Springer Proceedings in Physics 44

Springer Proceedings in Physics

Managing Editor: H. K. V. Lotsch

Volumes 1 – 29 are listed on the back inside cover

Optical Fiber Sensors

Proceedings of the 6th International Conference, OFS '89, Paris, France, September 18–20, 1989

Editors: H. J. Arditty, J. P. Dakin, and R. Th. Kersten

With 404 Figures

Springer-Verlag Berlin Heidelberg New York
London Paris Tokyo Hong Kong

Dr. Hervé J. Arditty
Photonetics, 52, avenue de l'Europe, F-78160 Marly-le-Roi, France

Professor John P. Dakin, Ph. D.
York Ltd., York House, School Lane, Chandler's Ford, Hampshire SO5 3DG, UK

Professor Dr. Ralf Th. Kersten
SCHOTT Glaswerke, Hattenbergstr. 10, D-6500 Mainz 1, Fed. Rep. of Germany

ISBN-13:978-3-642-75090-8 e-ISBN-13:978-3-642-75088-5
DOI: 10.1007/978-3-642-75088-5

2154/3150-543210 – Printed on acid-free paper

Preface

This conference is now the sixth of the Optical Fiber Sensors (OFS) series, the first having been launched in London in 1983. Progress over the last six years has been rapid and the number of laboratories working in the field has grown significantly. In addition, the number of commercially available sensors is growing, although, as usual, not as rapidly as the wilder optimists originally predicted. The subject is now maturing significantly, yet this volume shows that there is still no shortage of bright, new ideas.

The OFS series is devoted to the most recent research developments in the field and, in most respects, leaves commercially oriented conferences to cover items closest to production. However, as the technology progresses, it is appropriate to provide some coverage of applications-oriented research and development, and more applied sessions of this nature have been included.

The standard of papers contributed to the conference has been very high this year, and it has unfortunately been necessary to reject a number of interesting papers. Those that remain, invited, contributed and poster papers, are included in this volume. The poster papers were chosen for their greater suitability for presentation to smaller groups of people, giving more scope for detailed discussion around the poster display.

The conference this year has been somewhat unusual in having allowed a much shorter time between the initial announcement and call for papers, and the subsequent deadline, and the conference itself. This has, I believe, led to a much improved topicality of papers, most of those submitted being of a very up-to-date nature. Also I would care to speculate that laboratories with a large number of potential papers to submit have probably only had time to prepare papers covering their best results. It is becoming too common an experience nowadays to attend a conference having already seen a large number of the papers in the rapid-publication monthly journals. We all hope that this will be less likely for this conference.

Whatever the outcome, I am sure you will appreciate the quality of the papers contained in this volume, and will join me in thanking heartily all the authors and members of the organising and technical committees who have contributed to the success of the conference. I owe a particular debt to Professor Ralf Kersten, who, as the European co-chairman, has relieved me, by voluntarily taking on far more than his normal share of a considerable workload, during a period when I changed my post to join a new company. This would normally be a major activity in its own right, without the additional effort of organising the technical program of an international conference at the same time!

Of course, I must not forget the tremendous efforts of Dr. Hervé Arditty, chairman of the International Steering Committee, who has dealt with the enormous administrative load of running the conference, whilst, at the same time, directing his company. Special credit is also due to the other continental chairmen, Professor Ohtsuka and Dr. Kim, who have, in a smooth and efficient way, dealt with all the duties asked of them.

With the greatest thanks to all concerned, I hope that you enjoy the conference and find these proceedings valuable as a reference book for many years to come.

Chandler's Ford, England
July 1989

J.P. Dakin,
Chairman of the
Technical Program Committee

International Steering Committee

Dr. Hervé J. Arditty (Chair), Photonetics, France
Prof. Brian Culshaw, University of Strathclyde, UK
Dr. Anthony Dandridge, Naval Research Laboratory, USA
Prof. Shaoul Ezekiel, Massachusetts Institute of Technology, USA
Prof. Ralf Th. Kersten, Schott Glaswerke, FRG
Dr. Frederick J. Leonberger, United Technologies Research Center, USA
Prof. Takanori Okoshi, University of Tokyo, Japan
Dr. Scott Rashleigh, Australian Optical Fibre, Australia
Prof. Anna-Maria Scheggi, IROE-CNR, Italy
Dr. Yoshito Ueno, Nippon Electric Company, Japan

Technical Program Committee

Dr. Anastasious J.A. Bruinsma, TNO, The Netherlands
Dr. Anthony Dandridge, Naval Research Laboratory, USA
Dr. John P. Dakin (Chair), York Ltd., UK
Dr. Gordon W. Day, NIST, USA
Dr. Alan L. Harmer, Battelle, Switzerland
Dr. Masamitsu Haruna, Osaka University, Japan
Dr. Kazuo Hotate, University of Tokyo, Japan
Dr. Mark Johnson, York Ltd., UK
Dr. J.D.C. Jones, Heriot-Watt University, UK
Prof. Ralf Th. Kersten (Europe Co-chair), Schott Glaswerke, FRG
Dr. B. Yoon Kim (America Co-chair), Stanford University, USA
Dr. Rainer Kist, Fraunhofer Institute, FRG
Dr. Kazuo Kyuma, Mitsubishi Electric Corporation, Japan
Dr. Hervé C. Léfèvre, Photonetics, France
Dr. Walter F. Love, Corning Glass Works, USA
Dr. P.B. Lyons, Los Alamos National Laboratory, USA
Dr. Mario Martinelli, CISE, Italy
Dr. Gordon Mitchell, Metricor, USA
Dr. K. Nishizawa, Nippon Sheet Glass Co., Japan
Prof. Yoshihiro Ohtsuka (Asia Co-chair), Hokkaido University, Japan
Dr. Olivier Parriaux, CSEM, Switzerland
Prof. Colin Pask, University of New South Wales, Australia
Prof. Anna-Maria Scheggi, IROE-CNR, Italy
Dr. Elias Snitzer, Polaroid Corporation, USA
Prof. Toshihiko Yoshino, University of Tokyo, Japan

Contents

*Invited paper

Part VII Systems and Applications

Part VIII Current and Magnetic Field Sensors

Part IX Poster Papers

Part XIV Distributed Sensors

Part XII **Pressure Sensors**

Part XIII **Multiplexed Sensors**

Part I

Components and Special Fibers

101 Uses for Single Mode Fiber Directional Couplers

H.J. Shaw

Edward L. Ginzton Laboratory, Stanford University, Stanford, CA 94305, USA

We will review the saga of the humble but ubiquitous single mode fiber directional coupler which is the principal means of fiber interconnection in single mode fiber circuits. Its use in interferometric sensors and sensor networks will be described. Its use in other applications areas and the interplay between these and sensor applications will also be considered. The types of single mode fiber couplers now available and their prospects for the future will be discussed. Speculations will be made on evolving or possible new forms of couplers to circumvent limitations of present types and to open new applications.

Summary

Single mode fiber directional couplers are used to couple a propagating mode in one fiber to a propagating mode in another fiber. They are more important to the field of single mode fiber optics than the ancestors from which they derived, namely microwave long slot directional couplers, are to the field of microwaves. This is partly because in microwaves there are several types of couplers and junctions, while in single mode fiber optics the directional coupler is the only available all-fiber choice at this time.

In the beginning, about a decade ago, there were in some quarters strong concerns about the prospects for making very good single mode fiber directional couplers, even though integrated optic couplers and multimode fiber couplers had been successfully made during the preceding decade. Multimode couplers were made by mechanical lapping which, applied to the single mode case, would require working to submicron tolerances in spacings measured from the nearly invisible surface of the very small fiber core. Some ingenious methods were devised in an attempt to avoid this. However, by carefully establishing precise reference surfaces from which to measure, very high performance units were demonstrated using a lapping approach.

Today, single mode fiber directional couplers can be made routinely, although making state of the art units requires considerable skill. Of the various possible formats for single mode couplers, such as (1) etched types for simplicity of fabrication, (2) polished types for precise variable tuning and minimum power loss, (3) clamped, bonded or fused polished types for precise fixed coupling ratios and low loss, and (4) biconical fused types for high performance with minimum size in both fixed and tunable formats, the second and fourth types have found the widest use to date, with the latter predominating. Following the lead of the microwave directional coupler art, all of these couplers can be analyzed in terms of even and odd normal modes, leading to an understanding of the mechanisms of directivity, dispersion, polarization behavior and power loss.

Single mode fiber optic directional couplers were first developed to satisfy the needs of fiber interferometers and resonators for sensors, an application to which they are ideally

Springer Proceedings in Physics, Vol. 44 Optical Fiber Sensors
Editors: H.J. Arditty · J.P. Dakin · R.Th. Kersten

suited. Sensors of this type were the first devices to employ single mode fibers because of the need for spacial coherence for interferometry, and single mode fiber components were developed for this field before a real need for them arose in the communication field. Here single mode directional couplers are used to couple waves coherently to produce interference fringes having high sensitivity to environmental measurands. The fit of single mode directional couplers to the needs of these devices was profound, allowing the construction of all fiber circuits in which the light never left the fiber, giving high stability as well as sensitivity. Here the directional coupler was a replacement for the familiar bulk optic beam splitter, which preceded microwaves, and which is also a directional coupler but not a single mode device.

With high quality directional couplers on hand, it became possible to also construct complex fiber optic networks for the interconnection of multiple arrays of sensors. A variety of ladder and lattice topologies have been employed, allowing one to make design tradeoffs between power efficiency and cross talk. Single mode fiber components and techniques for sensor networks have some features in common with fiber circuits and networks in other applications areas. These include fiber optic systems for microwave signal processing and fiber optic communication systems. As these areas develop there is cross fertilization between them and the sensor area.

The marriage of microwaves and fiber optics was pleasing in view of the historic microwave background for the directional coupler itself. One purpose was to transform and process the microwave signal, in either time or frequency domains, at higher speeds than could be done electronically. This was done by modulating the microwave signal onto an optical carrier and injecting this composite signal into a complex single mode fiber processing network. The same basic network topologies as used for sensor networks can also be employed as multi-element transversal filters for this purpose. These systems can involve cascaded arrays of large numbers of directional couplers. Information bandwidth which is difficult or impossible to handle in rf circuits becomes narrowband information and easily processed in these optical circuits, leading to enhanced levels of computations per unit time proportional to the square of the optical bandwidth. This, together with the ability to form large numbers of parallel fiber paths and join them all together with directional couplers leads to new possibilities for high data throughput rates. Filter characteristics can be synthesized using the classical general methods of electronic filtering. Fiber optic recirculating memories, another group of devices based on directional couplers, are complementary to the above ladder and lattice systems. They are transient memories with storage times from microseconds to milliseconds and beyond, which can function as buffer memories for the former devices.

Directional couplers find their way into the fiber optic communication area in several ways. They are applicable as taps and power splitters for the interconnection of multiple users in fiber optic local area networks. For serial distribution systems multiple users can be distributed along the length of a fiber bus using one or more directional couplers per user which act as taps onto the bus. For parallel distribution systems single mode tree couplers and star couplers can be constructed as cascades of individual directional couplers. The wavelength dependence of the beat length in directional couplers provides a basis for design of wavelength division multiplexers, where long interaction lengths are used in the couplers to provide wavelength resolution. Directional couplers can also be employed as multiport hybrid junctions for combining signals in diversity receivers and for coupling local oscillators to signal channels in coherent communication receivers. In all of these applications fiber couplers offer benefits of compatibility with the system fibers and very low internal dissipative signal loss. Prospects for directional couplers in communications should grow with the emergence of fiber networks for broadband services to homes and businesses, handling data or information modulated either directly on the optical carriers or on microwave sub-carriers.

Other advances in the art of single mode fiber directional couplers, both linear and nonlinear have been made, some of which are important for fiber sensors. Directional couplers in various special forms can act directly as sensor elements. As circuit elements, couplers constructed with polarization maintaining fiber have been demonstrated which have very small polarization cross coupling. Fiber analogs of bulk wave polarizing beam splitters, which send orthogonally polarized beams into separate channels, have been developed. Y junctions, an alternate form of directional coupler, are also under investigation using single mode fibers.

Directional coupler halves made by mechanical polishing provide efficient access to the guided waves in a single mode fiber, opening up possibilities for modification of the propagation characteristics of these waves without removing them from the fiber. Planar elements such as birefringent media, optical amplifier media, nonlinear media and gratings can be placed on the surface of a directional coupler half so as to couple optically to waves propagating in the fiber. A variety of so-called evanescent wave devices using this approach have been under study for some time. Practical polarizers of high extinction ratio are an example of devices of this kind.

Single mode fiber optic couplers are of course reciprocal devices and in some applications this is an important advantage. For example in the fiber optic rotation sensor (fiber gyro), the most stable and sensitive of fiber sensors, reciprocity is the key property. In fiber sensor and communication networks reciprocity can lead to extra attenuation. Nonreciprocal couplers can be made in multimode and active forms, both of which are being studied. Generalized directional couplers in the form of active or nonlinear couplers, which function as binary switches, show promise for coping with this problem. Such switches could also accomplish message routing in communication systems. Eventually, arrays of such switches could constitute switchyards through which signals could self-route in response to address codes in their headers, which could be read and distributed to the binary switches. Ultimately all segments of the process could be carried out optically in fiber circuits for maximum speed. Directional couplers could play key rolls in all segments of the process.

Directional couplers are available from a number of suppliers. Two major potential market areas, communications and sensors, are felt by some to be destined to reach demands for very large quantities, with prices falling to a fraction of their current levels. These applications areas are being watched with interest in attempts to predict just when they will actually make a strong upward turn.

$LiNbO_3$ and $LiTaO_3$ Integrated Optic Components for Fiber Optic Sensors

F.J. Leonberger, T.K. Findakly, and P.G. Suchoski

United Technologies Research Center, East Hartford, CT 06108, USA

Over the last several years, there has been a growing interest in the use of integrated optic (IO) components and circuits in fiber sensor systems[1-3]. Initial work on integrated optic sensors focused on use of the IO chips themselves as sensors for temperature[4] and pressure[5]. More recently, integrated optics has generally been employed to perform key optical signal processing functions in a fiber sensor or as a signal transduction mechanism for an electrooptic sensor system. This paper will focus on these latter applications. Emphasis will be on chips for fiber optic gyros, laser vibration sensors, and electromagnetic field sensors. In many cases, recent advances in chip technology such as annealed-proton-exchange waveguides in $LiNbO_3$ and $LiTaO_3$, has enhanced circuit performance and made IO technology a viable choice for the fiber sensor designer.

Perhaps the primary focus for IO chips for sensors has been the fiber optic gyro. In the interferometric gyro[6], the IO chip can play the role of power splitter, polarizer, and phase modulator/frequency shifter, either as individual components or as a monolithic circuit. A strong motivator for IO circuitry for gyros is the ability to greatly reduce the size and separate component count of a sensor. For single-dimensional sensors, the savings are significant, but the largest payoff will come with multi-dimensional sensors (e.g., three-axis gyro) where many of the components for the several sensors can be formed on one chip. This chip could be nearly the same physical size as a single-dimension sensor chip. For a variety of reasons centered on gyro performance, a Y-junction circuit containing a polarizer and phase modulators, as shown in Fig. 1, is most commonly used. The Y-junction[7,8] is a well-behaved wideband reciprocal power splitter. The electrooptic phase modulators can be driven with typically 1-GHz bandwidth and < 10 V drive for a variety of modulation formats. A commonly used approach is serrodyne modulation of the phase modulator which produces an effective frequency shift. Phase modulator performance for gyros is controlled by the capacitance of the electrode structure, packaging effects and the quality of the electronics. For example, in the serrodyne modulation format, a high linearity ramp generator must be constructed with .04 fall to rise time ratio to achieve -50 dB sideband suppression levels in frequency shifting applications[9].

Excellent polarization control is essential for good gyro performance. Many types of polarizers have been demonstrated in Ti-diffused $LiNbO_3$, many involving metal/dielectric film overlays. We have recently reported a depolarization effect in Ti-diffused $LiNbO_3$ circuits[10] which limits the polarization retention of a chip to about 32 dB. To improve this value significantly, annealed-proton-exchange (APE) guides have been

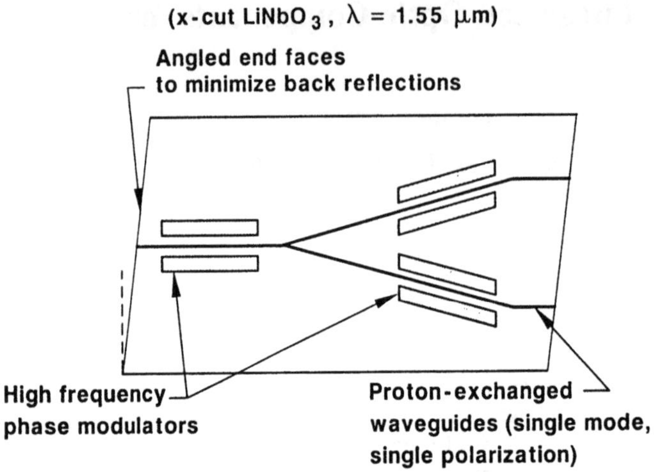

Fig. 1 Schematic of Y-junction IO circuit for fiber optic gyro.

developed[11]. These devices have exhibited polarization extinction of 60 dB (measurement limited) with no degradation in electrooptic properties.

Overall gyro chip performance is also influenced by reflections and residual intensity modulation in the phase modulators. An APE $LiNbO_3$ chip designed for 1.55 micron operation of the configuration shown in Fig. 1 has recently been reported[12] with -60 dB back reflection, 60 dB polarization extinction, 10^{-3} intensity modulation with a diode laser source, 1-GHz bandwidth, 4 dB insertion loss, and V_π of 10V.

More complex gyro chips have also been built. Recently, chips containing 11 components, as shown in Fig. 2, have been achieved using both Ti-indiffused technology[13] and annealed-proton-exchange technology[14]. This IO circuit permits the use of a special internal optical beat detection interferometric circuit to monitor the Sagnac-nulling phase shift, thereby forming a rate-integrating gyro[15].

Integrated optic chips have also been demonstrated for laser-based fiber heterodyne sensors[16,17] for use in velocity, vibration and displacement measurement. These sensors are based on laser heterodyne Doppler measurement techniques. The device is basically a Mach-Zehnder interferometer with associated optics and electronics required to detect a Doppler-shifted signal. Figure 3 shows one such circuit which replaces three beamsplitters, frequency shifter and polarization filter. This device has been fabricated in APE $LiNbO_3$.

Another class of sensors that utilizes IO chips is electromagnetic field sensors. Here, a linear IO modulator is used to convert a remotely received electromagnetic signal to a linearly modulated optical signal for transmission to an electronic receiver. The IO requirement is for a modulator with sensitivity to microvolt signals and a large linear dynamic range. Devices of this type have been reported utilizing asymmetric Mach-Zehnder modulators and synchronous directional couplers. For the Mach-Zehnder, a dynamic range of 84 dB, 1.1 microvolt sensitivity, 3 kHz detec-

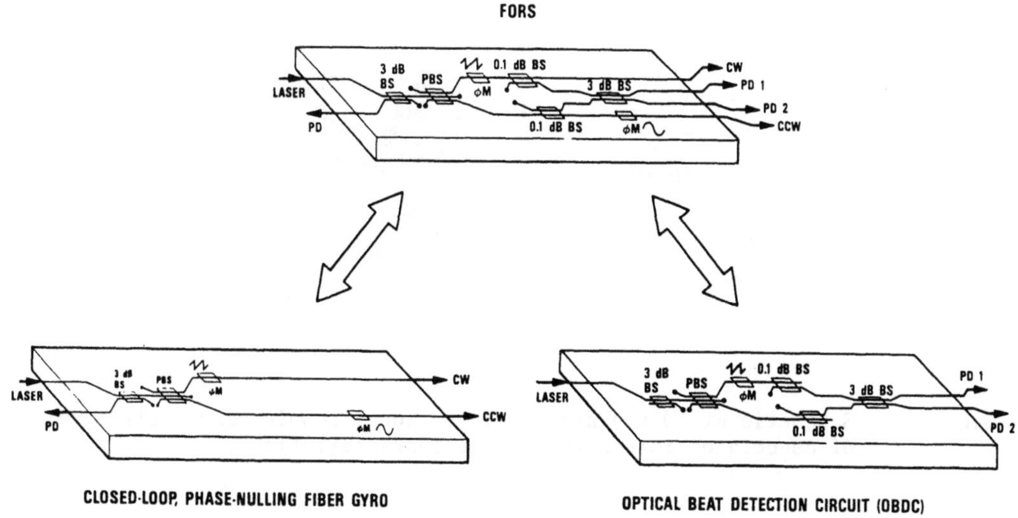

Fig. 2 Schematic of fiber gyro circuit incorporating both conventional closed-loop phase nulling and optical beat detection (Ref. 15).

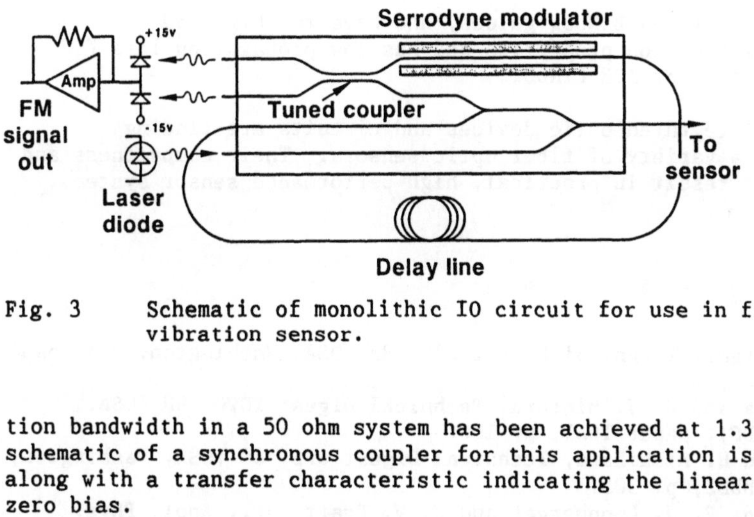

Fig. 3 Schematic of monolithic IO circuit for use in fiber optic laser vibration sensor.

tion bandwidth in a 50 ohm system has been achieved at 1.3 microns[18]. A schematic of a synchronous coupler for this application is shown in Fig. 4, along with a transfer characteristic indicating the linear response about zero bias.

In this sensor, as well as in gyro and heterodyne sensors, there is a performance advantage to be gained by operating at a relatively short wavelength. This advantage is due to the enhancement of the electrooptic efficiency which scales with wavelength squared (fourth power of the wavelength for required power). The obvious choice is the GaAs laser wavelength rather than the 1.3-1.5 wavelength band of GaInAsP. However, Ti-indiffused $LiNbO_3$ is known to exhibit photorefractive (optical damage) effects at low power levels at 0.8 microns[19] that can produce temporal instabilities in the transmitted optical signal. We have found that substantial improvement in power handling capability without these effects can be achieved in APE $LiNbO_3$ and especially in APE $LiTaO_3$.

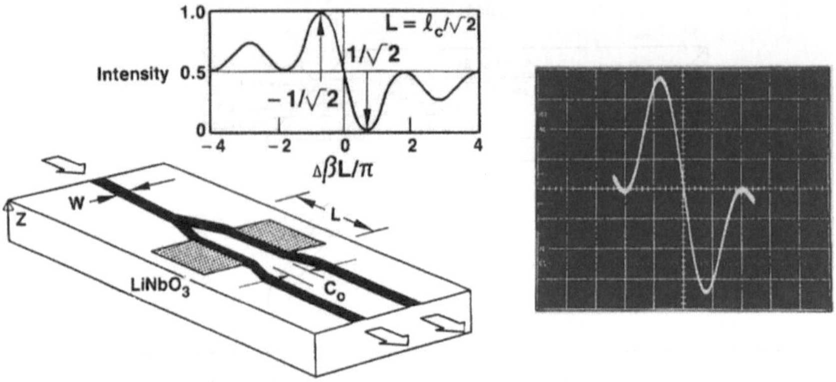

Fig. 4 Schematic of IO synchronous directional coupler and oscillogram of associated intensity-voltage characteristic.

In $LiTaO_3$, we have demonstrated 70 mW of stable power transmission[20]. Detailed studies of the negligible photorefractive effects in such guides have been reported[21]. This high power handling capability of $LiTaO_3$ coupled with the ready availability of high power GaAs laser diodes should lead to improved sensitivity and performance of sensor systems. APE $LiTaO_3$ devices exhibit similar electrooptic properties as conventional Ti-indiffused $LiNbO_3$ guides, and have the high (60 dB) polarization extinction properties and has low propagation loss (.2 dB/cm) characteristics of APE $LiNbO_3$.

In summary, integrated optic devices and circuits are finding increasing use in a variety of fiber optic sensors. Their compactness and performance should result in practical, high-performance sensor systems.

REFERENCES

1. F. J. Leonberger, Technical Digest OFS '85 (OSA, Washington, DC), paper ThAA3, p. 124.
2. R. P. DePaula and W. J. Minford, Technical Digest IGWO '88 (OSA, Washington, DC), paper TuA1, p. 186.
3. M. Haruna and H. Mishihara, Technical Digest OFS '88 (OSA, Washington, DC), paper ThDD2, p. 305.
4. L. M. Johnson, F. J. Leonberger and G. W. Pratt, Jr., Appl. Phys. Lett., 41, 134, (1982).
5. M. Izutsu, A. Enokihara, N. Mekada and T. Sueta, 2nd European Conf. Integrated Opt., Tech. Digest, p. 144, Firenze, Italy, October 1983.
6. See, for example, S. Ezekial and H. J. Arditty, in Fiber Optic Rotation Sensors and Related Technologies (Springer-Verlag, Berlin, 1982), p. 2.
7. H. Yajima, Appl. Phys. Lett., 22, 647 (1973).
8. R. H. Rediker and F. J. Leonberger, IEEE J. Quantum Electron., QE-18, 1813 (1982).
9. L. M. Johnson and C. H. Cox, J. Lightwave Tech., LT-6, 109 (1988).
10. P. Suchoski, T. Findakly and F. Leonberger, Electronics Lett., 23, 1357 (1987).
11. P. Suchoski, T. Findakly and F. Leonberger, Opt. Lett., 13, 1050, (1988).

12. P. Suchoski, T. Findakly and F. Leonberger, SPIE Proceedings, Vol. 993, p. 240, 1988.
13. W. Minford, F. Stone, B. Youmans and R. Bortman, SPIE Proceedings, Vol. 1169, 1989.
14. T. Findakly, P. Suchoski, F. Leonberger, M. Bramson, SPIE Proceedings, Vol. 1169, 1989.
15. K. R. K. Bartman, B. R. Youmans and N. M. Merheim, SPIE Proceedings, Vol. 719, p. 122, 1986.
16. H. Toda, M. Huruna, H. Nishihara, J. Lightwave Tech., LT-5, 901 (1987).
17. J. Waters, SPIE Proceedings, Vol. 1162, 1989.
18. C. Bulmer, Appl. Phys. Lett., 53, 2368 (1988).
19. R. Becker, Appl. Phys. Lett., 45, 121 (1984).
20. T. Findakly, P. Suchoski and F. Leonberger, Opt. Lett., 13, 797 (1988).
21. T. Findakly, J. Farina, P. Suchoski and F. Leonberger, IGWO '89 Technical Digest (OSA, Washington, DC), Paper WCCZ, p. 246.

Broader Spectral Width 1.5μm InGaAsP Superluminescent Diode of Stacked Active Layers

O. Mikami, Y. Noguchi, H. Yasaka, and K. Magari

NTT Opto-electronics Laboratories, 3-1 Morinosato Wakamiya, Atsugi-shi, 243-01, Japan

We have reported fabrication and characteristics of broader spectral width 1.5 μm InGaAsP superluminescent diodes having a novel structure of <u>st</u>acked <u>ac</u>tive layers (STAC-SLDs). The emission spectral width as broad as 140 nm is obtained, which is larger than the twice that of conventional 1.5 μm SLDs.

1. Introduction

Superluminescent diodes (SLDs)[1] are expected to be the optimum light sources in optical fiber gyroscope[2] and OTDR (Optical time-domain reflectrometry)[3] applications. The broad-band characteristics and the short coherent length of SLDs significantly improve the performance of systems. Various structures of SLDs[4-9] have been demonstrated to suppress the lasing mode due to reflection at the cleaved facets. However, there have not been any reports on broadening the emission spectral width of SLDs. As the spectral width is broadened, the coherent length is reduced. Therefore broader spectral width SLDs are preferred for obtaining the ultimate sensitivity in these systems.

In this paper, we report the successful fabrication of broader spectral width 1.5 μm InGaAsP SLDs. The SLDs proposed here have a structure of <u>st</u>acked <u>ac</u>tive layers (STAC-SLDs). We obtained emission spectral width as broad as 140 nm, which is larger than twice that of conventional 1.5 μm SLDs.

Springer Proceedings in Physics, Vol. 44 Optical Fiber Sensors
Editors: H.J. Arditty · J.P. Dakin · R.Th. Kersten

2. Device structure

The same design concept as laser diodes (LDs) and light emitting diodes (LEDs) has been applied to SLDs. Consequently, the spectral width is limited by the band-gap energy of the single active layer used. Conventional 1.3 μm SLDs have a spectral width of 30-40 nm[6-9], and the 1.5 μm SLDs have 60-70 nm width. The present device has an active structure consisting of stacked twin active layers, Q1 and Q2, whose band-gap wavelengths λ 1 and λ 2 are slightly different. The band structure of the active region is schematically shown in Fig. 1. InGaAsP active layers with λ 1 of about 1.50 μm and λ 2 of about 1.55 μm are used. The thickness of both layers is 50 nm. They have three InGaAsP layers whose band-gap wavelengths are 1.3 μm in-side as well as out-side of the twin active layers in order to obtain larger optical confinement.

The SLDs, having the active region mentioned above, were fabricated using LPE grown InP/InGaAsP wafers with a doubly channeled planar buried heterostructure (DCPBH)[10]. A new lasing suppression structure consisting of a straight buried active stripe and a bent buried absorbing stripe, shown in Fig. 2, was developed[9]. This structure makes it possible to suppress the lasing oscillation with a

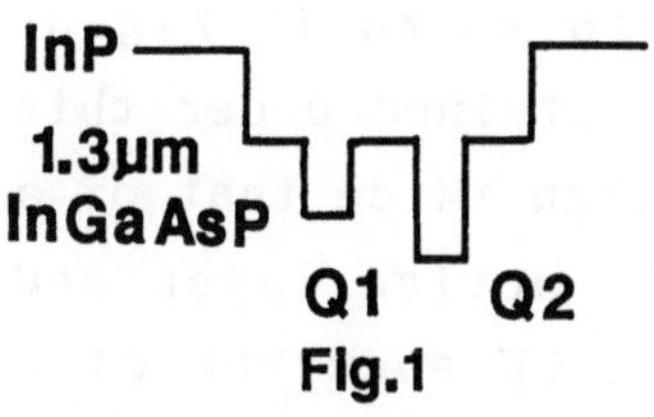

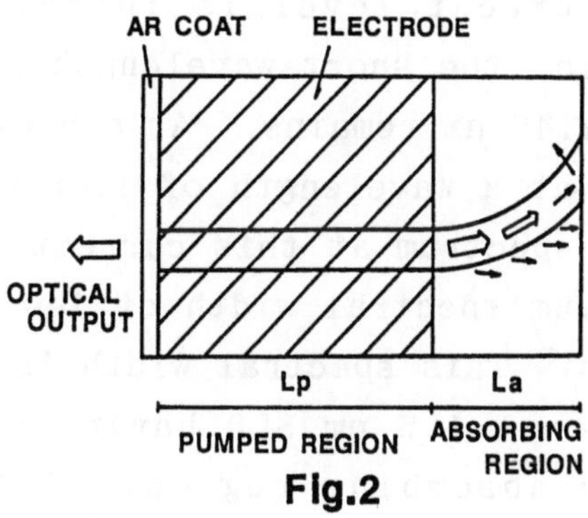

Fig.1 Band structure of active layers.
Fig.2 Buried bent absorbing guide structure.

short absorbing region. Typical device dimensions are as follows; a stripe width of W=2 μm, a pumped region length of Lp=400 μm, an absorbing region length of La=200 μm, and a bend guide radius of R=500 μm. No anti-reflection coating was deposited.

3. Characteristics

We first investigated characteristics of LDs fabricated by cleaving the pumping region at both ends. The threshold current at room temperature for a 200 μm long cavity is about 20 mA. This threshold current is as low as that of conventional DCPBH-LDs which have a single active layer. A few kinks due to mode jumping are observed in L-I curves. Lasing occurs in the long wavelength region at a low current level. When the bias is increased, lasing mode in the short wavelength region is also observed, and two peaks exist in different wavelength regions. As the bias is further increased, only one lasing mode in the short wavelength remains. This spectrum behavior is explained by the twin active layer structure.

Spectra of the fabricated SLDs having a stack of twin active layers under a pulsed condition were observed, using the injection current as a parameter. At a bias current of 200 mA, the emission peak is at a wavelength of 1.45 μm. As the current level is increased, a new emission peak appears in the short wavelength region while the emission peak at 1.45 μm remains. At a bias of 350 mA, two peaks are observed at a wavelength of 1.45 μm and 1.38 μm. A typical emission spectrum at this current level is shown in Fig.3. The maximum spectral width of 140 nm is obtained under this condition. This spectral width is more than twice that of a conventional 1.5 μm SLD having a single active layer and the same absorbing region. A typical CW spectrum of a conventional SLD is shown in Fig. 4. When the current is over 350 mA, the emission peak at the shorter wavelength

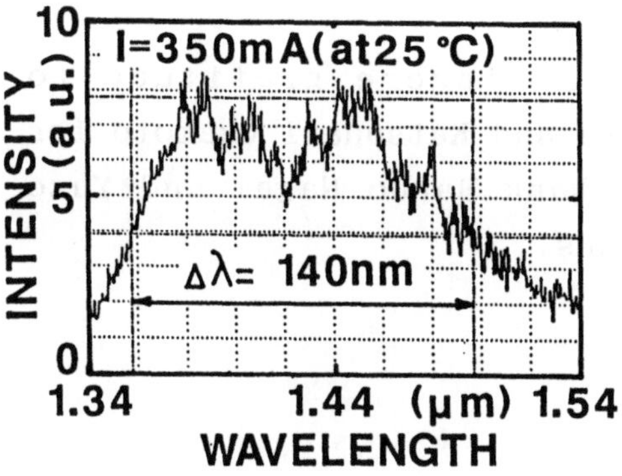

Fig.3

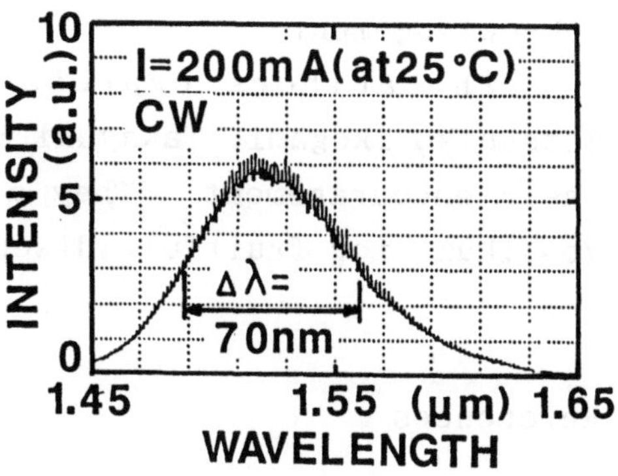

Fig.4

Fig.3 Emission spectrum of a fabricated SLD having stacked twin active layers (STAC-SLD).

Fig.4 CW emission spectrum of a conventional 1.5 μm SLD.

becomes dominant. The broader spectral width obtained in this device apparently comes from the stacked structure of twin active layers whose band-gap wavelengths are slightly different.

The output power at a current of 350 mA is about 1 mW. This value is somewhat lower than conventional 1.5 μm SLDs. However, this can be overcome by optimizing the structure of the active layers. The active layers are located very close so that high coupling efficiency to optical fibers are obtained.

4. Summary

Novel structure of broader spectral width InGaAsP SLDs having stacked active layers (STAC-SLDs) is proposed and characteristics in the 1.5 μm wavelength region have been reported. This structure is found to be quite effective for broadening the spectral width, although optimum design is necessary for the active layer structure. The broader spectral width SLDs are expected to improve the sensitivity in OTDR applications. This structure also seems very useful as a "white light source" in near-inrared spectroscopy.

Acknowledgement

The authors express their sincere gratitude to Tetsuhiko Ikegami, Takayuki Sugeta and Katsunari Okamoto for their encouragement. They also thank Haruo Nagai and Yuzo Yoshikuni for fruitful discussions.

References

[1]L.N.Kurbatov, S.S.Shakhidzhanov, L.V.Bystrova, V.V.Krapukhin and S.I.Kolonenkova: Soviet Phys.Semicon. 4, 1739 (1971).

[2]W.K.Burns, C.L.Chen and R.P.Moeller: IEEE/OSA J.Lightwave Technol. LT-1, 98 (1983).

[3]K.Takada, I.Yokohama, K.Chida and J.Noda: Appl.Opt. 26, 1603 (1987).

[4]T.P.Lee, C.A.Burrus and B.I.Miller: IEEE J.Quantum Electron. QE-9, 820 (1973).

[5]N.S.K.Kwong, K.Y.Lau, N.Bar-Chaim, I.Ury and K.J.Lee: Appl.Phys.Lett. 51, 1879 (1987)

[6]I.P.Kaminow, G.Eisenstein, L.W.Stulz and A.G.Dentai: IEEE J.Quantum Electron. QE-19, 78 (1983).

[7]K.Gen-ei, A.Tanioka, H.Suhara and K.Chinen: Appl.Phys. Lett. 53, 1138 (1988).

[8]N.S.K.Kwong, N.Bar-Chaim and T.Chen: Appl.Phys.Lett. 54, 298 (1989).

[9]H.Nagai, Y.Noguchi and S.Sudo: Appl.Phys.Lett. 54, 1719 (1989).

[10]I.Mito, M.Kitamura, K.Kobayashi and K.Kobayashi: Electron.Lett. 18, 953 (1982).

Third-Harmonic Measurement of Surface Temperature of a Thermal Fibre-Modulator

J.B. Brown [1], *J.D.C. Jones* [1,*], *S.J. Rogers* [1], *R.K.Y. Chan* [2], *and H.H. Wong* [2]

[1]Physics Laboratory, University of Kent at Canterbury, UK
[2]Hong Kong Baptist College, 224 Waterloo Road, Kowloon, Hong Kong
*Present address: Physics Department, Heriot Watt University, Riccarton, Edinburgh, EH144AS, UK

A method is presented for measuring the instantaneous temperature of a fibre undergoing thermal modulation by passing a heating current through a thin-film metallic jacket whose resistance is placed in an electrical bridge circuit. The temperature is recorded by a signal at the third harmonic of the driving frequency.

Introduction

Fibre-optic phase-shifters and modulators have been proposed in which the heat is applied by electrically heating a thin-film jacket of metal evaporated on to the fibre, either with or without its plastic coating intact. They offer the possibility of a light weight, sensitive device with response to high frequencies, comparing favourably in all respects with the more usual piezoelectric modulator.

An early report of the frequency response of an aluminium-coated fibre was made by Dandridge, Tveten and Giallorenzi [1], in which they were interested in the device as a current detector. Petuchowski, Sigel and Giallorenzi [2] measured the response of a similar fibre apparently firmly anchored thermally by its supports. More recently White et al. [3] demonstrated the utility of a gold covered fibre as a phase modulator and stabilizer for an interferometer at the quadrature point.

Schuetz et al. [4] analyzed the expected optical response to external temperature changes for fibres with different jacket materials and made clear the importance of the axial strain term associated with the expansion of a metal jacket, as compared with the core temperature term, particularly at higher frequencies.

As pointed out by Maurer, Cole and Bucaro [5], a major impediment to understanding the response has always been the difficulty associated with the direct measurement of the fibre surface temperature. Even a thin film thermocouple or other device added to the metal jacket will seriously perturb the result. Our method presented here simply uses the metal jacket itself as its own resistance thermometer by placing it in an a.c. bridge circuit during the application of the heating power to it. An interesting feature of the method is that the signal produced by the bridge, operating from a generator supplying current at frequence f, with measurement of the voltage appearing across an element whose resistance is changing at 2f, appears at a frequency 3f.

Experimental Method

Our first measurements have been made with a 10.6 cm section of single mode fibre of 125 μm cladding diameter and 5 μm core with an aluminium coating of somewhat less than 1 μm. It was mounted across paper at its ends with its active portion about 0.2 mm clear of a horizontal length of varnished aluminium channel.

The modulator is in a conventional Mach-Zehnder interferometer used with a He-Ne laser. The thermal measurement circuitry is as in Fig.1.

In the limit of small temperature changes one may take the heat supplied to the film as $R_4 I_0^2[1 + \cos(2\omega t)]/2$, where I_0 is the peak electrical current supplied at angular frequency ω and t is the time. The temperature rise of the film can then be given as

$$T = c_1 W + c_2 W \cos(2wt - \phi).$$

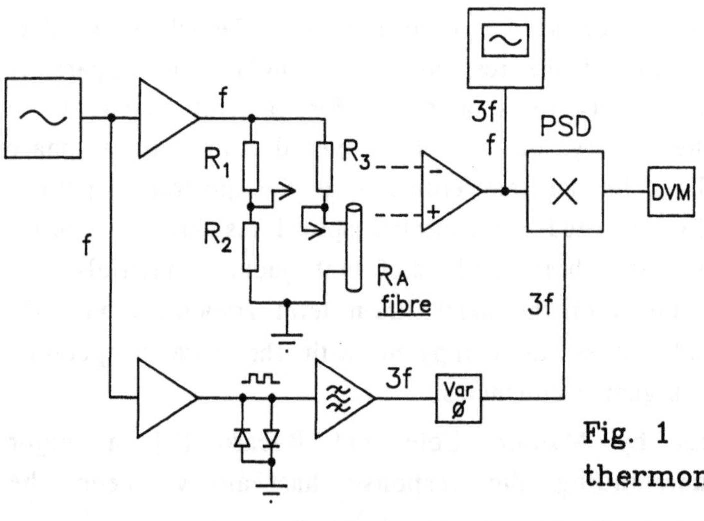

Fig. 1 Third harmonic thermometry circuit.

Here c_1 is a constant giving the d.c. temperature rise while c_2 is a frequency-dependent thermal impedance, ϕ is a phase lag and $W = I_0^2R_4/2$. The resistance of the film can then be represented in this range as

$$R = R_0[1 + \alpha T] = R_0[1 + \alpha c_1 W + \alpha c_2 W\cos(2\omega t - \phi)] \ .$$

Substituting the above expression into the bridge balance equation gives the unbalanced signal appearing for small temperature excursions as:

$$v = [V_0R_3R_0\alpha W/(R_0 + R_3)^2][c_1\cos(\omega t) + (c_2/2)\cos(\omega t - \phi)]$$
$$+ [V_0R_3R_0\alpha c_2 W/2(R_0 + R_3)^2].\cos3\omega t - \phi).$$

Here $V_o \cos\omega t$ is the total voltage across the bridge.

Measurement of the last term above gives c_2W, the amplitude of the a.c. temperature swing.

Thermal Impedances

Heat supplied to the metal film can be likened to that from a constant current generator in the analogous electrical situation. The heat capacity of the film itself can then be seen as an electrical capacitance in parallel with the admittance of two transmission lines, one being that into the interior of the fibre itself, where the a.c. component of heat flow is dissipated, the film becoming isothermal for d.c. temperature rise if the loss through the ends can be neglected. The other path, outwards from the fibre might be expected to be dominated by the thermal diffusivity represented by a transmission line at higher frequencies, but it may present thermally resistive behaviour at low frequencies, particularly if gas convection past the fibre carries air a fibre diameter distance in the time of a heating period, or if there is thermal conduction through the supports such as was observed by Petuchowski et al. [2].

For the internal path into the fibre we have examined the thermal diffusion equation in cylindrical coordinates for our fibre parameters. As noted by Hughes and Priest [6], the solution for the temperature internally in terms of that given at the surface may be expressed in terms of Kelvins ber and bei functions. The methods are very similar to those encountered in the case of electrical eddy currents treated by

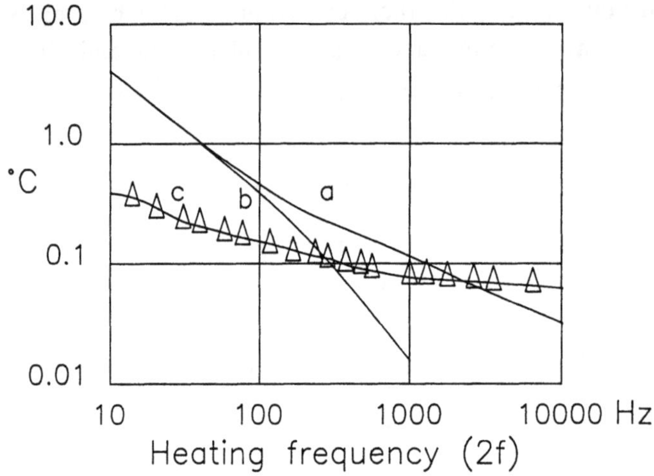

Fig. 2 Computed surface (a) and core (b) a.c. temperature based on fibre thermal impedance only, and observed signal (c).

Smythe [7]. Using a small computer, however, with series expansion it is now simple to use a direct solution in the form

$$T(r) = \frac{Q_0 I_0[(jp)^{1/2} r] e^{j\omega t}}{(jp)^{1/2} I_1[(jp)^{1/2} a] K}$$

Here I_0 and I_1 are modified Bessel functions of the zeroeth and first order, a is the fibre radius, r the distance from the centre, K is the thermal conductivity of the quartz and $p = \omega/D$, where D is the thermal diffusivity. We have obtained the temperature rise, $T = T_0 e^{j\omega t}$, in terms of the inward flux of heat, $Q = Q_0 e^{j\omega t}$, at the surface, rather than the temperature there, as done by Hughes and Priest. This approach yields the thermal impedance, T_0/Q_0, and hence, for the case of no heat loss outwards, the temperature rise at the surface when r = a, and the corresponding temperature rise at the core when r = 0. Both these quantities are plotted in fig.2 for an input of 2.09 W peak power. We have taken $K = 0.014$ W °C^{-1} cm^{-1} and $D = 8.45 \times 10^{-3}$ cm^2 s^{-1}.

Results

Preliminary experimental results using the third harmonic method of temperature measurement for this fibre are also plotted on the same scale in fig.2, to compare the temperature rise with that which would obtain if all the a.c. heat produced were to be absorbed by the fibre, the difference then being accounted for by that carried outwards. The relative flatness of the observed response variation with frequency and its absolute lower value at low frequencies suggest a component of resistive-like heat flow outwards. We have not yet explained the crossing over of the two curves at the highest frequencies. Tests with a dummy carbon resistor in place of the film show that background signals do not vary strongly with frequency and are less than 5% of that observed even at the higher frequencies. It was noted, however, that a dummy resistor in the form of a large rheostat gave a strong third harmonic signal traced to its being wound from magnetic wire. Because measurements are taken with the bridge at balance for the fundamental frequency, there is strong discrimination against any harmonics arising from residual distortion in the driving oscillator.

It is concluded that the new method offers promise in the measurement of fibre surface temperatures and the elucidation of the mechanisms involved. It is proposed to carry out comparisons with optical frequency response in the same fibres and to investigate different mounting and cooling arrangements.

References

1. A.Dandridge, A.B.Tveten and T.G.Giallorenzi, Electron Lett., **17**, 523 (1981)
2. S.J.Petuchowski, G.H.Sigel and T.G.Giallorenzi, Electron. Lett., **18**, 814 (1982).
3. B.J.White, J.P.Davis, I.C.Bobb, H.D.Krumboltz and D.C. Larson, J.Lightwave Techn. LT-5, 1169 (1987).
4. L.S.Schuetz, J.H.Cole, J.Jarzynski, N.Lagakos and J.A.Bucaro, Appl. Opt.22, 478 (1983).
5. G.S.Maurer, J.H.Cole and J.A.Bucaro, Optics Lett., **11**, 561 (1982).
6. R.Hughes and R.Priest, Appl.Optics, **19**, 1477 (1980).
7. W.R.Smythe, <u>Static and Dynamic Electricity</u> (McGraw-Hill, New York 1968).

Fluoride Glass Fiber-Optic Sensors: Realizations and Prospects

G. Mazé, V. Cardin, F. Chiquet, and M. Poulain

Z.I. du Champ Martin, F-35770 Vern-sur-Seiche, France

Since the discovery of zirconium-fluoride glasses, in 1974, by Michel and Marcel Poulain, fluoride glasses have been extensively studied because of their predicted ultra low attenuation level, 10^{-2}-10^{-3} dB/km in the mid-IR, around 2.5 µm [1]. To date, these glasses are the most promising candidates for high bit rate repeaterless submarine fiber-optic telecommunication links.

Fluoride glass fibers have still to be improved by several orders of magnitude before reaching the performances needed for telecommunications. The figures of merit of the single-mode and multimode fluoride glass fibers, available on the market-place, compared with that of the "telecommunication" fiber, are given in Table 1.

Table 1

Figure of merit of fluoride glass fibers (1989)
(Length -m-/Attenuation -dB/km-)

Fiber	Objective (single-mode)	IRguide SM1 (single-mode)	IRguide 150/200 (multimode)
Length (m)	3000	1000	500
Min.attenuation (dB/km)	.1	50	10
Figure of merit (FOM)	30 000	20	50
Normalized FOM (NFOM)	1	.67 10^{-3}	1.67 10^{-3}

Fiber-optic sensors do not usually necessitate ultra low loss fibers. For most applications, a few tens of dB/km are acceptable. These requirements are met by current fluoride glass fibers, either multimode or single-mode . Both types are used in a number of fiber-optic sensing devices. This paper presents some of them and discusses a few prospects.

Springer Proceedings in Physics, Vol. 44 Optical Fiber Sensors
Editors: H.J. Arditty · J.P. Dakin · R.Th. Kersten

Fluoride glass fibers technology and properties

Among the thousands of fluoride glasses which have been studied in the past years, only a few compositions are suitable for making fibers [2]. This is the case for the glasses of the ZBLAN family (Z stands for Zirconium Fluoride [ZrF_4], B for Barium Fluoride [BaF_2], L for Lanthanum Fluoride [LaF_3], A for Aluminium Fluoride [AlF_3] and N for Sodium Fluoride [NaF]).

In fluoride systems, a small percentage of almost any fluoride can be incorporated, without damage to the stability of the glass. Glasses doped with rare earths and transition metal elements are quite easy to make. In the ZBLAN family, Lanthanum [La], can be partly or completely replaced by any of the other rare earth elements.

Compared with silica-based materials, fluoride glasses exhibit a low softening temperature, a large thermal expansion coefficient and a similar refractive index. To some extent, these characteristics can be modified, by changing the composition, within the boundaries of the glass family. Aluminium [AlF_3] and Sodium Fluoride [NaF], for instance, are known to give a low refractive index, while Barium fluoride [BaF_2] is responsible of the high value of the coefficient of liner expansion. These properties are used to obtain the refractive index difference betwenn the core and the cladding, which is responsible of the guiding properties of the fiber, and to match the expansion coefficients of the glasses (Table 2).

TABLE 2

Typical fluoride glass compositions for core and cladding (molar %) and physical properties

	ZrF_4	BaF_2	LaF_3	AlF_3	NaF	Tg(°C)	ß(K^{-1})	n_d
Core	59	34	5	3	0	300	180 10^{-7}	1.52
Cladding	55	30	5	5	5	300	180 10^{-7}	1.51
Silica						1120	5.4 10^{-7}	1.46

Owing to the low viscosity of fluoride glasses the core can be cast in a cladding tube. The step index preform obtained that way is drawn into a fiber at a temperature close to 400 °C. The diameter of the fiber, controlled through the pulling speed, can be made with very tight tolerances. Either single-mode or multimode fibers, with core diameters ranging from 11 to 1000 microns, over a few tens of meters to 1 km, and are produced currently at LeVF.

The main property of fluoride glass fibers consist in their large optical window which expands from 300 nm in the UV to 5 µm in the IR. Compared with silica fibers, their infrared edge

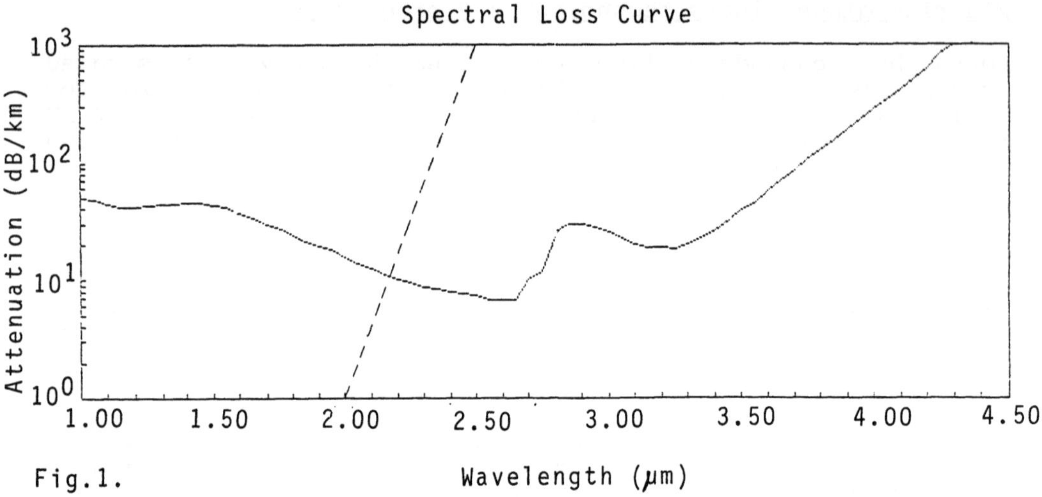

Fig.1.

is shifted 3 µm further (Fig 1). This large window opened on the mid-infrared part of the spectrum is at the root of most of the fluoride glass fiber-optic sensors which are made nowadays.

The absorption bands, at 1.5 and 2.9 µm, on the attenuation curve of a typical multimode IRguide fiber (Fig. 1), are due to impurities. At wavelengths longer than 3.5 µm the attenuation rises continuously because of the IR absorption of the matrix. The lowest the level of attenuation the largest the window of the fiber (Table 3).

TABLE 3

Low attenuation windows in multimode fluoride glass fibers (IRguide 1989)

Wavelength range (µm)	Attenuation (dB/km)
.4 - 4.3	< 1000
1 - 3.7	< 100
1.7 - 3.5	< 50
2.6	< 10

Those recent years, a lot of fluoride glass fiber-optic sensors have been constructed. Their applications are for remote near and mid-IR spectrometry, low temperature pyrometry, photothermal processes and infrared imaging.

Remote infrared spectrometry

Many diatomic gaseous molecules and most organic molecules exhibit characteristic absorption bands, linked to vibrational

TABLE 4

Absorption bands of some gaseous molecules

Molecule	Absorption wavelength (µm)
HO_2	6.8
H_2O	2.7 - 6.2
H_2O_2	7.8
CO	4.65
CO_2	4.2
HF	1.25
HCl	3.5
HBr	3.8
HI	4.4
HNO_3	5.8
NO	5.2
N_2O	7.7
NO_2	6.2
SO_2	4
CH_4	7.5 - 3.3
C_2H_6	3.4
CH_2O	3.6
CH_3Cl	3.3

and rotational transition in the infrared, between 1 and 14 microns (Table 4). These molecules are commonly identified and their concentration measured by IR transmittance spectrometry.

Since many of these bands are located in the low attenuation window of fluoride glasses, it becomes possible to guide a polychromatic light, down the sampling cell, and pick up the signature of the molecules via fluoride glass fibers. The damping undergone by a monochromatic light in this loop can be expressed as follows :

$$\beta = \beta_f (l_{f1} + l_{f2}) + 4,35\ 10^5\ a\ l_c + C + R \qquad (1)$$

where β_f stands for the attenuation of the fiber [dB/km], l_{f1} and l_{f2} for the lengths of the upstream and downstream fibers [km], a for the absorption coefficient of the sample [cm^{-1}] and l_c the cell's length [cm], C and R the coupling losses in the cell and the reflection losses [dB], respectively.

The longer the distance between the cell and the spectrometer, the longer the fiber length, the lower its attenuation should be (Eq. 1). The <u>Spectrosysteme 3100</u> developed by the company Fluidystème, of France, for the control and monitoring of organic compounds in the chemical plants, is the first system, operanting in the range of 1.5 to 2.5 microns, allowing distances as long as 100 m between the spectrometer and the cell [3].

Various cell designs are used according to the nature of the sample and its absorption coefficient. For most gases and some liquids, with low absorption coefficient, a long intercation length is necessary to obtain an acceptable signal/noise ratio. D. Pruss et al. constructed a miniature multipass reflection cell offering 54 cm total pathlength and a coupling efficiency of 60 % [4].

In the case of highly absorbing liquids, the interaction length is so short and the sample cell so narrow that the liquid cannot flow. Such drawback does not exist with an evanescant wave probe because the interaction between the incident beam and the absorbing medium occurs at the outer surface of the fiber. Since the intensity of the evanescent wave decreases exponentially with the distance from the core boundary, a substantial amount of power can be coupled in the surrounding medium only where the cladding has been reduced to a few microns thickness . It is well known also that in multimode fibers, only the modes near cut-off are likely to loose substantial power through the evanescent wave mechanism. In short-length probes, single-mode fibers should be preferred to multimode. With the evanescent wave probe, the term C in Eq. 1 vanishes and the reflection losses can be restricted to those occurring at the output ends of the fiber. The damping equation in the loop becomes then :

$$\beta = \beta_f (l_{f1} + l_{f2}) + c\, l_c\, a + R \qquad (2)$$

where c stands for the coupling coefficient of the probe and l_c for its length.

The spectrum of water and alcohol (Fig. 3a-b) has been recorded using a zirconium fluoride fiber probe, 150 µm core, which cladding was reduced to 10 µm thickness over 3 mm length (Fig. 2).

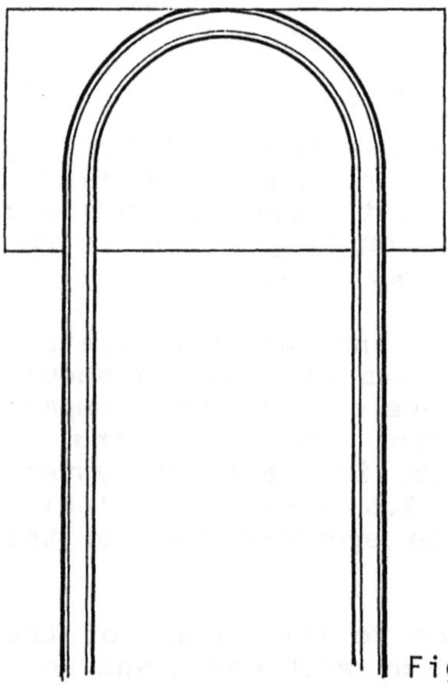

Fig.2

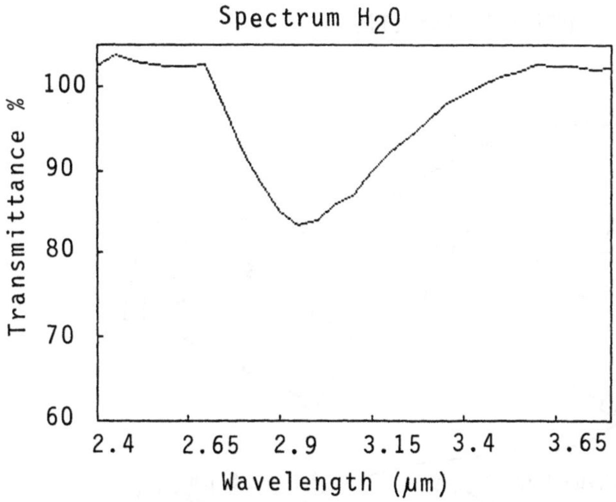

Fig.3a

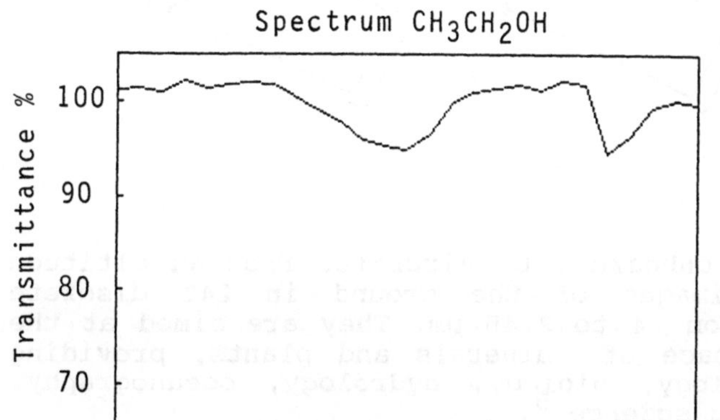

Fig.3b

For the analysis of solid substances, the transmittance and the evanescent wave techniques cannot be employed. They are replaced by IR reflectance spectrometry. In this approach, the solid sample is enlighted by the monochromatic beam coming out of an IR fiber and part of the light scattered by its surface is collected by an integrating sphere or a fiber-optic bundle.

The imaging spectrometer AVIRIS (Airborne Visible/Infrared Imaging Spectrometer)

AVIRIS was constructed in 1985 and 1986 at the Jet Propulsion Laboratory in Pasadena (CA), under contract with the NASA. This is the first civilian airborne IR imaging spectrometer incorporating fluoride glass fibers. AVIRIS started flying at

AVIRIS OPTICS SCHEMATIC

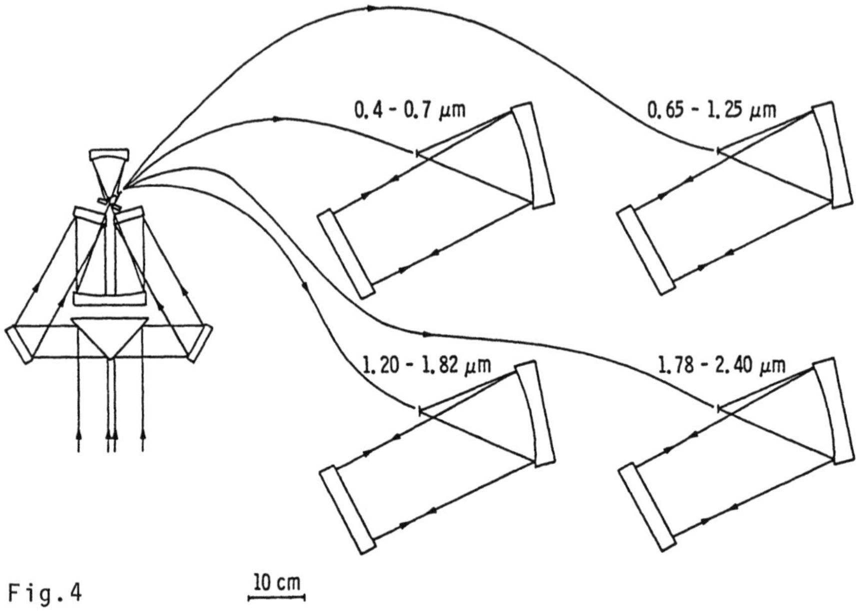

Fig.4

the beginning of 1987 onboard a U2 aircraft. From an altitude of 20 km it takes images of the ground in 242 discrete wavelengths ranging from .4 to 2.45 µm. They are aimed at the identification from space of minerals and plants, providing data relevant to geology, biology, hydrology, oceanography, atmospheric and sensing science [5].

The instrument includes 4 spectrometers, each one covering a spectral band [.4-.7], [.68-1.27], [1.25-1.86] and [1.84-2.45] µm (Fig. 4). The first high numerical aperture (.45) fluoride glass fiber was developed by LeVF to connect the 3d and 4th spectrometers to the foreoptics.

Low temperature fiber-optic pyrometry

Radiation thermometry is widely used in the industry to measure surface temperatures. The measurement relies on Planck's law which links the spectral radiance of the blackbody to its temperature.

$$L_l = 2\ n\ h\ c^2\ l^{-5}\ (\exp\ (hc/klT)-1) \qquad (3)$$

In a radiation thermometer the light emitted by a surface is filtered, modulated and focused on the detector. Instead of viewing directly the radiating surface, through the objective of the instrument, one can use optical fibers to convey the thermal radiation.

The early fiber-optic radiation thermometers, which appeared some 20 years ago, used silica-based fibers [6]. Their transmittance is limited in the IR to the wavelengths shorter than 2 µm, while the low temperature backbody radiates most of its power in the IR, at wavelengths longer than 2 um. As a consequence, it is necessary to use large core fiber-optic bundles to catch enough power to measure low temperatures. In spite of that it is not possible to measure temperatures lower than 100 °C through 1 m long fiber-optic bundle. For longer distances, the limit is still higher.

These temperature and length limitations, inherent to conventional glass fiber systems, are repelled further with the use of fluoride glass fibers as stated in the low temperature fiber-optic radiation pyrometers reported recently [7,8,9].

Using a 350 µm fluoride glass fiber, coupled to a nitrogen-cooled Indium Antimonide (InSb) detector, we succeeded, in our laboratory, to measure a temperature as low as - 20°.

Assuming the radiating surface to be a lambertian source, the power $W_{\Delta h}$ escaping from the pyrometer in the wavelength interval Δh is expressed by :

$$W_{\Delta h} = \int_{\Delta h} e_{h,T}\, R_{h,T}\, G\, D_{h,L} \prod_{1}^{n} (r_{i,h})\, S_h\, g\, dh \qquad (4)$$

where $e_{h,T}$ stands for the emissivity of the source, $R_{h,T}$ the spectral radiance of the surface, $D_{h,L}$ the damping undergone over the fiber length L, $r_{i,h}$ the Fresnel losses on the ith dielectric surface crossed by the beam, S_h the spectral sensitivity of the detector and g the gain of the amplifier.

The power transmitted through the fluoride fiber corresponds to the steep side of the spectral radiance curves of the blackbody. These instruments are likely to give an excellent precision, because a small temperature change results in a large variation of the radiated power.

The response of the pyrometer is proportional to the emissivity $e_{h,T}$ of the surface. In the case of gases and substances of weak emissivity, the power radiated does not suffice to allow the measurement. The blackbody tip was proposed recently to overcome these problems. The entrance of the fiber is covered with a lightweight high-emissivity substance which converts the temperature of the surrounding medium into a detectable radiation.

Photothermal processes

The physical changes induced in solids and fluids heated by a light radiation are at the basis of photothermal techniques. In a solid surface submitted to a pulsed laser beam, the point of impact becames the source of thermal waves which propagate into the medium. Their decay, with time and direction, contains information relevant to the physical state of the substrate. Owing to the fact that the over-heating is very small, only a non-contact technique is allowed to measure the time-space distribution of the temperature. Fluoride glass fibers are then very useful to pick-up the infrared radiation.

A. Lothon and co-workers, proposed recently to apply a photothermal technique to measure the temperature of the vanes of steam turbines [11]. A NdYAg laser pulse of 100 mJ and about 10 ns, is shot on the vane. The time variation of the infrared flux collected on the detector is due only to the radiation of the impact surface. This technique allows to separate in the radiated flux the contribution of the vane from those of the surroundings. A fluoride glass fiber is used to convey the radiated flux to the detector.

Conclusion

The large infrared transmittance of fluoride glasses fibers have been used to construct fiber-optic sensors for IR spectrometry, low temperature pyrometry and infrared imaging. Some of these instruments are already on the market-place. New sensors using single-mode and active fluoride glass fibers are in sight.

References

1 S. Shibata M. Horiguchi, K. Jinguji, S. Mitachi, T. Kanamori, T. Manabe, Prediction of loss minima in infrared optical fibres, Electronic Letters, 1981, Vol. 17.

2 M. Poulain, Halide glasses, in Preparation properties and structure of special glasses, pp. 1-14, 1983.

3 Fluidystème, Spectrosysteme 3100 technical data-sheet, 1989.

4 D. Pruss: Application of IR-glass fibers for remote spectroscopy, Ext. Abstacts 5th Int. Symp. on Halide Glasses, Shizuoka, (1988), pp. 76-84.

5 G. Vane, First results from the Airborne Visible/Infrared Imaging Spectrometer (AVIRIS), JP.L. Publication 87-37 (1987), pp. 89-97.

6 R. Vanzetti and A.C. Traub, Uses of fiber-optics in industrial radiation thermometers. In : Radiation Thermometry, DeWitt and Nutter ed., 1988, pp. 625-649.

7 Y. Denayrolles, B. Trouville : Pyromètre à fiber optique basse température. JNOG, Montpellier 2 - 3 avril 1987.

8 M. Shimizu, M. Shimoishizaka, S. Yoshida, Radiometric temperature measurement using infrared optical fibers, OFS Stuttgart, 1985.

9 S.R. Mordon, A.H. Cornil, J.-M. Brunetaud : Temperature measurement with a zirconium fluoride glass fiber, Applied Optics Vol. 26, 1987, pp. 607-609.

10 G. Mazé, V. Cardin, F. Chiquet, M. Poulain, Applications and prospects, in Fluoride Glasses, A.E. Comyns ed., 1989.

11 A. Lothon, Y. Denayrolles, A. Kleitz, J.J. Courant, Pyrométrie photothermique impulsionnelle appliquée aux aubages basse-pression des turbines à vapeur d'E.D.F., J.N.O.G., Résumé des communications, pp. 36-37, 1988.

Novel Single Mode Channel Waveguide Fabricated by Photopolymerization for Integrated Optic Sensors

H. Hosokawa, J. Takagi, N. Horie, S. Aoyama, S. Ogata, and T. Yamashita

Central R&D Lab., OMRON Tateisi Electronics Co.,
20 Igadera, Shimokaiinji, Nagaokakyo, Kyoto 617, Japan

A novel single mode channel waveguide fabrication method by photopolymerization has been developed. This waveguide can be widely applied to integrated optic sensors because of its low cost and single mode characteristics. It is expected that practical use in integrated optic sensors will be made of this polymer waveguide.

INTRODUCTION

Optical sensors using integrated optics technologies are being developed because of their possibility of integrating several components upon a common substrate. In addition, they are compact, reliable and relatively inexpensive. In particular, integrated optic sensors, utilizing an interferometer configuration, have achieved highly accurate measurements. For example, an integrated optic microdisplacement sensor(1)(2), as shown in Fig. 1, is able to detect both the direction and the quantity of displacement with 0.2 μm resolution. However, mass production of single mode channel waveguides, necessary for practical use in the integrated optic sensors, has not yet been achieved.

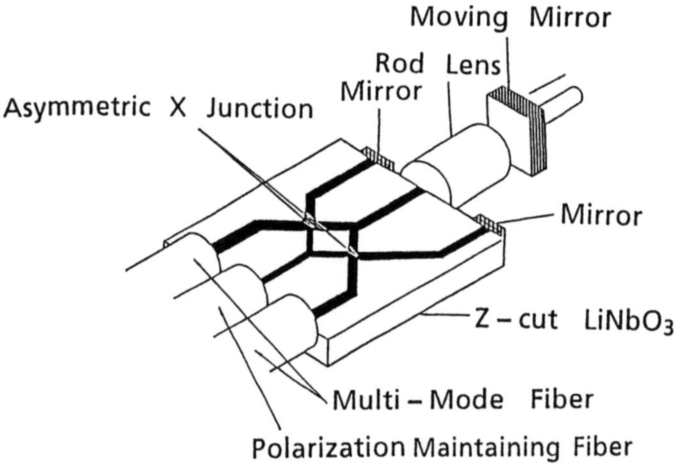

Fig.1 Integrated optic microdisplacement sensor

Springer Proceedings in Physics, Vol. 44 **Optical Fiber Sensors**
Editors: H.J. Arditty · J.P. Dakin · R.Th. Kersten

In this paper, we report a novel single mode channel waveguide, fabricated by photopolymerization, which is suitable for mass production.

DESIGN

A novel single mode channel waveguide fabrication process is shown in Fig. 2. The process is as follows: (1) a ridged stripe channel waveguide is fabricated by electron beam(EB) lithography, which allows highly accurate control of the width and height of the pattern profile; (2) a stamper for the waveguide mold is fabricated from an EB resist pattern using an electric gilding technique; (3) UV resin as a waveguide material is injected and pressed on the glass substrate which has two grooves to obtain smooth cleaved end faces of the waveguide; (4) UV resin is cured by UV light; (5) the waveguide on the glass substrate is removed from the stamper; and (6) the glass substrate is cleaved, and smooth end faces of the waveguide are obtained. Utilizing this method, it is possible to replicate a single mode waveguide easily by repeating steps (3) through (6).

Figure 3 shows a schematic illustration of the fabricated rib waveguide. To obtain a single mode waveguide, the width W of the waveguide should be within the range of 2.1 - 6.2 μm.

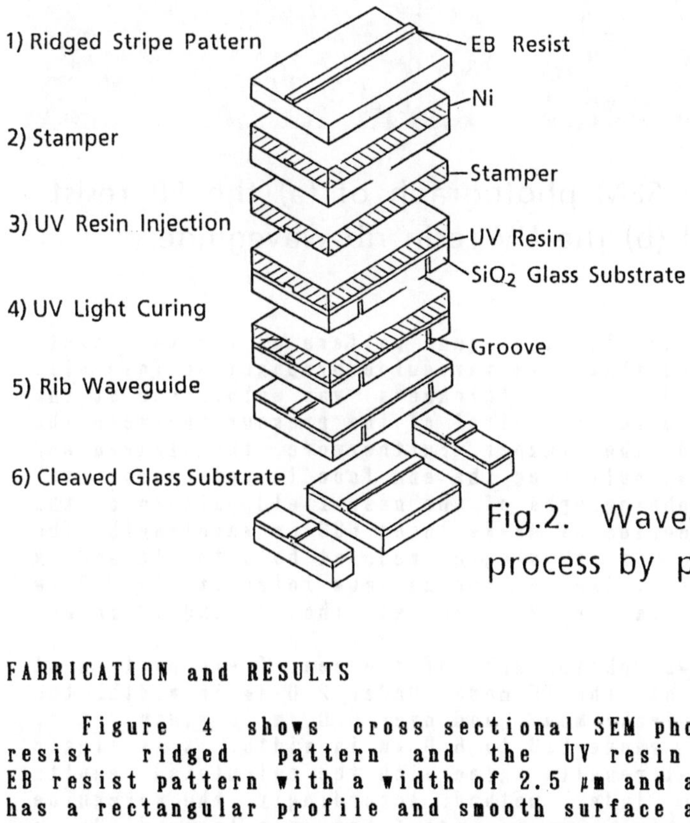

Fig.2. Waveguide fabrication process by photopolymerization

FABRICATION and RESULTS

Figure 4 shows cross sectional SEM photographs of the EB resist ridged pattern and the UV resin rib waveguide. The EB resist pattern with a width of 2.5 μm and a height of 1.5 μm has a rectangular profile and smooth surface and side walls. The UV resin rib waveguide replicated exactly from the EB resist

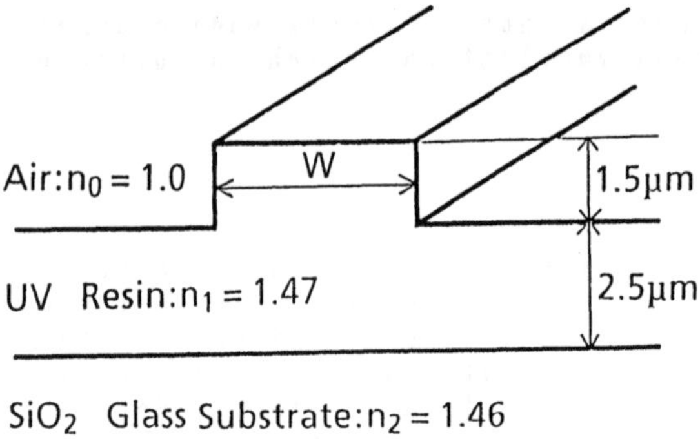

Fig.3. Schematic illustration of the rib waveguide

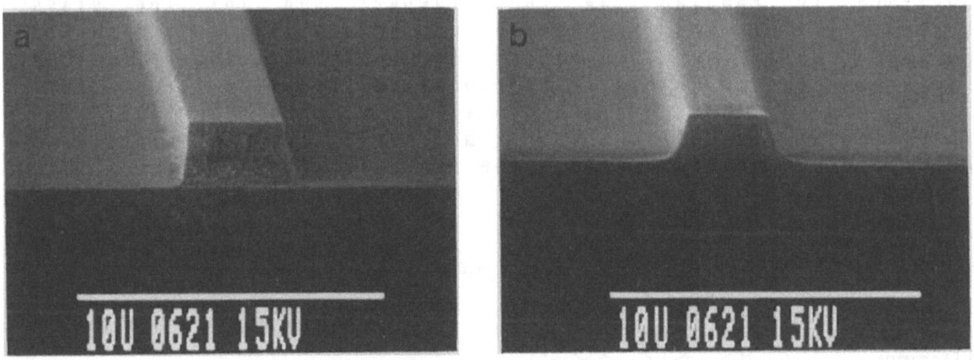

Fig.4.Cross sectional SEM photograph of (a) the EB resist ridged pattern and (b) the UV resin rib waveguide.

pattern has approximately the same pattern as the EB resist. Thus, it is expected that this waveguide propagation loss will be very low. The UV resin thickness was maintained at the desired value of 2.5 μm by control of the pressure between the glass substrate and the stamper. Furthermore, the cleaved end face is so smooth that polishing the end face is not necessary.

Figure 5 shows photographs of the near field pattern of the fabricated rib waveguide at He-Ne laser 633 nm wavelength. The fabricated waveguide is a single mode at both the TE and TM modes. The beam size at $1/e^2$ of the maximum intensity is 2.9 μm × 2.8 μm and 3.0 μm × 2.8 μm at the TE and TM modes, respectively.

Figure 6 shows photographs of the near field pattern at varied rib width at the TE mode. Under 2.0 μm in width, the guided wave light is leaky mode, and over 6.0 μm in width, it is multi-mode. In the range 2.0 to 6.0 μm in width, it is single mode. The experimental results agree with the calculated results using the effective index method. Accordingly, the waveguide coupling efficiency to a single mode fiber or a laser diode is

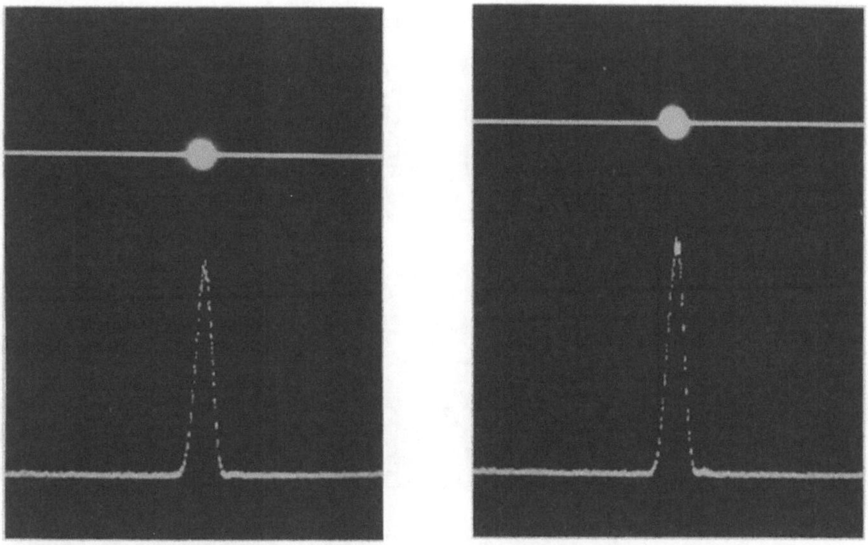

(a) $W_{//} = 2.9\mu m$ $W_{\perp} = 2.8\mu m$

(b) $W_{//} = 3.0\mu m$ $W_{\perp} = 2.8\mu m$

Fig.5.Photograph of the near field pattern at (a) TE mode and (b) TM mode.

expected to be very high, because various beam patterns can be easily controlled by varying the rib width and resin thickness.

APPLICATIONS

This polymer waveguide can be applied to a wide variety of the integrated optic sensors because of its low cost and single mode characteristics. In particular, applications to interferometric sensors are expected, for example, microdisplacement sensors using the Michelson interferometer as shown in Fig. 1, and temperature[3], humidity[4] and pressure sensors[5] using the Michelson or Mach-Zehnder interferometer. Furthermore, applications will include not only channel guided wave sensors, but also optical planer guided wave sensors such as optical pick up heads[6] because of the simultaneous fabrication of a waveguide and grating couplers.

CONCLUSION

A novel single mode channel waveguide fabrication method by photopolymerization has been developed. This waveguide can be widely applied to integrated optic sensors because of its low

(a) W = 2.0μm

(b) W = 2.5μm

(c)W = 4.0μm

(d)W = 5.5μm

(e)W = 6.0μm

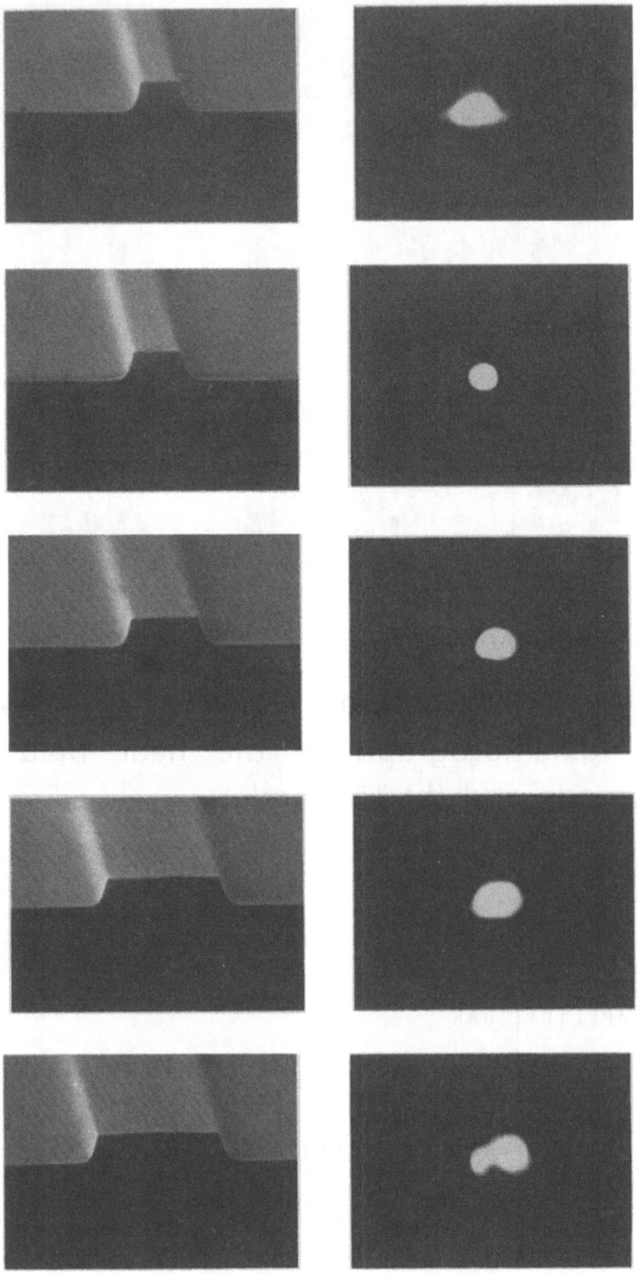

Fig.6.Measured guided wave beam size parallel to the substrate as a function of the rib width.

cost and single mode characteristics. It is expected that practical use in integrated optic sensors will be made of this waveguide fabrication method.

REFERENCES

1)T.Yamashita, and J.Takagi, "Integrated optic microdisplacement sensor using two asymmetric X junctions and a rod lens", Proceedings of OFS'86, 19, Tokyo, (1986) .

2)H.Hosokawa, J.Takagi, and T.Yamashita, "Integrated optic microdisplacement sensor using a Y junction and a polarization maintaining fiber" , Proceedings of OFS'88, ThAA8, 137, New Orleans, (1988) .

3)L.M,Johnson, and F.J.Leonberger, "Integrated optical temperature sensor" , Appl. Phys. Lett., 41(2), 134,(1982).

4)M.Izutsu, A.Enokihara, and T.Sueta, "Integrated optic temperature and humidity sensors" , IEEE J.Lightwave Tech., LT-4, 833, (1986) .

5)M.Izutsu, A.Enokihara, N.Mekada, and T.Sueta,"Optical-waveguide pressure sensor," in Tech. Dig. 2nd ECIO, 144, (1983) .

6)S.Ura, T.Suhara, and H.Nishihara, "An integrated-optic disk pickup device," IEEE J.Lightwave Tech., LT-4, 913,(1986) .

Part II

Interferometer Signal Processing I: Narrow Band Sources

Passive Quadrature Phase Detector for Coherent Fiber Optic Systems

S.Y. Huang, H.G. Park, and B.Y. Kim

Ginzton Laboratory, Stanford University, Stanford, CA94305, USA

We report a novel approach to achieving optical interferometric signals that are in phase quadrature, eliminating the need for modulators and/or active feed-back operations for the elimination of signal fading in fiber sensors and coherent optical communications. The passive device utilizes modal interference in a two-mode optical waveguide and a differential phase shift of $\pi/2$ between the near and far field radiation patterns.

1. INTRODUCTION

In coherent fiber-optic systems for sensing and telecommunication applications, the well known signal fading problem has to be overcome in order to achieve high sensitivity and stability. The interferometric response of the output optical intensity to the optical phase shift critically depends on the phase bias between the two interfering waves, which is sensitive to environmental parameters. In particular, when the phase difference between the two interfering waves is $N\pi$ (N=integer), the output intensity in interferometric systems has zero sensitivity to small optical phase shifts.

Several approaches have been devised to overcome the signal fading problem, including actively stabilizing the phase bias(1), heterodyning(2,3), and using a passive device that produces quadrature phase information(4-7). In this paper, we report a simple new device that directly provides two optical signals carrying quadrature phase information.

2. PRINCIPLE OF OPERATION

An optical fiber with highly elliptical core geometry can be operated as a two-mode fiber that supports only the fundamental symmetric LP_{01} mode and the second order antisymmetric LP_{11} mode with a stable cross-sectional mode intensity distribution, as illustrated in Fig. 1-a(8). As the two modes propagate along the fiber, a differential phase shift ϕ accumulates between them due to the difference in their propagation constants. When both the LP_{01} and LP_{11} modes are excited with approximately equal intensity, the output radiation pattern from a two-mode fiber varies as a function of the phase difference between the two modes as shown in Fig. 1-b. Detection of upper or lower half of the output radiation pattern provides an intermodal interference signal, from which the differential phase shift between the optical waves in the LP_{01} and LP_{11} modes can be determined.

Springer Proceedings in Physics, Vol. 44 Optical Fiber Sensors
Editors: H.J. Arditty · J.P. Dakin · R.Th. Kersten

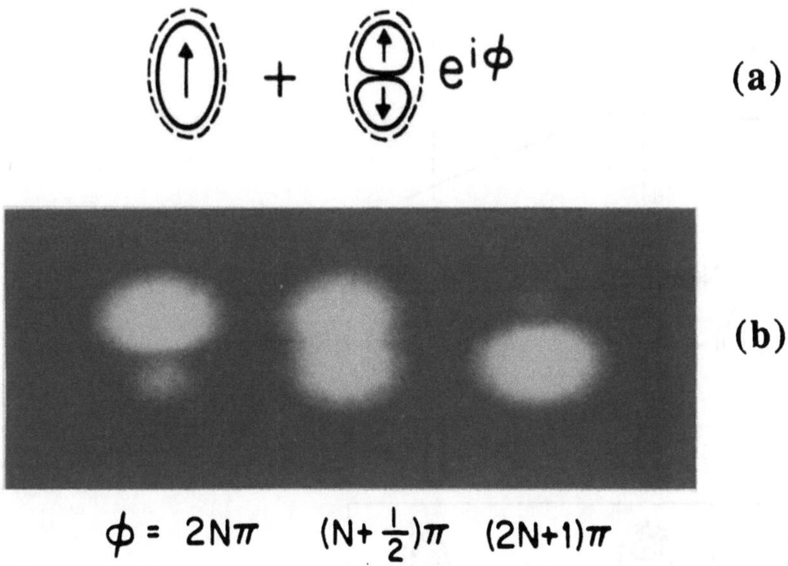

Figure 1. Interference of LP_{01} and LP_{11} modes in a two-mode optical fiber with a highly elliptical core geometry as a function of phase difference ϕ between the two modes.

The two-mode fiber interferometer is the key element of the passive quadrature phase detector described in this paper, along with a little known phenomenon called the Guoy effect(9). The Guoy effect introduces an additional phase shift to a Gaussian beam passing through a focal region or a beam waist. The Gaussian modes (TEM_{lm}) diverging away from its beam waist can be expressed as(10),

$$E_{lm}(x,y,z) = A_{lm}(x,y,z)\exp\{i[-kz - \frac{k(x^2+y^2)}{2R(z)} + (1+m+l)\eta(z)]\} \qquad (1)$$

where $A_{lm}(x,y,z)$ is the amplitude, k is the wavenumber, and the beam waist is at z=0. R(z) and $\eta(z)$ are given by,

$$R(z) = z(1 + z_0^2/z^2) \qquad (2)$$

$$\eta(z) = \tan^{-1}(z/z_0) \qquad (3)$$

where $z_0 = (1/2)w_0^2k$ is the minimum beam radius for the TEM_{00} mode at z=0, at which the optical field amplitude is down by 1/e compared to its value at its maximum. Notice the value of $\eta(z)$ becomes $\pi/2$ when $z>>z_0$, leading to differential phase shifts of $\pi/2$ between Gaussian modes having a difference of 1 between their values of m+l, as the beam makes a transition from its waist to the far field radiation region. This mode number dependant phase shift provides a simple and reliable way of getting quadrature optical phase information in interferometric systems if two spatial modes propagating away from a beam waist are used as two arms of an interferometer.

In view of the fact that the two guided spatial modes in a two-mode optical fiber (LP_{01} and LP_{11} modes) can be accurately approximated as Gaussian modes (TEM_{00} and TEM_{10} modes), the radiating output from an end of a two-mode optical fiber is

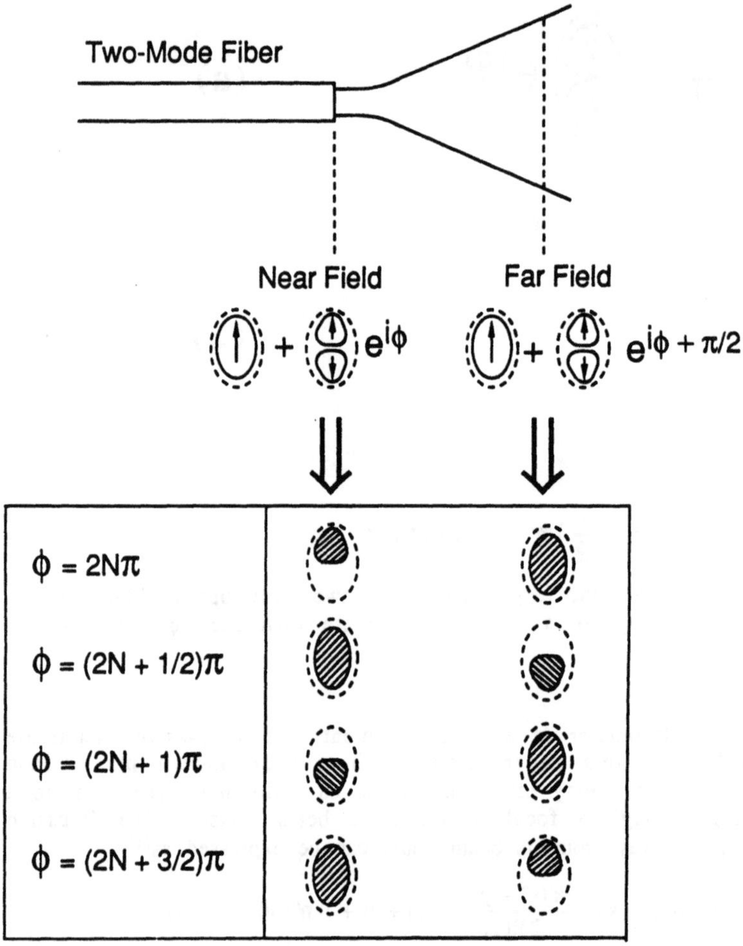

Figure 2. Modal interference pattern of a highly elliptical core two-mode fiber at near and far field radiation region.

an ideal medium for implementing the $\pi/2$ Guoy phase shift for quadrature phase detection. The fiber output end face represents the beam waist or the focal point, and as the LP_{01} and LP_{11} modes radiate away from the fiber end face a $\pi/2$ phase difference is introduced between them. As illustrated in Fig. 2, the resultant modal interference patterns in the near and far field region carry interferometer signals that are in phase quadrature. The implementation of this device in coherent single-mode fiber interferometers requires transfer of the optical signals in the two separate single-mode fiber arms of the interferometer into the two spatial modes in a two-mode optical fiber. This can be easily accomplished as described later in this paper.

3. EXPERIMENT

The experimental setup to demonstrate the passive quadrature phase detection is shown in Fig. 3-a. A two-mode fiber with a highly elliptical core (4.1×2.2 μm, provided by Polaroid Corp.) and with a LP_{11} mode cutoff wavelength of 633 nm was

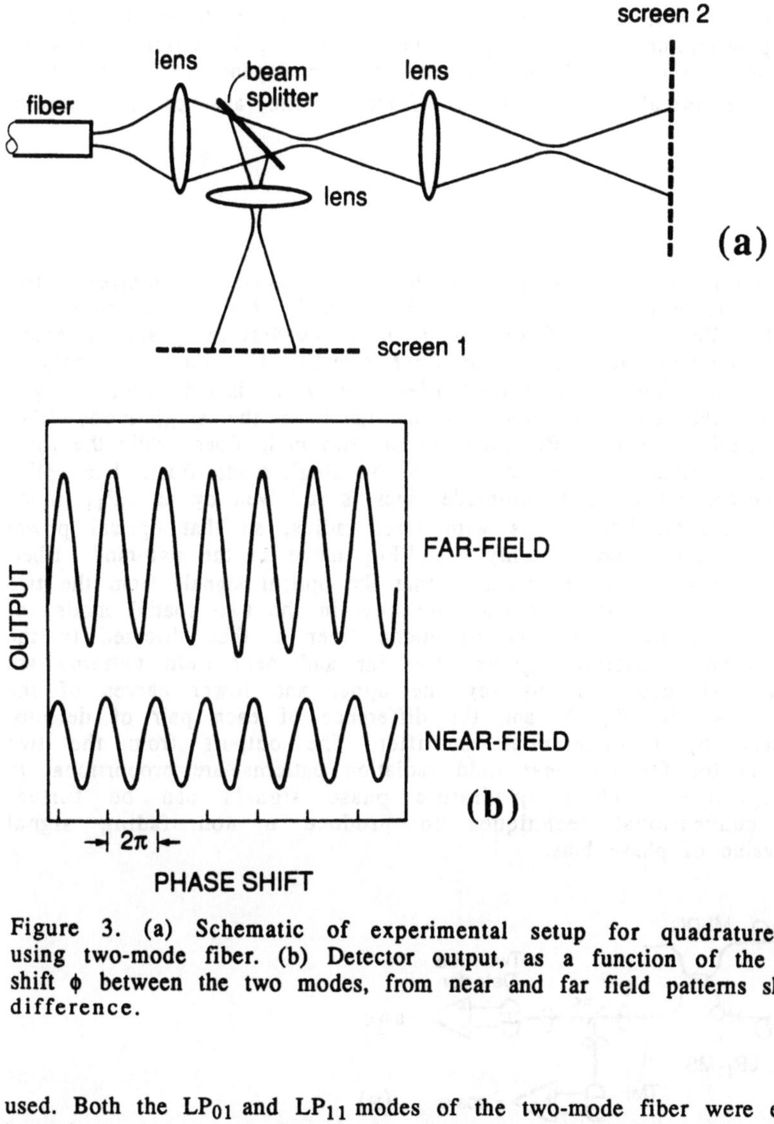

Figure 3. (a) Schematic of experimental setup for quadrature phase detection using two-mode fiber. (b) Detector output, as a function of the differential phase shift ϕ between the two modes, from near and far field patterns showing $\pi/2$ phase difference.

used. Both the LP_{01} and LP_{11} modes of the two-mode fiber were excited with equal intensity using an Ar-ion laser operating at 514.5 nm wavelength. The output from the fiber was converged by a 20× microscope objective and split into two beams by a beamsplitter. The near field pattern was then imaged on screen 1 by placing a lens at the beam waist of the converging beam. The far field pattern was projected on screen 2 and its size was adjusted by a lens located far away from the beam waist as shown in Fig. 3-a. A short section of the two-mode fiber was stretched to vary the phase difference between the fundamental and second order modes as described in reference 11. A clear observation of changing far and near field patterns with a $\pi/2$ phase difference was made as predicted in Fig. 2. The two screens were then replaced by two photodetectors offset from the optic axes of the two beams in order to get a direct measurement of the quadrature phase relationship between the two signals. Figure 3-b shows the output of the two detectors as a function of differential phase shift between the two modes introduced by the stretching. A $\pi/2$ phase difference between the response of the two detector signals is clearly seen.

It should be mentioned that we did not observe a significant dependence of the device on the polarization state of light, making it a polarization insensitive device. This is expected due to the negligible polarization dependence of the modal interference in an elliptical core two-mode fiber in the absence of external perturbations(12).

4. APPLICATIONS

The passive quadrature phase detector can be implemented in interferometric systems with a simple modification of the fiber circuit. Figure 4-a shows an example of a Mach-Zehnder interferometer with a quadrature phase detector, where one of the conventional single-mode fiber directional couplers is replaced by a mode selective directional coupler(MSDC)(13). The MSDC is built with a single- and two-mode fiber pair such that the optical signal in the single-mode fiber experiences 100% coupling to the LP_{11} mode of the two-mode fiber, while the LP_{01} mode in the two-mode fiber does not couple to the single-mode fiber. The splice (×) in Fig. 4-a between single- and two-mode fiber is followed by an LP_{11} mode stripper, formed by several fiber loops with tight radius, so that optical power from the single-mode fiber produces only the LP_{01} mode in the two-mode fiber. The result of this fiber circuit arrangement is that the optical signals from the two single-mode fiber arms of the interferometer are now in the two spatial modes of the two-mode fiber. The output of the two-mode fiber is then directed to the quadrature phase detector described above. The far and near field patterns are each detected with two detectors covering the upper and lower halves of the radiation patterns shown in Fig. 1, and the difference of each pair of detector outputs is produced by a differential amplifier. The outputs from the two differential amplifiers for far and near field radiation patterns are proportional to $\sin\phi$ and $\cos\phi$ respectively. These quadrature phase signals can be further processed using conventional techniques to produce a non fading signal regardless of the value of phase bias.

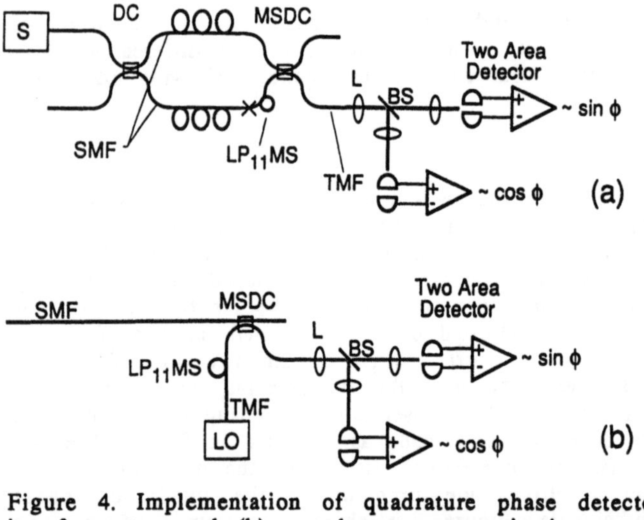

Figure 4. Implementation of quadrature phase detectors in (a) a Mach-Zehnder interferometer, and (b) a coherent communication receiver with a local oscillator. S; light source, DC; directional coupler, SMF; single-mode fiber, MSDC; mode selective directional coupler, TMF; two-mode fiber, L; lens, BS; beamsplitter, LO; local oscillator.

Figure 4-b illustrates the use of a quadrature phase detector to eliminate signal fading in a coherent homodyne communication system with a local oscillator operating at the same frequency as the incoming signal in a single-mode fiber. Light from the local oscillator is coupled to the LP_{01} mode of the two-mode fiber and unwanted excitation of LP_{11} mode is eliminated by a mode stripper. The incoming signal is coupled to the LP_{11} mode of the two-mode fiber through a mode selective directional coupler, resulting in inter-modal interferometer in the two-mode fiber. The rest of the detection system is identical to that for the example in Fig. 4-a.

5. CONCLUSION

We have described a novel passive quadrature phase detector for the elimination of signal fading in fiber-optic interferometric sensors and coherent communication systems. The device is based on modal interference in a highly elliptical core two-mode fiber and on the Guoy phase shift. We have successfully demonstrated the principle of operation of the device.

This work was supported by Litton Systems, Inc.

REFERENCES

1. D. A. Jackson, R. Priest, A. Dandridge, and A. B. Tveten, "Elimination of drift in a single-mode optical fiber interferometer using a piezoelectrically stretched coiled fiber," Appl. Opt. vol. 19, 2926 (1980)
2. J. H. Cole, B. A. Danver, and J. A. Bucaro, "Synthetic -Heterodyne Interferometric Demodulation," IEEE J. Quantum Electron., vol. QE-18, 694 (1982)
3. B. Y. Kim, and H. J. Shaw, "Phase -Reading All-Fiber-Optic Gyroscope," Opt. Lett., vol. 9, 378 (1984)
4. Th. Niemeier, and R. Ulrich, "Quadrature Outputs form Fiber Interferometer with 4×4 Coupler," Opt. Lett., vol. 11, 677 (1986)
5. S. K. Sheem, "Fiber-Optic Gyroscope with [3×3] Directional Coupler," Appl. Phys. Lett., vol. 37, 869 (1980)
6. D. W. Stowe, and T. Y. Hsu, "Demodulation of Interferometric Sensors Using a Fiber-Optic Passive Quadrature Demodulator," IEEE J. Lightwave Technol., vol. LT-1, 519 (1983)
7. L. G. Kazovsky, "Phase- and Polarization-Diversity Coherent Optical Techniques," IEEE J. Lightwave Technol., vol. 7, 279 (1989)
8. B. Y. Kim, J. N. Blake, S. Y. Huang, and H. J. Shaw, "Use of Highly Elliptical Core Fibers for Two-Mode Fiber Devices," Opt. Lett., vol. 12, 729 (1987)
9. A. E. Siegman, "Lasers," Chapter 17, University Science Books, Mill Valley, California (1986)
10. A. Yariv, "Quantum Electronics," Second Edition, John Wiley and Sons, Inc., New York (1975), p. 118.
11. J. N. Blake, S. Y. Huang, B. Y. Kim, and H. J. Shaw, "Strain Effects on Highly Elliptical Core Two-Mode Fibers," Opt. Lett., vol. 12, 732 (1987)
12. S. Y. Huang, J. N. Blake, and B. Y. Kim, "Perturbation Effects on Mode Propagation in Highly Elliptical Core Two-Mode Fibers," to be published in IEEE J. Lightwave Technol.
13. W. V. Sorin, B. Y. Kim, and H. J. Shaw, "Highly Selective Evanescent Modal Filter for Two-Mode Optical Fibers," Opt. Lett., vol. 11, 581 (1986)

Polarimetric Resonator Sensor with High Sensitivity and Large Dynamic Range

F. Maystre and R. Dändliker

Institute of Microtechnology, University of Neuchâtel,
Breguet 2, CH-2000 Neuchâtel, Switzerland

Abstract: A new opto-electronic system which converts the variations of birefringence induced inside an optical resonator into an optical phase difference with a linear response over many periods is presented. This system has been applied to measure static force by stress induced birefringence in a fiber-optic Fabry-Perot resonator.

Introduction: In optical polarimetric sensors the sensitivity of the induced retardation to the external measurand is usually low (lateral force on a 125 µm diam. fiber: 0.35 deg/g , Faraday rotation: $1.7 \cdot 10^{-4}$ deg/A·turn for silica at 780 nm [1]). The net phase shift between two eigenpolarizations which pass through a birefringent medium can be enhanced by placing it into an optical resonator. This technique has been recently demonstrated for optical fiber sensors [1]. However, the phase response is non-linear and the dynamic range of direct phase detection is essentially limited to the width of the resonance. This paper presents a new opto-electronic system, which converts the variations of the birefringence (linear or circular) inside an optical resonator (fiber or bulk, Fabry-Perot or other) into an optical phase difference with a linear response over a range of many periods. It combines high sensitivity and large dynamic range and it is therefore well suited for applications to high resolution transducers.

Principle: The system consists of three parts to perform the following steps.

1) Locking the frequencies of two lasers to two resonances of the two eigenpolarizations of the birefringent resonator [1]. As shown in Fig. 1, this procedure converts the relative position of the two resonances into an optical frequency difference.

2) Comparing the two optical frequencies in one and the same two-beam reference interferometer (e.g. Mach-Zehnder) with an appropriate optical path difference. This interferometer converts the optical frequency variations of each laser into phase variations with linear response and identical scale factors. Direct electronic detection of the beat frequency between the two lasers would avoid the reference interferometer, however, the dynamic range is then limited by the bandwidth of the photodetector and the accuracy depends on the spectral width of both lasers.

Springer Proceedings in Physics, Vol. 44 **Optical Fiber Sensors**
Editors: H.J. Arditty · J.P. Dakin · R.Th. Kersten

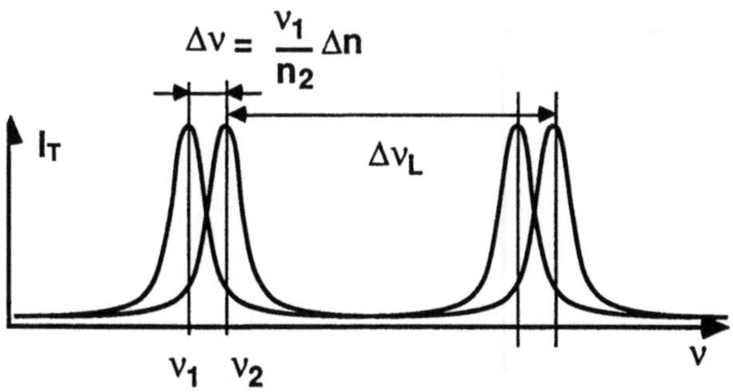

Fig. 1. Intensity transmitted through a birefringent Fabry-Perot resonator as a function of the optical frequency. The retardation of a single pass inside the cavity is $\Delta n 2\pi d/\lambda$, where $\Delta n = n_1 - n_2$ is the difference of the refractive indices of the eigenpolarizations and d is the length of the resonator.

3) Measuring the difference of the phase variations for the two optical frequencies at the output of the reference interferometer. A precise determination of the phase variations for both optical frequencies is accomplished by counting the periods and interpolating between them.

Experiments: To demonstrate the principle of the transducer, the birefringence induced by a transverse force applied on a bare optical fiber was measured. A Fabry-Perot resonator with a free spectral range of $\Delta\nu_L$ = 850 MHz and a finesse of 10 was constructed from a 125 µm diam. fiber with low intrinsic birefringence (York LB600) by gluing gold coated mirrors on both ends [2]. Two well isolated single-mode laser diodes (Hitachi HL 7801, $\lambda \cong 780$ nm) provided the two orthogonal linear polarizations corresponding to the eigenpolarizations of the resonator. The linewidth of each laser was measured to be about $\delta\nu$ = 20 MHz at 5 mW, which gives a coherence length $L_c = c/\pi\delta\nu$ of 5 m. To reduce the effect of dispersion in the resonator and also in the reference interferometer, the lasers were carefully selected and operated at temperatures where they have the same wavelength. This was verified with a grating spectrum analyzer of 0.05 nm resolution. PI controllers with about 20 kHz bandwidth maintained the optical frequencies within 1/100 of the resonance width (i.e. 0.85 MHz) under normal laboratory conditions. The unbalanced Mach-Zehnder reference interferometer uses polarization maintaining fibers (York HB600). At the output, the cross-talk between the two lasers was –20 dB, which limits the accuracy of the interpolation to $2\pi/600$. The optical path difference is 3 m, which gives a sensitivity for the frequency-to-phase conversion of $2\pi/100$ MHz. Fringe counting and interpolation to $2\pi/1000$ is performed by two synchronously triggered fringe counters (FT-Technologies 612 AS).

Figure 2 shows experimental results obtained with this setup. Open circles represent measured values of phase difference at the output of the Mach-Zehnder as a function of the

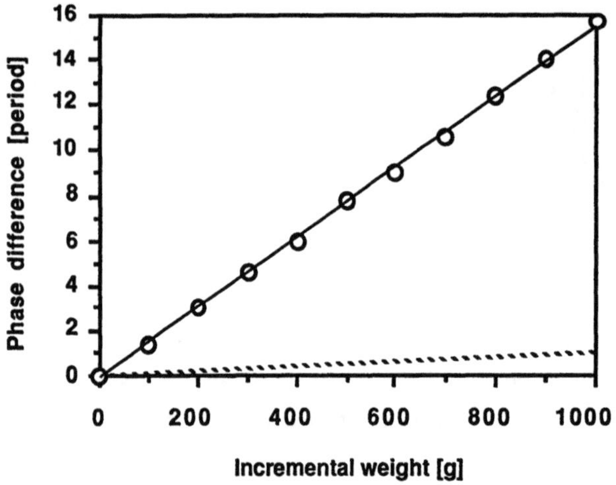

Fig. 2. Experimental results. The dashed line corresponds to the induced retardation inside the cavity for a 125 μm silica fiber. The solid line is the theoretical response for the system with 3 m optical path difference of the reference interferometer. Open circles are measured values.

weight applied to the fiber. The straight line corresponds to the theoretical value for a silica fiber with a diameter of 125 μm [1] and a Mach-Zehnder with an optical path difference of 3 m. The dashed line corresponds to the retardation induced inside the cavity. It shows that the frequency-to-phase conversion system increases the sensitivity by a factor 15. The resolution was only about $2\pi/20$, mainly limited by mechanical instabilities of the recombination of the two beams at the output of the Mach-Zehnder. This problem can be avoided by using an all-fiber polarizing coupler instead of the polarizing beam splitter PBS1.

Conclusion: The presented concept of the opto-electronic detection scheme can be applied to any transducer based on the change of birefringence inside an optical resonator. For transducers with precision in the ppm range, all-fiber systems using high birefringence fibers and all-fiber components can provide the required mechanical stability.

References

[1] Maystre F. and Dändliker R., "Polarimetric fiber optical sensor with high sensitivity using a Fabry-Perot structure", to be published in Appl. Opt., 1989, June issue.

[2] Stone J., "Optical-fibre Fabry-Perot interferometer with finesse 300", Electron. Lett., 1985, **21**, pp. 504-505.

Absolute Fiber-Optic Interferometric Strain Sensor with Insensitive Leading Fiber

L. De Maria, V. Gusmeroli, and M. Martinelli

CISE Tecnologie Innovative S.p.A., Via Reggio Emilia 39, I-20090 Segrate, Milano, Italy

A non-incremental fiber-optic interferometric sensor suitable for remote measurements of strain is reported. The absolute measurement is obtained by joining a continuous sinusoidal frequency modulation of the laser source with a heterodyne process. The use of a low-coherence laser source permits the implementation of an unbalanced sensing cavity and of a noise-free leading fiber. A resolution better than 10 $\mu\varepsilon$ is achieved.

The absolute measurement of physical parameters by means of interferometric techniques represents one of the present research targets. The extension of the fiber-optic interferometric sensors to field applications is limited by its incremental behaviour: all information about the "zero" is lost when the measurement is interrupted. Different techniques (1/2/3) have been proposed in recent years to overcome this restriction. In particular, the continuous sinusoidal frequency modulation (CSFM) of the laser source(4) appears attractive. It joins a simplicity of the scheme with a potential high resolution of the measurement.

The aim of this contribution is to present a remote absolute fiber-optic measurement of strain by means of a non-incremental interferometric technique. The original scheme, based on a low-coherence source, isolates the leading fiber from any environmental noise source. The realized sensor applies the CSFM techniques to a fiber-optic loop cavity.

In Fig.1 the basic characteristics of CSFM techniques are shown. In a normal interferometric sensor (Fig.1a) a single-frequency laser source is used to detect the dynamical physical parameter (the measurand). If the time behaviour of the measurand is a pure sinusoid, the signal from the phase demodulator is again a sinusoid at the same frequency of the measurand and with amplitude proportional to the measurand strength. Let us suppose now to want to detect only the DC component (Fig.1b) of the measurand (static or absolute measurement). In this case, the laser line is sinusoidally modulated at constant frequency. The output from the interferometer is again a sinusoid at the same frequency of the laser modulating signal and with amplitude proportional to

Editors: H.J. Arditty · J.P. Dakin · R.Th. Kersten

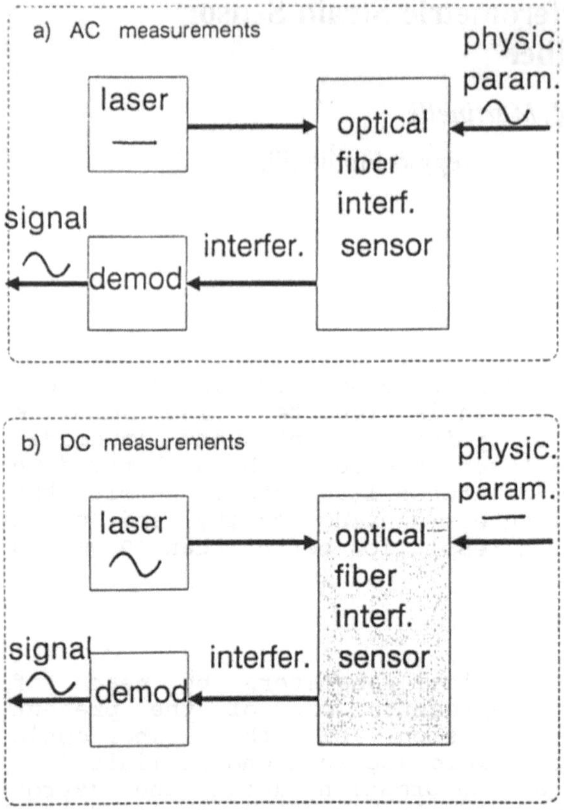

Fig.1.The principle of the Continuous Sinusoidal Frequency Modulation.a) AC measurement:the laser source is single frequency,the measurand is sinusoidal modulated.b) DC measurement:the laser source is sinusoidal frequency modulated, the measurand is constant.

the unbalance $\Delta\ell$ introduced by the DC measurand on the interferometer arms. The demodulated phase signal results:

$$\Delta\phi = \frac{2\pi}{c}\,\Delta\nu\,\Delta\ell \qquad (1)$$

where $\Delta\nu$ is the laser frequency sweep.

In Fig.2 the realized sensor is shown. The interferometric layout is based on the concept of double unbalanced interferometric cavities (5/6). The first unbalanced interferometer is constitued by the sensor (remote). A loop of single-mode fiber, whose optical length nL1 is much longer than the coherent length of the source (nL1>>Lc), is closed on a single-mode fused coupler. The optical path unbalance of the sensor loop is compensated by means of a second unbalanced

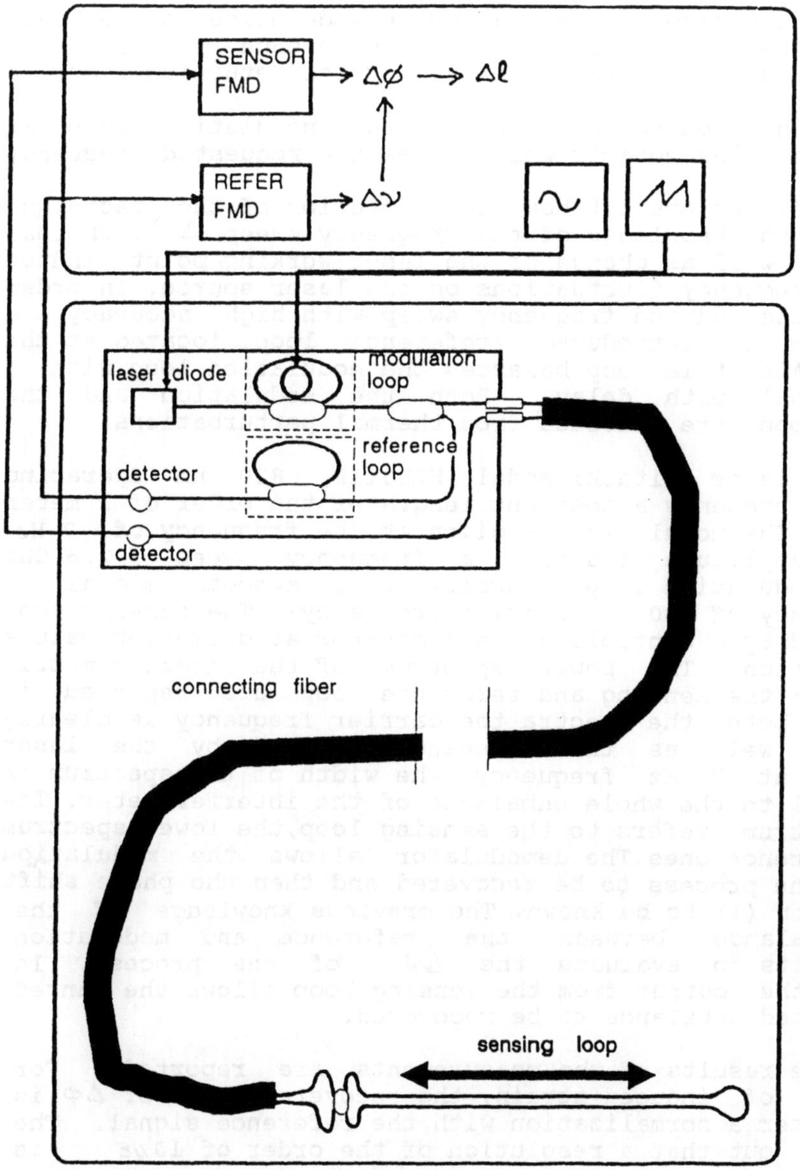

Fig.2.The realized setup.The connecting fibers are insensitive because the use of a low-coherence laser source.The sensing and reference loops are compensated by the modulation loop.

loop (modulation loop, located at the receiver) in order to recover the interferometric signal.If the length of this compensating loop is exactly equal to the sensor length, a high visibility signal is detected. The fiber connecting the two loops guides both the delayed and the direct beams coming from the modulation loop. Any environmental perturbation

acting on this fiber induces a common-mode noise on the beams and it is rejected during the interference process. The interference signal is then conveyed by the returning fiber to the receiver.
In order to heterodyne the scheme, the modulation loop is wound on a piezoceramic which gives the requested frequency shift.
Expression (1) points out how the resolution of the technique depends on the knowledge of the frequency sweep $\Delta\nu$. Thermal phenomena as well as change of the laser working point induce unwanted frequency fluctuations on the laser source. In order to measure the induced frequency sweep with high accuracy, a second loop is introduced (reference loop, located at the receiver). Also this loop balances the modulation loop with a known optical path delay. Both the modulation and the reference loops are shielded from thermal perturbations.

The adopted laser (Hitachi model HL8311E, 820 nm operating wavelength) presents a coherent length of the order of 1 meter and a half. The modulation is given at the frequency of 7 Hz and with amplitude inducing a frequency sweep of 2.6 GHz peak. The modulation loop is excited by a sawtooth signal at the frequency of 300 Hz (carrier frequency). The sensing loop is stretched by a controlled piezoactuator at different values of deformation. The power spectrum of the interferometric signals from the sensing and reference loops are reported in Fig.3. In both the spectra the carrier frequency is clearly visible as well as the sidebands induced by the laser modulation at 7 Hz frequency. The width of the spectrum is proportional to the whole unbalance of the interferometer. The upper spectrum refers to the sensing loop,the lower spectrum to the reference ones.The demodulator allows the modulation index of the process to be recovered and then the phase shift of expression (1) to be known. The previous knowledge of the whole unbalance between the reference and modulation loops, permits to evaluate the $\Delta\nu$ of the process. In parallel, the output from the sensing loop allows the wanted strain-induced unbalance to be recovered.

In Fig.4 the results of the measurements are reported. For each value of induced-strain, the recovered value of $\Delta\phi$ is reported after a normalization with the reference signal. The data point out that a resolution of the order of $10\mu\varepsilon$ is achieved.

In conclusion an interferometric setup suitable for remote measurements of absolute strain is described. The novel scheme permits to remotize the sensing loop while maintaining a noise-rejection of the leading fiber. A reference loop permits fine knowledge of the laser modulation depth and improves the measurement process.
The developed scheme is realized by all-fiber components, thus guaranteeing a high field-detection reliability. Moreover, the whole scheme can be suitable for Integrated Optics implementation. The adopted low-coherence technique permits the development of a multiplexed version of this scheme.

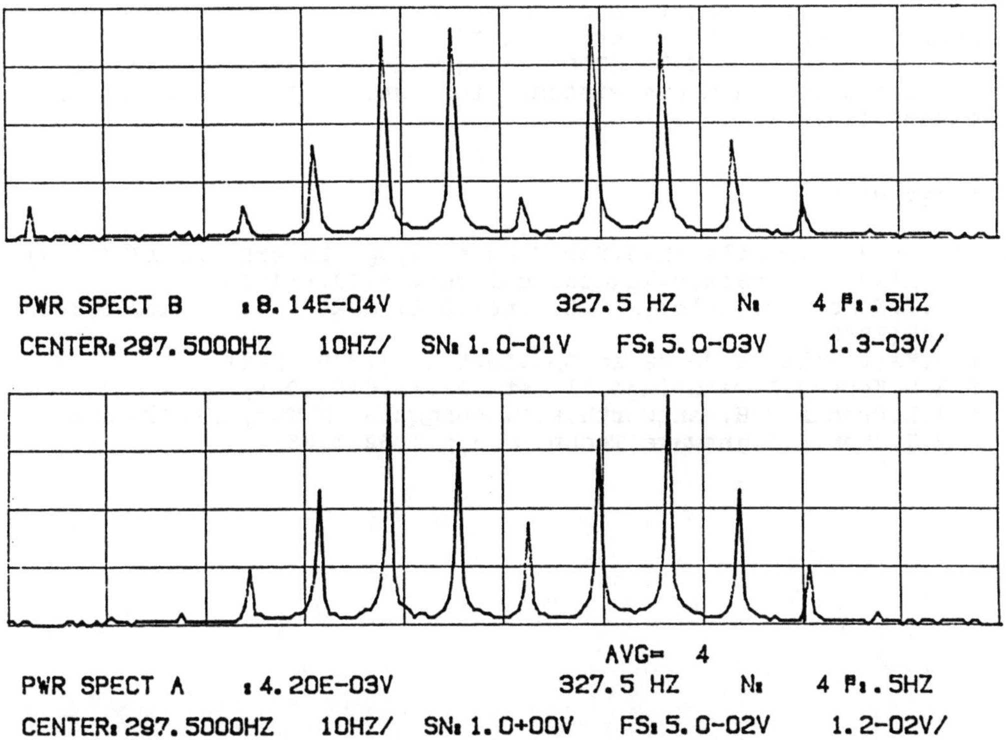

Fig.3.The spectra of the interference signal.The upper spectrum refers to the reference output,the other one to the sensing output.

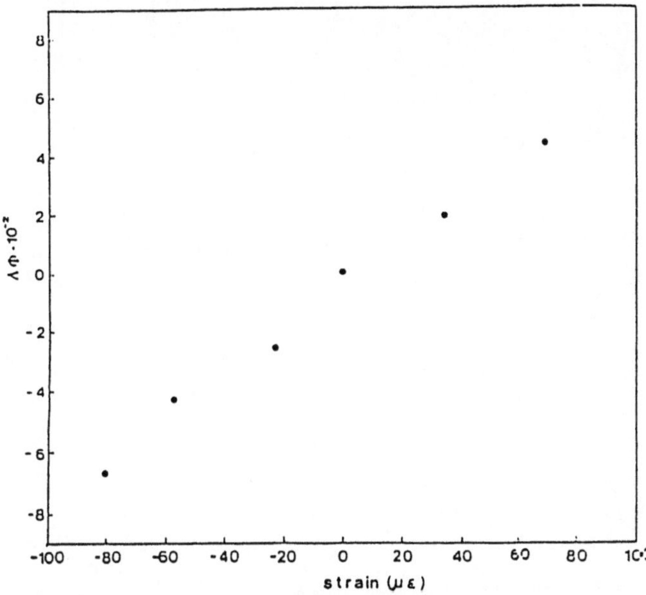

Fig.4.The demodulated sensing output versus the imposed strain.The signal has been normalized by the reference output.

This work was performed under the ENEL-CRIS (Structure and Hydraulic Research Center)"CEMOPT" contract. The authors wish to thank the ENEL-CRIS Management, who authorized this communication, and C.N.Pistoni for stimulating and helpful discussion.

REFERENCES:

1) D.Uttam,I.P.Giles,M.S.Ner,B.Culshaw,Opt.LaserEng.5,173(1984)
2) H.Kikuta,K.Iwata,R.Nagata,Appl.Opt.25,2976(1986)
3) D.J.Webb,R.M.Taylor,J.D.C.Jones,D.A.Jackson,Opt.Comm.66,5-6, 245(1988)
4) M.Martinelli,L.De Maria,Opt.Lett.13,9,779(1988)
5) A.D.Kersey,A.Dandridge,El.Lett.22,11,616(1986)
6) J.L.Brooks,R.H.Wentworth,R.C.Youngquist,M.Tur,B.Y.Kim and H.J.Shaw,J.Lightwave Techn.LT-3,5,1062(1985)

Multimode Fiber Interferometry and Several-Mode Fiber Polarimetry via Phase Conjugation

P.R. Forman, F.C. Jahoda, B.L. Mason, and C.B. Netterfield

Los Alamos National Laboratory, P.O. Box 1663, CTR-8, MS K639, Los Alamos, NM 87545, USA

A multimode fiber interferometer using self-pumped phase conjugation and high frequency heterodyne detection is reported. Lower frequency piezo-electric stretcher modulation has been demonstrated for several channels of multimode interferometry using only a single conjugator. Initial results of Faraday rotation measurements on several-mode fiber with phase conjugation are also given.

INTRODUCTION

While the simplest fiber optics sensors are realized by transmission intensity modulations impressed on multimode fibers, great enhancements in sensitivity are achieved by interferometric or polarimetric methods with light propagated through single-mode fibers. In this paper we report preliminary experiments on retaining the sensitivity of interferometric and polarimetric measurements while using only multimode fiber components. The motivation is greater experimental simplicity (light injection and alignment tolerances), lower cost (cheaper fiber optics components), use of multimode components not yet available in single mode (e.g. high Verdet fibers), or larger intensity transmission limits (which can be translated to higher time resolution for a given signal-to-noise ratio).

It is not an intrinsic requirement for interferometric or polarimetric measurements that a fiber be single mode, since all modes of a multimode fiber are affected by an externally applied stimulus. Single mode operation is, however, generally a practical necessity because of the averaging over the exit phase or polarization, significantly different for each mode, that occurs even in the quiescent state of a multimode fiber before the application of the stimulus.

We use phase conjugation for unscrambling of the spatially superposed modes at the exit of a multimode fiber, prior to deriving the common information carried by all the modes. The phase conjugation is achieved with self-aligning, self-pumped phase conjugation in $BaTiO_3$ [1], requiring no auxiliary beams. While the original phase conjugate, faithful to the transverse spatial distribution exiting the fiber in one polarization, forms only slowly (seconds) in the photorefractive crystal, the response to subsequent spatially uniform phase shifts along the direction of light propagation does not have a sign reversal [2]. It also follows from this that the response to both this kind of phase shift or to any change in the incident intensity is instantaneous. We consider interferometry and polarimetry in separate sections below.

MULTIMODE FIBER INTERFEROMETRY

A. Single Channel

It has been demonstrated by Fischer and Sternklar [3] that a spatially uniform interference pattern can be obtained with a multimode fiber and self-pumped phase conjugator in one arm of a Michelson interferometer configuration. Quite recently Kwong [4] has demonstrated time resolved interferometrically determined temperature changes impressed on the multimode fiber in such an arrangement. He also demonstrated improved utilization of the available light with the polarization-preserving modification for self-pumped $BaTiO_3$ phase conjugation introduced by McMichael et al [5].

In [4] the data were obtained by fringe counting in the basic homodyne arrangement. This does not permit fractional fringe counts, accuracy better than integral fringe number count, nor unambiguous determination of maxima and minima in the measured stimulus, i.e. the true integer value of successive fringes if the time derivative of the measurand changes sign close to a fringe extremum. In our first modification of this basic arrangement we used a heterodyne method, e.g. [6], by placing a 40 MHz acousto-optic Bragg cell in a Sagnac loop at the end of the reference path and measuring the beat signal between the interferometer output and the driver frequency of the Bragg cell.

A 15-cm section of the multimode fiber (50 μm core) in the signal arm was passed through an electrically heated oven on its way to the phase conjugator. The results over a three minute time span with a changing temperature in the oven gave about 100 fringes for a 20^0 temperature change. The average sensitivity, obtained by normalizing to conventional thermocouple readings, is 0.4 fringes/cm/^{0}K. This value is consistent with published values for the thermal expansion of glass. However the ambient noise drift over three minutes with no temperature change in the oven was separately determined to be about 1 to 2 fringes.

The ambient noise was reduced to about 0.2 fringes over a three-minute interval by routing two multimode fibers closely together along their entire length. The fibers had a length difference of 10 cm within the oven, but were made nearly equal total length by increasing the path length of the other fiber outside the oven. Both fiber outputs are phase conjugated on the same crystal, vertically displaced to avoid beam overlap within the crystal, and the return beam of each is recombined with the reference beam. The two approximately 40MHz beat patterns are then electronically processed and computer analyzed for their relative phase differences. The results during changing oven temperature were consistent with the one fiber measurement and are displayed in Fig. 1.

The drift noise decreases nearly an order of magnitude further if the total time duration examined is reduced to only several seconds. Fig. 2 shows the ambient noise (when the oven temperature is held constant) for three different time intervals. The residual value is not much more than the 0.02 fringes obtained independent of time duration when two identical fibers are intertwined along their entire length. The latter is not a useful configuration for an interferometric measurement, but gives an approximation of our electronic noise limit.

An extension of this phase of our work, deferred for that reported below, would also incorporate a fiber in the reference arm with the acousto-optic cell and use self-aligning phase conjugation on the same or a second self-pumped crystal. A by-product of this approach would be a doubled heterodyne frequency due to the forth and back transit of the Bragg cell.

When the very high frequency response of the Bragg cell heterodyne arrangement is not needed (e.g. temperature measurements) it is simpler and more economical to eliminate the seperate reference path altogether and to put a piezo-electric stretcher modulation on one of the two multimode fiber phase conjugator inputs. This reverts to a conventional Michelson two arm interferometer except that both arms terminate in the same phase conjugator crystal. The piezo-electric modulation in one arm achieves the same non-integer fringe count capability and unambiguous determination of sign reversal in time of the signal as described in the beginning of this section.

This type of signal processing has been described before, e.g. [7]. A sinusoidal length modulation gives an equivalent sinusoidal modulation to the phase factor in the exponential of the electric field of the light propagating through the fiber. After recombining the coherent light in the two arms on a square-law intensity detector, a mathematical identity permits the transmitted intensity to be written as an infinite series of Bessel functions in integer harmonics of the modulation frequency. The odd and even terms are amplitude modulated respectively with the sine and cosine of other slowly varying phase factors (e.g. desired signal) in either interferometer arm. Processing the detected intensity electronically to determine separately the first and second harmonic terms for the modulation frequency impressed by the stretcher then gives the sine and the cosine and hence the unambiguous value itself of the signal induced phase shift. If the amplitude of the modulation frequency is reasonably near 2.4 radians phase shift the intensity of the zero order Bessel term is suppressed, and both the 1st and 2nd harmonics have suitably large intensities.

Using a 20 kHz piezo-driver modulation frequency results similar to Fig. 1 were obtained. This was chosen as the detection method in the multiple channel measurements reported below.

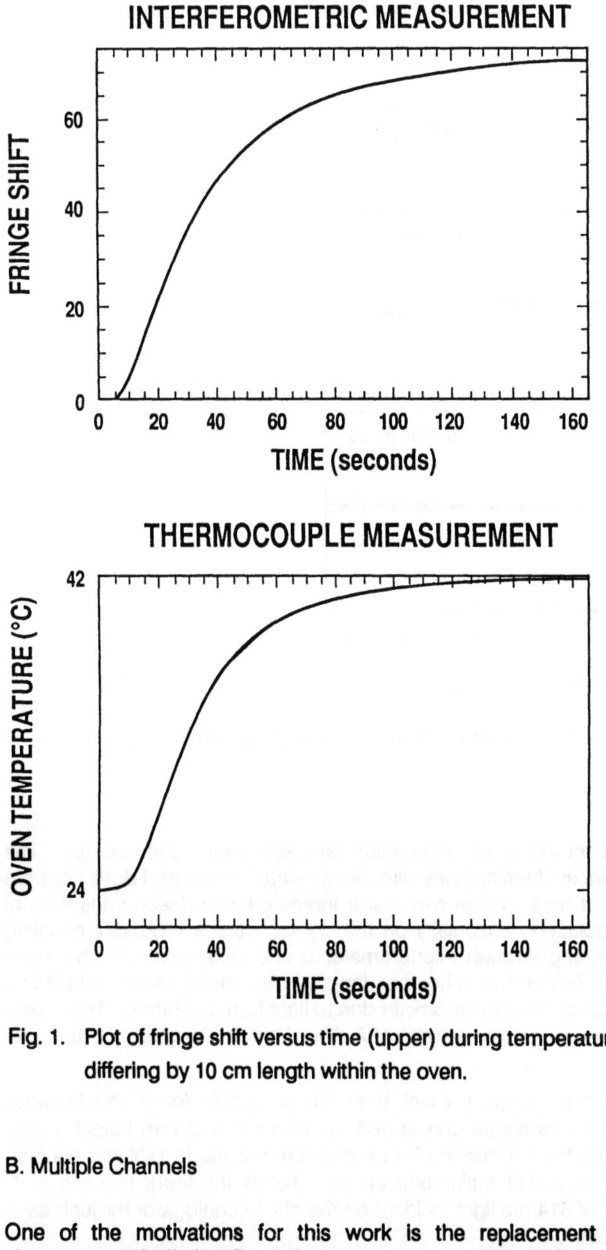

Fig. 1. Plot of fringe shift versus time (upper) during temperature change of oven (lower) for two fibers differing by 10 cm length within the oven.

B. Multiple Channels

One of the motivations for this work is the replacement of numerous thermocouples in the electromagnetically hostile environment of a magnetic confinement fusion experiment. This would gain the well-known fiber optic EMI advantages as well as finite length averaging instead of individual spatial point temperatures. To make this anywhere near cost competitive it is desirable to use the same $BaTiO_3$ phase conjugator for as many separate interferometer pair inputs as possible. (One 5mm x 5mm x 5mm crystal from Sanders, Inc. costs about $3.4K).

The c-axis of the crystal is in the horizontal plane and all the beam fanning and four-wave mixing is also in the horizontal plane. With GRIN lenses terminating the fibers, which limits the vertical spread of the light in the conjugator, it has been possible to separate the two single channel inputs by only 0.8 mm vertically without cross-talk. This should permit six beams or three interferometer pairs in a vertical column on the 5 mm high crystal. A 15 mm vertical height crystal for $6.0K is on order.

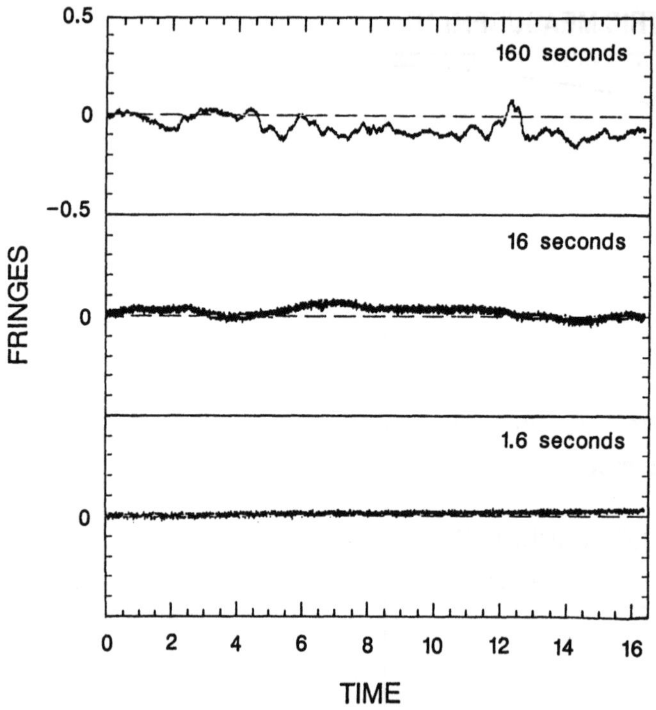

Fig. 2. Observed ambient noise for three different time intervals at constant oven temperature.

For horizontal displacement of beams on the crystal interaction between beams often occurs. The literature on this subject, [8] and references therein, deals with steady state images and does not treat the rapid phase-only changes of interest here. When two of our interferometers (each consisting of two vertically displaced inputs) are displaced horizontally on the crystal there is invariably coupling (some light from interferometer A returning through interferometer B and vice versa). If the equal length arms in A have the same length (within the coherence length of the laser) as those in B this coupling results in cross-talk (modulation of one interferometer due to light from the other). This cross-talk can be eliminated despite partial coupling if the lengths of A differ from the lengths of B by more than the coherence length.

We summarize the above by stating that at the present time the prospects for 6 simultaneous independent interferometer pairs on a 5 mm height crystal and 18 pairs on a 15 mm height crystal appear reasonable. We use an argon-ion laser furnishing light through a multimode 1xN star coupler, each output arm spliced to one arm of a 2x2 multimode coupler that is the input to each of N interferometers. It requires about 2 mw of 514 nm light incident on the phase conjugator through each interferometer arm to get good conjugation.

There are two directions for future research which might increase the number of inputs in a horizontal plane possible on a single phase conjugator crystal. The first is using different incommensurate piezo-stretcher frequencies for each vertical column of interferometer pairs and electrically filtering out the cross-talk. The second is using different wavelength light inputs (multiple laser diodes replacing the argon-ion laser source). No coupling between incoherent beams differing by more than 1 nm has been reported. However, self-pumped phase conjugation in $BaTiO_3$ becomes more difficult and less efficient at the longer wavelengths characteristic of laser diodes [9].

SEVERAL-MODE FIBER POLARIMETRY

The circular birefringence imposed on the two degenerate orthogonal polarization modes of a "single mode" fiber by a longitudinal magnetic field is the basis of Faraday rotation. Via Ampere's law the polarization rotation through single mode fiber loops measures the total current enclosed by the loops. The present status of this type of measurement is reviewed in these proceedings [10].

The biggest problem in Faraday rotation measurements with fibers is the suppression of intrinsic or externally induced linear birefringence. Intrinsic birefringence results from departures from perfect cylindrical symmetry introduced in the manufacturing process, and externally induced birefringence is most commonly the result of the bending radii and asymmetric pressure introduced in the fiber deployment. The fiber leads to and from the sensor section can be a further source of problems, e.g.[11]. The effects of linear and circular birefringence are not simply additive, and the magnitude of the linear birefringence relative to the circular birefringence can either distort or completely obliterate the Faraday rotation. Solutions to suppressing the deleterious effect of linear birefringence include a) twisting the fiber along its axis to induce a large bias circular birefringence b) annealing in situ to relieve stress, and c) Spun- HiBi fiber (nearly circularly polarized eigenmodes) [12].

The Faraday effect is non-reciprocal, i.e. simple reflection at the exit point and re-traversal of the incoming path to the entrance point doubles the rotation angle of linearly polarized light. This occurs because both the magnetic field and the sense of rotation (clockwise or counter-clockwise viewed into the light direction) change sign. A perfect phase conjugator with instaneous response acts exactly like a reflector as far as rotation sense is concerned.

We report below our initial results on making Faraday rotation measurements with the $BaTiO_3$ self-pumped phase conjugator at the 514 nm argon-ion laser wavelength using a Corning fiber that is single mode at 1.3 μm. Based on the V-number there are 10 (doubly degenerate) modes at 514 nm [13]. Linearly polarized light entering the approximately 5 meter long fiber section emerges almost completely unpolarized. The fiber is loosely wound in 6 turns of 25 cm diameter around a damped, oscillatory capacitor discharge current. As expected, a two-polarization analyzer measurement (45° extinction separation) at the fiber exit yielded nothing.

By using the two-orthogonal polarization input method on the $BaTiO_3$ phase conjugator of [5], (subsidiary optics put both polarizations into the plane required by the crystal), we made the two-analyzer-at-45° measurements following a non-polarizing beam splitting at the entrance end of the fiber. The conventional processing gave the result presented in Fig. 3a. While this is far from an accurate representation of the true current time history measured with a Rogowski loop and given in Fig. 3b, the qualitative resemblance is extremely encouraging to pursue further improvement. (Note that in conventional single-mode one-way transits we have previously achieved excellent agreement with Rogowski coils [6]).

The reasons for the distortions in the current waveform are not clear at present. Either or both of two distinct problems may be involved. The first is that we have no knowledge of the amount of residual linear birefringence in our ten mode fiber. As stated above linear birefringence is known to cause distortions. If this is the dominant problem methods used with single mode (particularly annealing) may improve the results and open the way for truly multimode (large core) fiber to be used.

The other open question is more fundamental. We do not know how well the exquisitely precise unscrambling of modes in a fiber by phase conjugation works when the polarization of the return wave, due to polarization rotation away from the conjugator, creates different polarization directions everywhere along the fiber path for the forward and return beams. There is, for instance, the Goos-Hanchen Effect [14], a differential shift along the propagation direction for ordinary and extra-ordinary rays undergoing total internal reflection. (This will be suitable for experimental test when we bring together a non-birefringent single mode fiber at a wavelength amenable to phase conjugation and of more than 2 mw intensity exiting the fiber.) If the distortion problem is indeed inherent to phase conjugation we also do not know if increasing the number of modes to true multimode will increase the distortions or decrease them by further averaging.

In any case it seems likely that smaller rotation angles than the ~ 90 degrees maximum involved in Fig. 3a, either due to smaller currents or fewer fiber loops, will improve the fidelity of the measurement.

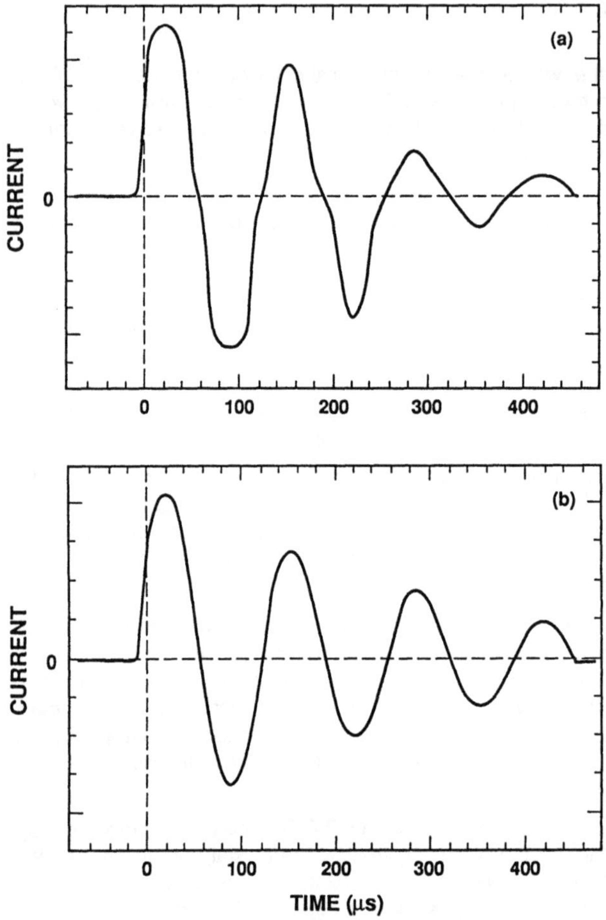

Fig. 3 a. Current vs time deduced from Faraday rotation of light in a 10 mode fiber.

b. Current vs time measured by a current transformer simultaneously with a.

ACKNOWLEDGEMENTS

We gratefully acknowledge the experimental assistance of D. K. Miller and C. R. Hall, and the donation of a piece of Corning SMF-28 fiber by D. L. Franzen, NIST, Boulder, CO, USA.

REFERENCES

[1] J. Feinberg, "Self-pumped, continuous-wave phase conjugator using internal reflection," *Opt. Lett.* **7**, 486 (1982).

[2] J. Feinberg, "Interferometer with a self-pumped phase-conjugation mirror," *Opt. Lett.* **8**, 569 (1983).

[3] B. Fischer and S. Sternklar, "Image transmission and interferometer with multimode fibers using self-pumped phase conjugation," *Appl. Phys. Lett.* **46**, 113 (1985).

[4] N. S. K. Kwong, "Fiber-optic sensor using a tandem combination of a multimode fiber and a self-pumped phase conjugator," *Opt. Lett.* **14**, 590 (1989).

[5] I. McMichael, M. Khoshnevisan, and P. Yeh, "Polarization-preserving phase conjugator," *Opt. Lett.* **11**, 525 (1986).

[6] G. I. Chandler, P. R. Forman, F. C. Jahoda, and K. A. Klare "Fiber-optic heterodyne phase-shift measurement of plasma current," *Appl. Opt.* **25**, 1770 (1986).

[7] D. A. Jackson, A. Dandridge, and S. K. Sheem, "Measurement of small phase shifts using a single-mode optical-fiber interferometer," *Opt. Lett.* **5**, 139 (1980)

[8] P. Yeh, "Coupled-mode theory of hologram sharing in mutually pumped phase conjugators," *Appl. Opt.* **28**, 1961 (1989).

[9] B. T. Anderson, P. R. Forman, and F. C. Jahoda "Self-pumped phase conjugation in $BaTiO_3$ at 1.06μm," *Opt. Lett.* **10**, 627 (1985).

[10] G. W. Day, "Recent advances in Faraday effect sensors," *OFS-6 '89*, these proceedings.

[11] P. R. Forman and F. C. Jahoda, "Linear birefringence effects on fiber-optic current sensors," *Appl. Opt.* **27**, 3088 (1988).

[12] Experimental product of York Ventures and Special Optical Products, Ltd.

[13] S. E. Miller, E. J. Marcatili, and T. Li, "Research toward optical fiber transmission systems. Part I: The transmission medium," *Proc. IEEE* **61**, 1703 (1973).

[14] A. W. Snyder and J. D. Love, in *Optical Waveguide Theory* (Chapman and Hall, 1983).

Part III

Interferometer Signal Processing II: White Light Interferometry and Coherence Modulation

Interferometric and Polarimetric Sensors Using White-Light Interferometry

R. Ulrich

Technische Universität Hamburg-Harburg,
D-2100 Hamburg 90, Fed. Rep. of Germany

In optical fiber sensors using white-light interferometry the information-transmitting light-flux has a broad ("white-light") spectrum with a channelled structure. Its envelope of width B is determined by the optical source, by the transmission spectrum of the fiber, and by the response of the detector. Under this envelope there exist a large number of narrow lines, typically spaced equidistantly. Either that spacing δf or the absolute spectral position of one line of this "comb" is used to represent the measurand x. In these sensors, therefore, the transducer is an optical comb filter, typically an interferometer that is tuned by the measurand. The comb spectrum is decoded, i.e. the value of x is recovered, in an evaluation unit that contains a receiving interferometer. The decoding process involves a parameter y that is equivalent to x, and some scheme of scanning, tuning, or dithering to determine that value of y for which the detector signal becomes maximum. The corresponding output value of the sensor is denoted here by x_o, so that $\delta = x_o - x$ represents the error of the sensor.

The main advantages of these sensors are:

- A high insensitivity of the output value x_o to loss variations in the fiber-optic transmission line. This requires a sufficiently large number $M = B/\delta f$ of lines in the comb. Experimentally errors as small as $\delta / x_{max} \leq 10^{-6}$ for a 5dB loss variation have been observed.
- A very high possible resolution, expressed in the number $R = x_{max}/\delta$ of resolvable spots. $R = 10^6$ has been reached.
- The possibility to operate the sensors with an LED source, with multimode fibers and with inexpensive connectors.

As a consequence, entire sensor systems can be based on white-light interferometry, with tranducers designed for a variety of measurands and for different ranges, providing connector-compatible interchangeability without recalibration. The development of such systems proceeds presently in the directions of improved simplicity, stability, and reliability for applications in adverse industrial environments.

In the talk, the principle of operation of these sensors is analyzed in some detail, in the spectral domain as well as in the time domain. The problem of index dispersion in the interferometer arms is addressed. It is shown how this problem may

Springer Proceedings in Physics, Vol. 44 **Optical Fiber Sensors**
Editors: H.J. Arditty · J.P. Dakin · R.Th. Kersten

either be avoided, or how it may be used advantageously in the encoding process. Various kinds of interferometers are compared for the use as fiber-optic displacement sensors and as refractometers. Polarization-optical arrangements are shown to be just special implementations of the general interferometric principle, permitting the compact design of sensors for temperature, force, pressure, electrical and magnetic measurands.

Reflectometry with High Spatial Resolution and No Moving Parts by Means of Source-Coherence Modulation

K. Hotate and O. Kamatani

RCAST, Research Center for Advanced Science and Technology,
The University of Tokyo, 4-6-1 Komaba, Meguro-ku, Tokyo 153, Japan

Novel reflectometry with high spatial resolution and a simple setup for evaluating the optical components or the total system are proposed, in which the source coherence is intentionally modulated through the direct frequency modulation of the laser diode. Several modulation schemes are evaluated and a successful preliminary experiment using a simplified model is reported.

1. Introduction

An optical system for communication or sensing is being actively developed for practical use, exploiting recent progress in optical devices. The need to evaluate the characteristics of each component or total system along the optical path is steadily increasing. In spite of the demands on such a measurement system, there are a few efficient techniques to meet the requirements of high resolution and dynamic range with an uncomplicated setup. So far, methods using a low coherent source [1,2] and modified optical frequency domain reflectometry [3] have been reported. However, these methods have mechanical moving structures.

In this paper, we propose a new reflectometry by means of source-coherence modulation for high resolution measurements of optical waveguide devices using a laser diode light source. This method has no mechanical moving parts. The oscillation frequency of the laser diode is directly controlled by the injection current with an appropriate waveform, and then its time-averaged spectrum shape or the coherence function can be modulated to obtain the distribution of the backscattering intensity. This method is an Optical Coherence Domain Reflectometry: OCDR [4]. We present the principle of the OCDR, considering three types of modulation waveform. The basic experiment has been successfully carried out using a simplified model.

2. Principle of the OCDR

The fundamental setup for the OCDR is shown in Fig.1. This is the Michelson interferometer with a frequency shifter, acousto-optical modulator(AOM) driven by frequency f/2. The

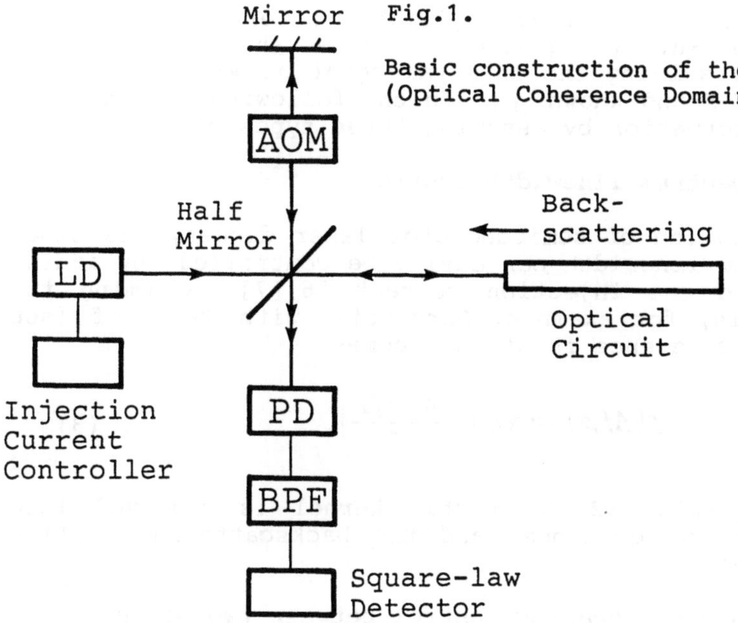

Fig.1.

Basic construction of the proposed OCDR (Optical Coherence Domain Reflectometry).

interference is detected as an intermediate signal of frequency f by a photo detector(PD). We assumed to first order that the signal caused in the optical circuits is Rayleigh backscattering [5], and carried out the theoretical analysis. Interference component I(t) of the frequency f in the PD output is expressed as

$$I(t) = -j\left(\frac{E_0 n 2\pi f_0}{c}\right)\exp(-j2\pi f t) \int R(z)\gamma(F,z)\Delta n(z)\exp(-j2kz)\,dz \quad , (1)$$

where f_0 is the average oscillation frequency of the light source, and R(z) is the backscattering amplitude profile along the distance z. (The position where the path length difference is zero corresponds to z=0.) γ is the coherence function of the light source. In Eq.(1), F is the parameter that expresses the power spectrum shape of the source and $\Delta n(z)$ is the normalized refractive index fluctuation [5]. Input light amplitude, effective refractive index of the waveguide, and the wave-number are denoted as E_0, n, and k, respectively.

Then the output signal P(F) of the square-law detector (SQD) becomes

$$P(F) = \left(\frac{E_0 n 2\pi f_0}{c}\right)^2 \alpha_R' \int R^2(z)\gamma^2(F,z)\,dz \quad , (2)$$

where α_R' is the normalized Rayleigh scattering coefficient corresponding to $\Delta n(z)$. Equation (2) shows that we can obtain $R^2(z)$ profile by solving the integral equation with measured P as a function of F and known γ. In the following, we show the practical transformation by assuming three types of γ.

(a) OCDR with spectrum linewidth control

In the situation of semiconductor laser diodes, the time-averaged spectrum linewidth can easily be controlled by adding white noise into the injection current [6][7]. Assuming the spectrum shape is, for example, Lorentzian with HWHM Δf (see Fig.2-(a)), its coherence function becomes

$$\gamma(\Delta f, z) = \exp\left(-\frac{4\pi \Delta f z}{c}\right) \quad . \quad (3)$$

The integral equation (2) with this kernel is resolved into linear simultaneous equations, and the backscattering profile $R^2(z)$ is obtained.

(b) OCDR with spectrum spacing control between two peaks

Assuming the spectrum has two peaks as shown in Fig.2-(b), the coherence function is expressed as

$$\gamma(\Delta f, f_s, z) = \exp\left(-\frac{4\pi \Delta f z}{c}\right)\cos\left(\frac{2\pi f_s z}{c}\right) \quad , \quad (4)$$

where f_s is the spacing, and each shape is assumed Lorentzian. The spectrum with two peaks is easily obtained by injecting a current of a square waveform with time averaging (see Fig.2-(b)). This modulation has been confirmed to give the coherence function of Eq.(4) by detailed calculation. Using Eqs. (2) and (4), we have a Fourier cosine transformation equation. The backscattering profile $R^2(z)$ is given, by inverse calculation with measured P as a function of f_s, as

$$R^2(z) = K\int\left\{P(f_s) - \frac{1}{2}P(0)\right\}\cos\left(\frac{4\pi f_s z}{c}\right)df_s \quad , \quad (5)$$

where K is a constant. The spatial resolution δz is determined, by its total frequency shift f_m, as

$$\delta z = \frac{c}{2f_m} \quad . \quad (6)$$

Assuming the use of a DFB laser, for example, the frequency shift 600 GHz corresponds to a resolution of 250 μm (in air).

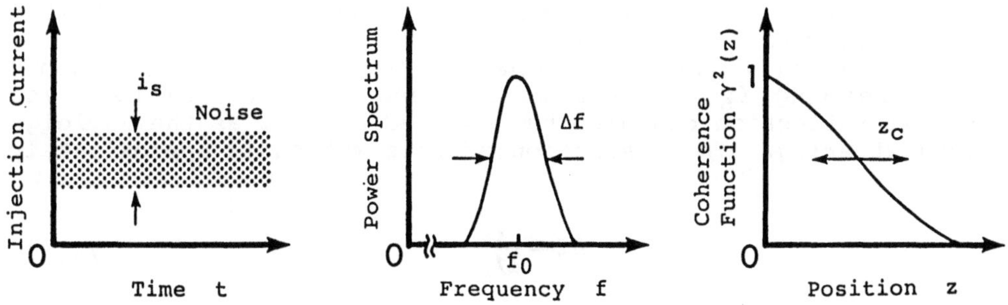

(a) OCDR with spectrum linewidth control

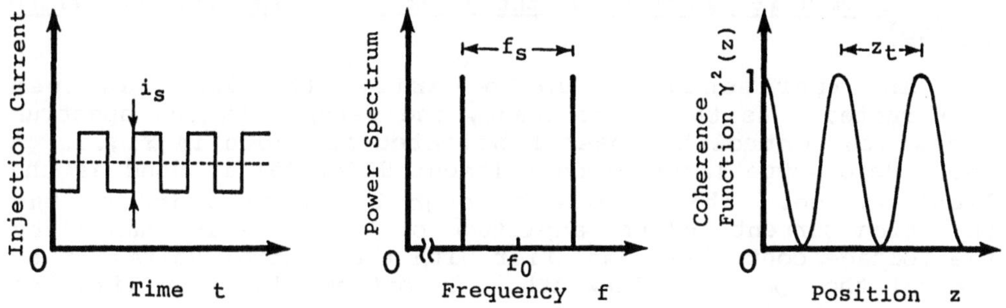

(b) OCDR with spectrum spacing control between two peaks

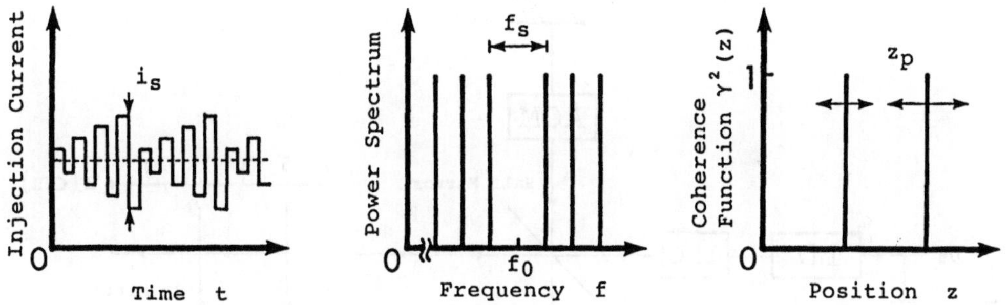

(c) OCDR with spectrum spacing control with many peaks

Fig.2. Three types of the OCDR techniques

(c) OCDR with spectrum spacing control with many peaks

In the case where the spectrum shape has 2N peaks, as shown in Fig.2-(c), which has also been confirmed to be achieved by injecting a waveform shown in the figure with time averaging, the coherence function is derived as

$$\gamma(\Delta f, f_s, z) = \exp\left(-\frac{4\pi \Delta f z}{c}\right) \sum_{k=1}^{N} \cos\left(\frac{2\pi k f_s z}{c}\right) \quad , \quad (7)$$

where the summation term approximately gives a periodic delta function assuming relatively large N (See Fig.2-(c)). Choosing the parameters so that only one delta-like function is located at the portion to be measured, the distribution of the backscattering $R^2(z)$ is obtained directly as the SQD output with sweeping f_s. There is no need for calculation, and Fresnel reflection can also be measured as well as the Rayleigh backscattering. The resolution of this method is approximately given by

$$\delta z = \frac{c}{2f_m} \tag{8}$$

3. Experimental results of the OCDR with spectrum linewidth control

An experimental setup to verify the OCDR has been constructed. As the first stage, the setup with the spectrum linewidth control has been fabricated as shown in Fig.3. A Fabry-Perot-type laser diode (Hitachi HLP-1400) is used as the light source. White noise through 2 GHz is added to the injection current and the amplitude of the noise is changed by the voltage controlled amplifier with the control voltage V to control the spectrum linewidth. An optical isolator with the

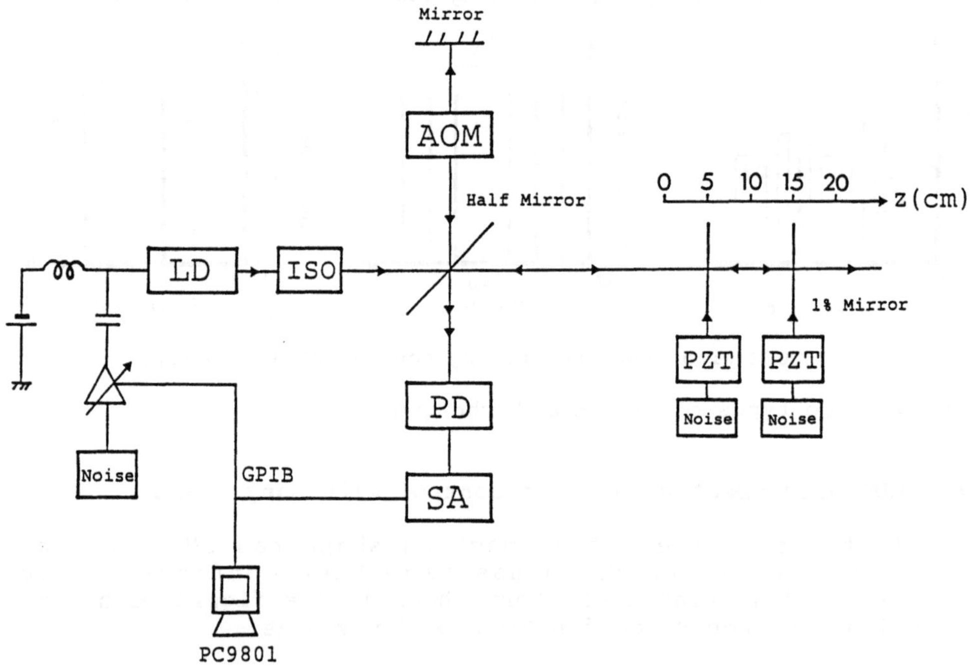

Fig.3. Experimental setup to verify the OCDR with spectrum linewidth control.
LD:laser diode, ISO:isolator, AOM:acousto-optical modulator, PD:photodetector, SA:spectrum analyzer, PZT:piezoelectric transducer

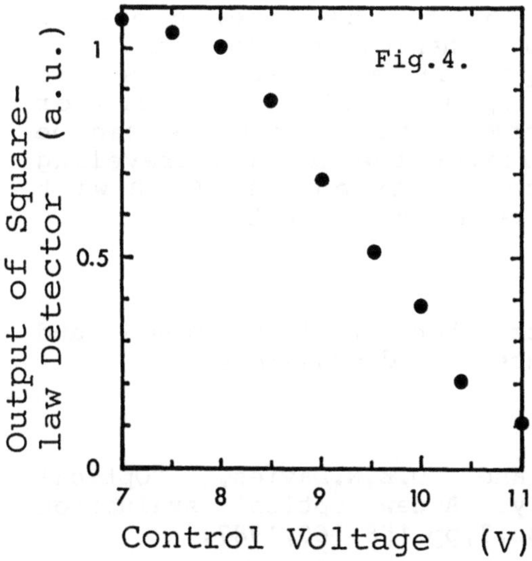

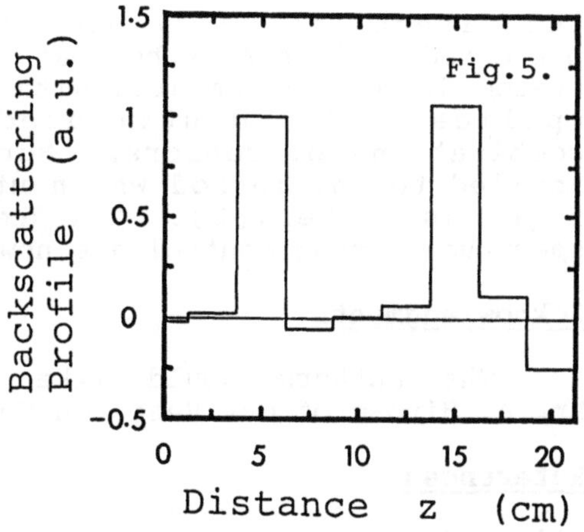

Fig.4. The relation between the output of the square-law detector and the noise control voltage when 1% mirrors are placed at distances of 5 and 15 cm.

Fig.5. Backscattering profile calculated from the experimental data in Fig.4. Two peaks correspond to the two 1% mirrors.

isolation of 50 dB is inserted in the path to prevent optical feedback-induced intensity noise. Two 1% mirrors are set as the scatterer. To simulate the characteristics of the Rayleigh scattering $\Delta n(z)$, each mirror is fixed on a piezoelectric transducer(PZT), driven by a noise source. The AOM is driven at 40 MHz, and the interference signal is measured by a spectrum analyzer(SA), as a 80 MHz signal. The mirrors are placed at distances of 5 cm and 15 cm from the point where the path difference is zero.

The relationship between obtained P and the control voltage V is shown in Fig.4. Because the laser linewidth increases due to the rise of the noise control voltage V, the degree of coherence is reduced. As a result, the output of SQD decreases. The kernel $\gamma(F,z)$ in Eq.(2), that is $\gamma(V,z)$, has been measured and used in solving the equation. From the result in Fig.4, the backscattering profile along the distance z has been calculated as shown in Fig.5. Equation (2) has been converted to the linear simultaneous equations with the distance increment of 2.5 cm. Two peaks in Fig.5 correspond fairly well to the mirror positions at 5 cm and 15 cm respectively.

4. Conclusion

We have proposed a new OCDR technique for high resolution measurements of optical circuits, and also presented the preliminary experimental results, which confirm the theoretical analysis. In the OCDR, the source-coherence modulation through the direct frequency modulation of the laser diode is performed, and this method has no mechanical moving parts. The system is simple compared with previous methods. It is also applicable to distributed optical fiber sensors or combinations of sensors. Moreover, this technique can be applied to the method which utilizes the forward traveling light as in Ref.[2]. The experiments on the OCDR with spectrum spacing control are now being carried out.

Acknowledgments

The authors would like to thank Prof. T. Okoshi and Mr. A. Hirose of our Research Center for discussions.

References

[1] R.C.Youngquist, S.Carr, and D.E.N.Davies, "Optical coherence-domain reflectometry: A new optical evaluation technique,"Opt.Lett.,vol.12,no.3,pp.158-160,1987.

[2] M.Tsubokawa, T.Higashi, and Y.Sasaki, "Measurement of mode couplings and extinction ratios in polarization-maintaining fibers,"J.Lightwave Technol.,vol.7,no.1,pp.45-50,1989.

[3] H.Barfuss and E.Brinkmeyer, "Modified optical frequency domain reflectometry with high spatial resolution for components of integrated optic systems," J.Lightwave Technol.,vol.7,no.1,pp.3-10,1989.

[4] The name OCDR has already been used in Ref.[1], in which the moving part is inherently required. The principle of the technique that we report here is completely different from that shown in Ref.[1].

[5] M.Nakazawa,"Rayleigh backscattering theory for single-mode optical fibers,"J.Opt.Soc.Am.,vol.73,no.9,pp.1175-1180,1983.

[6] K.Hotate, N.Okuma, M.Higashiguchi, and N.Niwa, "Rotation detection by optical heterodyne fiber gyro with frequency output,"Opt.Lett.,vol.7,no.7,pp.331-333,1982.

[7] D.T.Jong and K.Hotate, "Optical fiber interferometric sensor with immunity from environmental disturbance induced in the lead-fiber," submitted to OFS-6'89.

Interferometric Spectrally Encoded Sensor Using a Superradiant Diode

P. Sansonetti, M. Lequime, and H. Giovannini

Bertin & Cie., BP 22, F-13762 Les Milles Cedex, France

A spectrally encoded displacement sensor that only uses a superradiant diode as source and a Michelson interferometer as sensor is presented.The diode Fabry-Pérot spectral density modulation and short coherence length create a correlation signal when the interferometer path length difference is equal to a multiple of the optical round trip of the diode cavity.The displacement is measured either when the diode injection current is constant or varied to lock the diode optical round trip to the interferometer.

INTRODUCTION

Interferometric spectrally encoded sensors are very interesting for their potential unsensitivity to source output power or in-line transmission variations as well as their ease of multiplexing (1).

Usually, such a sensor is based on a broad spectrum source (for instance a LED) which is launched, through a multimode optical fibre, into a sensing interferometer; the path length difference of the latter is greater than the source coherence length and depends on the parameter to measure.The output spectral density therefore exhibits a periodic variation in frequency.When this signal is injected, again through a multimode fibre, into another receiver interferometer,a correlation signal appears when the difference between the path length differences of both interferometers is smaller than the source coherence length.The measurement can then be made either with a static receiving interferometer or by phase-locking it to the sensing one.

Here, we present a new scheme that requires only the sensing interferometer and a Superradiant Diode (SRD) as the source (2).It takes advantage of two SRD properties: a Fabry-Pérot modulation of the spectral density and a short coherence length.Therefore, a correlation signal is obtained when the sensing interferometer path length difference is close (within the coherence length) to a multiple of the SRD cavity optical round trip.

PRINCIPLE

The principle of the sensor is represented in Figure 1.Here, the sensing interferometer is of the Michelson type but another one could be used.The correlation signal is detected at the other port but a coupler could also allow detection at the input port.

The SRD spectral density can be written as :

$$S(k)=S_0(k)\,(1+2\,\Sigma_{p=1,\infty}\,(R_1 R_2 G_i^2)^{p/2}\cos(p\,2\,k\,n_{eff}\,L))$$

where $k=2\Pi/\lambda$, λ being the wavelength.The second term corresponds to Fabry-Pérot modulation;it was calculated using the autocorrelation formalism (3). $S_0(k)$ is the envelope of the SRD spectral density;it is given by (4):

$$S_0(k)=(1-R_1)/(1-R_1 R_2 G_i^2)\,S_s(k)$$

$S_s(k)$ is the spontaneous emission power integrated along the cavity ; R_1 and

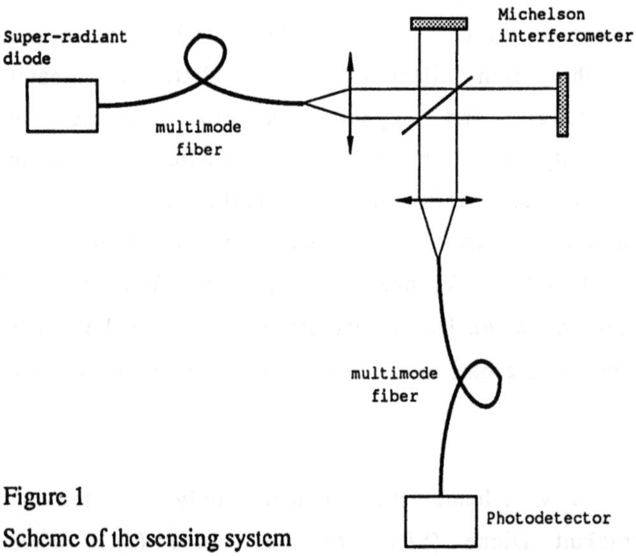

Figure 1
Scheme of the sensing system

R_2 respectively are the power reflectance of the front and rear facets, G_i is $(G_+G_-)^{1/2}$ where G_+ and G_- are the single-pass power gains in opposite directions, n_{eff} the real part of the guided mode effective index and L is the SRD cavity length.The round trip guided mode overall power gain is given by $R_1R_2G_i^2$;it is smaller than one in a SRD, since it is operating below laser oscillation.This gives SRD a broad spectrum containing Fabry-Pérot modulation.The power of the signal coming out of the interferometer is:

$$P= {}_{\Delta k}\int S(k)\,(1+\cos k\Delta)/2\;dk$$

where Δk represents the SRD spectral width and Δ is the interferometer path length difference.If Δ and $2\,n_{eff}\,L$ are both greater than $2\Pi/\Delta k$ which is the SRD coherence length, a correlation signal is obtained when Δ is close to a multiple of $2\,n_{eff}\,L$ (within the SRD coherence length);the output power of the pth order is therefore given by:

$$P_p= {}_{\Delta k}\int S_0(k)/2\;\left(1+(R_1R_2G_i^2)^{p/2}\cos k\,(p\,2\,n_{eff}L-\Delta)\right)dk$$

If we take the simple model of a gaussian distribution for S_0 and G_i, that is: $S_0(k)=S_0^M\exp\left(-4\left((k-k_0)/\Delta k\right)^2\right)$ and $G_i(k)=G_i^M\exp\left(-4\left((k-k_0)/\Delta k\right)^2\right)$, the value of P_p is finally given by:

$$P_p=P_0/2(1+(p+1)^{-1/2}(R_1R_2(G_i^M)^2)^{p/2}\exp(-(\Delta k\,\delta/4(p+1)^{1/2})^2)\cos k_0\delta)$$

where $\delta = p\,2\,n_{eff}\,L-\Delta$ and P_0 is the SRD output power.To measure Δ, the value of $n_{eff}\,L$ can be kept constant or varied through the injection current to lock $2\,p\,n_{eff}\,L$ on Δ.

EXPERIMENTS

The SRD is of the GaAs/GaAlAs type from General Optronics.At an injection current of 160 mA, its whole emission spectrum is represented in Figure 2a.This broad spectrum is the cause of the SRD short coherence length (50 µm).The Fabry-Pérot modulation is clearly seen in Figure 2b.

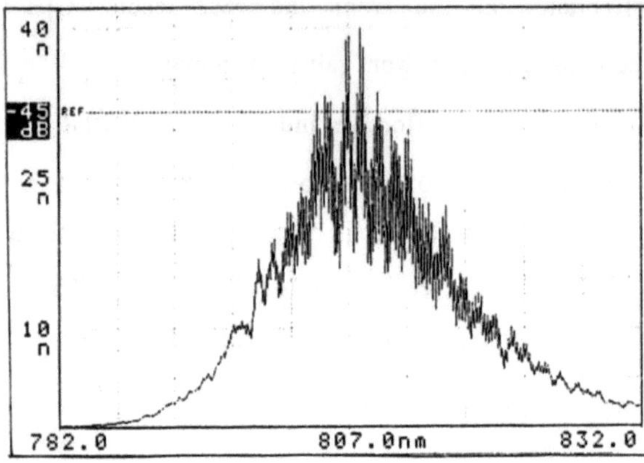

Figure 2a

Wavelength emission spectrum of the SRD at 160 mA

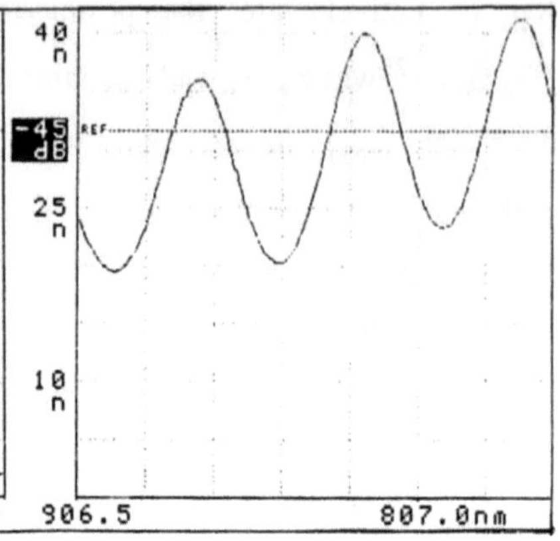

Figure 2b

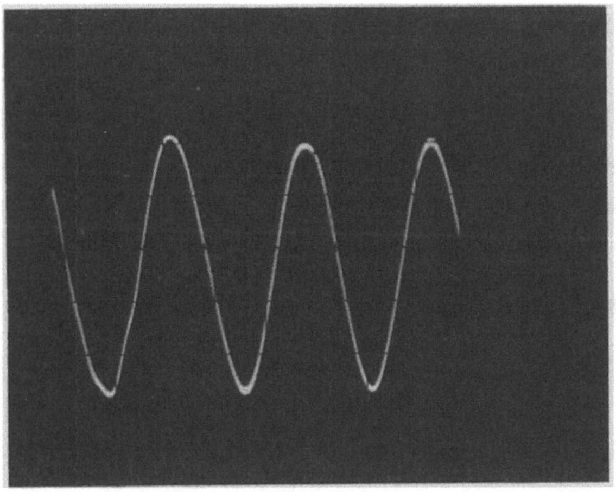

Figure 3 Oscilloscope trace of the correlation signal vs mirror translation(i=160 mA)

The SRD output was launched into the sensor system according to the experimental set-up represented in Figure 1.When Δ increases from zero,the classical interference term of the Michelson disappears.Then, the first order correlation term appears and has its maximum value when Δ equals 2.740 mm.Taking a value of 3.6 for n_{eff}, which is a typical value for GaAs/GaAlAs diode, a value of 380 μm is obtained for L.When one Michelson mirror has a ramping movement created by a piezoelectric translator, the oscilloscope trace of the signal versus the piezoelectric voltage is represented in Figure 3;the injection current was 160 mA in that case.A full fringe represents a mirror displacement of a half- wavelength.

Therefore, we see that a simple spectrally encoded displacement sensor is obtained.We shall see, further in this paper, that it also exhibits the multiplexing features of this type of sensor.

In Figure 4, the measured modulation index of this signal is represented versus the SRD injection current.The variation is due to the one of G_i.Even at 190 mA, the index is 12.9%, which means that the diode still exhibits a superradiant behaviour.Thus, higher injection currents could still be used.

Then the sensing system was operated with n_{eff} L locked on Δ through the SRD injection current, by taking advantage of the variation of $n_{eff}L$ (mainly through n_{eff}) with the injection current.Such an experiment was carried out for the first and second order correlation signals.We have represented in Figure 5 the resulting current variation, from 170 mA, versus the Δ variation.As predicted, the second order has twice the sensitivity of the first order. Δ was varied slowly and therefore mainly the thermal effect is responsible for the increase of n_{eff} L with current.A negative variation would be obtained at higher frequencies where the free carrier effect is dominant.In

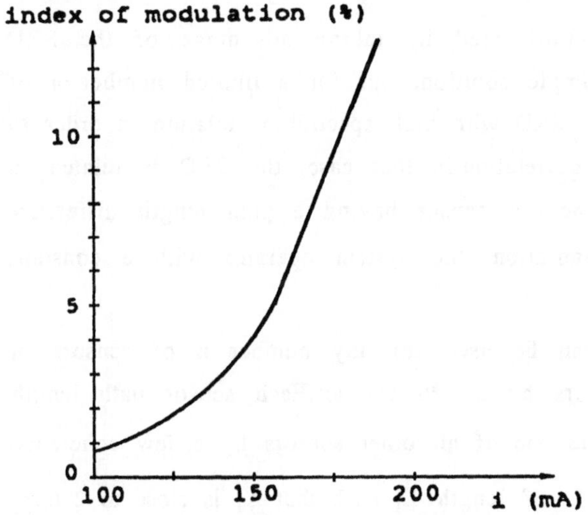

Figure 4 Modulation index of the correlation signal versus the SRD injection current

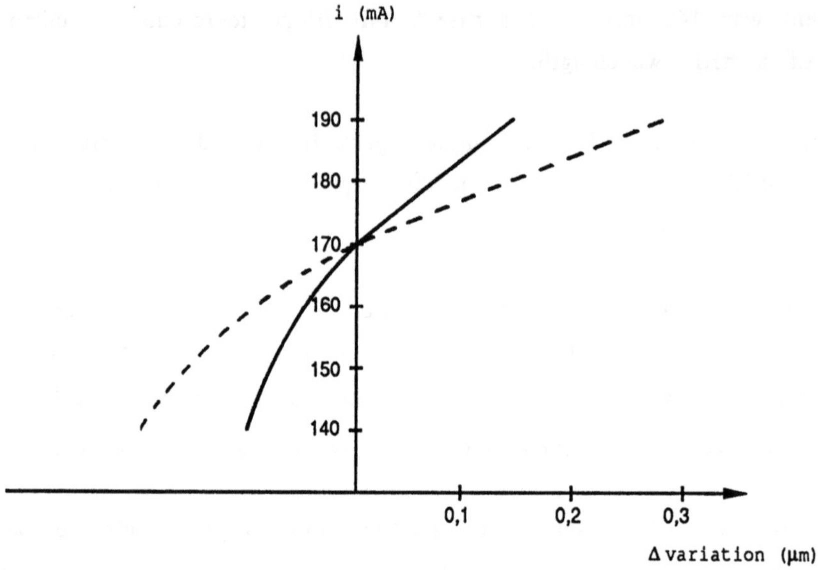

Figure 5 Injection current variation versus mirror translation in a phase locked scheme (—— first order; - - - second order)

that case, very high measurement bandwidth could be obtained. To check that the phase variation is not due to wavelength shift, the experiment was carried out for $\Delta=0$ (zero order interference term); no phase variation was observed.

POTENTIAL MULTIPLEXING

These sensors can be multiplexed by taking advantage of the SRD short coherence length. A first simple solution, but for a limited number n of sensors, would be to use a single SRD with high spectral modulation in order to have power in several orders of correlation.In that case, the SRD is related to the n sensors by a 1*n coupler, the pth sensor having a path length difference close to p 2 n_{eff} L.In that configuration, the system operates with a constant SRD injection current.

Another multiplexing scheme can be used for any number n of sensors; n SRDs are coupled to the n sensors by a n*n coupler.Each sensor path length difference Δ_i is separated from the one of all other sensors by a few coherence lengths, and corresponds to a SRD of length L_i such that Δ_i is close to 2 n_{eff} L_i.Therefore, the multiplexing scheme is simply obtained by using n SRDs (of different lengths) correlated to n sensors.Such a scheme can operate with either constant SRD injection current or in phase locked configuration.

CONCLUSION

An interferometric spectrally encoded displacement sensor that only uses a superradiant diode as source and a Michelson interferometer as sensor has been presented.The diode Fabry-Pérot spectral density modulation and short coherence length create a correlation signal when the interferometer path length difference is equal to a multiple of the optical round trip of the diode cavity.Displacement was measured by using either the first or second order correlation signal. The injection current was either constant or varied to allow the locking of the diode cavity optical round trip on the interferometer path length difference.In the former case, the dynamic range is twice the SRD coherence length (around 100 μm).In the latter case, mainly thermal effects were present.A dynamic range of 0.48 μm was obtained for the second order.In the phase locking configuration, the free carrier effect offers the possibility of a very simple and high bandwidth operation.

Finally, potential multiplexing schemes have been presented that use either a single SRD with each sensor accorded on one correlation order or as many SRD as sensors, each sensor being accorded to one SRD.

REFERENCES

(1) Th. Bosselmann, R. Ulrich, 'High accuracy position-sensing with fibre-coupled white-light interferometers', OFS' 84 , Stuttgart, pp 361-364.

(2) P. Sansonetti, French patent No. 89 02388.

(3) M. Lequime, private communication.

(4) I. P. Kaminow, G. Eisenstein, L.W. Stulz, 'Measurement of the modal reflectivity of an antireflection coating on a superluminescent diode', IEEE J. Quant. Elect., Vol. QE-19, No. 4, 1983, pp 493-495.

Interferometric Displacement Sensing by Visibility Modulation

T.A. Berkoff[1] *and A.D. Kersey*[2]

[1]Sachs-Freeman Associates, Landover, MD 20785, USA
[2]Optical Techniques Branch, Code 6570, US Naval Research Laboratory, Washington, DC 20375, USA

An interferometric technique for the measurement of displacement is described which is based on fringe-visibility modulation. The approach utilizes a synthesized dual-mode source with variable mode spacing to modulate the interference visibility in an unbalanced Mach-Zehnder fiber interferometer which includes an air path in one arm.

Introduction

A range of approaches to monitor quasi-static displacement using interference-based techniques with single-mode optical fibers have been demonstrated over the past decade, using direct interferometry [1], frequency chirped (FMCW) interferometric ranging [2], sub-carrier RF interferometry [3], and dual-wavelength interferometry [4]. These various approaches cover a wide spectrum of measurement ranges, from sub-nm to mm resolutions, and thus are suitable for use in a diverse range of application areas. In this paper we report preliminary results obtained using a new displacement sensing technique based on the modulation of the interference fringe visibility function in an unbalanced fiber interferometer.

Principle

The principle of operation is based on the well-known dependence of the interference fringe visibility on optical path difference (OPD) between the arms of an interferometer illuminated by a multimode source. In particular, the dependence of the visibility on OPD for a dual-mode source is of interest here; Figure 1 shows the dependence for a single-mode (single frequency) and a dual mode. In the case of the dual-mode source, the visibility function or interferogram envelope is characterized by a series of nulls, superimposed on the decay due to the finite linewidth (coherence length) of each mode. The OPD at which these nulls occur is given by

$$L_n = (n - 1/2).\{c/\Delta\nu\}, \qquad (1)$$

where, $n \geq 1$ is the null order, $\Delta\nu$ is the spacing between the modes (frequency) and c is the velocity of light. The principle of operation of the displacement sensing technique described

Springer Proceedings in Physics, Vol. 44 Optical Fiber Sensors
Editors: H.J. Arditty · J.P. Dakin · R.Th. Kersten

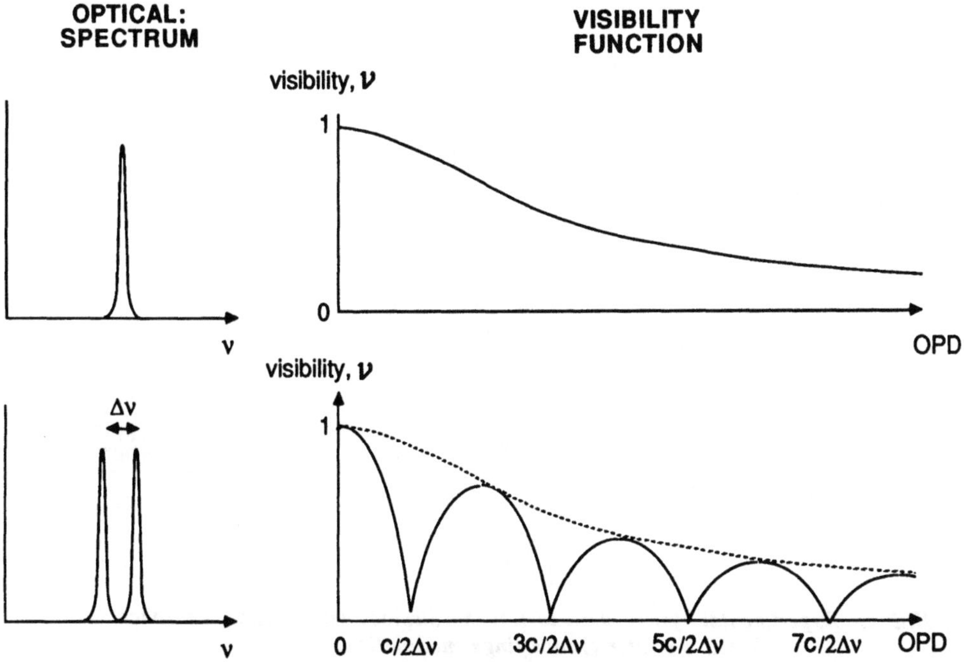

FIGURE 1. Visibility function vs. OPD for a quasi-monochromatic source and a dual mode source with mode spacing Δν (note intrinsic mode linewidth assumed equal in each case)

here relies on the determination the OPD of an unbalanced interferometer by variation of the mode spacing $\Delta\nu$. Tracking of this visibility null in the manner used in the sub-carrier filtering approach is then used to monitor changes in the OPD of the interferometric system.

In order to generate a dual-mode source with variable mode spacing we used a technique described by Auracher and Keil [5] for use in the determination of bandwidth of Mach-Zehnder modulators. The experimental arrangement is as shown in Figure 2. Light from a single frequency diode-laser is coupled through an integrated-optic Mach-Zehnder modulator, biased at a differential phase of π rads. and driven by a RF modulation $\Delta\phi \cos(\omega_c t)$. Under these modulation and bias conditions, it can be shown that the optical frequency spectrum of the oputput light from the modulator comprises a suppressed optical carrier (ω_o) and two symmetric sidebands at $\pm\,\omega_c$ (mode spacing $\Delta\nu = 2f_c$, where $f_c = \omega_c/2\pi$). The relative power in these sidebands (to the input power P_o) is given by

$$P_{\pm 1}\,(\omega_o \pm \omega_c) = P_o\,\{J_1^{\,2}(\Delta\phi\,/2)\}, \qquad (2)$$

which for $(\Delta\phi/2) =$ 1.8 rad, the ratio P_s/P_o is maximized at ~ 0.3. Figure 3 shows the type of optical spectra observed using a scanning Fabry-Perot interferometer. Figure 3.a shows the modulator output without biasing or RF modulation, whereas Figure 3.b shows the result obtained with a bias phase $\phi_o = \pi$ rad and a modulation at 80 MHz applied to the modulator.

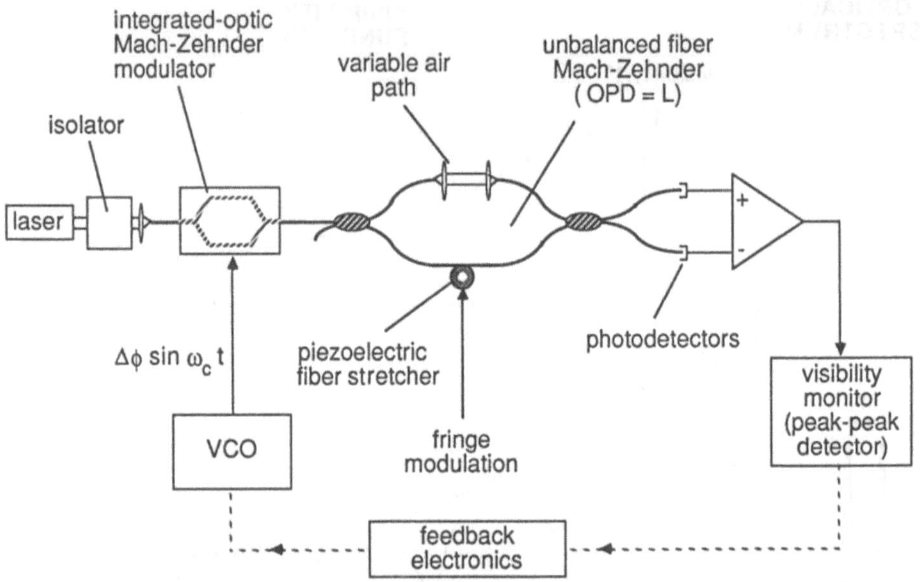

FIGURE 2. Experimental system used to test the displacement sensing approach based on interference-finge modulatiom

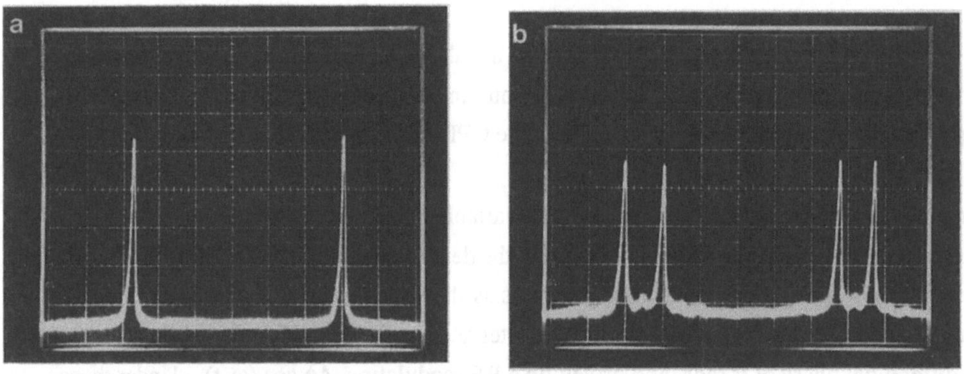

Figure 3. Output optical spectrum of the IO-MZ modulator (a) without RF modulation and no bias, and (b) with a bias phase of π and RF modulation at 80 MHz. Free spectrtral range of Fabry-Perot ≈ 1 GHz.

The two sideband modes at ± 80 MHz, about the suppressed optical carrier are clearly seen in this Figure. Variation of the modulation frequency applied to the MZ modulator confirmed that the sideband (mode) spacing could be tuned. In the experimental sen'sor system, the light at the output of the MZ modulator was fed to an unbalanced fiber Mach-Zehnder interferometer which incorporated an air-path in one arm as shown. This air path was used to alter the effective OPD of the interferometer. The output visibility was monitored by modulating the phase of the interferometer over a number of fringes at several kHz using a small piezoelectric fiber stretcher included in the all-fiber arm of the interferometer and detecting the fringe amplitude using a peak-to-peak detector.

Results

Figure 4 shows the recorded output visibility function of the interferometer when the frequency of the modulation applied to the input IO-MZ was swept from 10 to 100 MHz. In this case the air path was eliminated by butting the fiber ends together in a fiber splicing machine. The visibility function displayed nulls at frequencies ~ 27 and 81 MHz, as expected for the OPD of the system which was ~ 1.9 m of fiber (≈ 2.8 m air). In order to make displacement measurements, we locked the center frequency of the VCO is the first-order null by dithering the VCO at a frequency f_d~ 100 Hz, detecting the resulting modulation in the visibility at f_d to derive an error signal which was then fed back to the VCO control input. Changes in the OPD of the interferometer then gave rise to an automatic shift in the locking frequency of the VCO. From Eqn 1 (subsituting $\Delta v = 2f_c$) the change of the VCO frequency, Δf_c , with path length change , ΔL, (assuming $\Delta L << L$) is given by

$$\Delta f_c = \{-c/4L^2\}\ \Delta L \tag{3}$$

Figure 5 shows the first results obtained using this approach. The OPD of the interferometer was changed using the variable air-path by ± 7 cm. (the nominal overall OPD was ~ 1.4 m). The VCO was locked to the first-order null in the visibility. As seen by the results, the VCO frequency was found to be proportional to the path change over the displacement range used.

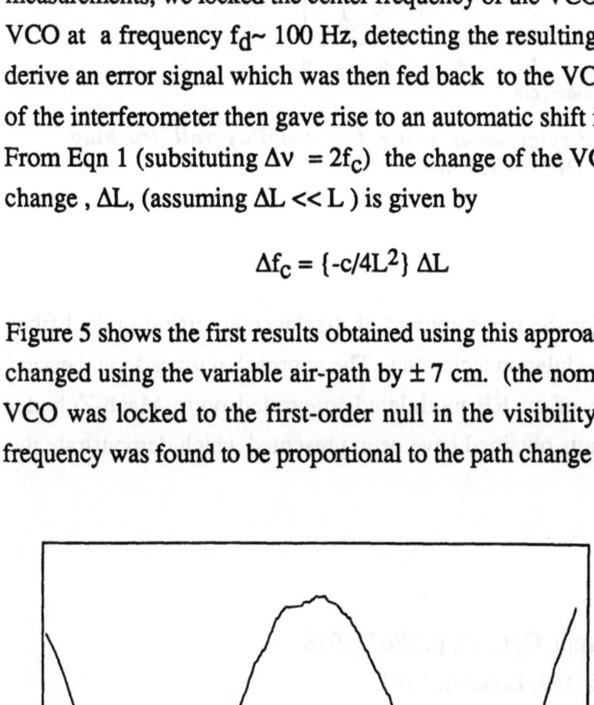

Figure 4. Variation in visibility observed using the unbalanced interferometer with an OPD of 2.75 m (air equivalent) when the frequency of the modulation applied to the IO-MZ was swept from 10 to 100 MHz.

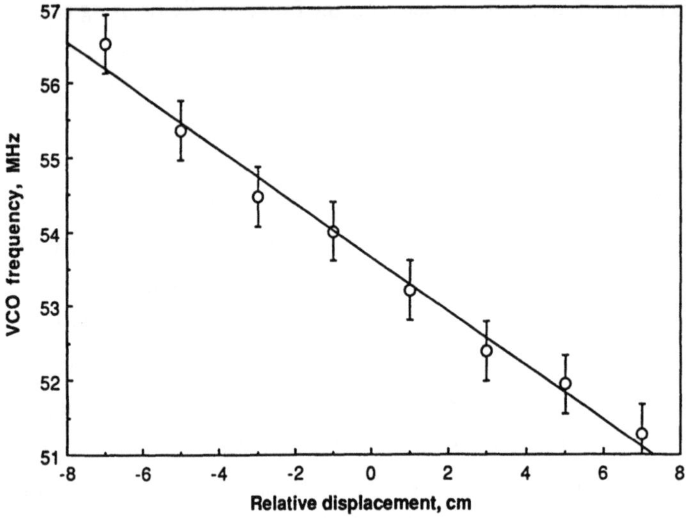

Figure 5. VCO frequency shift vs displacement using the visibility-null tracking feedback system. (Nominal overall OPD ≈ 1.4 m)

Conclusions

We have desribed a new technique for the measurement of displacement using optical fiber interferometry based on a visibility-modulation technique. The approach utilizes a synthesized dual-mode source based on the use of an RF modulated integrated optic Mach-Zehnder intensity-modulator. Preliminary results obtained have been presented which demonstrate the principle of the approach.

References

1. C. D. Butler and G. B. Hocker, Appl. Opt., 17, p. 2867, 1978.
2. D. Uttam *et al.*, Proc 1 st. OFS, p. 182, London, 1983.
3. C. A. Wade *et al.*, Proc. 'Fiber Optic Sensors', Cannes, SPIE Vol. 586, p. 223, 1985.
4. A. D. Kersey and A. Dandridge, Proc. 'Fiber Optic Sensors II', The Hauge, SPIE Vol. 798, p. 176, 1987.
5. F. Auracher and R. Keil, Appl. Phys. Lett., 36, p. 626, 1980.

Part IV

Ring Resonators

Fiber-Optic Resonant Ring Sensors: System and Device Technology

K. Kyuma, S. Tai, K. Kojima, and M. Takahashi

Central Research Laboratory, Mitsubishi Electric Corporation,
8-1-1 Tsukaguchi-Honmachi, Amagasaki-City, Hyogo 661, Japan

The operation principle and the current status of the fiber-optic ring-resonator sensors are briefly reviewed. The fiber-optic passive ring-resonator gyroscope under development in our laboratory is introduced, together with the narrow linewidth MQW-DFB laser diodes for coherent optical sensors. Several unique applications of the fiber-optic ring-resonators are also described.

1. Introduction

Fiber-optic ring-resonators are expected to bring a new class of optical sensing/processing systems such as rotation sensors and optical spectrum analyzers[1~4]. However, there are several problems to be solved to realize the high performance optical systems. Among them, it is very important to remove the reflections[4] in the optical path and to suppress the crosstalk in a coupler between the two orthogonally polarized modes. In addition to these problems, since most optical systems using the ring-resonators are based on the sharp resonant characteristics (high finesse), high coherent (narrow linewidth) light sources are required. In the first part of this paper, we report on the solutions to these problems, and describe the current status of the fiber-optic passive ring-resonator gyro. In the second part, we propose and demonstrate several unique applications of fiber-optic ring-resonators, including a nondestructive measurement for the misaligned angle of polarization maintaining fiber-couplers, and a measuring and controlling technique for the spectral linewidth of laser diodes.

Springer Proceedings in Physics, Vol. 44 Optical Fiber Sensors
Editors: H.J. Arditty · J.P. Dakin · R.Th. Kersten

2. A Fiber-Optic Passive Ring-Resonator Gyro

2.1 System technology

The principle of operation for the passive ring-resonator gyro is based on the shift in the resonant frequency of the fiber ring-resonator in the presence of rotation due to the Sagnac effect[1]. The system configuration under development in our laboratory is shown in Fig. 1. The resonator was constructed of a PANDA-type polarization-maintaining fiber and a low-loss fiber-directional-coupler. The two counterpropagating laser beams in the fiber-ring are needed as a sensing light and a reference light, respectively. The output of the reference light is fedback to the piezoelectric transducer [PZT(1)] in order to compensate for the fluctuations in the resonant peak caused by the external disturbances. The fiber-length, the ring-diameter, and the finesse were 4 m, 10 cm, and 50, respectively. An AlGaAs/GaAs external-cavity laser diode whose spectral linewidth was 100 KHz was used as a coherent light source.

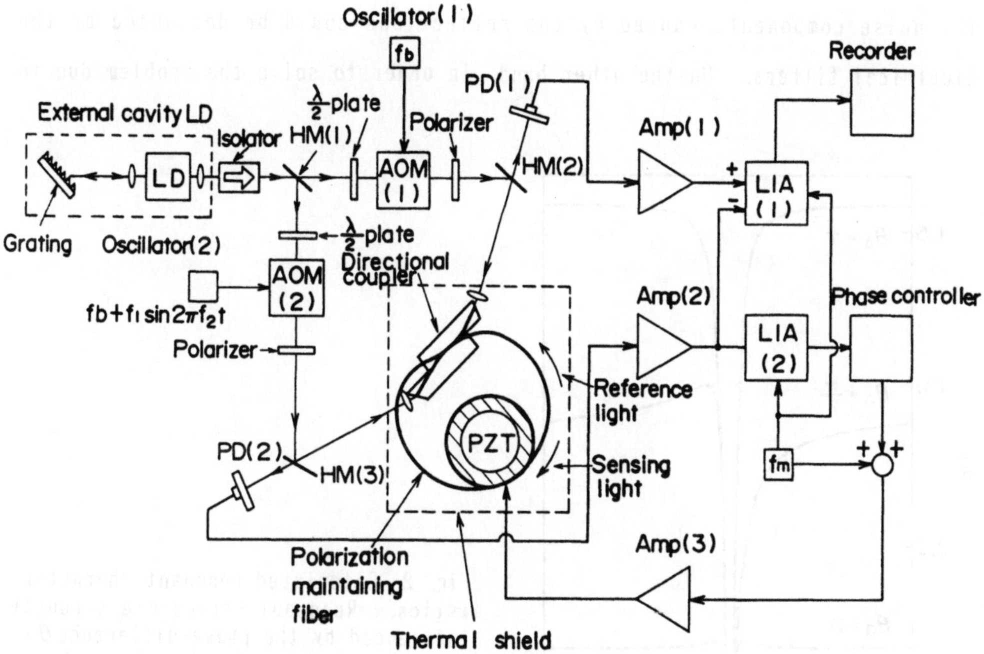

Fig. 1 Schematic configuration of the constructed fiber-optic ring-resonator gyroscope.

The shot-noise-limited detection sensitivity is theoretically 10^{-7} rad/sec when the output power on the photodiode for the sensing arm is $5\mu W$.

However, there are several noise sources which degrade the resonant characteristics and then limit the rotation detection sensitivity. The most dominant noise factor which we made clear is the reflected light at the fiber-ends, discrete optical components, and/or the Rayleigh scattered light in the fiber[4) . These stray lights propagate back to interfere with the sensing and the reference light on the detectors. Figure 2 is the typical observed resonant curves which show that the degree of the degradation is also dependent on the phase difference between the incident light and the reflected light at the fiber-end. The second dominant factor is the crosstalk between the orthogonally polarized modes, which is caused in the fiber-coupler. In the system shown in Fig. 1, the reflection at the fiber-ends was reduced less than 10^{-4} by polishing them at an oblique angle of 7° . In addition, the optical phase- or frequency-modulation technique was introduced so that the reflected light was shifted in frequency from the sensing and reference light, and then the noise components caused by the reflections could be decoupled by the electrical filters. On the other hand, in order to solve the problem due to

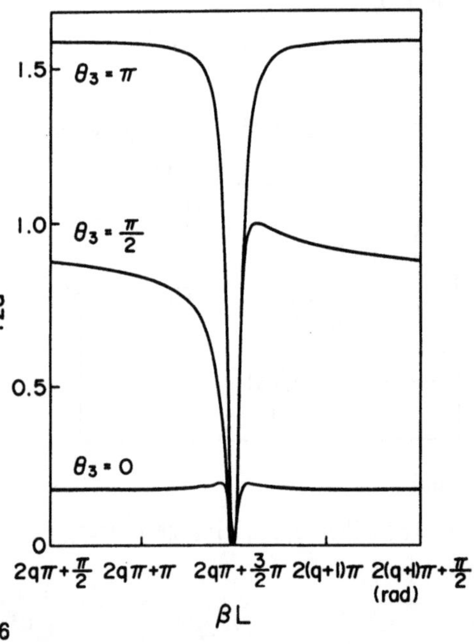

Fig. 2 Calculated resonant characteristics. Resonant curves are strongly influenced by the phase difference θ_3 between the incident light and the reflected light at the fiber-ends.

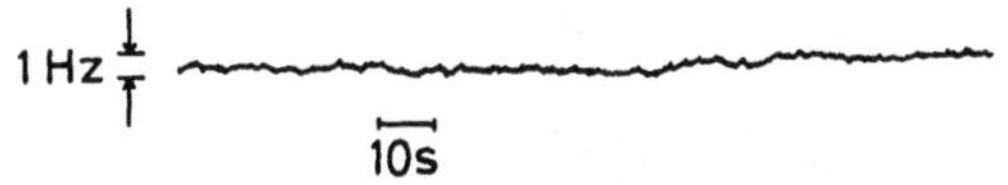

Fig. 3 Measured drift characteristics of the gyro output. The time constant was 30 sec.

the crosstalk, the operating point of the main resonance was ajusted at the optimum point with respect to the sub-resonance. As a result, the detection sensitivity as low as 2×10^{-6} rad/sec was obtained with the time constant of 30 sec, as shown in Fig. 3.

2.2 Narrow linewidth semiconductor lasers

Narrow spectral linewidth distributed feedback (DFB) laser diodes are strongly required for the ring-resonator gyro as well as other coherent optical fiber sensors[5]. The typical linewidth required for such applications is as narrow as a few hundred KHz. It is well known that the linewidth of a single mode laser is expressed as[6]

$$\Delta \nu = \frac{v_g \Gamma g n_{SP}}{4 \pi I} (1 + \alpha^2)$$

where v_g the group velocity of light, Γ is the confinement factor, g is the threshold gain, n_{SP} is the spontaneous emission factor, I is the photon number in the cavity, and α is the linewidth enhancement factor called α-parameter. Equation (1) shows that the reduction of α-parameter is very important to reduce the linewidth. We have already experimentally demonstrated for the first time in 1987[7,8] that α-parameter and then linewidth are dramatically reduced by introducing a multiple-quantum-well (MQW) active layer because the differential gain with respect to the carrier density is enhanced.

In this paper, we have designed the MQW layer in an optimum structure[9]. The AlGaAs/GaAs MQW-DFB laser fabricated by two-step molecular-beam-epitaxy (MBE) growth technique is schematically shown in Fig. 4. It is found from Eq. (1) that the linewidth is decreased with decreasing n_{SP} and Γ. This means that the linewidth can be narrowed with a small number of wells because n_{SP} is decreased with increasing the population inversion and because Γ is decreased.

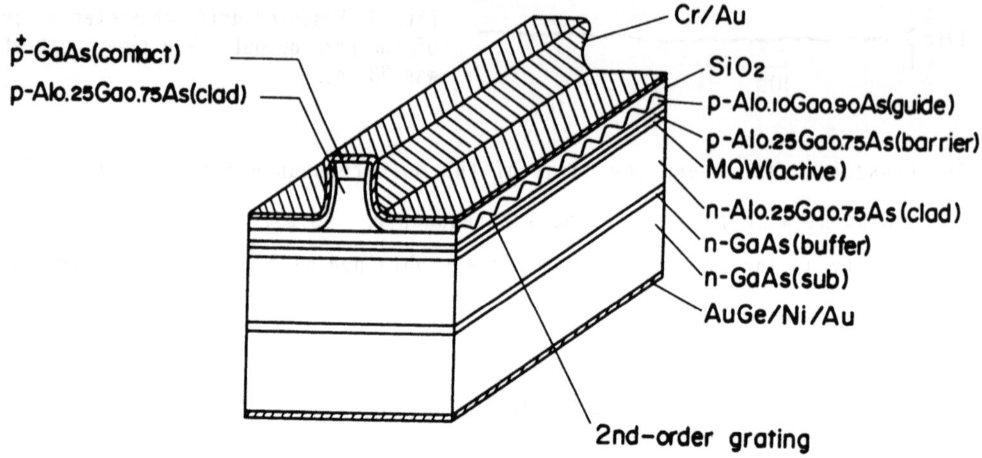

Fig. 4 Schematic configuration of the AlGaAs/GaAs MQW-DFB laser.

However, α-parameter is increased with increasing the carrier density. Therefore, in the fabricated device, the MQW active layer was designed with 3 undoped GaAs wells (l_w=100Å) and 2 undoped $Al_{0.2}Ga_{0.8}As$ barriers (l_b=50 Å) so that the linewidth was minimized. The measurement of α-parameter using the Chirp-Halfwidth Product method[10] indicated that the α-parameter for the current device(-3.3) is about half of that of the DH laser(-4.6~-6.0), as expected theoretically. The cavity length was 830μm, and the oscillation wavelength was 860 nm. Figure 5 shows the measured output power dependence of

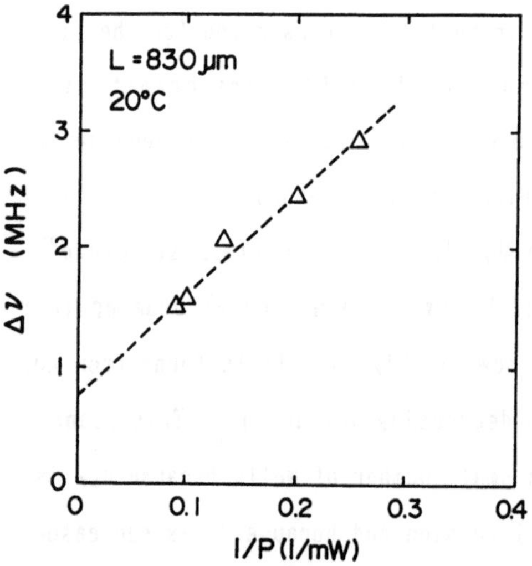

Fig. 5 Measured spectral linewidth as a function of the inverse output power.

the linewidth. The linewidth as narrow as 1.5 MHz was obtained at the output power of 11 mW. This value is, to our knowledge, the narrowest ever reported for AlGaAs lasers. However, the obtained linwidth is still broard, as compared to the requirement from the ring-resonator gyro. One of the methods to further reduce the linewidth is the p-doping in the MQW structure. The possibility of such a device technology will be also discussed at the conference.

3. Other Applications

The resonant characteristics of the fiber-optic ring-resonators are strongly dependent on the spectral purity of the light source, the loss and the polarization crosstalk in the fiber-coupler. These facts indicate that the ring-resonators are able to be utilized for monitoring and/or controlling the optical properties of such components.

3.1 Nondestructive measurement of misaligned angle in a polarization-maintaining fiber-coupler

The polarization coupling (crosstalk) in the fiber-coupler generally degrades the performance of the optical systems based on the polarization- and phase-modulation. The polarization coupling is originated from the fact that the principal axes of the two fibers are mutually deviated somewhat from the desired parallel condition by a misalignment in the assembling process. This misaligned angle ϕ_d can be nondestructively measured by observing the deformed resonant curve of the fiber-optic ring-resonator[11].

Figure 6 shows the calculated resonant curves for several η_d [$=(\beta_x - \beta_y)L$] values under the appropriate conditions, where β_x and β_y are propagation constants of the X-polarized and Y-polarized lights, respectively, and L is the fiber-length. Solid and broken lines correspond to the cases for ϕ_d =0.1 rad and ϕ_d =0, respectively. Figure 7 shows the observed resonant curves of the constructed resonator. η_d was intentionally changed by altering the temperature of the fiber-ring. We can see that the close agreement between the observed and the calculated curves is obtained. The most important feature

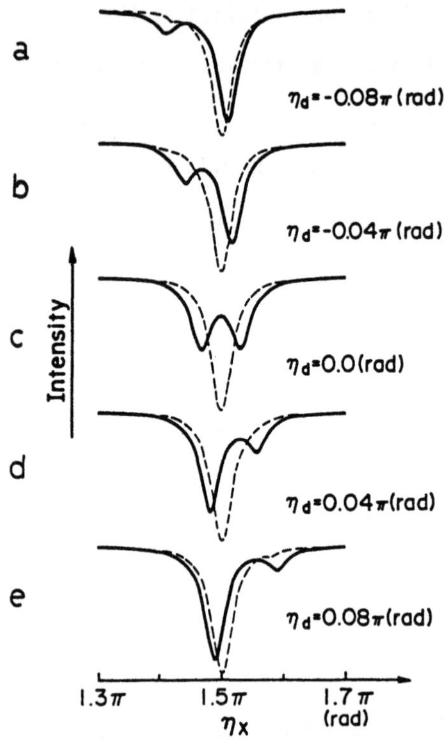

Fig. 6 Calculated dependence of the resonant characteristics on η_d[ϕ_d= 0.1 rad(solid line), 0.0rad(broken line), $\Delta\beta L$=-0.08π rad(a), -0.04π rad(b), 0(c), 0.04π rad(d), and 0.08π rad(e)].

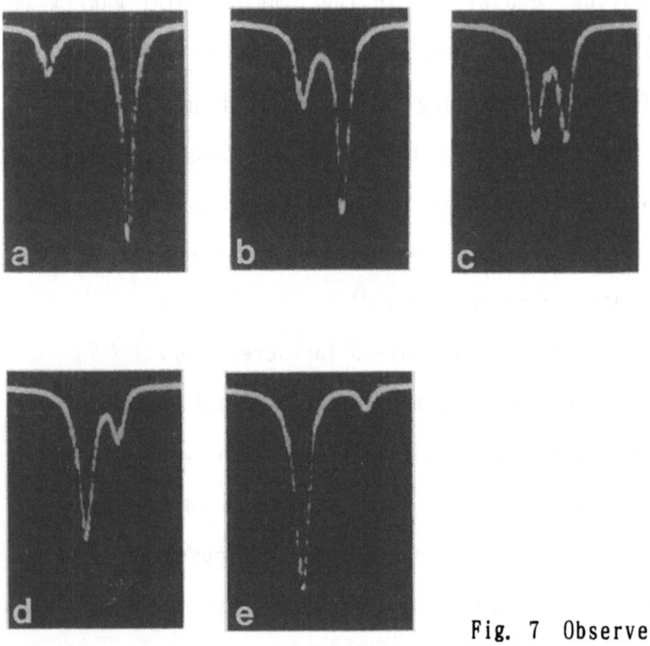

Fig. 7 Observed resonant characteristics.

is that the resonant curve becomes a symmetric and double-robed curve in the case of $\eta_d = 0$. Since the distance between these two resonant points is given by $2\phi_d$, the misaligned angle ϕ_d of the coupler can be deduced by measuring the distance. In the case of the example in Fig. 7, ϕ_d is estimated to be approximately 0.1 rad (6°).

3.2 Monitoring and controlling of the spectral linewidth of an external-cavity laser

An external-cavity laser diode is widely ulilized as a coherent light source at the present time because of its simple structure and excellent spectral purity. However, it is required to control the optical feedback condition for the stable operation. We have reported on the stabilized external-cavity laser[12] which is in principle based on the automatic phase control (APC) technique of feedback light using the fiber-optic ring resonator as a linewidth discriminator.

The experimental arrangement is shown in Fig. 8. The GaAs/AlGaAs single mode laser oscillating at 830 nm was used, together with the external mirror mounted on a PZT plate placed 15 cm away from the laser diode. The linewidth of this laser was about 10 MHz in the free running state. The ring resonator was constructed of a standard single-mode fiber (1 m-length), a low loss fiber-coupler and a phase modulator. Since the resonant characteristics of the resonator depends strongly on the linewidth, the resonator acts as the linewidth discriminator if the cavity finesse is high enough (finesse=60 in this work). According to our theoretical estimate, the linewidth from 100 KHz to 100 MHz can be measured with the accuracy less than ±10 to ±20 % by this technique. The optoelectronic APC loop operates as follows. The peak value of the resonant output is compared with the predetermined reference voltage by the differential amplifier. By applying the error signal to the PZT plate, the position of the mirror is moved before and behind to satisfy the phase maching condition of the feedback light. The bandwidth of the APC loop is about 1 KHz. According to the experimental results, the stable spectral linewidth of 500 KHz was obtained for the weak feedback external cavity laser.

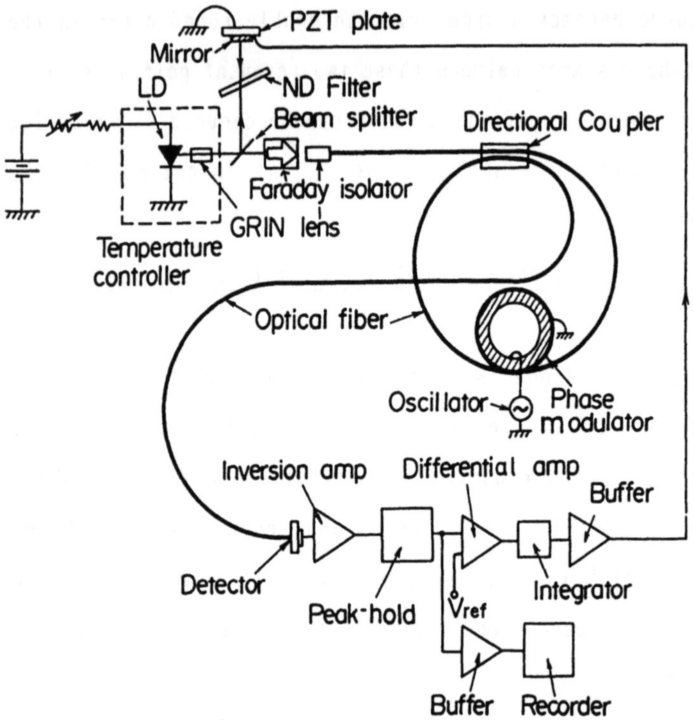

Fig. 8 Experimental arrangement for the spectral linewidth stabilization using the fiber-optic ring-resonator as a linewidth discriminator.

4. Conclusion

In conclusion, we have reported the state-of-the-art performance of the fiber-optic ring-resonator gyro. Several techniques to remove the noise sources, which degrade the rotation sensitivity, are introduced. The narrow linewidth MQW-DFB laser diodes developed in our laboratory are also reported. They are expected to be widely employed in the coherent optical sensing systems including the ring-resonator gyroscopes. Furthermore, we have proposed and demonstrated several applications of the fiber-optic ring-resonators. The fiber-optic ring-resonators will find unique application fields in the near future by further improving the performance of the constitute optical components.

REFERENCES

(1) R. E. Meyer, S. Ezekiel, D. W. Stowe, and V. J. Tekippe, Opt. Lett. 8, 644 (1983).

(2) K. Iwatsuki, K. Hotate, and M. Higashiguchi, Appl. Opt. 23, 3916 (1984).

(3) M. Takahashi, S. Tai, K. Kyuma, and K. Hamanaka, Opt. Lett. 13, 236 (1988).

(4) M. Takahashi, S. Tai, K. Kyuma, and K. Hamanaka, Opt. Lett. 13, 413 (1988).

(5) K. Kojima and K. Kyuma, Electron. Lett. 20, 869 (1984).

(6) C. H. Henry, IEEE J. Quantum Electron. QE-18, 259 (1982).

(7) S. Noda, K. Kojima, K. Mitsunaga, K. Kyuma, K. Hamanaka, and T. Nakayama, Appl. Phys. Lett. 48, 1867 (1986).

(8) S. Noda, K. Kojima, K. Kyuma, K. Hamanaka, and T. Nakayama, Appl. Phys. Lett. 50, 863 (1987).

(9) K. Kojima, K. Hara, M. Kameya, and K. Kyuma, Electron. Lett. 25, 240 (1989).

(10) M. Osinski, D. F. G. Gallagher, and I. H. White, Electron. Lett. 21, 981 (1985).

(11) M. Takahashi, S. Tai, and K. Kyuma, Electron. Lett. 25, 600 (1989).

(12) S. Tai, K. Kyuma, and T. Nakayama, Opt. Acta, 33, 1539 (1986).

Bias of an Optical Passive Ring-Resonator Gyro due to the Misalignment of the Polarization Axis in the Resonator Formed by the Polarization-Maintaining Fiber

*K. Hotate, K. Takiguchi, and M. Murakami**

RCAST, Research Center for Advanced Science and Technology,
The University of Tokyo, 4-6-1 Komaba, Meguro-ku, Tokyo 153, Japan
*Present address: NTT Laboratories

Misalignment of the polarization axis in the resonator formed by the polarization-maintaining fiber is analyzed as a bias source in Optical Passive Ring-Resonator Gyro. Large bias occurs when the phase difference between the polarization modes is $2m\pi$. However, this bias can be suppressed by fiber polarizers set at the resonator leads.

1. Introduction

Optical Passive Ring-Resonator Gyro (OPRG), which utilizes a high finesse fiber ring-resonator, is expected to achieve a high performance with a much shorter length of a fiber loop than an interferometer type fiber gyro. In our group, behaviors of noise sources in OPRG have been analyzed in detail [1]-[5], and some experimental verification of the noise reduction methods has been carried out [2], [6], [7]. Polarization fluctuation is one of the dominant noise sources, and we have also analyzed its effect generally [4]. We pursue, in this paper, the detail of the polarization effect in OPRG using a practical model, and show that the polarizers set at the lead portions of the resonator have the role to reduce the bias.

In OPRG, the ring-resonator must be composed of the polarization-maintaining fiber [4]. Relatively easy way to construct the resonator is to connect two lead portions of the coupler by the splice. Therefore, the misalignment of the polarization axis in the resonator takes place, which is analyzed to induce a bias of OPRG. We have already evaluated this bias by using a quite simple model [8].

In this paper, we evaluate the bias caused by the misalignment of the polarization axis by using the more exact model than before [8], in which the twist caused by the misalignment is also taken into account. Moreover, the model includes the fiber polarizers at the lead portions of the

Editors: H.J. Arditty · J.P. Dakin · R.Th. Kersten

resonator, which have been pointed out to be effective in suppressing the excitation of the unwanted polarization mode [4].

It is clarified that a large bias takes place when the round phase shift difference between the two polarization modes of the fiber is equal to $2m\pi$(m;integer). The fiber polarizers are analyzed to make the bias flat when the misalignment is relatively small, and they are effective in suppressing the large bias mentioned above. This result is significant to achieve a high performance gyro. Similar analysis has been presented in Ref.[9]. However, it is not the analysis on the misalignment at the splice but that on the coupler. Moreover, it does not include the effect of the polarizers.

2. Analysis

The model for the analysis is shown in Fig.1. The ring-resonator is composed of the polarization maintaining fiber, in which there exists one misalignment of the polarization axis caused by the splice connection. The fiber is twisted uniformly due to the misalignment. Fiber polarizers are set at the lead portions of the resonator for suppressing the excitation of the unwanted polarization mode [4].

In general, two eigenstate of polarizations (ESOPs) exist in the fiber ring-resonator [4]. Each ESOP generates the resonance independently [4]. The resonance point of each ESOP is concerned with the eigenvalue of the transfer matrix of the whole fiber loop, and each ESOP corresponds to the eigenvector of the matrix.

First of all we calculate the transfer matrix of the whole fiber loop, which is expressed as the multiplication of the fiber transfer matrix taking the twist into account by the rotation matrix corresponding to the misalignment. The eigenvalues and eigenvectors are calculated using the

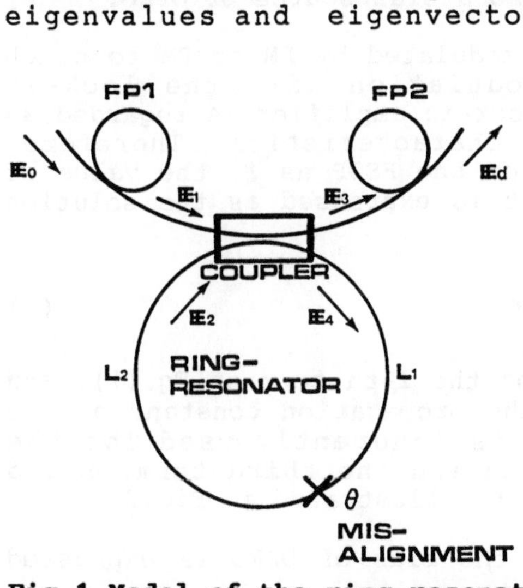

$$L = L_1 + L_2$$
$$\Delta L = L_1 - L_2$$

Fig.1 Model of the ring-resonator of OPRG for the analysis.

transfer matrix of the whole fiber loop. Then the field vector of the input lightwave is expressed as linear combination of the two eigenvectors. Next, we obtain the transfer matrix of the whole system as the multiplication of the transfer matrix of the ring-resonator by the matrix of the fiber polarizer. Multiplying the transfer matrix of the whole system by the field vector of the input lightwave, the ratio of the output power to the input power $|E_d/E_0|^2$ is expressed as

$$|E_d/E_0|^2 = K_1|U_1|^2 + K_2|U_2|^2 + K_3, \tag{1}$$

where

$$|U_j|^2 = K_4\{1 - \frac{K_5}{(1-K_6^{1/2})^2 + 4K_6^{1/2}\sin^2(\beta_j L/2)}\}, \tag{2}$$

K_i is a variable which is decided by the characteristics of the polarization maintaining fiber, the coupler, and the fiber polarizers as well as the position and the angle of the misalignment. A propagating constant of each ESOP is denoted as β_j.

In Eq.(1), the first and the second term represent the resonance characteristics corresponding to the two ESOPs. The third term is a cross-term of the two resonances, which appears because the transfer matrix of the fiber polarizer set at the output lead portion does not show the unitary characteristics.

Each resonance peak corresponding to each ESOP is observed at the different frequency of the lightwave. When one resonance peak used for the rotation detection is approached by the other peak due to the thermal fluctuation, the bias of OPRG occurs. Moreover, the third term of Eq.(1) shows the resonance at the same frequencies as the two ESOPs. Therefore, the third term is also a bias source of OPRG.

The incident lightwave is modulated by FM or PM to catch the resonance point by demodulation with the lock-in amplifier. The output of the lock-in amplifier is regarded as a differential of resonance characteristics. Therefore, denoting the round phase shift of the ESOP as ζ, the value of ζ describing the resonance point is expressed as the solution of the following formula:

$$D_1(\zeta) + D_2(\zeta - \Delta\beta' L) + D_3(\zeta) = 0, \tag{3}$$

where $D_i(\zeta)$ is a differential of the i-th term of Eq.(1), and $\Delta\beta'$ is the difference between the propagation constant of the two ESOPs. In Eq.(3), $D_1(\zeta)$ is inherently used for the rotation detection. The second and the third term, D_2 and D_3, induce the gyro bias, which is illustrated in Fig.2.

Denoting n as an integer, the bias of OPRG is expressed as follows [4].

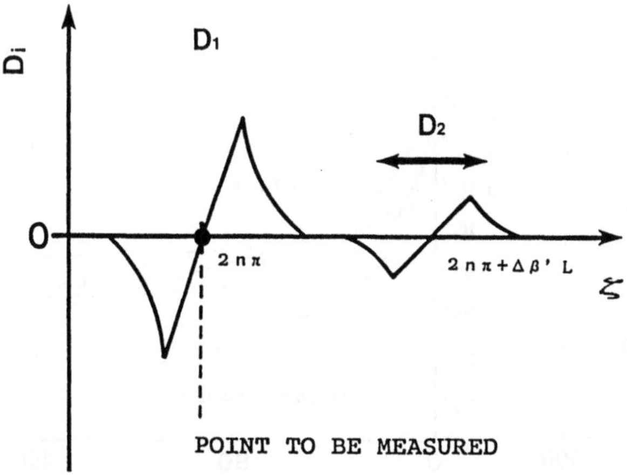

Fig.2 Bias due to the unwanted resonance characteristics. The value of n is an integer, and $\Delta\beta' L$ is the round phase shift difference between the two ESOPs.

$$\Delta\Omega = \frac{C_0 \lambda}{4\pi L r} (\zeta - 2n\pi), \tag{4}$$

where C_0, λ, and r are the light speed in vacuum, a wavelength of the light source, and a radius of the ring-resonator, respectively.

3. Numerical Evaluation

We have evaluated the bias caused by the misalignment of the polarization axis in the ring-resonator, based on the consideration in the former chapter. Here, the loss of the fiber and the coupling coefficient of the coupler are assumed to be the same for the two ESOPs.

Fig.3 shows the bias as a function of the round phase shift difference $\Delta\beta L$ between the two polarization modes of the fiber. The main parameters for evaluation are the angle of the misalignment of the polarization axis θ at the splice, the angle of the misalignment of the input lightwave polarization into the fiber θ_i, and the existence of the fiber polarizers. One of the parameters $\Delta\beta\Delta L$ is chosen so that the bias becomes the maximum. Besides, ε^2 is a extinction ratio of the fiber polarizers, and F is a finesse of the ring-resonator.

The conclusion derived from Fig.3 is as follows.

(1) When $\Delta\beta L = 2m\pi$, the bias becomes very large. It is because that at this condition each ESOP approaches the ellipse or the linear polarization which is inclined at 45 degrees. This phenomenon has been pointed out by our group at the Ref.[8]. Besides, the similar phenomenon has been observed when the effect of the polarization coupling at the coupler is analyzed [9].

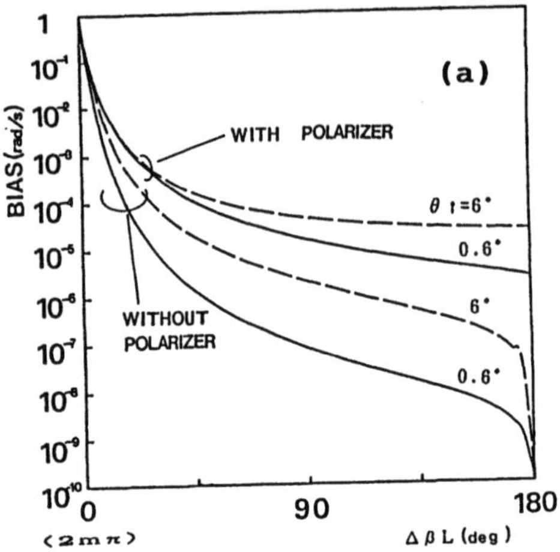

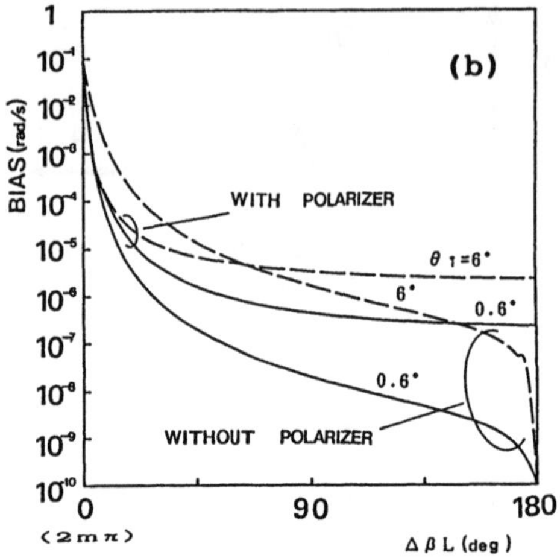

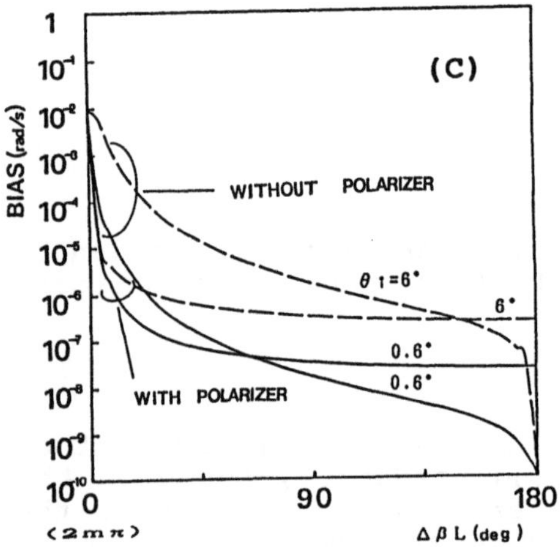

*birefringence of the fiber $\Delta\beta$=5000rad/m

*fiber length L=15m

*extinction ratio of the polarizer ε^2=40dB

*finesse F=50

*incident polarization misalignment θ_i

*misalignment at the splice θ

Fig.3 Bias of OPRG as a function of the round phase shift difference $\Delta\beta L$ between the two polarization modes of the polarization maintaining fiber. (a) θ=1° , (b) θ=0.1° , (c) θ=0.01°

(2) Though the fiber polarizers are effective in suppressing the bias caused by the second term in Eq.(1) [4], they are not so effective in suppressing the bias caused by the third term in Eq.(1). When the angle θ becomes larger, ESOPs deviate from the ideal polarization modes of the polarization maintaining fiber at any value of $\Delta\beta L$. Therefore, because the effect of the third term is promoted at the above condition, the bias becomes larger. Consequently, when the angle θ is large and the angle θ_i is small, the fiber polarizers enhance the bias. On the contrary, when the angle θ is small and the angle θ_i is large, the fiber polarizers are effective in making the bias flat.

Table1 Bias of OPRG around $\Delta\beta L=2m\pi$ as a function of the incident polarization misalignment θ_i. (a) with fiber polarizers, (b) without fiber polarizers. The unit of the bias is "rad/sec". Using the polarizers, the bias does not change even when θ_i changes.

$\Delta\beta L$ (deg) \ θ_i (deg)	0.6	1.8	DIFFERENCE
0	7.2179×10^{-2}	7.2195×10^{-2}	1.5631×10^{-5}
2	2.7536×10^{-3}	2.7672×10^{-3}	1.3649×10^{-5}
4	8.1069×10^{-4}	8.2113×10^{-4}	1.0439×10^{-5}
6	3.2168×10^{-4}	3.2970×10^{-4}	8.0285×10^{-6}
8	1.5472×10^{-4}	1.6113×10^{-4}	6.4066×10^{-6}
10	8.5178×10^{-5}	9.0468×10^{-5}	5.2906×10^{-6}

(a)

$\Delta\beta L$ (deg) \ θ_i (deg)	0.6	1.8	DIFFERENCE
0	7.3620×10^{-2}	7.6746×10^{-2}	3.1263×10^{-3}
2	3.0250×10^{-3}	5.6159×10^{-3}	2.5908×10^{-3}
4	7.0705×10^{-4}	1.8538×10^{-3}	1.1468×10^{-3}
6	2.1557×10^{-4}	7.0750×10^{-4}	4.9193×10^{-4}
8	8.2869×10^{-5}	3.1877×10^{-4}	2.3590×10^{-4}
10	3.8012×10^{-5}	1.6461×10^{-4}	1.2660×10^{-4}

(b)

(3) In this paper, we have evaluated the bias caused by one of the two counterpropagating lightwaves. Table1 shows the bias around $\Delta\beta L=2m\pi$ in case with or without fiber polarizers. From Table1 (a), we can conclude that the bias changes little, in spite of the change of the incident polarization angle θ_i,when we use the fiber polarizers. On the contrary, without polarizers, the bias changes large as the angle θ_i changes. It is because that the misalignment of the polarization direction of the input lightwave is decreased enough with the

fiber polarizer. Moreover, because each ESOP approaches the ellipse or the linear polarization which is inclined at 45 degrees around $\Delta\beta L=2m\pi$, the excitation ratio between the two ESOPs is almost unity. In closed loop operation in practical OPRG, the output is the difference of the resonance frequency between the two counterpropagating lightwaves. Therefore, Table1 (a) suggests that, in a practical gyro system, the large bias around $\Delta\beta L=2m\pi$ can be extremely reduced with the fiber polarizers.

4. Conclusion

We have evaluated the bias of OPRG caused by the misalignment of the polarization axis, in which one misalignment caused by the splice connection in the ring-resonator and the uniform twist caused by the misalignment are taken into account. Moreover, we have investigated the validity of the fiber polarizers at the lead portions of the resonator.

Consequently, it is clarified that the fiber polarizers can make the bias flat when the misalignment is relatively small, and they are effective in suppressing a large bias, which occurs when the round phase shift difference between the two polarization modes of the fiber is equal to $2m\pi$. The results are significant to achieve a high performance OPRG.

The authors would like to thank Prof. T. Okoshi of our Research Center for his encouragement.

References

[1] K. Iwatsuki, K. Hotate, M. Higashiguchi, Appl. Opt. 23, 3916 (1984).
[2] K. Iwatsuki, K. Hotate, M. Higashiguchi, Appl. Opt. 25, 4448 (1986).
[3] K. Iwatsuki, K. Hotate, M. Higashiguchi, J. Lightwave Technol. 4, 645 (1986).
[4] K. Iwatsuki, K. Hotate, M. Higashiguchi, Appl. Opt. 25, 2606 (1986).
[5] K. Hotate, M. Murakami, OFS '88 (New Orleans) FBB6-1, 405 (1988).
[6] K. Iwatsuki, K. Hotate, M. Higashiguchi, Electron. Lett. 22, 135 (1986).
[7] K. Hotate, K. Takiguchi, A. Hirose, IOOC '89 (Kobe) 20D3-2, (1989).
[8] M. Murakami, Master's Thesis in Electrical Engineering, Postgraduate school of the University of Tokyo, Feb. 1988 (in Japanese).
[9] M. Takahashi, S. Tai, K. Kyuma, Paper of Technical Group, IEICE Japan, OQE88-63, Sep. 1988 (in Japanese).

An All Fibre Ring Resonator Gyroscope Using a Low Coherence Length Source

F. Farahi, K. Kalli, and D.A. Jackson

Physics Laboratories, University of Kent, Canterbury, Kent CT2 7NR, UK

An all fibre gyroscope based on white light interferometry is described. The sensing loop comprises an all fibre ring resonator which acts as dual cavity for two counter propagating beams. A comparison between this configuration and other fibre gyroscopes is presented and the components required to improve the performance of the system are assessed.

Introduction

The all fibre Sagnac gyroscope has the advantage of increased stability and simplicity over the bulk optic Sagnac system [1]. However, there are problems associated with the all fibre gyroscope which limit its performance. It has been shown that one of the most important of these is the phase error produced by coherent back scattered Rayleigh light [2]. This effect can be substantially reduced by decreasing the coherence length of the source; superluminescent diodes have been found to be near ideal for such applications [3].

There are however different approaches for increasing the sensitivity of the fibre gyroscope. One approach is to use a high sensitivity fibre ring resonator as the rotation sensor. The basic configuration of a ring resonator gyroscope is shown in figure 1, where the ring is illuminated from both directions thus producing two corresponding output signals. In this case the sensitivity is directly proportional to the finesse of the fibre ring resonator (F_R) in comparison with Sagnac interferometer ($F_S = 2$) where the sensitivity increases in the ratio $\frac{F_R}{F_s}$ [4,5]. An all fibre ring

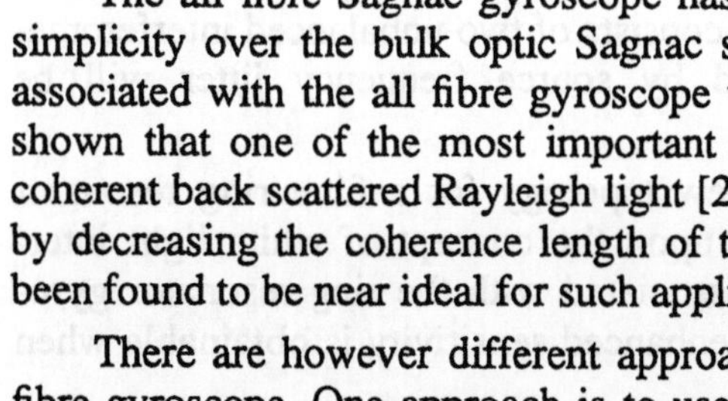

Figure 1. Schematic diagram of fibre optic ring resonator gyroscope.

resonator with a finesse greater than 1200 has recently been reported [6]. The use of such a resonator would substantially improve the sensitivity of the ring resonator gyroscope relative to the Sagnac interferometer. There are however problems associated with the fibre optic resonator which currently limit the obtainable resolution in gyroscope applications, these are outlined bellow:

i) it requires a highly coherent source, hence the effect of the Rayleigh back scattered signal is important

ii) non - linear effects, such as stimulated Brillouin scattering have recently been reported for high finesse ring resonators at very low input powers [7]. This signal and Rayleigh back scattered signal introduce undesirable interference signals which both increase the noise floor and degrade the linearity of the fibre ring resonator gyroscope.

iii) the system consists of two independent non - reciprocal ring resonators corresponding to two directions of propagation (Figure 1) which are highly sensitive to environmental perturbation. Although the perturbation effect is the same and the effect of the rotation is of opposite sign for the two rings it would be difficult in practice to measure a small differential signal in the presence of large noise signals.

iv) the Sagnac interferometer is a zero path imbalance configuration in contrast with the resonator of figure 1 which consists of two unbalanced interferometers, therefore phase noise caused by source frequency jitter will be significantly greater.

In this presentation we describe a new topology for a fibre ring resonator operating as a gyroscope which is based upon the concept of white light interferometry. We show that the problems associated with the ring resonator gyroscope are resolved and that the predicted enhanced sensitivity is obtainable when compared with Sagnac interferometer.

Theory and experimental

Figure 2 shows schematically the operation of the low coherence source fibre ring resonator in a reciprocal configuration. Light is launched into the system via one of the input arms of a directional coupler producing two guided beams which illuminate the fibre ring from two opposite directions thus forming two indepen-

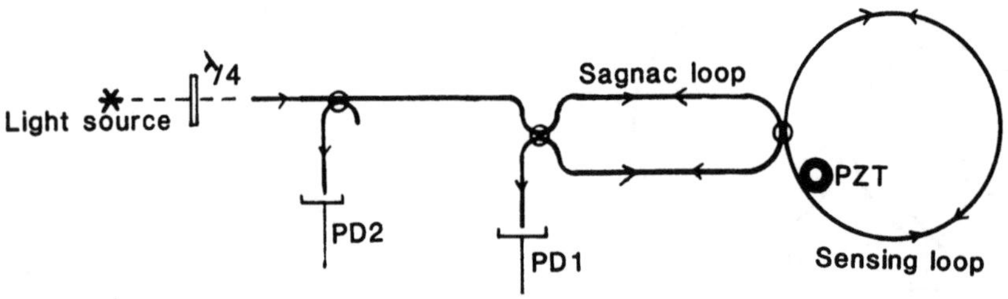

Figure 2. Schematic diagram of an all-fibre ring resonator gyroscope with a low coherence length source.

dent ring resonators. The outputs are then further combined at the first directional coupler and detected at the output of its forth port (PD1) or preferably PD2. In this system the fibre links between the input coupler and the ring coupler form a Sagnac interferometer (Figure 2). The transfer function of the system may be written as;

$$I = S\, R_1\, R_2 \tag{1}$$

where S is the transfer function of Sagnac interferometer and R_i corresponds to ith ring resonator (i=1,2 for two opposite directions) and is given by [8]

$$R_i = (1-\gamma_0)\left[1 - \frac{(1-k_r)^2}{(1+k_r)^2 - 4k_r \sin^2\left(\frac{\beta L_i}{2} - \frac{\pi}{4}\right)}\right] \tag{2}$$

where γ_0 is the intensity loss and k_r is the resonant coupling coefficient.

The Sagnac interferometer is a zero path imbalance interferometer and when operated in a fully reciprocal configuration is relatively insensitive to many physical quantities such as temperature, strain,...; its sensitivity to rotation is directly proportional to the length of fibre loop. When a short fibre loop is used for the Sagnac interferometer its sensitivity approaches zero hence one can assume S to be constant for the very short fibre links used between the two couplers in the configuration shown in figure 2. Equation (1) can then be rewritten as;

$$I \,\alpha\, R_1\, R_2 \tag{3}$$

We next consider the case when the system is illuminated by a source with a coherence length short in comparison with the cavity length of the ring resonator such that no optical interference is observable at the output of each individual resonator. Assuming that the spectrum of the light source can be represented by a Lorentzian (L) function equation (2) finally reduces to

$$I \,\alpha \int R_1\, R_2\, L\, d\phi \tag{4}$$

where ϕ is the interferometric phase. This has been previously calculated for different resonant coupling coefficients and it was shown that high finesse output could be achieved for two balanced ring resonators [9].

An arrangement similar to that of figure 2 was used to demonstrate the operational principle of the presented configuration. Light from a laser diode source, (780 nm wave length) with a coherence length of about 5 m was launched into the input port of a 50-50 fused tapered directional coupler. A quarter wave plate was placed between the light source and the input fibre with its axis at 45° with respect to the polarisation azimuth of the laser beam such that circularly polarised light was injected into the coupler. The output ports were fusion spliced to a second directional coupler with a similar coupling ratio. The output ports of the second

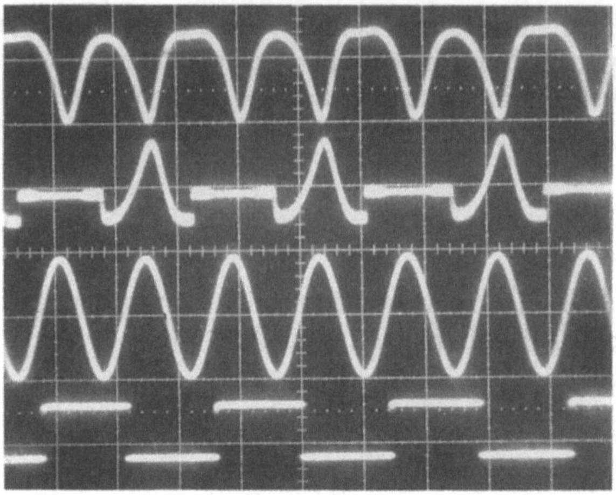

Figure 3. Photo-detector output (trace a), the gated carrier signal (trace b), band pass filter output (trace c) and square wave signal (trace d).

coupler were spliced to a 450 m length of fibre which formed the ring resonator. A phase modulator formed by wrapping part of the resonator fibre around a piezo electric cylinder was inserted close to one of the output ports of coupler. This modulator is used to enable pseudo-heterodyne signal processing [10] to be used to recover the phase shift induced by the rotation of the ring. The output signal was monitored with the photodetectors and patterns similar to the output of two beam interferometer were observed with no increase in finesse. This was due to the excess loss in the splicing joints between the two couplers and the splicing joints within the sensing loop, in addition the coupling ratio of the coupler used for the ring resonator was not appropriate for high finesse configuration ($k = 0.5$).

In order to check the reciprocity of the system the phase modulator was sinusoidally driven at frequency of 1.35 kHz to induce a phase modulation amplitude of ~ 2.82 radians (Figure 3.a) [10]. Alternate half cycles of the output were gated and bandpass filtered at 2.7 kHz (Figure 3.b,c). Square waves of equal duty cycle were obtained from the resulting electronic carrier via an automatic gain control and a divide-by-two bistable (figure 3.d) [11]. The phase of this signal was compared with a derived reference signal using a lock-in analyzer. The sensing loop was placed on a rotatable platform and driven sinusoidally with a peak to peak value of 3.18 $°s^{-1}$ (~0.055 $rad\ s^{-1}$), and the output of lock-in analyzer was simultaneously plotted (Figure 4). During the experiment the gyroscope was subject to environmental perturbation and no random phase modulation was observed, but like other gyroscope configurations it could be influenced by externally induced birefringence.

The sensitivity of this system could be substantially improved by using a low loss coupler for the sensing ring resonator with a coupling ratio equal to resonant coupling coefficient [9]. To avoid signal fading due to the polarisation change, an

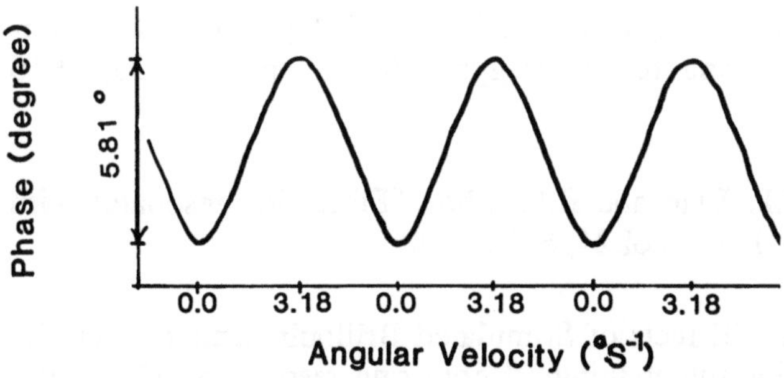

Figure 4. Phase change of the all-fibre ring resonator gyroscope subject to a sinusoidal angular velocity.

all fibre polarisation maintaining ring resonator is required [7] this automatically eliminates the possible drift caused by Faraday effect [12]. Obviously as the finesse of the ring resonator increases the required fibre length in the sensing loop would decrease. This short length and high finesse output impose new conditions in selecting the signal processing. Research is presently in progress to compare the merits of different signal recovery schemes suitable for this new configuration.

The system presented in this paper is potentially more sensitive than Sagnac interferometer, and unlike the ring resonator gyroscope it has a reciprocal format and experiences no Rayleigh and stimulated Brillouin signal caused by the use of a highly coherent source. It is also a balanced configuration hence no excess noise from source frequency jitter is to be expected.

Finally, we have demonstrated a novel configuration based on fibre ring resonator in conjunction with a short coherence source. The system can be operated reciprocally, it is potentially stable, not affected by back scattered signals and promises higher sensitivity than the Sagnac interferometer.

References

1) R.B. Bergh, H.C. Lefevre and H.J. Shaw "An overview of fibre-optic gyroscopes" J.Lightwave Tech., Vol. LT-2, 91 (1984).

2) C.C. Cutler, S.A. Newton and H.J. Shaw "Limitation of rotation sensing by scattering" Opt. Lett., Vol.5, 448 (1980).

3) K. Bohm, P. Martin, K. Petermann, E. Weidel and R. Ulrich "Low-drift fibre gyro using a superluminescent diode" Electron. Lett. Vol.17, 352 (1981).

4) R.E. Meyer, S. Ezekiel, D.W. Stowe and V.J. Tekippe "Passive fibre-optic ring resonator for rotation sensing" Opt. Lett., Vol.8, 644 (1983).

5) K. Iwatsuki, K. Hotate and M. Higashiguchi "Effect of Rayleigh back scattering in an optical passive ring resonator gyro" Appl. Opt., Vol.23, 3916 (1984).

6) G.Y. Yue, J.D. Peng, Y.P. Liao and B.K. Zhou "Fibre ring resonator with finesse of 1260" Electron. Lett., Vol.24, 622 (1988).

7) R. Kadiwar and I.P. Giles "Effects of Stimulated Brillouin scattering on the performance of polarisation maintaining all fibre ring resonators" Opt. Lett., Vol.14, 332 (1989)

8) L.F. Stokes, M. Chodorow and H.J. Shaw "All single mode fibre resonator" Opt. Lett, Vol.7, 288 (1982).

9) F. Farahi, N. Takahashi, J.D.C. Jones and D.A. Jackson "Multiplexed fibre optic sensors using ring interferometers" J. Modern Opt., Vol.36, 337 (1989).

10) A.C. Lewin, A.D. Kersey and D.A. Jackson "Two wide dynamic range signal recovery schemes for the fibre optic gyroscope" IEE Proceedings, Vol.132, 271 (1985).

11) R.P. Tatam, J.D.C. Jones and D.A. Jackson "Opto-electronic processing schemes for the measurement of circular birefringence" Optica Acta, Vol.33, 1519 (1986).

12) K. Hotate and K. Tabe "Drift of an optical fiber gyroscope caused by the Faraday effect: influence of the earth's magnetic field" Appl. Opt., Vol.25, 1086 (1986).

Evaluation of a Sawtooth Generator in a Closed-Loop Fiber-Optic Gyroscope

A. Kurokawa[1], *K. Kajiwara*[1], *N. Usui*[1], *Y. Hayakawa*[1], *M. Haruna*[2], *and H. Nishihara*[2]

[1]Mitsubishi Precision Co., Ltd., 345 Kamimachiya, Kamakura, Kanagawa 247, Japan

[2]Department of Electronics, Faculty of Engineering, Osaka University, 2-1 Yamada-oka, Suita, Osaka 565, Japan

Abstract: A closed-loop fiber-optic gyroscope(FOG) with an integrated-optic modulator(IOM) was successfully developed. Optimum characteristics of a sawtooth generator which drives the IOM were verified experimentally regarding a scale factor stability and linearity. Under the optimum condition of the sawtooth generator output, the scale factor stability of 13.8ppm(1σ) and the linearity of 82ppm of full scale(100°/sec) were achieved.

1. Introduction

Fiber-optic gyroscope(FOG) is being actively developed as a next generation gyroscope. In particular, closed-loop FOG with a sawtooth phase-modulated feedback has been vigorously investigated because of it's potential of reaching a excellent scale factor stability and linearity together with a wide dynamic range[1-3].

In Ref.1, the influence of the characteristics of a sawtooth phase modulation signal on gyro performances is discussed theoretically. In Ref.2 and Ref.3, the performance test results of a closed-loop FOG are presented.

However, the experimental data on the influence of a flyback time and amplitude of a sawtooth wave on gyro performances in the closed-loop FOG have not been presented yet.

This paper presents the experimental results of a sawtooth generator in the closed-loop FOG with an integrated-optic modulator(IOM). The experiments revealed

① the influence of the flyback time and amplitude of the sawtooth wave on the scale factor and linearity

and ② the influence of the amplitude fluctuation of the sawtooth wave on a scale factor stability.

2. Demodulator output of the closed-loop FOG

A schematic diagram of the closed-loop FOG using the IOM and the sawtooth genrator is shown in Fig.1. In this FOG, the sawtooth phase modulation signal, in addition to the sinusoidal phase modulation signal, is given to the counter-propagating light. Usually, this sawtooth phase modulation signal is the function

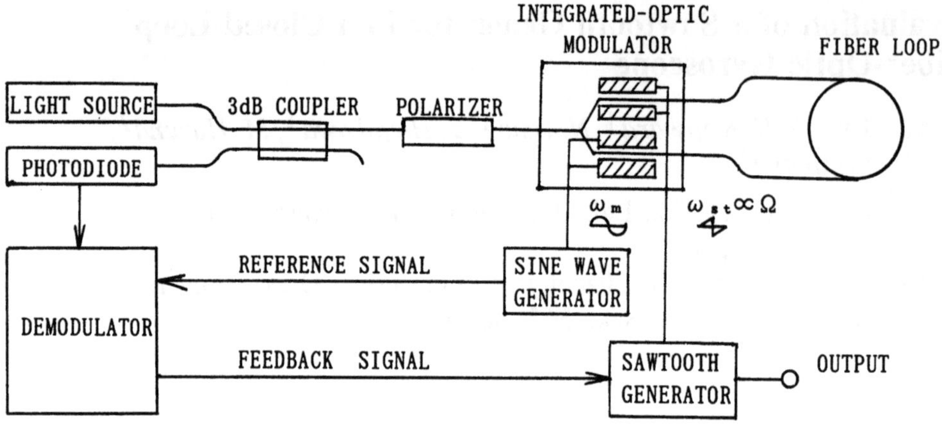

Fig.1. A schematic diagram of a closed-loop FOG

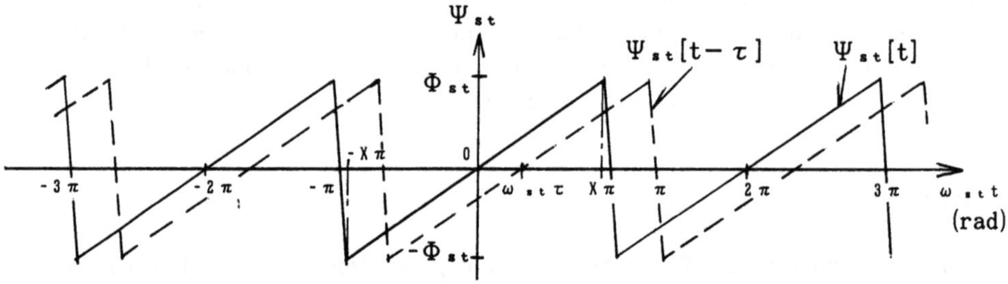

Fig.2. Sawtooth phase modulation function: Ψ_{st}

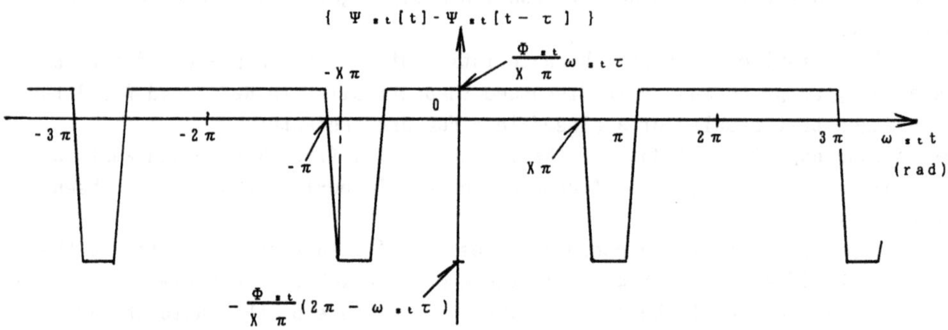

Fig.3. $\{\Psi_{st}[t]-\Psi_{st}[t-\tau]\}$

Ψ_{st} with a finite flyback time as shown in Fig.2. In this case, the phase shift between the clockwise (CW) and counterclockwise (CCW) light generated by the sawtooth wave is the function $\{\Psi_{st}[t]-\Psi_{st}[t-\tau]\}$ as shown in Fig.3, where Φ_{st} is the peak phase shift, $X=1-t_f\cdot f_{st}$, t_f is the flyback time, f_{st} is the sawtooth phase modulation frequency, and $\omega_{st}=2\pi f_{st}$.

The demodulator outputs the scalar product of the intensity of the interference light detected by a photodiode and the sinusoidal phase modulation signal.

So far, the terms related to the sawtooth wave in the demodulator output signal were expressed with the complex Fourier coefficients[1].
In this paper, for the convenience of caluculating the frequency f_{st} which nulls the demodulator output signal, the demodulator output signal is expressed with the real Fourier coefficients as follows:

$$V = 4G\sqrt{P_{cw}P_{ccw}} \cdot J_1(\eta)\cdot\{A_0/2 + \sum_{k=1}^{\infty}[A_k\cdot\cos(k\omega_{st}t) + B_k\cdot\sin(k\omega_{st}t)]\} \quad (1)$$

where G is the constant,, P_{cw} and P_{ccw} are the intensity of the CW and CCW light respectively, J_1 is the first order Bessel function of the first type, η is the phase modulation index, A_0, A_k and B_k are the real Fourier coefficients, and t is time.
Assuming that the ac components are eliminated by a low pass filter, the demodulator output signal by above Eq(1) is written as

$$V = 2G\sqrt{P_{cw}P_{ccw}} \cdot J_1(\eta)\cdot A_0 \quad (2)$$

where

$$A_0 = \frac{4(1-X)X}{\Phi_{st}}\cdot \sin\left(\frac{\Phi_{st}}{X}\right)\sin\left\{\phi_s - \frac{\Phi_{st}}{X} + \frac{\Phi_{st}}{X\pi}\omega_{st}\tau\right\} + \frac{1}{\pi}(2X\cdot\pi - \omega_{st}\tau)\cdot\sin\left\{\phi_s + \frac{\Phi_{st}}{X\pi}\omega_{st}\tau\right\} + \frac{1}{\pi}\{2(X-1)\cdot\pi + \omega_{st}\tau)\}\cdot\sin\left\{\phi_s - \frac{\Phi_{st}}{X\pi}(2\pi - \omega_{st}\tau)\right\}, \quad (3)$$

ϕ_s is the Sagnac phase shift, and τ is the delay time of the light traversing the fiber loop.
The closed-loop FOG has the feedback loop which controls the frequency f_{st} in order to null the demodulator output signal. Therefore, in case of $t_f=0$ (i.e., X=1) and $\Phi_{st}=q\pi$ ($q=\pm1$), the ideal gyro output is obtained as follows:

$$f_{st} = -\frac{\phi_s}{2\pi q\cdot\tau} = -\frac{1}{q}\cdot\frac{2R}{\lambda n_f}\cdot\Omega \quad (4)$$

where R is the fiber loop radius, λ is the wavelength in free space, n_f is the refractive index of the fiber, Ω is the rotation rate, and $\mathrm{sign}(\Omega/q)<0$.
In the practical sawtooth wave, however, $t_f\neq0$ (i.e., $X\neq1$) and Φ_{st} deviates from $q\pi$, so that the ideal gyro output cannot be obtained.
The following is the experimental results regarding the influence of the flyback time and amplitude in the sawtooth wave on the scale factor, non-linearity and scale factor unstability.

3. Experiments

3·1. Characteristics of the fabricated closed-loop FOG

The optical system of the fabricated closed-loop FOG consists of a super radiant diode with a temperature controller, two 3dB polished fiber couplers, a fiber polarizer, an IOM, a polarization maintaining fiber loop and an avalanche photo-

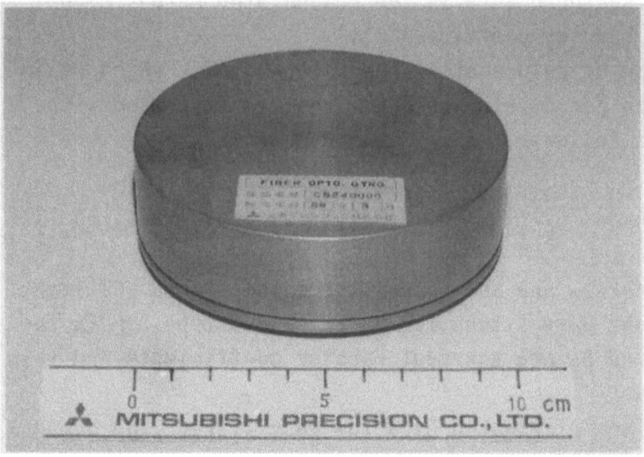

Fig.4. Closed-loop FOG

Table 1. Characteristics of the closed-loop FOG

Size		[mm]	:	ϕ110 × 30(H)
Wavelength		[nm]	:	816
Fiber	- Type		:	PANDA
	- Length	[m]	:	601.3
Sinusoidal phase modulation frequency		[kHz]	:	170
Optical scale factor		[deg/(°/sec)]	:	0.833

Table 2. Characteristics of the IOM

Size of $LiNbO_3$ substrate		[mm]	:	5(W) ×49(L) ×1(T)
Electrode length		[mm]	:	27
Excess loss	- waveguide	[dB]	:	2.5
	- waveguide to fiber	[dB]	:	1.3
Splitting ratio		[%]	:	44 : 56
Crosstalk		[dB]	:	≤ -17
Half-wave voltage		[V]	:	4.2
3dB bandwidth		[MHz]	:	≥ 800

diode. The IOM consists of a Y-junction waveguide and two phase modulators integrated on a Ti-diffused,X-cut Y-propagation $LiNbO_3$ substrate. The guide width of the waveguide is 5μm and the branching angle of the Y-junction is 1/50 rad. Characteristics of the closed-loop FOG and the IOM are listed in Table 1 and Table 2 respectively. The appearance of the closed-loop FOG is shown in Fig.4.

In order to investigate the influence of the characteristics of the sawtooth wave on the gyro performances,two sawtooth generators were fabricated. One had 40nsec flyback time,the other had 520nsec flyback time,and output amplitude of both generators were variable. Characteristics of two sawtooth generators are listed in Table 3.

Table 3. Characteristics of the sawtooth generators

Parameters			(A)	(B)
Frequency	- Range	[Hz]	0.1mHz～200kHz	0.1mHz ～200kHz
	- Accuracy	[ppm]	± 5	± 5
	- Jitter	[nsec]	± 4.4	±75
Flyback time		[nsec]	40	520
Amplitude	- Range at 50 Ω	[Vp-p]	0 ～ 10	0 ～ 15
	- Accuracy	[%]	± 0.2	± 0.05
	- Frequency coefficient	[%/kHz]	− 0.0046	− 0.037

3·2. Experimental methods and results

(1) Scale factor and linearity

At first, in order to investigate the variations of the scale factor and the linearity due to the change of the amplitude Φ_{st}, we measured the scale factor and the linearity for the different amplitude in the closed-loop FOG with the generator (A) [t_f=40nsec]. The selected amplitude were 0.8π, 0.9π, π, 1.1π, 1.2π Next, in order to investigate the variations of the scale factor and the linearity due to the change of the flyback time, the same experiments were repeated for the FOG with the generator (B) [t_f=520nsec]. These experiments were performed at the ambient temperature of 23°C±2°C without temperature control of the FOG. The scale factor was taken as the slope of a regression straight line calculated by least squares method, where the output frequencies f_{st} were measued from -100°/sec to 100°/sec with 10°/sec increment . The linearity was taken as the maximum deviation from this line and expressed as the ratio to a full scale.

The data of the scale factor and linearity for the diffrent amplitude of the sawtooth wave are shown in Fig. 5 and Fig. 6 respectively together with the calculated values. The calculated values were obtained by introducing the data in Table 1 and Table 3 into Eq. (3).

The measured values of the scale factor and the linearity correspond to the calculated.

The following facts are recognized from Fig. 5.

① The change rate of the scale factor by the amplitude is minimum at $|\Phi_{st}|=\pi$.

② This change rate becomes small with the decrease of the flyback time.

The following facts are recognized from Fig. 6.

① The linearity error is minimum at $|\Phi_{st}|=\pi$.

② The linearity error becomes small with the decrease of the flyback time.

And, the measured linearities at $|\Phi_{st}|=\pi$ were 82ppm of full scale at t_f=40nsec and 132ppm of full scale at t_f=520nsec.

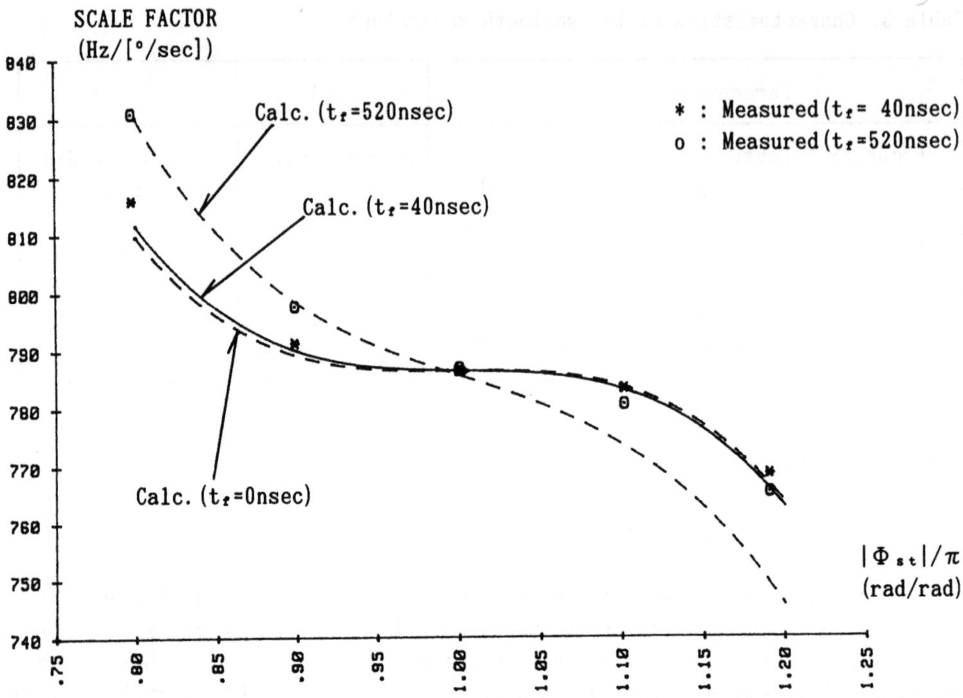

Fig. 5. The amplitude of the sawtooth wave vs. the scale factor

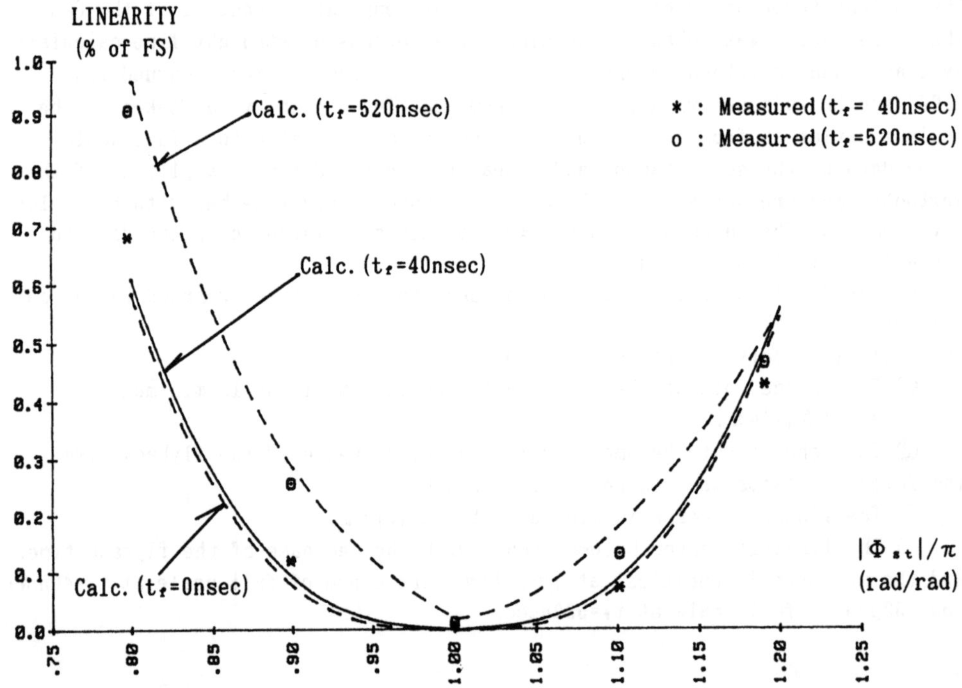

Fig. 6. The amplitude of the sawtooth wave vs. the linearity

(2) Scale factor stability

In order to evaluate the scale factor stability of the fabricated closed-loop FOG, the experiment under the similar conditions described in 3·1(1) was performed five times at $|\Phi_{st}|=\pi$ (nominal) using both the generator (A) and (B). The cooldown time between each measurement was more than 1hr. The scale factor stability was taken as the standard deviation of five scale factor values and expressed as the ratio to the average scale factor value.

The scale factor stability was 13.8ppm (1σ) with the generator (A) and was 38.2ppm (1σ) with the generator (B). The detailed data are listed in Table 4.

On the other hand, We calculated the values of the scale factor stability using Eq. (3) and the data in Table 1, Table 3 under the assumption that the amplitude fluctuation (i.e., the accuracy) of the generator output signal varied at random with the normal distribution.
The calculated results were 11.7ppm (1σ) with the generator (A) [the flyback time of 40nsec, the amplitude fluctuation of ±0.2%] and 40.3ppm (1σ) with the generator (B) [520nsec, ±0.05%] as shown in Table 5.

Table 4. Scale factor stability (measured value)

Data No.	Scale factor [Hz/(°/sec)]	
	Generator (A)	Generator (B)
1	786.2456	786.8230
2	786.2292	786.8151
3	786.2374	786.8254
4	786.2307	786.8033
5	786.2125	786.7439
Average value	786.2311	786.8021
Scale factor stability	13.8ppm (1σ)	38.2ppm (1σ)

Table 5. Scale factor stability (calculated value)

Generator	Flyback time	Amplitude fluctuation	Scale factor deviation (calculated)	Scale factor stability (calculated)
(A)	40nsec	±0.2 %	35.2ppm p-p	11.7ppm (1σ)
(B)	520nsec	±0.05%	120.8ppm p-p	40.3ppm (1σ)

The measured values are in good agreement with the calculated. Therefore, it is clear that the scale factor stability depends on the amplitude fluctuation of the sawtooth generator output signal.

4. Conclusions

In the closed-loop FOG with the IOM and the sawtooth generator, the relations between the characteristics of the generator and gyro performances were experimentally made clear, and it was verified that the experimental results are in good agreement with the calculated values.

In general, the scale factor error for the inertial grade gyroscope is required within 100ppm. According to the experimental results, it is concluded that the condition necessary for realization of the inertial grade close-loop FOG is to drive the IOM by the sawtooth wave which has the flyback time shorter than or equal to 40nsec together with $|\Phi_{st}|=\pi$. Besides, the bias stability of the fabricated closed-loop FOG was 0.43°/hr(1σ).

References

[1] A. Ebberg and G. Schiffner/OPTICS LETTERS, Vol. 10, No. 6, P. 300, 1985
[2] H. C. Lefèvre/Proc. of SPIE, Vol. 838, Fiber Optic and Laser Sensors V, P. 86, 1987
[3] G. A. Pavlath/Proc. of the NATIONAL TECHNICAL MEETING, P. 90, 1988

Polarisation Effects in Highly Birefringent Fibre Ring Resonators

Z.K. Ioannidis, R. Kadiwar, and I.P. Giles

University College London, Electronic and Electrical Engineering Department, Torrington Place, London WC1E 7JE, UK

The polarisation stability of highly birefringent all fibre ring resonators is investigated both experimentally and theoretically, and shown to be a major problem. Tolerance levels for the polarisation isolation of the highly birefringent fibre coupler are defined, and solutions to avoid polarisation instabilities are demonstrated. The performance of such resonators in gyroscope applications is also discussed.

Introduction: One of the major problems in Optical Fibre Ring Resonator[1], fig.1, applications arises from the fact that real single mode fibres support two orthogonal non-degenerate polarisation modes. This creates a second resonance dip in the intensity transmission of the resonator, fig.2, the level ρ_2 and phase separation θ of which drift with environmental parameters. However, in applications the performance of a resonator should be independent of environmental perturbations and ideally only one resonance should be present It is therefore evident that such a resonator should use fibre that shows polarisation preserving properties.

In this communication we show both theoretically and experimentally that such a goal is very difficult to achieve using common polarisation maintaining fibre due to the unavoidable coupling between the fibre polarisation modes. We also show that this problem can be overcome if the fibre+coupler cavity exhibits sufficient differential loss and/or coupling constant between the fibre modes. The performance of such resonators in gyroscope applications is also discussed.

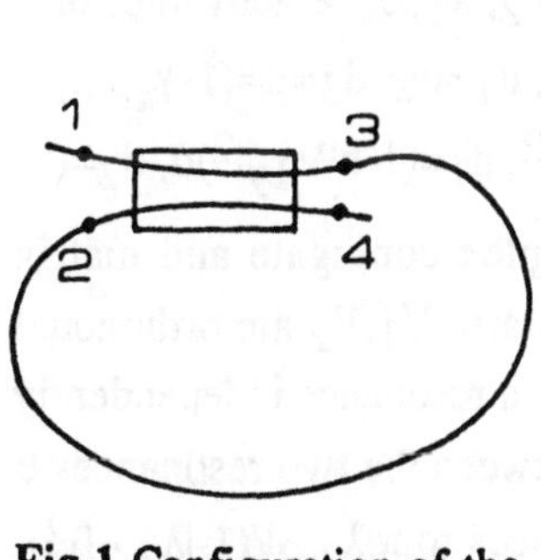

Fig.1 Configuration of the all-fibre ring resonator

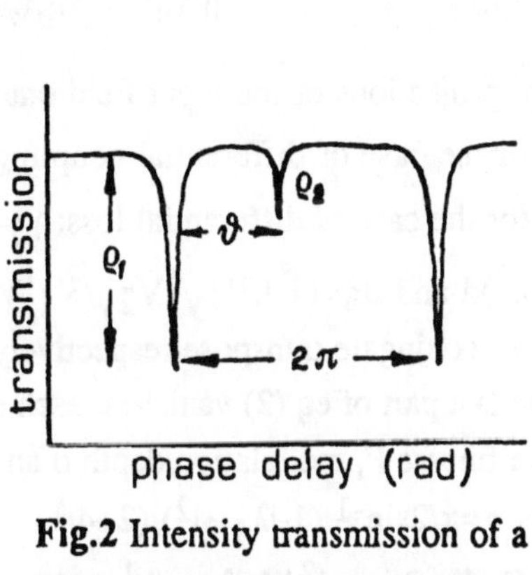

Fig.2 Intensity transmission of a resonator for a 2π optical path change

Theory: The polarisation properties of the optical fibre ring resonator can be described by means of its eigenstates of polarisation[2], which are defined as the polarisation states that remain the same after one transit in the ring; an input field is analysed onto the two eigenstates, each of them producing a single resonance dip in the response. Let us consider the Polarisation Maintaining fibre Ring Resonator (PMRR) of fig.1, in which the fibre modes are allowed to exhibit differential coupling constant in the coupler and/or differential loss either in the coupler or the fibre loop. In a real PMRR there is always cross-talk between the fibre polarisation modes which may occur in the fibre loop but predominantly in the coupler due to misalignment of the fibre geometrical axes[3]. Let us now consider the Jones matrix representation of such a resonator. The field at each port i is expressed as $E_i=[E_x,E_y]^T$ where E_x,E_y are the complex amplitude projections of the field onto the fibre birefringence axes and T denotes matrix transpose; the roundtrip transfer matrix can then take the form:

$$S=\lambda_o\begin{bmatrix} \cos\alpha\ \exp(i\phi/2) & \sqrt{h}\ \sin\alpha \exp(-i\phi/2) \\ \sin\alpha\ \exp(i\phi/2) & \sqrt{h}\ \cos\alpha\ \exp(-i\phi/2) \end{bmatrix} \quad (1)$$

where $\lambda_o= k_x(1-\gamma_x)\exp[-fL+i(\beta L+\pi/2)]$ and $h=k_y(1-\gamma_y)(1-g)/k_x(1-\gamma_x)$. [$\gamma_x,\gamma_y$: coupler loss; k_x,k_y: coupling constants; f: fibre atten. coeff.; L: fibre length; $\beta=(\beta_x+\beta_y)/2$, $\phi=(\beta_x-\beta_y)L$, β_x,β_y : fibre propagation constants; g: differential loss between the fibre modes and angle α describes the level of isolation between the fibre polarisation modes $R=-20\log(\sin\alpha)$]. The eigenstates of polarisation are now calculated as the eigenvectors $\mathbf{V}_1,\mathbf{V}_2$ of S. Quantity h (so called anisotropy parameter) expresses the anisotropic properties of the resonator cavity and shows that any of the differential coupling constant, coupler loss or fibre loss can have equivalent effect on $\mathbf{V}_1,\mathbf{V}_2$. The intensity transmission can now be calculated as

$$T=(1-\gamma_0)-\frac{(1-d_1)|e_1|^2}{|1-\lambda_1|^2}-\frac{(1-d_2)|e_2|^2}{|1-\lambda_2|^2}-2\mathcal{Re}\left\{\frac{(1-d_3)\,e_1\dot{e}_2\,V_1^tV_2}{(1-\lambda_1)^*(1-\lambda_2)}\right\} \quad (2)$$

[e_1,e_2: projections of the input field onto the eigenstates $\mathbf{V}_1,\mathbf{V}_2$; λ_1,λ_2: eigenvalues of S and: for the case of differential coupling constant: $\gamma_0=\gamma_x=\gamma_y$, $d_1=d_2=d_3=d=(1-\gamma_x)\exp(-2fL)$; for the case of differential loss: $\gamma_0=\gamma_x+(\gamma_y-\gamma_x)|E_{1y}|/|E_1|^2$, $d_1=(1-h|V_{1y}|^2)d$, $d_2=(1-h|V_{2y}|^2)d$ and $d_3=(1-hV_{1y}{}^*V_{2y}/V_1{}^tV_2)d$ (*,t denote complex conjugate and matrix complex conjugate transpose respectively)]. When h=0, eigenstates $\mathbf{V}_1,\mathbf{V}_2$ are orthogonal and the last part of eq.(2) vanishes; each eigenstate then creates a resonance independently and the finesse F, modulation depth ρ and phase separation between the two resonances θ are: $F_{1,2}=\pi/2\sin^{-1}(1-|\lambda_{1,2}|^2)/(2+4|\lambda_{1,2}|)$ (when F>>2 it reduces to $\pi|\lambda_{1,2}|/(1-|\lambda_{1,2}|)^2$), $\rho_{1,2}=(1-d_{1,2})|e_{1,2}|^2/(1-|\lambda_{1,2}|)^2$ and $\theta=\arg(e_1)-\arg(e_2)$. When h=0, then $\mathbf{V}_1,\mathbf{V}_2$ are no more orthogonal and the last part of eq.(2) shows that the two resonances depend on its other. This part is more significant close to the resonance points and can be neglected away from them; however, for small excitation of the unwanted eigenstate, its magnitude is very small and can still be neglected unless extreme phase sensitivity is required, such as in gyroscope applications as will be highlighted below.

Discussion / Experiment: Although the polarisation holding ability of highly birefringent fibres is not affected by usual external perturbations (eg slight bends,twists), the differential phase delay between the fibre polarisation modes ϕ is very sensitive even to small environmental changes.

Fig.3 uses the Poincare' sphere representation of polarisation states to illustrate the drift of V_1, V_2 of a PMRR with ϕ, as calculated from eq(1). In an ideal PMRR with no cross-coupling (α=0), the eigenstates are fixed at points H and V (fibre birefringence axes). Curve (a) shows the drift of V_1, V_2 assuming R=25dB isolation level between the fibre modes, which is a typical value obtained in our polarisation preserving couplers. When $\phi=\pi$, the eigenstates are linear, very close to points H and V. A linear input polarisation state along a fibre axis excites the unwanted resonance at a level of -28dB; however as ϕ approaches zero, the eigenstates drift quickly to the poles R,L (right, left circular

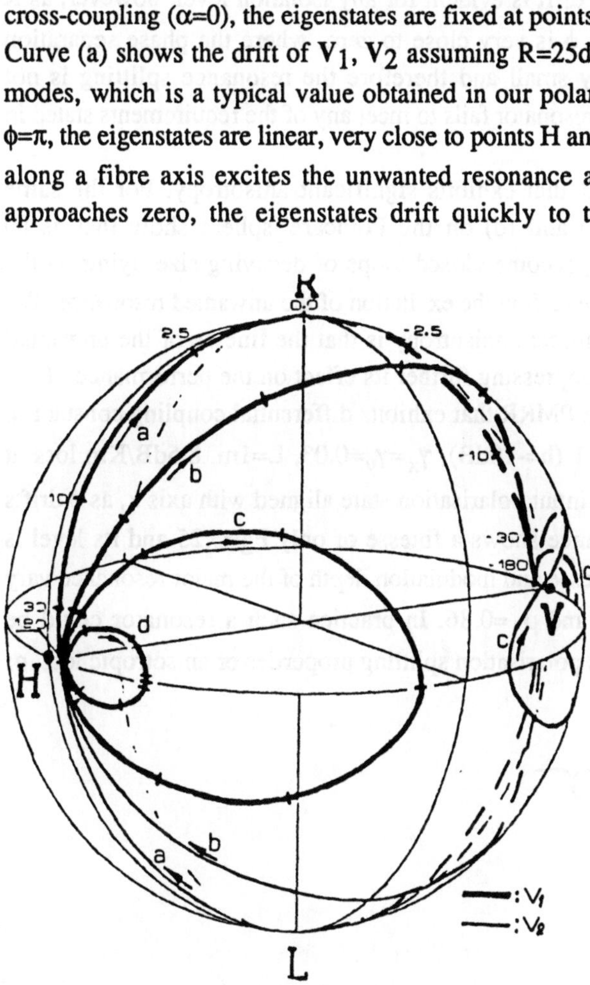

Fig.3 Poincare' sphere representation of the drift of the polarisation eigenstates of a PMRR with ϕ. Parqameter h is : a)0, b)0.5, c)1, d)3dB.

polarisation states) and the same input state excites them now at equal levels. At $\phi=-\pi$ the eigenstates exchange positions on the sphere. Curve (b) shows that similar drift takes place even when a small anisotropy ($|h|<<1dB$) is present; it is easy to verify in this case that V_1,V_2 do not follow orthogonal trajectories on the sphere any more. A 1m PMRR was fabricated using York bow-tie fibre (l_b=1mm), single mode at λ=633nm. The finesse and modulation depth were measured: $F_1=120$, $F_2=115$, $\rho_1=0.97$, $\rho_2=0.91$, yielding an anisotropy parameter of just h=-0.05dB. The intensity transmission for an input linearly polarised along a fibre axis is shown in fig.4. As predicted, when ϕ approaches zero the depth of the minor resonance increases and at $\phi\sim0$ the response consists of a pair of resonance dips. Eq(2) produces exactly the same response assuming R=23dB. It is important to note that the described drift is evident for any isolation level; however, as R rises, the drift occurs mainly when ϕ is very close to zero, where the phase separation between the resonance dips is very small and therefore the resonance splitting is not distinguished easily. Clearly, such a resonator fails to meet any of the requirements stated in the beginning.

Let us now consider a PMRR that exhibits significant anisotropy. For the same isolation level of 25dB, curves (c) and (d) on the Poincare' sphere show that as $|h|$ increases, the trajectories of V_1,V_2 become closed loops of decaying size, lying on the locus of H and V, thus minimizing the drift in the excitation of the unwanted resonance. The second benefit of introducing a significant anisotropy is that the finesse of the unwanted eigenstate is now greatly reduced, suppressing further its effect on the performance of the resonator. As an example, consider a PMRR that exhibits differential coupling constant in the coupler and has $k_x=0.97$, $k_y=0.1$ (h=-10dB), $\gamma_x=\gamma_y=0.03$, L=1m, 8.5dB/Km loss at λ=633nm and R=25dB. For a linear input polarisation state aligned with axis x, as ϕ drifts between $-\pi$ and π, the minor resonance shows a finesse of only $F_2=3.25$ and its level is now kept below -48dB, while the finesse and modulation depth of the major resonance vary only 4% around the values $F_1=94$ and $\rho_1=0.86$. In practice such a resonator could be realized using a coupler that exhibits polarisation splitting properties or anisotropic loss, or

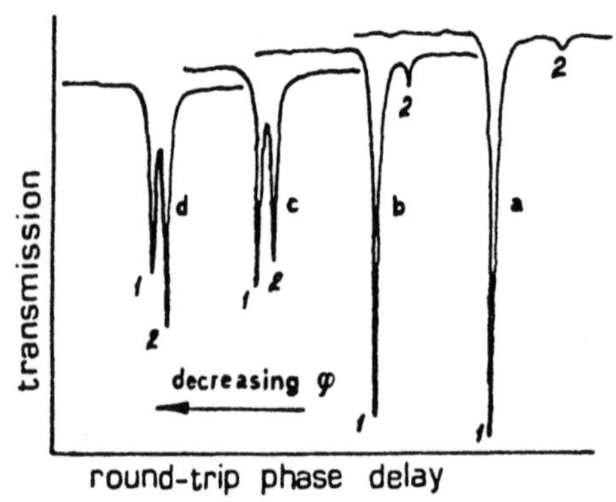

Fig.4 Experimental transmission of a PMRR for different values of ϕ.

inserting an in-line polariser in the fibre loop (alternatively, a resonator made of polarising fibre[4] could serve the same purpose). An attempt to fabricate such a resonator was tried by inserting a thin film of nematic liquid crystal (NLC) in between the two polished fibre blocks that form the coupler of the polished type, instead of the usual refractive index matching liquid. The cigar-like molecules of the NLC were aligned homogeneously (ie lying on the block surfaces), perpendicular to the fibre axes; thus a TM-like field faces the lower (ordinary) refractive index which for the NLC used (BDH mixture 14616[5]) was n_o=1.458, while a TE-like field experiences the higher (extraordinary) one n_e=1.5. Due to this anisotropic interlayer the coupler exhibits different coupling constants and loss for the two fibre polarisation modes; the anisotropy parameter h was estimated to be h=-8dB. The fibre used was the same as before and had a length of L=0.8m. Fig.5 shows the intensity response of this resonator under linear optical path modulation, for a linear input polarisation state inclined at 45deg to the fibre axes. Clearly the two polarisation eigenstates exhibit vastly different finesses (F_1=60,F_2: almost sinusoidal modulation). When the input was aligned with the fibre axis that creates the higher finesse, the response consisted of a single resonance dip and was unaffected by normal external perturbations. To our knowledge this is the first demonstration of a resonator that exhibits vastly different finesses and polarisation stabilized response.

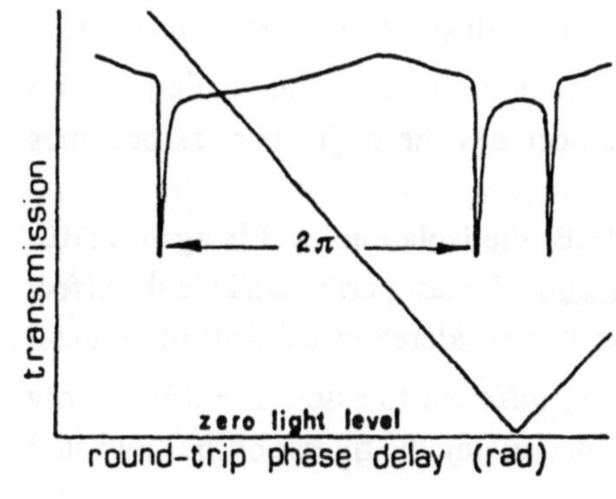

Fig.5 Experimental transmission of an anisotropic PMRR (h=-8dB).

Gyroscope applications: In gyroscope applications, due to reciprocity,the excitation of the unwanted eigenstate is not a problem itself as long as the eigenstates are exited on both directions identically. However, in practice there is always mismatch between the two input polarisation states, therefore the shape distortion of the major resonance due to the presence of the minor is different in each direction. In what follows, the PMRR of the previous example is considered and the resonance drift ω_d (ie the differential phase shift of the resonance point between the two counter propagating fields-in the absence of rotation) is calculated using eq(2). For one direction the polarisation state is assumed linear along the x fibre axis, while for the other a -30dB component is also present on the y axis. Curve (a) of fig.6 shows the variation of ω_d with ϕ in the absence of cross-talk (α=0). Curves (b) and

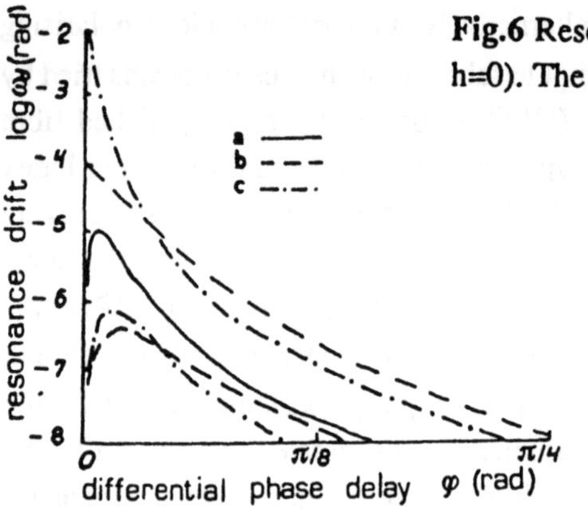

Fig.6 Resonance drift of a PMRR gyro (F=100, h=0). The isolation ratio is: a)0, b)25, c)33dB.

(c) show the drift for R=25 and 33dB respectively (the upper line corresponds to max. ellipticity of the mismatched input, while the lower one to min.). Clearly, in the presence of cross-talk the drift has increased significantly. It is important to note that the drift is bigger in the case of the higher isolation level (R=33dB); recalling fig.4, when ϕ=0, both eigenstates are equally exited but their phase separation decreases with the level of isolation, therefore the two resonance dips overlap to a greater extent and the resonance shift is enhanced. For the PMRR under consideration (F~100), the drift is maximized for R=33dB (curve (c)) and reduces therein. Another important effect illustrated by fig.4 is that a gyro system locked on resonance dip 1 -traces (a) or (b)- could even face signal fading when ϕ changes sign, since in this case, resonance dip 2 becomes the major one, as becomes evident from trace (d).

Let us now introduce an anisotropy of h=-10dB; the isolation level is again 25dB. Curve (a) of fig.7 shows the hypothetical drift neglecting the last part of eq(2),ie the effect of nonorthogonality of V_1,V_2. This curve can also be considered as the drift of an ideal PMRR (α=0), the eigenstates of which exhibit vastly different finesses. The drift is now several orders of magnitude reduced, indicating that decreasing the finesse of the unwanted

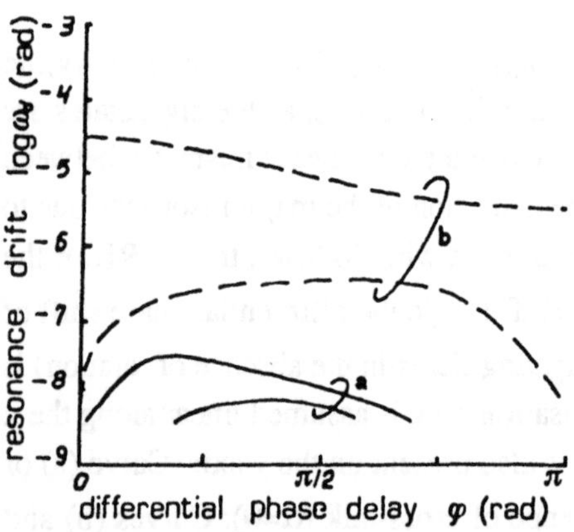

Fig.7 Resonance drift of the same PMRR as in fig.6 but h=-10dB. In curve (a) the non-orthogonality part of eq.(2) is neglected.

eigenstate can be a useful tool to tolerate the accuracy in matching the two input polarisation states. However, in the presence of cross-talk, the appearance of the nonorthogonality term of eq(2) is the price to pay for the suppression of the unwanted resonance. Although this term is very small ($<10^{-3}$), its sharp variation close to the major resonance point can still be the reason of significant drift,as shown by curve (b) which was calculated using the full eq(2); therefore, in applications where extreme phase sensitivity is required, the benefit of introducing anisotropy in the resonator cavity is eliminated by the fact that the polarisation eigenstates become non-orthogonal, indicating that the isolation level between the fibre modes is still a crucial parameter in such systems. The effect of this non-orthogonality on gyro applications and possible ways to overcome it, are still under examination.

In conclusion, we have showed that the task of a polarisation drift free fibre resonator cannot be reached by a PMRR, unless the isolation level between the fibre modes is kept at excellent levels (eg >>30dB for a finesse of ~100), which we find very difficult to realize in practice. The effect of the cross-talk was found to be eliminated by inserting a moderate anisotropy in the resonator cavity and the first resonator exhibiting such a property was demonstrated.

Acknowledgments: The authors would like to thank A. Kerr for his invaluable help in fabricating the resonators,and their colleagues in the Optical Fibre Systems group at UCL for their support, C. Bowry for her help in using the NLC and the members of the Telecommunication Section of the Electrical Engineering department. of University of Thessaloniki, Greece for providing the opportunity to one of the authors (Z.I.) to complete a part of this work with them.

References

1) L. F. Stokes, M. Chodorow and H. J. Shaw, *Opt. Lett.* **7**, 288 (1982)
2) K. Iwatsuki, K. Hotate and M.Higashiguchi, *Appl. Opt.* **25**, 2606 (1986)
3) A. W. Snyder and A. Stevenson, *J. Light. Techn.* **6**, 450 (1988)
4) M. P. Vahnam, D. N. Payne, A. J. Barlow and E. J. Tarbox, *Opt. Lett.* **9**,306 (1984)
5) I. Sage and D. Chaplin, *Electr. Lett.* **23**, 1192 (1987)
6) R. E. Meyer, S. Ezekiel, D. W. Twone and V. J. Tekippe, *Opt. Lett.* **8**, 644 (1983)

Part V

Optical Fiber Gyroscopes and Special Sources

Evolution of the Fiber Optic Gyroscope

H.C. Léfèvre

Photonetics, 52, Avenue de l'Europe, F-78160 Marly-le-Roi, France

This paper reviews the technical evolution of the interferometric fiber gyroscope over the last decade. Today a psychological barrier has been passed, and it is now accepted that this new technology will find many applications during the 90's.

1 - INTRODUCTION

Over the last decade, the fiber-optic gyroscope (FOG) research and development has evolved from an promising physics experiment to a practical device that is now close to production. This has been made possible by a refined analysis of the system architecture and of the possible signal processing schemes, but also by fundamental improvements of the various optoelectronic technologies, as, in particular, integrated optics, semiconductor sources, polarization preserving fibers, and in-line fiber components.

Today, a psychological barrier has been passed, and it is now accepted among inertial guidance and control specialists that this new technology will be a strong contender for many military and civilian applications of the next decade. Most leading companies in the inertial guidance field are heavily involved in R & D programs on the fiber optic gyroscopes, and prototypes have been flight tested [1]. A fiber-optic gyroscope has even already been selected to fulfil the Attitude and Heading Reference System (AHRS) function of a new regional airliner program [2].

We are now going to describe the technical state-of-the-art and trends of present FOG R & D. We will also present the applications that are foreseen because of the specific advantages of the FOG due to its solid-state configuration : high dynamic range, high bandwidth, rapid start-up, and ability to cope with a severe environment (in particular shocks and vibrations).

2 - BASIC ARCHITECTURE OF THE FIBER OPTIC GYROSCOPE [3]

The interferometric fiber-optic gyroscope is a passive two-wave ring interferometer. The input light is split and propagates in opposite directions along a multiturn single mode fiber coil. Because of the Sagnac effect, the waves experience a phase difference when the whole system is rotated with respect to inertial space.This phase difference is proportional to the rotation rate component Ω parallel to the coil axis :

$$\Delta\Phi_S = (2\pi\, L.D / \lambda.c).\, \Omega$$

where L is the length and D the diameter of the coil, λ is the source wavelength and c is the light velocity in vacuo.

Practical fiber gyroscopes usually work over plus or minus half a fringe, i.e. $\pm\, \pi$ radian of phase difference. The theoretical sensitivity (limited by photon noise) is on the order of 10^{-6}

Springer Proceedings in Physics, Vol. 44 **Optical Fiber Sensors**
Editors: H.J. Arditty · J.P. Dakin · R.Th. Kersten

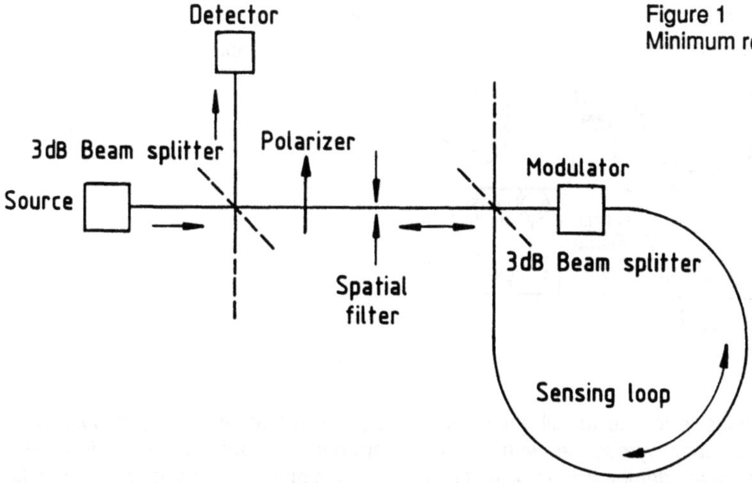

Figure 1
Minimum reciprocal configuration

radian or less of phase shift. Compared to the absolute phase accumulated by the waves when they propagate along the coil, this theoretical limit is extremely small, but reciprocity of light propagation in single-mode waveguides makes it detectable. This requires the use of the so-called minimum reciprocal configuration [4,5] [Figure 1], which is now universally employed. This architecture needs a single spatial mode and polarization filtering at the common input-output port of the interferometer and a biasing phase modulation at one end of the coil that serves as a delay line. This brings a drastic enhancement to the ring interferometer stability by making both opposite paths identical in the absence of rotation. It is also accepted that to further improve noise and drift, it is important to use a broadband source [6] and a polarization preserving fiber [7].

As a matter of fact the system comprises two primary waves sensitive to rotation but also a lot of parasitic waves (backreflected, backscattered or coupled in the crossed polarization) that yield secondary interferometers. The short coherence length of a broad band source suppresses the contrast of these spurious interferometers, and then limits their parasitic signals. The use of polarization preserving fibers is essential because such fibers are also birefringent by principle and one takes advantage of depolarization of cross-coupling with a broadband source. This limits the required polarizer rejection to values that are compatible with present technology [8,9]. Notice that white light interferometry has been found as a very powerful tool to carefully control the polarization and birefringence problems of the ring interferometer [Figure 2] [10], when one wants to get the best possible performances.

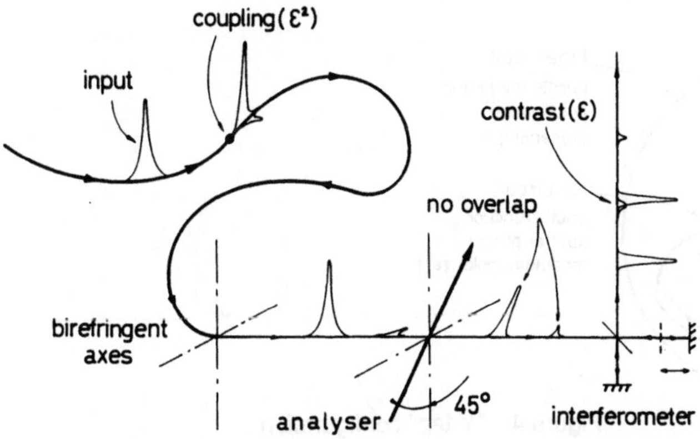

Figure 2
Testing of polarization coupling with white light interferometry

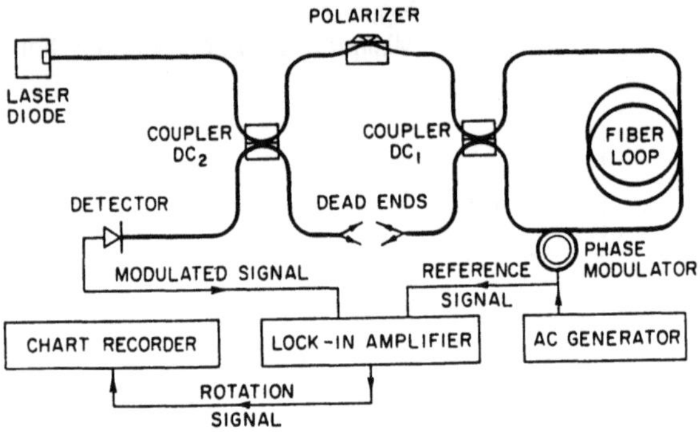

Figure 3
All-fiber gyroscope

Another point of concensus is to use an all-guided technology for ruggedness. In particular it is possible to make an all-fiber gyroscope with in-line components [Figure 3] [11]. However, this approach has only been demonstrated with an open-loop signal processing which yields problems of linearity and stability of the scale factor. This can be good enough for certain applications, but most teams are now working on a closed-loop approach that allows one to linearize and stabilize the scale factor. This was first proposed with the use of bulk acousto-optic frequency shifters [12,13], but the general trend is now to use integrated optic modulators to implement specific closed-loop phase modulation schemes as serrodyne modulation (analog phase ramp) [14] or digital phase ramp [15]. Such schemes require a broad and flat modulation bandwidth that is the main technical advantage of integrated optics. Notice that the digital phase ramp could simply look like a quantified analog ramp, but it is actually fundamentally better because the digital approach is synchronized with the modulation/demodulation biasing scheme. This solves automatically the problem of the finite fly-back of the ramping by synchronous gating and it eliminates the fiber index dependence in the scale factor [16].

Integrated optics is also a favored technology for implementing several functions on the same circuit for improving compactness and simplifying the connections. In particular the optimal simplicity is obtained with the so-called "Y-tap" configuration [Figure 4] [16], which uses a three-function integrated optics composed circuit of a Y junction for wavesplitting, a polarizer on the common base trunk and two phase modulators on both branches.

Such a circuit, now widely accepted and used as the " gyro circuit", has a parallelogram shape to avoid backreflections induced by the index mismatch between the substrate and the fibers [Figure 5]. As can be seen, the assembling of the components has a relative simplicity,

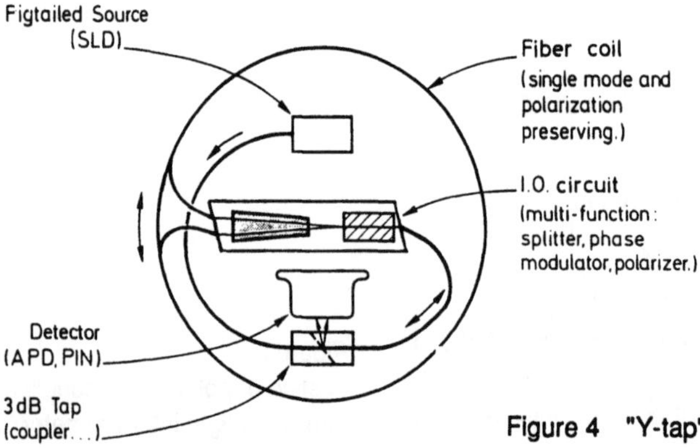

Figure 4 "Y-tap" configuration

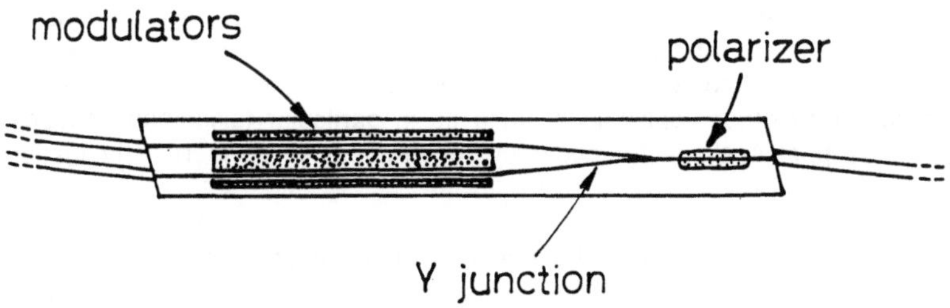

Figure 5 The "gyro circuit"

requiring only a pigtailing of the source and three pigtailings on the I.O. circuit. These connections can be ruggedized to withstand a rough environment.

For sake of completeness, one should not forget two effects that are intrinsically non-reciprocal as the Sagnac effect : the magneto-optic Faraday effect and the non-linear Kerr effect [17]. They could have been a drastic limitation to high performance because they cannot be distinguished from the Sagnac effect, but it was found that they can be eliminated very simply, and they are not a problem anymore. The use of a polarization preserving fiber eliminates the Faraday non-reciprocity if care is taken to avoid twist in the fiber [18], and the use of a broadband source averages out the Kerr non-reciprocity [19]. We find here a very nice property of the interferometric FOG : there are not any parasitic effects that require incompatible solutions. On the contrary, the same solution solves nicely several different problems at the same time : polarization preserving fibers apply to the Faraday effect but also to birefringent non-reciprocity ; broad band sources eliminate backscattering and polarization coupling noise but also the Kerr non-reciprocity !

Finally thermal transients and vibrations can produce parasitic signals because reciprocity is strictly valid only for time invariant systems [20]. Winding techniques must symmetrize the coil to get common mode rejection of the perturbations. The so-called quadrupolar technique [21] looks like the most efficient method. Adequate potting is also necessary to define a stable pointing accuracy of the sensing coil.

3 - PRESENT SUBJECTS OF CONCERN AND APPLICATIONS

FOG prototypes with very good performances have been reported by various R & D teams. Their bias stability is in the 10 to 0.1 degree per hour range for compact devices (100 to 50 mm diameter) using 100 to 500 meters of polarization preserving fiber, and a superluminescent diode as a broadband source. Scale factor accuracy and linearity are in the 100 ppm range or less, for closed-loop FOG's using integrated optics and in the 1000 ppm range, for open-loop FOG's. These performances exceed what is needed for AHRS as has been demonstrated in flight tests [1].

Present performances do fit tactical application requirements (0.1 to 10 °/h range) where they bring in addition a very high dynamic range (more than 1000 °/s) and a very wide measurement bandwidth (several kHz) that cannot be reached with "classical" technologies. Interest is particularly strong for agile missiles for example. The good behaviour of FOG solid state technology under shocks and vibrations is also making it a strong candidate for smart ammunitions. Besides military applications, the FOG looks also very promising for use in a hybrid navigation, attitude and heading system with a GPS (Global Positioning System) receiver. For these applications, efforts are directed towards ruggedization and reduction of cost which is expected to be very competitive once the FOG enters production. Use in cars is even seriously envisaged in Japan [22].

Advanced reseach on components is still continuing for high performance applications. It concerns superfluorescent rare earth doped fiber sources [23] that will improve the wavelength stability of the broadband source, compared to semiconductor sources. This research field is very promising for many other applications, such as in particular telecommunications, and it should help fulfil the ppm scale factor accuracy requirement of an inertial navigation grade FOG.

Another important topic is the improvement of the rejection of the integrated optic polarizer. The usual technique is Ti-indiffused waveguides on a lithium niobate substrate and polarization extinction is obtained with a metallic overlay. It is known however that proton exchange on a lithium niobate circuit yields single polarization waveguidance that gives a very high polarization extinction ratio (more than 60 dB instead of 30-40 dB with the overlay technique). Great progress in the annealing technique of proton exchange has yielded gyro circuits with very attractive performances [24] that should further improve the bias stability and get into the 0.01 °/h range.

4 -CONCLUSION

The fiber optic gyroscope is now widely accepted as the new gyro technology of the 90's and it will get a significant share of the market. Many companies, universities and governmental agencies are working on the subject : Litton, United Technologies/Hamilton Standard, Honeywell, Bendix, McDonell-Douglas, MIT, Stanford University, JPL and NRL, in the USA ; Mitsubishi, JAE, Hitachi and Tokyo University in Japan [22] ; SEL, Litef, Teldix, AEG-Telefunken, British Aerospace, Ferranti, Thomson-CSF, Sagem, Matra, Photonetics, University College and Strathclyde University in Europe. There is also an important effort on components.

Various demonstration prototypes have shown very good performances and several companies are close to production. The main advantage of the FOG compared to "classical" technologies is its solid-state configuration. This yields high dynamic range, high bandwidth, rapid start-up, good resistance to shocks and vibrations, which will extend the field of applications of inertial guidance techniques even further.

ADDITIONAL COMMENT

This paper is dedicated to the evolution of the interferometric FOG, sometimes called IFOG. There is however another possible approach to making an optical fiber gyroscope by using a resonant fiber ring cavity instead of a two wave interferometer. This approach, known as RFOG (R standing for resonant), has not reached the performances of IFOG so far but it also has interesting prospects, even if they seem technologically more difficult to achieve. A detailed description of this resonant approach is given in these OFS-6'89 proceedings by Dr. Kyuma in "Fiber optic resonant ring sensors - System and device technologies".

REFERENCES

[1] Litton in the USA and its German subsidiary Litef have both flight tested IMU with fiber gyros [Military Fiber Optic News, June 1989]

[2] Honeywell has been selected to make the avionics system for the new Dornier 328 program in Germany. The AHRS will use fiber gyros.

[3] For a review paper on interferometric FOG, one can see for example :
H.C. Lefèvre, ***"Fiber Optic Gyroscope"***
Fiber Optic Sensors, J. Dakin and B. Culshaw Editors, Volume 2, Chapter 11, Artech House, (1989)

[4] R. Ulrich, ***"Fiber optic rotation sensing with low drift"***
Optics Letters, 5, 173-175, (1980)

[5] S. Ezekiel and H.J Arditty, ***"Fiber-Optic Rotation Sensors : Tutorial Review"***
Springer Verlag Series in Optical Sciences, Vol. 32, 2-26, (1982)

[6] K. Böhm, P. Russer, E. Weidel and R. Ulbrich,
"Low drift fibre gyro using a superluminescent diode"
Electronics Letters, 17, 352-353, (1981)

[7] W.K. Burns, R.P. Moeller, C.A. Villaruel and M. Abebe,
"Fiber-optic gyroscope with polarization holding fiber"
Optics Letters, 8, 540-542, (1983)

[8] R.J. Fredericks and R. Ulrich,
"Phase error bounds of fibre gyro with imperfect polariser-depolariser"
Electronics Letters, 20, 330-332, (1984)

[9] H.C Lefèvre, J.P. Bettini, S. Vatoux and M. Papuchon,
"Progress in optical fiber gyroscopes using integrated optics"
NATO/AGARD Conference Proceedings, Vol. 383, 9A-1, 9A-13, (1985)

[10] H.C. Lefèvre, ***"Comments about fiber optic gyroscopes"***
SPIE Proceedings, Vol. 838, 86-97, (1987)

[11] R.A. Bergh, H.C. Lefèvre and H.J. Shaw,
"All single mode fiber gyroscope with long term stability"
Optics Letters, 6, 502-504, (1981)

[12] J.L. Davis and S. Ezekiel,
"Techniques for shot-noise-limited inertial rotation measurement using a multiturn fiber Sagnac interferometer"
SPIE Proceedings, Vol. 157, 131-136, (1978)

[13] R.F. Cahill and E. Udd, ***"Phase-nulling fiber-optic gyro"***
Optics Letters, 4, 93-95, (1979)

[14] C.J. Kay, ***"Serrodyne modulator in a fibre-optic gyroscope"***
IEE Proceedings, Vol. 132, 259-264, (1985)

[15] H.C. Lefèvre, Ph. Graindorge, H.J. Arditty, S. Vatoux and M. Papuchon,
"Double closed-loop hybrid fiber gyroscope using digital phase ramp"
Proceedings of OFS-3 ' 85, Post deadline paper PSP 7, (1985)

[16] H.C. Lefèvre, S. Vatoux, M. Papuchon and C. Puech,
"Integrated optics : a practical solution for the fiber-optic gyroscope"
SPIE Proceedings, Vol. 719, 101-112, (1986)

[17] S. Ezekiel, J.L. Davis and R.W. Hellwarth,
"Intensity dependent nonreciprocal phase shift in a fiberop gyroscope"
Springer-Verlag Series in Optical Sciences, Vol. 32, 332-336, (1981)

[18] K. Hotate and K. Tabe,
"Drift of an optical fiber gyroscope caused by the Faraday effect influence of the earth's magnetic field"
Applied Optics, 25, 1086-1092, (1986)

[19] R.A. Bergh, B. Culshaw, C.C. Cutler, H.C. Lefèvre and H.J. Shaw,
"Source statistics and the Kerr effect in fiber-optic gyroscope"
Optics Letters, 7, 563-565, (1982)

[20] D.M. Schupe,
"Thermally induced non-reciprocity in the fiber-optic interferometer"
Applied Optics, 19, 654-655, (1980)

[21] N.J. Frigo,
"Compensation of linear sources of non-reciprocity in Sagn interferometers"
SPIE Proceedings, Vol. 412, 268-271, (1983)

[22] Information about FOG R & D in Japan has been provided by Dr. Y. Hayakawa of Mitsubishi

[23] K. Liu, M. Digonnet, K. Fesler, B.Y. Kim and M.J. Shaw,
"Superfluorescent single mode Nd fiber source at 1060 nm"
Proceedings of OFS 5 ' 88, 462-465, New-Orleans, (1988)

[24] P.G. Suchoski, T.K. Findakly and F.L. Leonberger,
"$LiNbO_3$ integrated optical components for fiber optic gyroscopes"
SPIE Proceedings, Vol. 993, 240-244, (1988)

Fiber Optic Gyroscope with All-Digital Closed-Loop Processing

H.J. Arditty, P. Graindorge, H.C. Léfèvre, P. Martin, J. Morisse, and P. Simonpiétri

Photonetics, 52, Avenue de l'Europe, F-78160 Marly-le-Roi, France

1 - INTRODUCTION

After fifteen years, the fiber optic gyroscope, commonly abbreviated FOG, is now considered as a very advantageous technology for many civilian and military applications of the next decade. Several operating conditions are now well-admitted [1] as in particular the biasing modulation-demodulation. However, this method can suffer from the zero drift of the output DC amplifier of the analog demodulator that is commonly used in such a scheme. Temperature control and modeling of the electronics remain possible but digital processing brings a very good solution to that problem, as we are going to describe it.

2 - PRINCIPLE OF THE ALL-DIGITAL PROCESSING

The method of biasing modulation-demodulation uses a phase demodulator placed asymmetrically at one end of the fiber coil [Figure 1]. Then the FOG behaves like a delay line filter with respect to phase modulation. The phase difference is modulated about zero : at rest the interference signal contains only even harmonics of the modulation frequency, whilst in rotation odd harmonics are generated. By demodulation with a lock-in amplifier, one gets a biased sine signal of the rotation rate. In laboratory tests, good bias performance can be obtained, because one uses a high AC gain of the interference signal and RC filtering on the DC output. This allows one to test the intrinsic stability of the ring interferometer, but for real applications the system must

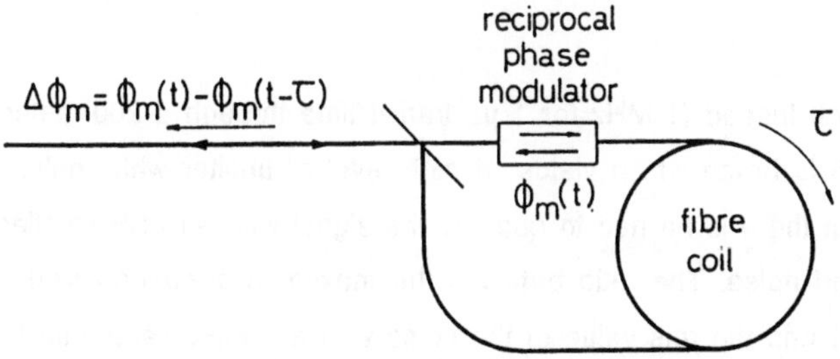

FIGURE 1 Principle of the phase modulation using the fiber coil as a delay line

Editors: H.J. Arditty · J.P. Dakin · R.Th. Kersten

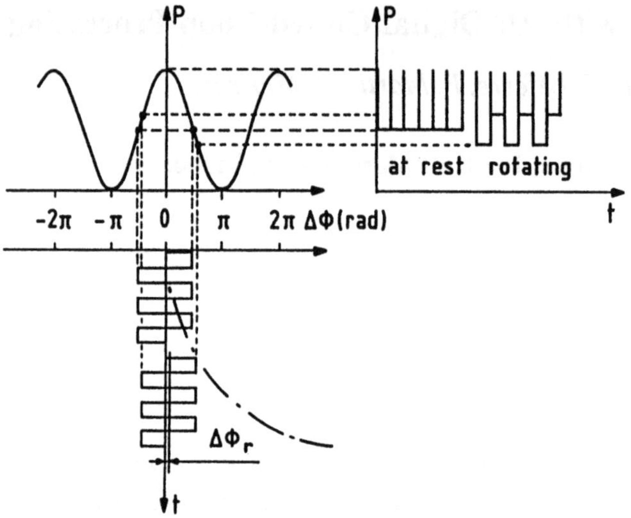

FIGURE 2
Square wave biasing modulation

withstand high rate, high angular acceleration with a short response time. This forbids the use of high AC gain and then bias is affected by the zero drift of the DC output amplifier of the demodulator. A temperature control and a rigorous design of this analog demodulator may be realized but this is a difficult engineering goal, especially when you must choose schemes that could be implemented in specific integrated circuits.

However, analog demodulation is not the only solution. One can directly digitalize the analog modulated signal and processes the converted signal with digital logic circuitry that can be easily integrated. As a matter of fact, it is known that one can use square wave biasing modulation instead of sine wave [Figure 2]. After having gated out the transient spikes, it is possible to digitalize the signal with a clock time equal to half the modulation period, i.e. the transit time through the fiber coil when the "proper" frequency is used [2]. At first thought, it seems that this approach is difficult because it would require an A/D converter with a very large number of bits to get the expected dynamic range, but when one looks more carefully at the problem, one realizes actually that this requires only a reasonable number of bits which is obtained with present technology.

The sampling frequency is quite high indeed (1 MHz for 1 μs transit time through a 200 meter fiber coil) and the signal bandwidth is broad which yields a high level of unfilter white noise. Signal processing theory shows that there is no ned to quantify the signal with an LSB smaller than the rms value of this unfiltered noise. The ratio between the maximum open-loop signal (for $\Pi/2$ radian induced phase shift) and the rms value of the noise with a 1 MHz bandwidth is only 10^{-3} for a typical noise level of 10^{-6} radian/$\sqrt{\text{Hz}}$. Now, once the high bandwidth signal is

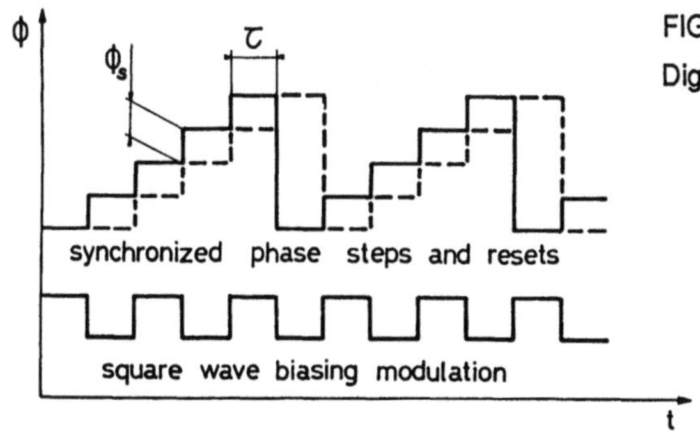

FIGURE 3

Digital phase ramp modulation

digitalized, digital filtering yields the same noise reduction as analog filtering, and one can get the digital value of the rotation rate with good resolution over a large number of significant bits. This all-digital processing yields the fundamental advantage not to suffer from any source of electronic drift of the bias, even while working over a large dynamic range, because it does not use any DC amplifier.

To take full advantage of this all-digital approach, one has to use a closed-loop processing, as in particular the so-called "digital phase ramp" that can be easily generated with digital logics and a D/A converter [3] (Figure 3). We realized such a processing electronics with three 7 cm x 9 cm cards. Their functions are respectively :

- optical signal detection, preamplification, transient spike gating and A/D conversion,
- digital logic with two specific integrated circuits,
- D/A conversion, and power amplification to drive the integrated optic modulator.

3 - FOG PROTOTYPES

This all-digital processing electronics was designed for FOG prototypes that we have fabricated. They have a diameter of 80 mm and a height of 28 mm. They use the well-admitted reciprocal "Y-tap" configuration [Figure 4].

They are composed of [Figure 5] :

- a 830 nm superluminescent diode (SLD) as the broadband source. It is pigtailed with a polarization preserving fiber and it is mounted on a Peltier element for temperature control,
- a side-polished 3 dB coupler fabricated directly on the source pigtail. This allows us to couple simply the output optical signal in a multimode fiber cable connected to the detection card of the processing electronics. This avoids the problem of ground loops between modulation and detection,

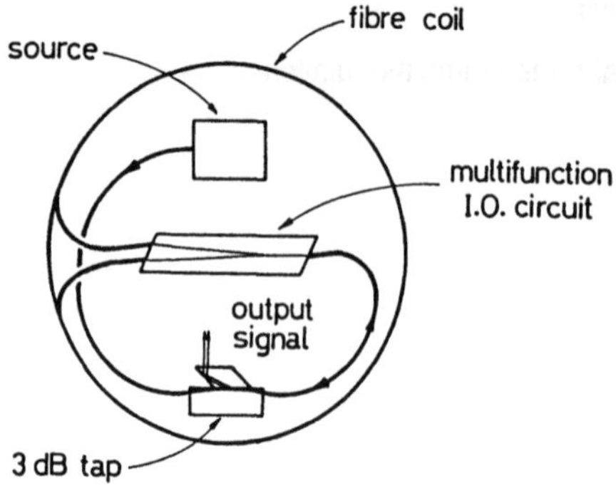

FIGURE 4

Reciprocal "Y-tap" configuration

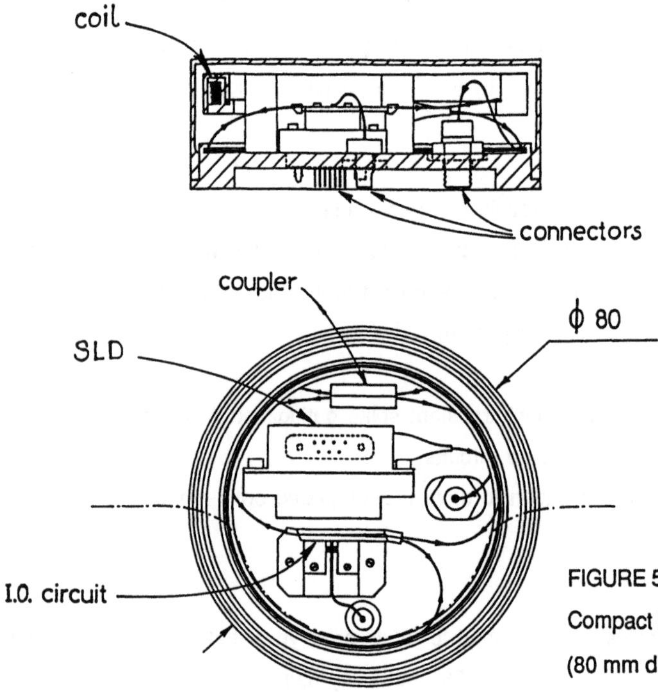

FIGURE 5

Compact FOG prototypes

(80 mm diameter)

- a multifunction integrated optic circuit (20 mm long and 2.3 mm wide) with fibers connected directly with small ferrules on the circuit egdes.

This is an X-cut Ti-indiffused $LiNbO_3$ circuit with a Y junction, a metallic overlay polarizer (35 dB rejection) and two push-pull phase modulators. The total V_Π efficiency is 1.8 V. The fiber to fiber additional loss is 8 dB.

- a polarization preserving fiber coil of 200 meters (i.e. a transit time of 1 µs) wound about a mean diameter of 69 mm. Notice that the polyimide coated fiber has a very thin coating

(105 μm diameter) that makes the coil very compact. The loss is 1 dB and the polarization conservation is 17 dB. For such a coil and a wavelength, there is a phase shift of Π radian for a rotation rate of 515 deg/s and a phase shift of 10^{-6} radian for 0.6 deg/h.

The three key components (SLD, I.O. circuit, and polarization preserving fiber) were purchased as developmental products, but in each case, we were able to select among several potential suppliers.

The energy budget is 26 dB from the source pigtail to the detector :

- 6 dB of intrinsic loss of the reciprocal configuration (the coupler loss is neglectable, < 0.1 dB),
- 9.5 dB of insertion loss of the pigtailed IO circuit after bonding of the fiber ferrules (instead of the original 8 dB),
- 1 dB for the fiber coil.

With 1 mW in the input pigtail, there is 2.5 μW back to the PIN detector. Figure 6 shows an example of the short-term bias when tested with an analog lock-in amplifier. The noise is evaluated at 0.5 (deg/h)/√Hz (8 x 10^{-7} radian/√Hz in terms of phase shift), i.e. a random walk of 8 x 10^{-3} deg/√h. Figure 7 shows a long term recording of this bias with an RC constant

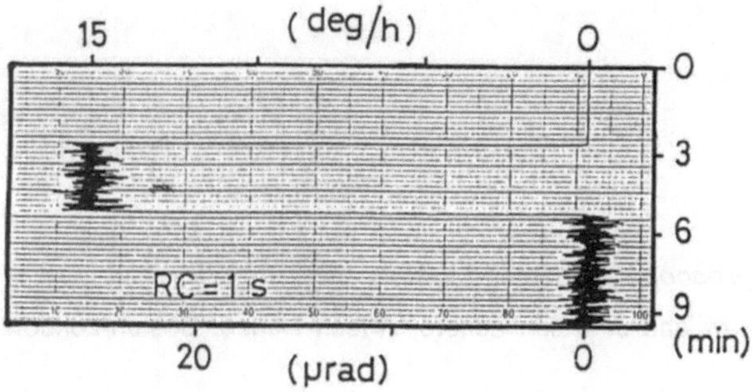

FIGURE 6
Short term bias

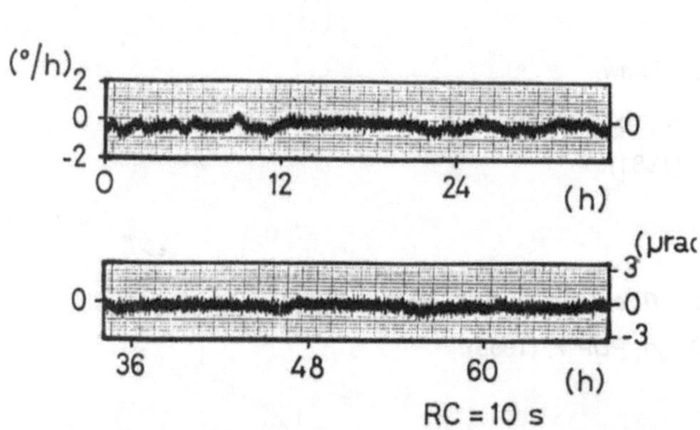

FIGURE 7
Long term drift

of 10 seconds. The peak to peak bias uncertainty is ± 0.3 deg/h which corresponds to a phase shift of $\pm 5 \times 10^{-7}$ radian.

Preliminary tests of our all-digital closed-loop electronics already show a linearity of 200 ppm over a dynamic range of 500 °/s, and more complete experimental results will be presented at the conference.

CONCLUSION

We have presented FOG prototypes with state-of-the-art performances and we have also described a very powerful method of all-digital processing that solves the difficult problem of the zero drift of an analog demodulator.

R and D efforts are now directed towards more compact FOG prototypes (50 mm diameter) with extanded dynamic range (2000 °/s) and high bandwidth (3 kHz).

ACKNOWLEDGEMENTS

Our present R and D program is sponsored by the DRET of the French Ministry of Defence.

REFERENCES

[1] See for example :
H.C. Lefèvre, ***"Fiber Optic Gyroscope"***, Chapter 11,
Optical Fiber Sensors, B. Culshaw and J. Dakin editors, Artech House, Boston-London, (1989).

[2] R.A. Bergh, H.C. Lefèvre and H.J. Shaw,
"All single-mode fiber optic gyroscope with long term stability",
Optics Letters, 6, 502-504, (1981).

[3] H.C. Lefèvre et al.,
"Double closed-loop hybrid fiber gyroscope using digital phase ramp",
Proceedings of OFS 85, San Diego, PDP 7, (1985).

Fiber Superfluorescent Sources for Fiber Gyro Applications

W.K. Burns[1], *I.N. Duling III*[1], *L. Goldberg*[1], *R.P. Moeller*[1], *C.A. Villarruel*[1], *E. Snitzer*[2], *and H. Po*[3]

[1]Naval Research Laboratory, Washington, DC20375, USA
[2]Rutgers University, Piscataway, NJ08855, USA
[3]Polaroid Corp., Cambridge, MA02139, USA

Double-clad fiber superfluorescent sources pumped by high power laser diode arrays are investigated for fiber gyro applications. Output powers of 40-80mW are demonstrated and fit parameters to a theoretical model are obtained. The sensitivity of the sources to feedback is investigated experimentally and compared to the model.

It is well known that broadband sources are beneficial for interferometric fiber optic gyroscopes due to a reduction in noise due to Rayleigh backscattering and polarization effects and in bias drift due to the Kerr effect. This need has primarily been filled by semiconductor superluminescent sources (SLD's) in the past, but the development of rare-earth doped fibers and Nd doped fiber lasers[1-3] and superfluorescent sources[4] has led to considerable interest in using these sources in interferometric fiber-optic gyroscopes. Advantages of the fiber sources include higher output power, longer life, and an improved wavelength stability[4]. The latter is critically important in considerations of scale factor stability. Prospects for these sources have improved even further with the development by workers at Polaroid Corp., of fiber configurations which contain the active single-mode core within a large multi-mode core [5,6] which confines the pump light, thus making practical end-fire pumping by high-power laser diode arrays or broad-stripe lasers[7,8]. Recently the use of a broadband fiber laser in a fiber gyro was reported[9]. This presentation will explore the potential of the Polaroid fiber as a superfluorescent source for fiber gyro applications.

The fiber cross sections are shown in Fig. 1. Fiber 1 (fig. 1a) is an all glass composition[5] and fiber 2 (fig. 1b) is a plastic clad composition[6] which achieved a higher multimode core-cladding

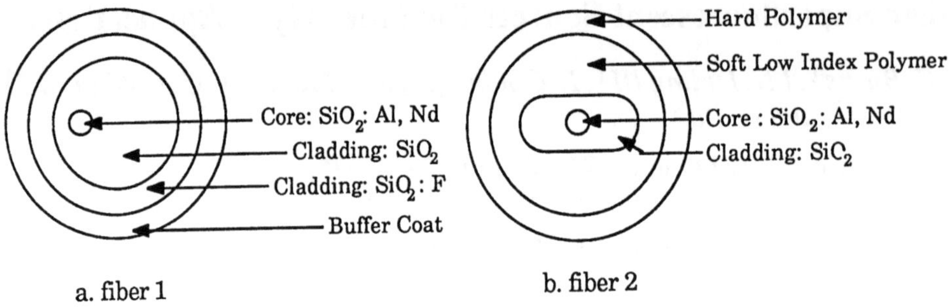

Figure 1.

numerical aperature (NA) than fiber 1 (fiber 2: NA = .4, fiber 1: NA = .1). The fibers were explored in somewhat different experimental conditions, fiber 1 being pumped with an injection locked 1 W broad stripe GaAlAs diode laser and fiber 2 by a 1 W GaAlAs laser diode array.

Superfluorescent output can be modeled using the theory developed by Digonnet[10]. If we allow for the possibility of reflection at the fiber ends, all cases (neglecting loss) can be treated by the solution of the equations

$$P^+(l)(1-R_l) + P^-(0)(1-R_o)+(P_s-2P_o)\ln\left(\frac{P^+(l)+P_o}{R_oP^-(0)+P_o}\right) = P_s g_o \qquad (1)$$

$$(R_o P^-(0) + P_o)(P^-(0)+P_o) = (P^+(l)+P_o)(R_lP^+(l)+P_o) \qquad (2)$$

where the powers $P^-(0)$ and $P^+(l)$ are defined in Fig. 2. A fiber length l and reflectivities R_o and R_l at z = o and l, respectively, are assumed. The output powers are given by $P^+_{out} = (1-R_l)\,P^+(l)$, copropagating with the pump and $P^-_{out} = (1-R_o)\,P^-(0)$, counterpropagating with the pump. In eq. 1 P_s is the saturation power, P_o is a constant accounting for

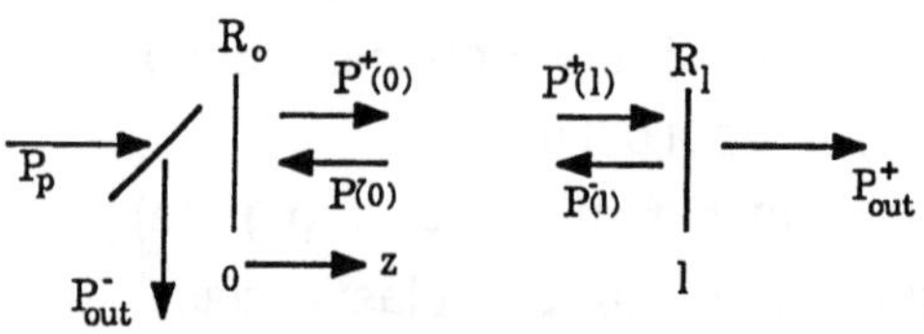

Figure 2.

spontaneous emission, and g_o represents the unsaturated gain. We take $g_o = k(l)P_p$ where P_p is the input pump power and k(l) is a length dependent constant. The important feature of eq. 1 is that gain saturation is explicitly taken into account. Equations 1 and 2 have analytic solutions only for special cases of reflectivity equal to zero or unity.

Using this model we can completely characterize each fiber by fitting the power in - power out characteristics, as is done in Figs. 3 and 4. In Fig. 3 for fiber 1 (l = 10m) the superfluorescent emission was taken out counterpropagating to the pump with reflectivities $R_o = 0$, $R_l = 1$. In Fig.4 for fiber 2 (l = 7.5m) emission was extracted copropagating with the pump with reflectivities $R_o = .02$, $R_l = 0$. The fit constants for these two cases are

	k(l) (mW^{-1})	P_s(mW)	P_o(mW)
Fiber 1	.051	8.48	.0165
Fiber 2	.051	6.23	.0051

which are quite close (although k(l) is length dependent and the pump wavelengths were slightly different), indicating the similarity of the

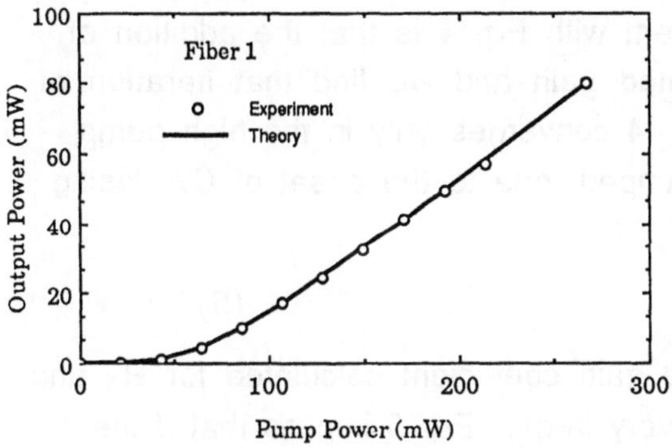

Fig. 3

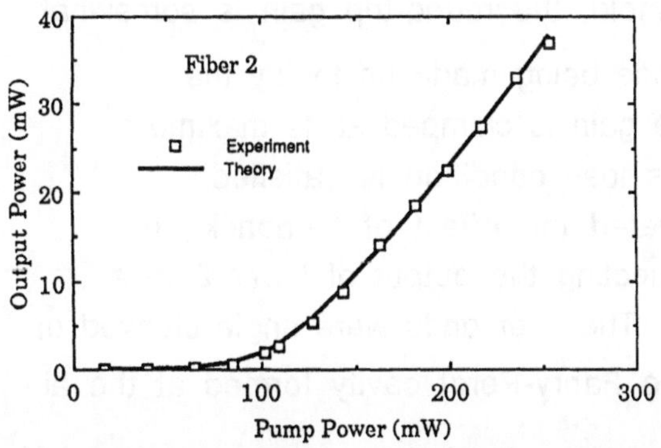

Fig.4

active singlemode cores. Note the large output powers achieveable (fiber 1: 80mW out with 258mW absorbed pump; fiber 2: 37mW out with 208mW absorbed pump) and the increase in power obtained using a 100% reflector.

It becomes quickly apparent experimentally that an important consideration in using these high gain sources is avoiding reflections that make them lase. We determine the onset of lasing by the observation of power spikes on a spectrum that was previously smooth. The first order approximation to calculating laser threshold is to set $P_o = 0$ in equations 1 and 2 and obtain the laser output equation which has its threshold at

$$R_o R_l = e^{-2g_o} \qquad (3)$$

which is the usual round trip gain equals loss condition. However, this result assumes zero optical output at theshold and so uses the unsaturated gain coefficient, which is too large for cases of large superfluorescent output. A better approximation might be to calculate the saturated gain for a superfluorescent source with no output reflector, and use that gain to estimate the output reflector that would satisfy the round trip gain equal loss condition.

$$R_o R_l\ (th) \simeq e^{-2g(l)\ (R_l = 0)} \qquad (4)$$

where g(l) is the saturated single-pass gain coefficient integrated over the fiber length. The problem with Eq. 4 is that the addition of the reflector changes the saturated gain and we find that iteration of the reflectivity R_l and gain in Eq. 4 converges only in the high pump power limit where the gain is clamped, due to the onset of CW lasing. We find that the equation

$$R_o R_l = e^{-2g_c(l)} \qquad (5)$$

holds where $g_c(l)$ is the clamped gain coefficient calculated for R_o and R_l as the pump power becomes very large. Eq. 5 implies that if the gain is not yet clamped at threshold, the round-trip gain is somewhat less than the loss, with the balance being made up for by the spontaneous emission. Once the gain is clamped at its maximum value, the round trip gain equals loss condition is satisfied.

We experimentally investigated the effect of feedback on superfluorescent sources by connecting the output of Fiber 2 to a fiber coupler as shown in Fig. 5. The fiber ends were angle cleaved or polished and arranged so that no Fabry-Perot cavity formed at the air-

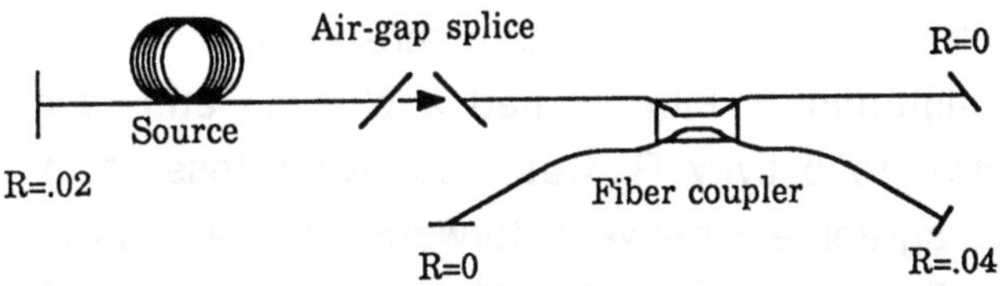

Figure 5.

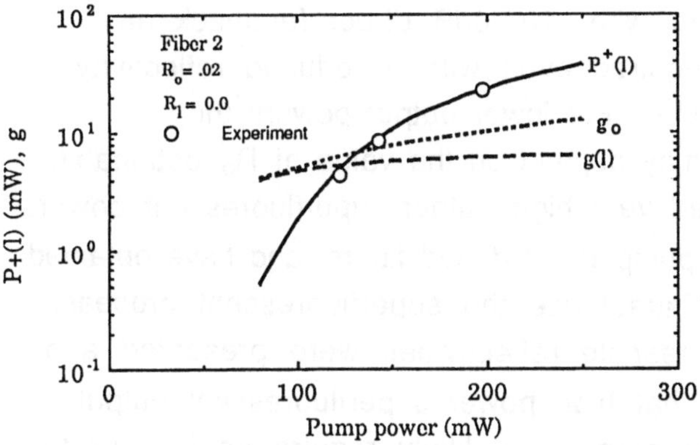

Figure 6.

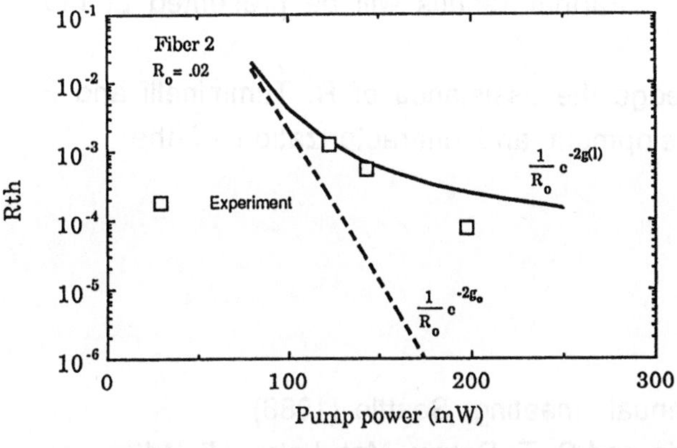

Figure 7.

gap connection between the source and the coupler. One coupler output fiber was cleaved to form a 4% reflector and the separation of the air-gap connection was varied to control the amount of feedback to the source. The results are shown in Fig. 6 and 7. Fig. 6 shows measured and modeled output power without feedback, and the saturated g(l) and unsaturated g_o single-pass gain coefficients. Fig. 7 shows the experimental feedback reflectivities that resulted in onset of lasing and the calculated models of Eq. 3 and 4. We see that the model using the saturated gain coefficient (Eq. 4) gives a good fit to experiment for output powers < 10mW. An improved fit over the entire range would require a more sophisticated model.

We can use the results of Fig. 7 to estimate threshold reflectivities for a gyro configuration. Fiber 2 had a cleaved end face at z = 0 so that the effective reflectivity R_o was somewhat less than .04. We achieved the best agreement between forward and backward output powers by choosing $R_o = .02$. For >20mW output the reflectivity product R_oR_l should then be less than ~1.6×10^{-6} to avoid lasing. For an interferometric fiber gyro with -10-20dB direct feedback an isolator probably will be required even with a reduced reflectivity angled face at the pump input. For lower output powers the requirement for an isolator may depend on the value of R_o obtainable.

We have demonstrated very high output superfluorescent powers with array and broad stripe pumped Nd doped fibers, and have obtained the fit parameters which characterize the superfluorescent process. Experimental results for threshold reflectivities were presented and the conclusion was drawn that high power superfluorescent output into an interferometric fiber gyro would likely require an isolator to avoid lasing from feedback. Further results will be presented at the conference.

The authors acknowledge the assistance of R. Tumminelli and F. Hakimi in the original development and characterization of the double-clad fiber.

References

1. H. Po, et. al., OSA Annual meeting, Seattle (1986).
2. B. J. Ainslie, S. P. Craig and S. T. Davey, Mat. Lett. 5, 143 (1987).
3. K. Liu, et. al., Electron Lett. 24, 838, (1988).
4. K. Liu, et. al., OFS '88, New Orleans (1988).
5. E. Snitzer, et. al.,OFS '88, PD5, New Orleans (1988).
6. H. Po, et. al.,OFC '89, PD07, Houston (1989).
7. I. N. Duling III, L. Goldberg, and W. K. Burns, CLEO '89, TUP4, Baltimore (1989).
8. I. N. Duling III, W. K. Burns and L. Goldberg, to be published.
9. K. A. Fesler, B. Y. Kim and H. J. Shaw, Electron Lett. 25, 534 (1989).
10. M. J. F. Digonnet, J. Lightwave Tech. LT-4, 1631 (1986).

Erbium-Doped Fibre Superfluorescent Source for the Fibre Gyroscope

P.R. Morkel

Optical Fibre Group, Dept. of Electronics and Computer Science, University of Southampton, Highfield, Southampton SO9 5NH, UK

The operation of an erbium-doped silica fibre as a superfluorescent source at 1.535μm pumped at 980nm is reported. The variation of superfluorescent output power with pump power and fibre length is characterised. The spectrum of the superfluorescent emission is seen to be dependent on fibre length , pump power and fibre temperature. However, under certain operating conditions, good spectral stability with pump power and fibre temperature can be obtained.

INTRODUCTION

A broadband optical source capable of providing several milliwatts of power in a single transverse mode is of interest for a number of optical sensor applications, in particular for the fibre-optic gyroscope (FOG). In the gyro application, use of a broadband source is a well-known means of reducing gyro bias introduced by coherent backscatter and the optical Kerr effect[1]. Superfluorescent emission from neodymium-doped silica fibres and its potential for a gyro source has been previously investigated[2,3]. As well as increased power, the advantage of using rare-earth doped fibres over semiconductor superluminescent diodes rests primarily in greater spectral stability with respect to temperature[3]. This factor directly determines gyro scale-factor stability (defined as output over rotation rate) and is a critical performance criterion for the FOG.

Potential advantages of the erbium system over neodymium include increased gain and hence higher output powers for a given pump power, and operation at a wavelength more compatible with telecommunications technology. In addition, the 1.54μm wavelength has advantages of decreased sensitivity to irradiation of the fibre, lower fibre/integrated optics loss and a higher integrated optics damage threshold. Coherent Rayleigh backscatter is also considerably reduced at the 1.54μm wavelength compared to 1.06μm, and the photon energy is lower so improving the shot noise limit.

A qualitative investigation of the amplified spontaneous emission (ASE) spectrum as a function of pump power and fibre length has previously been reported by Desuvire et al[4]. This work showed that the difference in absorption and emission spectra inherent in the 3-level erbium system gives rise to variations in emission spectra with fibre length and pump power. The latter effect suggests poor spectral stability and that the erbium-doped fibre superfluorescent source is poorly suited to the FOG.

In this paper we investigate the practicality of erbium doped fibres as a superfluorescent sources when pumped at 980nm. Particular attention is applied to spectral stability with pump power and temperature in order to determine whether a spectrally-stable device can be fabricated. Surprisingly, it is found that under certain operating conditions the source has excellent wavelength stability.

EXPERIMENTAL

Fig. 1 shows the experimental set-up used. A 10m length of Er^{3+} doped germano silicate fibre (NA 0.17, cutoff 925nm, Er^{3+} dopant concentration 60 ppm) was butted up to a

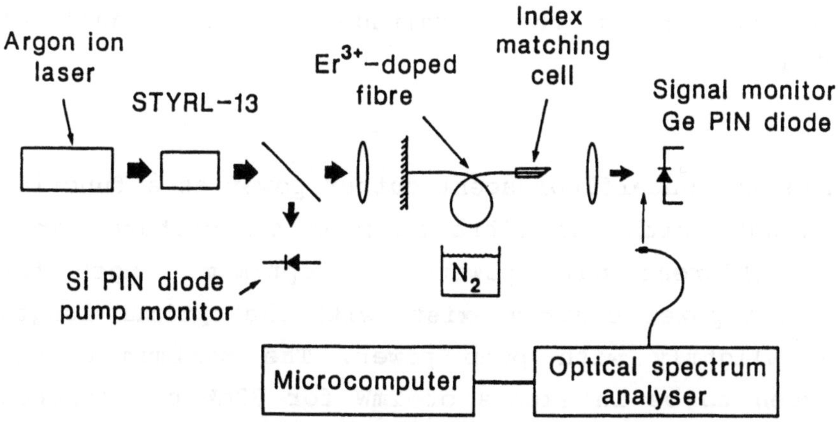

Fig. 1. Experimental set-up.

dielectric mirror which had greater than 99% reflectivity at 1.55μm, but transmitted 85% at 980nm. The other end of the fibre was carefully terminated in an index-matching cell in order to prevent laser oscillation in the fibre. The fibre was successively cut back and at each length the variation of output power with pump power at both room and liquid nitrogen fibre temperatures was recorded using a germanium PIN detector. It was found necessary to slightly angle the detector in order to prevent feedback and laser oscillation from the detector surface under conditions of heavy pumping. In addition, emission spectra were recorded using an Anritsu optical spectrum analyser at resolutions down to 0.1nm and the data stored on computer for processing. The pump source used in the experiments was a Styrl-13 dye laser tuned to 980nm pumped with a 6W argon-ion laser. Using this set up a maximum of approximately 35mW could be coupled into the fibre. The emission spectra were numerically integrated to give a value for the weighted mean wavelength according to the relation:

$$\overline{\lambda} = \frac{\int a(\lambda).\lambda \, d\lambda}{\int a(\lambda) \, d\lambda}$$

where $a(\lambda)$ represents the magnitude of the signal at wavelength λ.

RESULTS

Fig. 2 shows the superfluorescent output power as a function of fibre length with the fibre at room temperature for a number of different pump powers. An optimum length for maximum output power clearly exists with the optimum length increasing slightly with pump power. The maximum output power is seen to be in excess of 2mW for 33mW of launched pump power.

For the 6m fibre length, fig. 3 shows the variation of signal output power and optical bandwidth as a function of

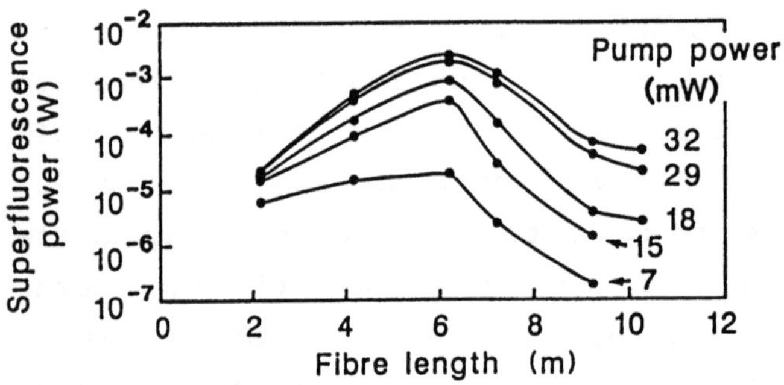

Fig. 2. Superfluorescent output power vs fibre length for a number of pump powers.

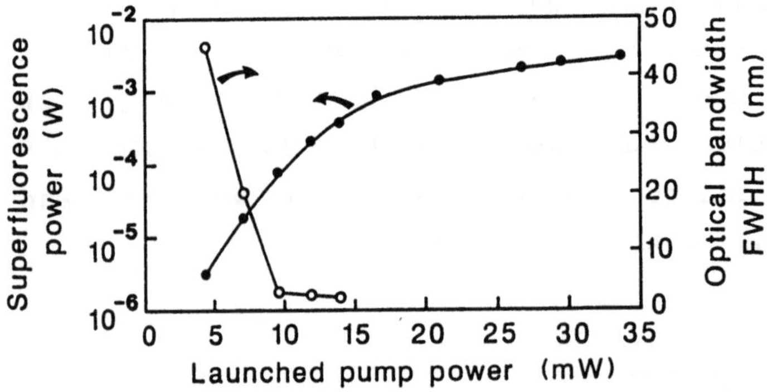

Fig. 3. Superfluorescent output power @ $1.54\mu m$ and optical bandwidth vs pump power. 6.2m fibre length.

launched pump power at room temperature. Above 15mW launched pump power the characteristic is seen to saturate from the expected exponential input/output relation and the optical spectrum tends asymptotically to around 2nm. The superfluorescent output of the fibre was observed to be randomly polarised within the limits of experimental error.

Fig. 4 indicates the variation of $\bar{\lambda}$ with pump power for the 6m fibre length. At low levels of pump power, the long wavelength edge of the gain spectrum dominates the ASE spectrum, consistent with quasi 4-level like behaviour[4]. However as the pump power is increased, $\bar{\lambda}$ is seen to tend towards 1.535μm, which corresponds with the peak of the fluorescence emission in this fibre type. Note also that at the lower pump-power levels the spectrum is seen to be significantly different for 77K and room temp. operation, consistent with redistribution of the populations of the Stark components of the energy levels. However, at the higher pump levels, the peak of the gain curve dominates and we may assume that this peak corresponds to only 1 Stark component of the transition. This effect clamps the spectrum at a particular wavelength relatively independent of temperature and pump power.

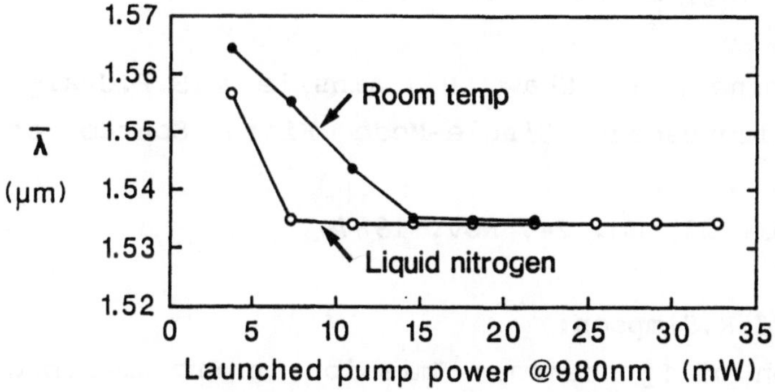

Fig. 4. Weighted mean wavelength vs pump power 6.2m fibre length. Fibre at room temp. & 77K.

CONCLUSIONS

The implementation of an erbium doped silica fibre superfluorescent source emitting >2mW @ 1.535μm when pumped at 980 nm has been shown. Examination of the emission spectrum shows a spectral width around 2nm for output powers in excess of approx 100μW, corresponding to one Stark transition. Cooling the fibre to 77K indicates that under high gain operating conditions temperature may only affect the output spectrum by ~10ppm/°C, a figure far superior to that of an ELED (~300 ppm/°C)..

REFERENCES

1) R.A.Bergh, B.Culshaw. C.C.Cutler, H.C.Lefevre & H.J.Shaw:
"Source statistics and the Kerr effect in fibre-optic gyroscopes",
Opt. Lett., Vol. 7, 11, Nov. 1982.

2) H.Po, F.Hakumi, R.J.Mansfield, B.C.McCollum, R.P.Tuminelli & E.Snitzer:
"Neodymium fiber laser at 0.905, 1.06 and 1.4μm",
Proc. OSA meeting, Seattle, Oct. 1986.

3) K.Liu, M.Digonnet, H.J.Shaw, B.J.Ainslie & S.P.Craig:
"10mW Superfluorescent Single-Mode Fibre Source at 1060nm",
El. Lett., Vol. 23, No. 24, Nov. 1987.

4) E.Desuvire & J.R.Simpson:
"Amplification of Spontaneous Emission in Erbium-Doped Single-Mode Fibers",
Journal of Lightwave Technology, Vol. LT-7, No. 5, May 1989.

Part VI

Noise in Interferometers

Limitations on the Dynamic Range and Sensitivity of Optical Interferometric Sensors Excited by Semiconductor Lasers with Non-Lorentzian Lineshapes

A. Arie and M. Tur

School of Engineering, Tel Aviv University, Ramat Aviv, Tel Aviv, Israel 69978

The phase induced intensity noise at the output of optical interferometers driven by semiconductor lasers whose lineshape exhibits relaxation-oscillation peaks is shown to be different from predictions based on a Lorentzian model. The role of the damping rate of the relaxation oscillations is discussed together with ways to partially overcome this problem in fiber-optic sensors.

The phase induced intensity noise at the output of optical interferometers driven by semiconductor lasers whose lineshape exhibits relaxation-oscillations peaks, is shown to be different from predictions based on a Lorentzian model. The role of the damping rate of the relaxation oscillations is discussed together with ways to partially overcome this problem in fiber-optic sensors.

Optical systems with multiple paths in which a semiconductor injection laser source is being split and after experiencing different time delays, recombined on an amplitude basis, are widely used in fiber optic interferometric sensors [1] and fiber optic signal processing systems [2], as well as in systems that characterize lasers for coherent communication applications [3, 4].

One of the major noise sources in these systems is an excessive amount of laser phase induced intensity noise (PIIN). Let us represent the output field of an intensity-noiseless, single-mode laser, emitting at a center frequency ω_0 by

$$E_{in}(t)=\sqrt{I_0}\exp[i(\omega_0 t+\phi(t))],\ (I_0 \text{ is the laser output power}) \tag{1}$$

If the laser light is injected into a multiple path system, such as a Mach-Zehnder interferometric sensor with a differential delay of τ between its arms, the output field will be

$$E_{out}(t)=K_1\sqrt{I_0}\exp[i(\omega_0 t+\phi(t))]+K_2\sqrt{I_0}\exp[i(\omega_0(t-\tau)+\phi(t-\tau))] \tag{2}$$

Where $K_1\sqrt{I_0}$ and $K_2\sqrt{I_0}$ are the coefficients of the undelayed and delayed output fields, respectively. The output intensity becomes

Springer Proceedings in Physics, Vol. 44 **Optical Fiber Sensors**
Editors: H.J. Arditty · J.P. Dakin · R.Th. Kersten

$$I_{out}(t)=|K_1|^2+|K_2|^2+2|K_1K_2|\cos[\omega_0\tau+\phi(t)-\phi(t-\tau)] \quad (3)$$

Since $\phi(t)$ is a random process, the third expression in Eq. 3 represents noise. This phase induced intensity noise is directly proportional to the input power and therefore it sets a floor to the available signal to noise ratio in the system. It may become much stronger than the source intensity noise and tends to severely limit the dynamic range and sensitivity of fiber optic phase sensitive sensors [5]. In order to eliminate the effect of the PIIN in fiber optic interferometric sensors such as the Mach-Zehnder, the differential delay τ must be minimized. For instance, in order to achieve sensitivities below 1 microradian with single-mode semiconductor lasers, the Mach-Zehnder paths should be matched to less than 1 mm.

Laser phase noise in optical interferometers was studied by many authors, e.g., [4]-[7]. Common to all their treatments is the assumption that the deviation of the laser instantaneous optical frequency, $\omega(t)$, from its mean ω_0, can be modelled by a stationary normal (Gaussian) random process with a flat spectrum [6]. Therefore, the laser phase noise $\phi(t)$ is a Wiener-Levy random process whose structure function $D_\phi(\tau)$ is proportional to $|\tau|$,

$$D_\phi(\tau)\equiv\langle[\phi(t+\tau)-\phi(t)]^2\rangle=2|\tau|/\tau_c \quad (4)$$

The angle brackets denote ensemble average and τ_c is the coherence time. Since the laser lineshape is the Fourier transform of the laser field autocorrelation function, $\langle E(t)E^*(t+\tau)\rangle=I_0\exp(-i\omega_0\tau)\exp[-0.5D_\phi(\tau)]$, the linear dependence of $D(\tau)$ on τ in Eq. 4 corresponds to a Lorentzian lineshape [8].

However, careful measurements of the lineshape of single-mode semiconductor lasers made by Vahala et. al [9] and by Daino et. al [10], showed that the lineshape is non-Lorentzian, due to side peaks at the relaxation-oscillation frequency. As a result, the structure function, $D_\phi(\tau)$, can no longer be assumed to be proportional to $|\tau|$ and it is the purpose of this paper to study the implications of the appropriately modified structure function, with its associated non-Lorentzian lineshape, on the power spectral density of the phase induced intensity noise at the output of multiple path optical systems.

The main intrinsic source of noise in lasers is the random spontaneous emission. Every such spontaneous event disturbs the steady state populations of the laser energy levels but the system quickly recovers in a short series of relaxation oscillations of both the output intensity and the carrier concentration. But since the refractive index of the semiconductor lasing medium is a relatively strong function of the carrier concentration, the spontaneous emission event generates intensity fluctuations, as well as phase fluctuations, through the non-coherent spontaneous photon and through the induced carrier concentration fluctuations [11]. Thus, Eq. 4, which only represents the random walk of the phases of the randomly emitted spontaneous photons, has to be modified to properly describe those phase fluctuations related to the relaxation-oscillations.

While several alternative formulations are available, we follow here that of C.H. Henry [8]

$$D_{\phi}(\tau) = \left(\frac{R}{2I}\right)\left[(1+\alpha^2)|\tau| + \frac{\alpha^2(\cos(3\delta)-\exp(-\Gamma|\tau|)\cos(2\pi f_{ro}|\tau|-3\delta))}{2\Gamma\cos(\delta)}\right] \quad (5)$$

R is the spontaneous emission rate into the lasing mode, I is the number of photons per mode, and α is the linewidth enhancement factor which relates the phase fluctuations with those of the carrier concentration: $\alpha=[\partial\mathrm{Re}(n)/\partial G]/[\partial\mathrm{Im}(n)/\partial G]$ (n is the complex refractive index of the lasing medium and G, the gain, is proportional to the carrier concentration). While in most other lasers $\alpha\simeq0$, semiconductor lasers are characterized by $4<\alpha<7$ [8]. Carrier injection in semiconductor diodes gives rise to band filling, which in turn shifts the optical absorption edge towards lower wavelengths and consequently also modifies the imaginary part of the material refractive index. According to the Kramers-Kronig relations, the imaginary part of the refractive index is coupled to its real part and since the lasing frequency depends on the optical length of the cavity, fluctuations in the carrier concentration result in excessive phase noise [11]. f_{ro} is the relaxation-oscillation frequency, Γ is the damping rate of relaxation-oscillation, and δ is defined by $\cos(\delta)=2\pi f_{ro}/[(2\pi f_{ro})^2+\Gamma^2]^{0.5}$. Eq. 5 ignores the relatively small effects of spectral hole burning and the shot noise related to the generation and recombination of minority carriers [8]. In the limiting case, where Γ reaches infinity: the relaxation oscillations are completely quenched, $D_{\phi}(t)$ is again proportional to τ and the coherence time becomes

$$\tau_L=4I/(R(1+\alpha^2)) \quad (6)$$

However, when we use a finite value of the damping rate, the coherence time may deviate from this value. For low damping rates, the coherence time becomes meaningless due to the oscillations in the structure function.

When the laser field of Eq. (1) is injected into a two beam interferometer (N=2) with a delay of τ, which does not alter the input polarization, the intensity covariance function, $\mathrm{Cov}(\Delta T)=<[I(t+\Delta T)-\langle I\rangle][I(t)-\langle I\rangle]>$, becomes [12]

$$\begin{aligned}\mathrm{Cov}(\Delta T)=&2|K_1K_2|^2I_0^2[\exp[0.5(-2D(\Delta T)-2D(\tau)+D(\Delta T+\tau)+D(\Delta T-\tau))]-\exp(-D(\tau))] +\\ &[(K_1K_2^*)^2\exp(2i\omega_0\tau)+(K_1^*K_2)^2\exp(-2i\omega_0\tau)]I_0^2\\ &[\exp[0.5(2D(\Delta T)-2D(\tau)-D(\Delta T+\tau)-D(\Delta T-\tau))] - \exp(-D(\tau))]\end{aligned} \quad (7)$$

According to the Wiener-Khinchine theorem, the power spectrum of the PIIN, S(f), is the Fourier transform of the autocovariance $\mathrm{Cov}(\Delta T)$. Fig. 1 shows the power spectrum of the PIIN, in case of in-quadrature ($\omega_0\tau=\pi/2+n\pi$) and maximum output ($\omega_0\tau=2n\pi$), for both Lorentzian (Fig. 1a) and non-Lorentzian (Fig. 1b-c) lineshapes. Both arms are identically excited ($K_1=K_2=1/2$). The coherence time for the Lorentzian case was chosen to be τ_L (Eq. (6)), which corresponds to a damping rate $\Gamma=\infty$ in the non-Lorentzian lineshape. For the non-Lorentzian lineshapes two values

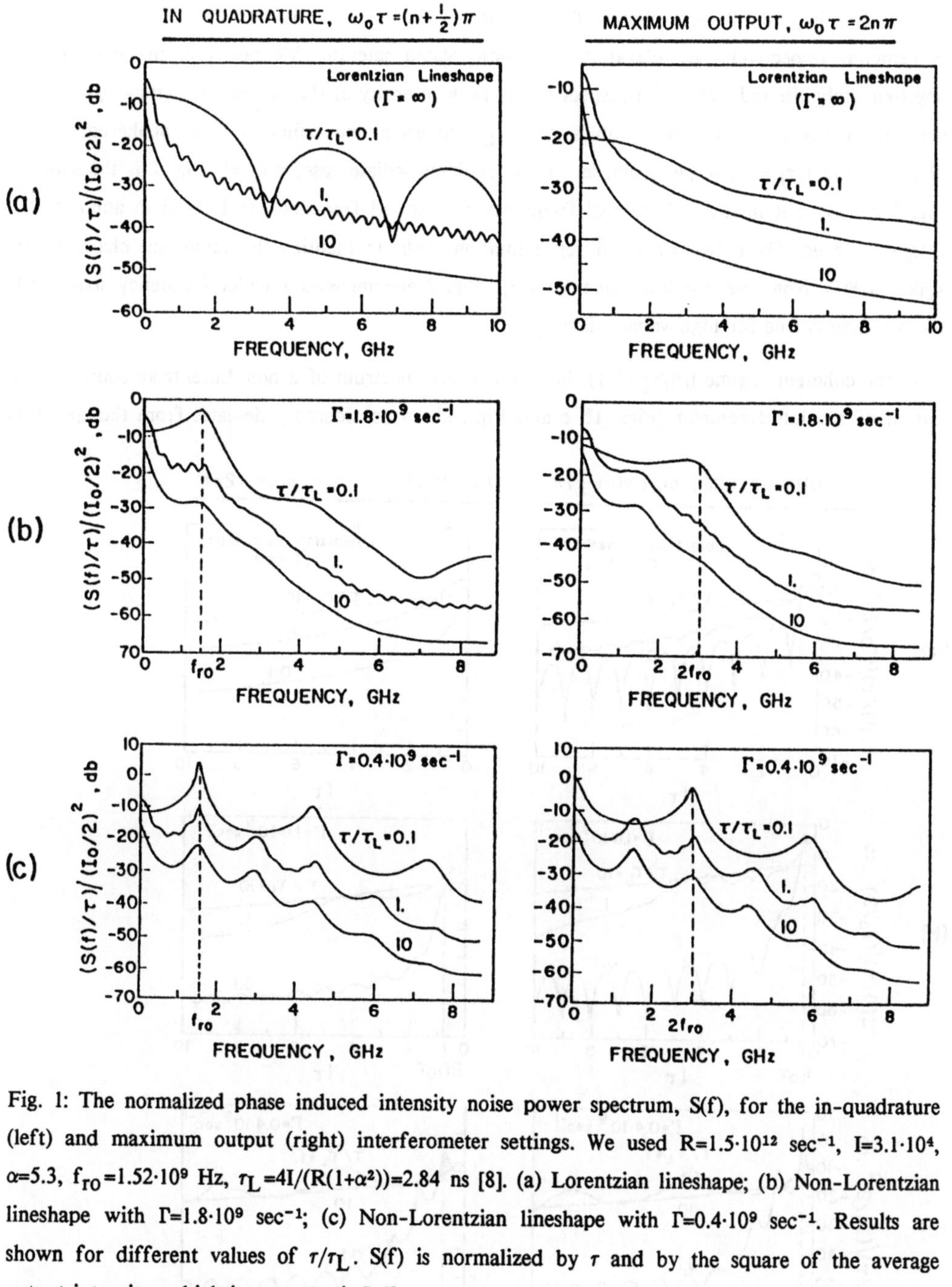

Fig. 1: The normalized phase induced intensity noise power spectrum, S(f), for the in-quadrature (left) and maximum output (right) interferometer settings. We used $R=1.5\cdot10^{12}$ sec^{-1}, $I=3.1\cdot10^4$, $\alpha=5.3$, $f_{ro}=1.52\cdot10^9$ Hz, $\tau_L=4I/(R(1+\alpha^2))=2.84$ ns [8]. (a) Lorentzian lineshape; (b) Non-Lorentzian lineshape with $\Gamma=1.8\cdot10^9$ sec^{-1}; (c) Non-Lorentzian lineshape with $\Gamma=0.4\cdot10^9$ sec^{-1}. Results are shown for different values of τ/τ_L. S(f) is normalized by τ and by the square of the average output intensity, which in our case is $I_0/2$.

of the damping rate Γ were chosen: $\Gamma=1.8\cdot10^9 sec^{-1}$, Fig. 1b, and $\Gamma=0.4\cdot10^9 sec^{-1}$, Fig. 1c, which correspond, respectively, to relaxation-oscillation-related satellite side peaks in the lineshape of approximately 1% and 10% (with respect to the peak intensity at the central frequency ω_0). Results are shown for three different values of τ/τ_L, covering the coherent and incoherent limits, $\tau/\tau_L=0.1$ and $\tau/\tau_L=10$, respectively, as well as an intermediate case, $\tau/\tau_L=1$. Fig. 2 is the same as Fig. 1 except that instead of the real frequency f, a scaled frequency $f\tau$ is used to annotate the frequency axes. These two different representations help to identify the important characteristic scales in S(f). Note that for low values of τ/τ_L Fig. 2 encompasses a wider frequency span while the converse is true for high values of τ/τ_L.

In the coherent regime ($|\tau|/\tau_L=0.1$) the noise power spectrum of a non-Lorentzian source at the output of the interferometer (Figs. 1b-c and Figs. 2b-c) considerably deviates from the spectrum

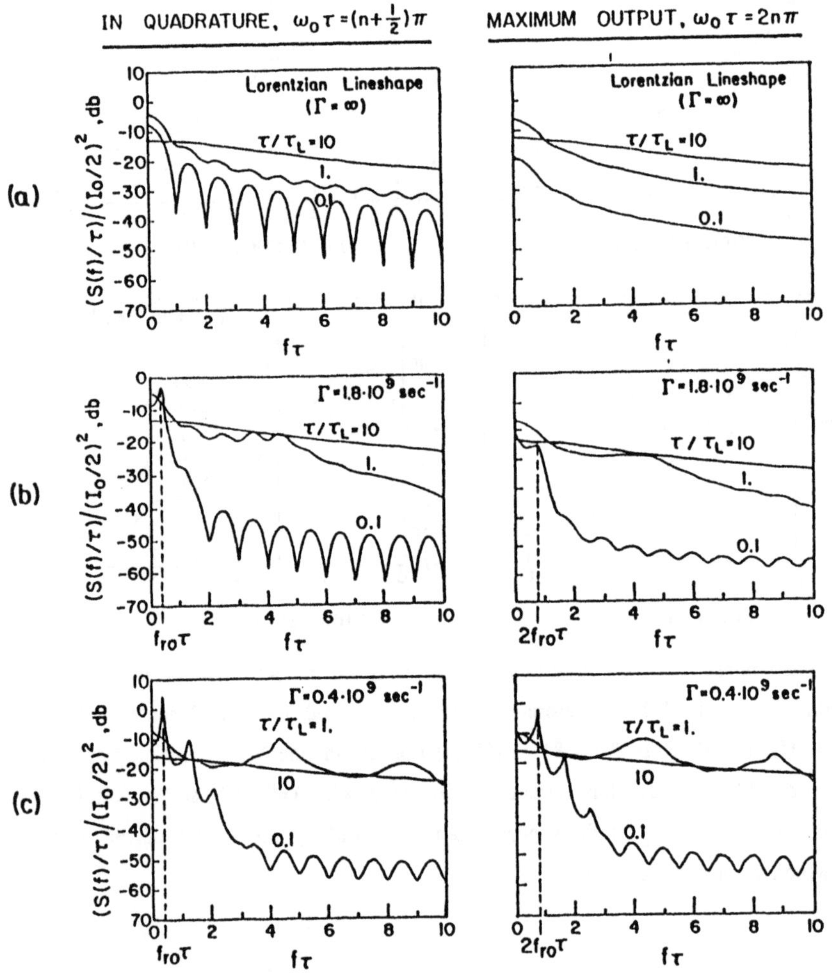

Fig. 2: The normalized phase induced intensity noise power spectrum, S(f), as a function of the scaled frequency $f\tau$. All other parameters are the same as in Fig. 1.

obtained under the assumption of a Lorentzian lineshape (Figs. 1a, 2a). At relatively low frequencies, the characteristic frequency in the Non-Lorentzian case is the relaxation-oscillation frequency, see Figs. 1b-c, whereas for higher frequencies it becomes $1/\tau$, as in the case of the Lorentzian lineshape, Figs. 2b-c. In the in-quadrature case, we get a peak in the PIIN spectrum at the relaxation-oscillation frequency f_{ro}, as observed by Daino et al. [10]. However, in case of **maximum** output, we get a peak at **twice** the relaxation-oscillation frequency. The existence of this peak is clearly supported by Fig. 3, which shows the autocovariance Cov(ΔT) for the in-quadrature and maximum output cases. To the best of our knowledge, this peak has not been

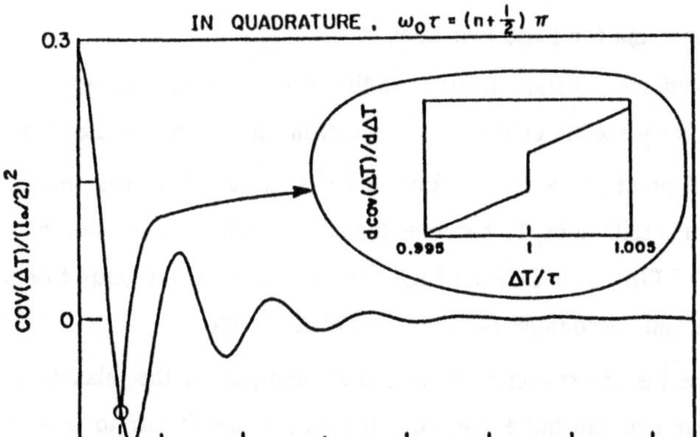

Fig. 3: The intensity autocovariance function (Eq. (7)) at the output of an symmetrically excited MZDL. We used $R=1.5\cdot10^{12}$ sec^{-1}, $I=3.1\cdot10^{4}$, $\alpha=5.3$, $f_{ro}=1.52\cdot10^{9}$ Hz, $\tau_L=4I/(R(1+\alpha^2))=2.84$ ns, $\Gamma=1.8\cdot10^{9}$ sec^{-1}, $\tau/\tau_L=0.1$. The insets show the discontinuity of the derivative of the autocovariance at $\Delta T=\tau$, which accounts for the $1/\tau$ period at the higher frequencies of the noise spectrum (See Figs. 2b-c). (a) In quadrature; (b) Maximum output.

observed so far. If confirmed it may serve as an indication for the state of the interferometer, being in-quadrature, out of quadrature or somewhere in between.

It is also interesting to note that the power spectrum of the Non-Lorentzian source in the high frequency range gets its minima at frequencies which are multiples of $1/\tau$ in case of in-quadrature, whereas in case of maximum output, maxima are obtained at the same frequencies, Figs. 2b-c. The $1/\tau$ periodicity, which is also evident in the coherent Lorentzian case, originates from discontinuities in the derivative $dCov(\Delta T)/d\Delta T$ at $\Delta T=\pm\tau$, see the insets in Fig. 3 ($dCov(\Delta T)/d\Delta T$ is antisymmetric in ΔT). The second derivative $d^2Cov(\Delta T)/d^2\Delta T$ will exhibit, therefore, Delta function discontinuities at $\Delta T=\pm\tau$: $+\delta(\Delta T-\tau)$ and $-\delta(\Delta T+\tau)$ for the in-quadrature case, but $-\delta(\Delta T-\tau)$ and $+\delta(\Delta T+\tau)$ at maximum output. These sign differences are responsible for the different locations of minima and maxima for these two cases.

Note that even relatively small relaxation-oscillation-related satellite side peaks (1% of the peak intensity at ω_0, $\Gamma=1.8\cdot10^9$ sec^{-1}) have a very sizable effect on the spectrum of the PIIN noise (Figs. 1b, 2b). The difference between the Lorentzian and Non-Lorentzian spectrum becomes smaller when the damping rate of the relaxation-oscillations, Γ, becomes higher. A high damping rate may explain the results of Ref. [4] where the PIIN of a DCPBH InGaAsP DFB laser at the output of a Mach-Zehnder interferometer could be fully accounted for by a Lorentzian model.

As we get into the incoherent regime the above mentioned peaks at multiples of the relaxation-oscillation frequencies tend to disappear and the noise spectrum becomes quite similar to that of the Lorentzian source. This result is consistent with the fact that the structure functions for the Lorentzian and Non-Lorentzian lineshapes are different only in the coherent regime, and become similar when $|\tau|$ goes to infinity [8]

The differences between the Lorentzian and Non-Lorentzian lineshapes in the incoherent regime may have an effect on the accuracy of the delayed self homodyne method for measurement of laser spectrum. In this method [3] the laser light is injected into an incoherent two beam interferometer, and the output phase induced intensity noise is measured by an RF spectrum analyzer. In case of a Lorentzian source, the output spectrum is also a Lorentzian [7] and its FWHM (Full Width at Half Maximum) is twice the FWHM of the source. Narrow linewidth sources such as DFB lasers can be characterized in this way. However, if the source is Non-Lorentzian, the output FWHM spectrum divided by two no longer reflects the source linewidth. Further more, as opposed to the Lorentzian case, the PIIN spectrum is not a scaled version of the source lineshape [13]. However, for the two specific non-Lorentzian sources under study in Figs. 1b and 1c, we get a very similar linewidth to the one predicted by the Lorentzian case, Fig. 1a.

The variance of the output intensity $\sigma^2_{I_{out}}(\tau)$, which represents the total amount of the PIIN noise, is obtained by evaluating the covariance function at its origin, $Cov(\Delta T=0)$, see Fig. 4. Again, there is a clear difference between the variance functions in the Lorentzian and Non-Lorentzian in the coherent regime.

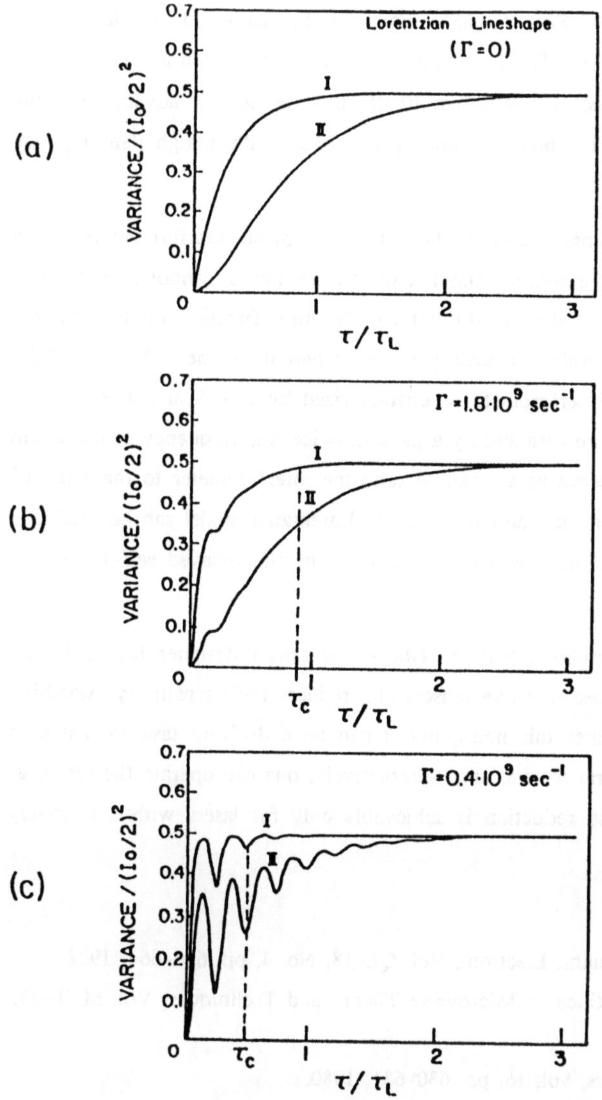

Fig. 4: The normalized variance of the phase induced intensity noise at the output of a two beam interferometer for the in-quadrature (curve I) and maximum output (curve II) settings. We used $R=1.5\cdot10^{12}$ sec^{-1}, $I=3.1\cdot10^4$, $\alpha=5.3$, $f_{ro}=1.52\cdot10^9$ Hz, $\tau_L=4I/(R(1+\alpha^2))=2.84$ ns. (a) Lorentzian lineshape. (b) Non-Lorentzian lineshape with $\Gamma=1.8\cdot10^9$ sec^{-1}. (c) Non-Lorentzian lineshape with $\Gamma=0.4\cdot10^9$ sec^{-1}. The coherence time for each case is indicated on the time axis.

Based on the PIIN spectrum of a Lorentzian source, it was suggested that operating a two beam interferometer near its first notch, at frequency $1/\tau$, instead of at low frequency, may reduce considerably the noise power [7] (up to 30 dB for $\tau/\tau_c=0.1$, see Fig. 2a). This suggestion was based on the deep notch in the noise spectrum of the coherent in-quadrature Mach-Zehnder at frequency $f=1/\tau$. However, the results of Figs. 1 and 2 show that for semiconductor lasers with Non-Lorentzian lineshape it may not be helpful. For instance, the noise reduction at $f=1/\tau$ for the source described in Fig. 2c ($\Gamma=0.4\cdot10^9$ sec^{-1}) is less than 10 dB. In order to take advantage of the lower noise at this frequency, one has to choose an appropriate laser, with a high damping rate value.

In summary, the phase induced intensity noise at the output of optical interferometers driven by semiconductor lasers whose lineshape exhibits relaxation-oscillations peaks, cannot be accurately predicted using a Lorentzian model. For the non-Lorentzian structure function of Eq. (5), the deviations from the Lorentzian-based results are mainly in the coherent regime, where the PIIN spectrum at the output of a two beam interferometer is characterized by a peak at the relaxation-oscillation frequency for the in-quadrature case and by a peak at twice this frequency at maximum output setting. These differences may serve as a mean to tune the interferometer to the states of in-quadrature or maximum output. In the incoherent limit, the Lorentzian model can be used as a fairly good approximation, although it may reduce the accuracy of the delayed-self-homodyne technique.

Phase induced intensity noise is a problem that the fiber-optic sensor designer has to handle since it limits the dynamic range of phase sensitive sensors and reduces their sensitivity. Matching the delays of an interferometer eliminates this noise, but it can be a difficult task to match to better than 1 mm. arms which are several meters long. Alternatively, one can operate the sensor at frequency $f=1/\tau$, but a substantial noise reduction is achievable only for lasers with a relatively high damping rate.

Bibliography

1. T.G. Giallorenzi et. al, IEEE J. Quant. Electron., Vol. QE-18, No. 4, pp. 626-664, 1982.
2. K.P. Jackson et. al, IEEE Transactions on Microwave Theory and Techniques, Vol. MTT-33, no. 3, pp. 193-210, 1986.
3. T. Okoshi et. al, Electronics Letters, vol. 16, pp. 630-631, 1980.
4. R.W. Tkach et. al, IEEE J. Lightwave Tech., vol. LT-4, no.11, pp. 1711-1716, 1986.
5. M. Tur et. al, IEEE J. Lightwave Tech. vol. LT-3, no.1 pp. 20-31, 1985.
6. J.A. Armstrong, J. Opt. Soc. Amer., Vol. 56, No. 8, pp.1024-1031, 1966.
7. B. Moslehi, IEEE J. Lightwave Tech., vol. LT-4, no.11, pp. 1704-1710, 1986.
8. C.H. Henry, IEEE J. Quantum Electron., vol. QE-19, no. 9, 1391-1397, 1983. See also C.H. Henry, Vol. LT-4, No.3, pp. 298-311, 1986.
9. K. Vahala et. al, Appl. Phys. Lett., **42** (3), pp. 211-213, 1983.
10. B. Daino et. al, IEEE J. Quant. Electron., Vol. QE-19, No. 3, pp. 266-270, 1983.
11. J. Manning et. al, IEEE J. Quant. Electron., Vol. QE-19, No. 10, pp. 1525-1530, 1983.
12. Ady Arie et. al, To be published in J. Lightwave Tech.
13. M. Nazarathy, Private communication.

Noise Characteristics of Optical Systems Driven by Quasimonochromatic Thermal Sources

Y. Weissman[1], *E. Shafir*[1], *M. Tur*[2], *and K. Bløtekjaer*[3]

[1]S.N.R.C., Yavne, Israel 70600

[2]Faculty of Engineering, Tel-Aviv University, Ramat-Aviv, Tel-Aviv, Israel 69978

[3]Division of Physical Electronics, Norwegian Institute of Technology, N-7034 Trondheim – NTH, Norway

Source induced noise in systems driven by recently introduced high power quasimonochromatic sources is analyzed, using a new formulation. This noise becomes dominant already for a few microwatts of optical power falling on the detector. Applications to the Mach-Zehnder and Fabry-Perot sensing architectures reveals interesting and significant features.

INTRODUCTION

Quasimonochromatic optical sources, like light emitting and superluminescent diodes are becoming increasingly popular in fiber optic sensors and LANS. In the fiber-optic gyro [1], as well as in certain types of optical measurement schemes [2], quasimonochromatic sources are required to enhance the performance characteristics. In LAN applications, the popularity of LEDs stem from their low cost, high availability, reliability and immunity to back-reflections. Recently introduced high power superfluorescent fiber sources, [3], will undoubtedly enhance this trend.

However, when incorporated in systems, these LEDs will give rise to fairly high detected optical powers and source noise may well become a dominant contributor to the system output noise. Indeed, it can be shown [4] that the intensity noise of LEDs exceeds both thermal noise and shot noise whenever the received optical power is in excess of a few microwatts.

It is by now well established that source noise in fiber structures driven by laser sources can pose severe problems [5], as its spectral range is in the RF band, thus overlapping the (electronic) spectral range of the signal itself. The present work intends to present guidelines for the evaluation of the source noise contribution in systems driven by quasimonochromatic sources. This treatment complements the previous treatments of the optical noise due to laser sources [5], however it is inherently different in view of the fundamental difference in the statistical properties of coherent and quasimonochromatic optical sources [6].

We start by exposing the main tools and concepts of a new method developed for the analysis of the quasimonochromatic source noise. The results are applied to the cases of Mach-Zehnder and Fabry-Perot architectures which are quite common structures in fiber optic sensors.

MATHEMATICAL FORMULATION

We present here a newly developed formulation for the treatment of source induced noise characteristics of systems driven by quasimonochromatic sources. The power spectral density of the source (optical) intensity noise is given by the Fourier transform of the optical intensity covariance function [6], which is of fourth order in the electric field E(t). However, in quasimonochromatic sources, which obey Gaussian statistics, it is possible to express fourth order moments in terms of second order moments [6]. Using this property, it can be shown that the source induced electrical noise power spectral density $P_{source}(f)$ is given by [7]

$$P_{source}(f) = \int_0^\infty d\nu \mathrm{Tr}[G(\nu)G(\nu+f)], \quad (1)$$

where Tr stands for the trace of a matrix. The elements of the matrix $G(\nu)$ are Fourier transforms of second order moments

$$G_{jk}(\nu) = \eta \int_{-\infty}^{+\infty} d\tau \exp[2\pi i\nu\tau]\langle E_j(t+\tau)E_k^*(t)\rangle, \quad (2)$$

where η is the detector responsivity, so that G is measured in units of Amp/Hz. The subscripts represent the two orthogonal polarizations and <> denotes ensemble averaging. Since the width of the spectral power density of a quasimonochromatic thermal source is of the order of 1000GHz or more, $P_{source}(f)$ has a similar width, resulting in a practically constant value for ordinary electronic frequencies.

We turn now to consider the effect of a linear network on the power spectral density of the optical noise. A linear network may be characterized in general by a Jones matrix $J(\nu)$. The network transforms $G(\nu)$ of Eq. (2) into [6]:

$$G_{out}(\nu) = J(\nu)G(\nu)J^\dagger(\nu), \quad (3)$$

where $J^\dagger$ is the Hermitian conjugate of J. Since a linearly filtered Gaussian process is also a Gaussian process, we can use Eq. (2) above to compute the power spectral density $P_{out}(f)$ of the electrical noise at the network output. The result is

$$P_{out}(f) = \int_0^\infty d\nu \mathrm{Tr}[H(\nu,f)G(\nu)H^\dagger(\nu,f)G(\nu+f)], \quad (4)$$

where $H(\nu,f) \equiv J^\dagger(\nu+f)J(\nu)$.

The above expression for $P_{out}(f)$ is most general and applies to any combination of source (provided it is of thermal nature) and linear system, represented by the matrices G and H respectively. We note that in many practical situations the source and the system dependent terms can be factorized. This factorization is feasible only if the matrices $G(\nu)$ and $H^\dagger(\nu,f)$ commute, for which case we can write

$$P_{out}(f) = \int_0^\infty d\nu \mathrm{Tr}[H\cdot H^\dagger \cdot G(\nu)G(\nu+f)]. \quad (5)$$

This factorization can be carried on for cases in which $H(\nu,f)$ is periodic in ν with some period Δf, which is much smaller than the spectral width of the source, so that $G(\nu)$ does not vary appreciably over Δf. The physical significance of this situation is that the coherence length of the source is much shorter than the optical path differences between the various paths that connect the source to the detector, a situation which is often referred to as the incoherent limit [5a]. This will be almost always true for quasimonochromatic sources in the near infrared, which are characterized by a coherence length of the order of approximately 10μm. The LED driven Mach-Zehnder and Fabry-Perot considered below are typical examples for this state of affairs.

To carry on the factorization we break the infinite integration interval in Eq. (5) into an infinite set of finite intervals of length Δf, and utilize the incoherent limit assumption, the periodicity of H and several obvious approximations to obtain

$$P_{out}(f) = \mathrm{Tr}\left[\sum_{k=0}^{\infty}\int_{k\Delta f}^{k\Delta f+\Delta f} d\nu\ \mathbf{H}\cdot\mathbf{H}^{\dagger}\cdot\mathbf{G}(\nu)\mathbf{G}(\nu+f)\right]$$

$$\simeq \mathrm{Tr}\left[\sum_{k=0}^{\infty}\mathbf{G}(k\Delta f)\mathbf{G}(k\Delta f+f)\int_{k\Delta f}^{k\Delta f+\Delta f} d\nu\ \mathbf{H}\cdot\mathbf{H}^{\dagger}\right]$$

$$\simeq \mathrm{Tr}\left[\mathbf{S_m}(f)\cdot\int_{0}^{\infty} d\nu\mathbf{G}(\nu)\mathbf{G}(\nu+f)\right], \tag{6}$$

where in the last step we have introduced the structure matrix $\mathbf{S_m}(f)$:

$$\mathbf{S_m}(f) = \frac{1}{\Delta f}\int_{[\Delta f]} d\nu\mathbf{H}(\nu,f)\mathbf{H}^{\dagger}(\nu,f). \tag{7}$$

The integration should be carried out over any frequency interval of length Δf (which is the period of $\mathbf{H}(\nu,f)$ in ν).

Equation (6) can be further simplified for two cases of practical interest. The first case is the polarization degenerate system, for which **H** is a scalar matrix (i.e. a function times a unity matrix) and the second is the case of the completely unpolarized source, for which **G** becomes a scalar matrix. It is easy to show that in both cases

$$P_{out}(f) = S(f)\cdot P_{source}(f), \tag{8}$$

where we have introduced the structure function S(f) [5a,b]

$$S(f) = \mathrm{Tr}[\mathbf{S_m}(f)]/2. \tag{9}$$

Since the electrical signal power is proportional to the average source-to-detector-transmittance squared, it is convenient to normalize the structure function S(f) by this quantity [4].

A question arises whether this source induced noise could be seen at all above the always present shot and detector thermal noise for practical sources and systems. Indeed, it can be shown [4] that when the detector load resistance is 100KΩ, source noise becomes dominant already for optical powers of 1-3μw falling on the detector, depending on the source spectral width. These optical powers are readily available today even from LEDs pigtailed to single-mode optical fibers. For a 10nm wide source with 100μw of received optical power, the overall source noise exceeds the shot+thermal noise (with a 100 kΩ load resistance) by 15dB.

We present below detailed treatments of some common fiber optic sensing architectures. As mentioned above our treatment is valid only in the incoherent limit.

THE MACH-ZEHNDER ARCHITECTURE

We start by considering the elementary case of the Mach-Zehnder system with equal powers in the two arms. This system consists of two paths connecting the source and the detector with a time delay T between them. Assuming that the system is polarization degenerate, its Jones matrix is given by:

$$J(\nu) = \frac{1}{2}[1 + \exp(2\pi i\nu T)]\cdot\mathbf{I}, \tag{10}$$

where **I** is the unity matrix. The corresponding **H** matrix and structure function S(f) are

$$H(\nu,f) = \frac{1}{4}[1 + \exp(2\pi i\nu T) + \exp(-2\pi i(\nu+f)T) + \exp(-2\pi i fT)]\cdot\mathbf{I}, \tag{11}$$

$$S(f) = \frac{1}{4}\left(1 + \frac{1}{2}\cos(2\pi fT)\right). \tag{12}$$

In this case the average system intensity transmittance is 0.5. Therefore S(f) normalized by the average transmittance squared is

$$S(f)\ (\text{Normalized}) = 1 + \frac{1}{2}\cos(2\pi ft). \tag{13}$$

It is instructive to compare this result to the case in which interference effects at the detector are eliminated. Such a system can be realized by a linearly polarized source, and a rotation of the plane of polarization in one of the arms by 90°. When this is done, then the two merging signals become mutually orthogonally polarized, and cannot interfere. The Jones matrix of such system is

$$J(\nu) = \frac{1}{2}[\mathbf{I} + \mathbf{Q}\exp(2\pi i\nu T)], \tag{14}$$

where **Q** is the rotation Jones matrix:

$$\mathbf{Q} = \begin{pmatrix} 0 & -1 \\ 1 & 0 \end{pmatrix}. \tag{15}$$

This system is not polarization degenerate and therefore the output noise must be calculated directly from Eq. (4). After some straightforward algebra it turns out that a structure function in the sense of Eq. (8) can be defined in this case as well, and it is given by

$$S(f)\ (\text{Normalized}) = \frac{1}{2} + \frac{1}{2}\cos(2\pi fT). \tag{16}$$

Here, in the absence of interference effects, Eq. (16) merely describes the RF frequency intensity filtering feature of the structure. In other words, here we get at the output only the filtered source intensity noise. However, when interference exists [Eq. (13)], an extra 1/2 appears, which seems to represent the interferometric conversion of the source phase fluctuations into additional intensity noise at the system output. The structure functions for both cases are plotted in Figure 1.

THE FABRY-PEROT ARCHITECTURE

Let us now apply the above treatment to a Fabry-Perot structure, both in the reflection and in the transmission modes. In addition to its importance for sensing applications [2], this structure also models reflections between connectors.

For the reflecting Fabry-Perot we have [7]

$$J(\nu) = r\,\frac{1+(t^2-r^2)\exp(2\pi i\nu T)}{1-r^2\exp(2\pi i\nu T)}\cdot\mathbf{I}\,, \tag{17}$$

where r and t are the field reflectances and transmittances of the end mirrors (which are assumed to be identical) and T is the circulation time. The structure function S(f), normalized to the square of the average device reflectance is given for this case by [5b]

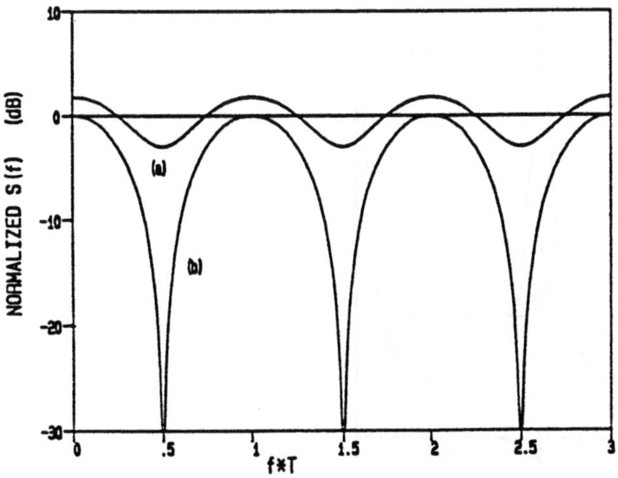

Figure 1: The normalized structure function S(f) (Normalized) for the Mach-Zehnder architecture with (a) parallel and (b) perpendicular polarizations in the two arms.

$$S(f)\ (\text{Normalized}) = 1+\frac{1}{2}(1-R)^2\,\frac{\cos(2\pi fT)-R^2}{1+R^4-2R^2\cos(2\pi fT)}\,, \tag{18}$$

where $R=|r|^2$ is the mirror intensity reflectance. Similarly for the transmitting Fabry-Perot we find [7]

$$J(\nu) = t^2\,\frac{\exp(\pi i\nu T)}{1-r^2\exp(2\pi i\nu T)}\cdot I\,, \tag{19}$$

and

$$S(f)\ (\text{Normalized}) = 1+2R^2\,\frac{\cos(2\pi fT)-R^2}{1+R^4-2R^2\cos(2\pi fT)}\,. \tag{20}$$

Here the normalization is with respect to the square of the average device transmittance. Equations (18) and (20) are plotted in Figs. 2a and 2b, respectively, for various mirror intensity reflectances R. It is clearly seen that in both modes the structure term S(f) (and thus practically also the whole resultant power spectral density) is again periodic with a period of 1/T. However, whereas in the transmission mode S(f) is characterized by very high (≃40dB) peaks, in the reflection mode the same term assumes a relatively flat shape, with spectral variations of less than 0.3dB. Obviously, in the transmission mode incautious choice of signal frequency may result in a significant degradation of the available dynamic range.

CONCLUSIONS

The random nature of the output field of a quasimonochromatic low-coherence source such as an LED, a superluminescent diode or a superfluorescent fiber source may significantly contribute to the output noise power of any optical system. It is found that the source induced noise becomes significant (i.e. above detector shot and thermal noise) for a few microwatts of optical power falling on the detector (with a 100KΩ load resistance). With the high power quasimonochromatic sources available today (e.g., 10mW for the fiber source [3]), the source induced noise at the system output may be a few orders of magnitude above detector shot and thermal noise.

A new formulation for analyzing the appearance of this noise in a general linear network was presented. It was applied to the incoherent Mach-Zehnder as well as to the incoherent

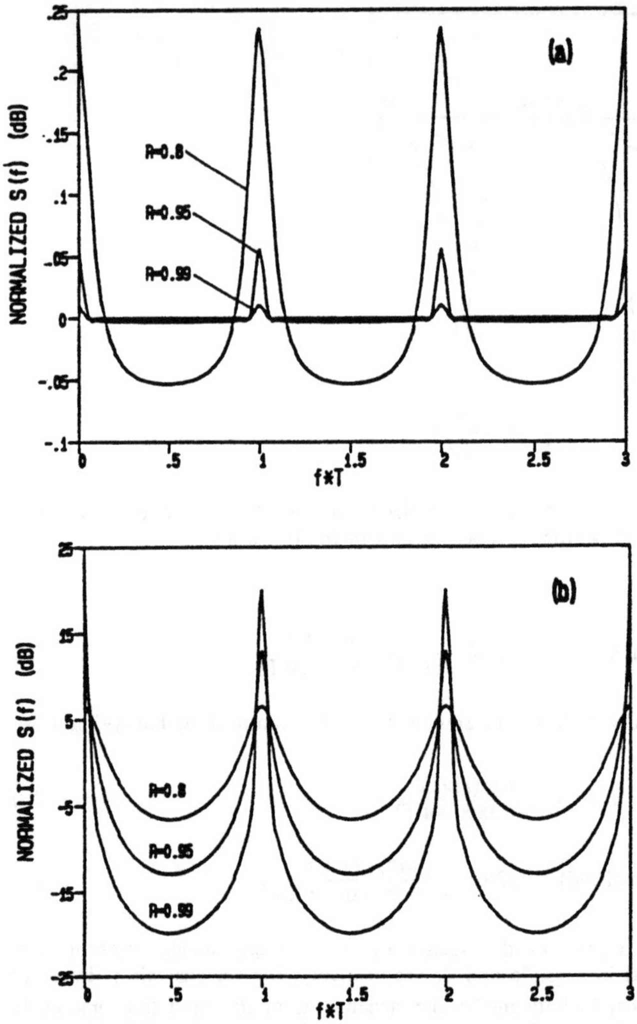

Figure 2: The normalized structure function S(f) (Normalized) for the Fabry-Perot architecture for (a) reflection mode and (b) transmission mode. Note the difference between the two vertical scales.

Fabry-Perot structure in both reflection and transmission modes. For the Mach-Zehnder it was found that deep notches appear at RF frequencies of 1/2T, 3/2T, etc. (where T is the propagation time difference between the two arms) provided interference is inhibited at the output of the device, where the two beams are superimposed. Here the mere function of the device is to filter the source intensity fluctuations. On the other hand for the ordinary Mach-Zehnder architecture, where interference takes place, the device also interferometrically converts the source phase fluctuations into an additional noise term at the output, raising the notches and flattening the noise spectrum.

For the Fabry-Perot it was found that whereas in the reflection mode the noise power spectral density is relatively flat, this is not the case in the transmission mode where very high peaks ($\simeq$40dB) may occur at integral multiples of 1/circulation time. The significance of these characteristics for a proper choice of the optimal signal frequency has been emphasized.

ACKNOWLEDGEMENT

One of the authors (YW) wishes to thank the Adler foundation for support.

REFERENCES

1. W. K. Burns and R. P. Moeller, J. of Lightwave Technology, *LT-2*, p. 430, 1984.
2. G. Beheim, Appl. Opt., *24*, p. 2335, 1985.
3. K. Liu et. al., Elect. Lett., *23*, p. 1320, 1987.
4. M. Tur, E. Shafir and K. Bløtekjaer, "Source induced noise in optical systems driven by quasimonochromatic sources", submitted for publication.
5. (a) M. Tur, B. Moslehi, and J. W. Goodman, J. of Lightwave Technology, *LT-3*, p. 20, 1985; (b) E. Shafir and M. Tur, J. Opt. Soc. Am. A, *4*, p. 77, 1987.
6. J. W. Goodman, *Statistical Optics*, John Wiley, New-York, 1985.
7. Y. Weissman, "Optical RF noise generation by a linearly filtered optical thermal source", submitted for publication.

Demonstration of Lead Sensitivity of a Fiber Interferometer due to Magneto-optically Induced Input Polarization Fluctuation

A.D. Kersey, M.J. Marrone, and A. Dandridge

Naval Research Laboratory, Optical Techniques Branch, Code 6570, Washington, DC 20375, USA

Abstract: We show that the dependence of the phase shift of an interferometric fiber sensor on the input state of polarization can lead to the generation of excess phase noise due to magnetically induced optical rotation in the input fiber. The use of depolarized source light to eliminate this source of excess phase noise is theoretically analysed and experimentally demonstrated.

Introduction: It is commonly assumed that interferometric fiber sensors are 'lead-insensitive', that is to say that perturbations of the input fiber to, or output fiber from an interferometer do not give rise to a shift in the phase of the interference signal generated at the detector. In a recent paper [1], we presented both a theoretical analysis and experimental results which showed that fluctuations in the input state of polarization (SOP) to an interferometric sensor can result in the generation of phase noise in the demodulated sensor output. In this paper we show for the first time that this mechanism can lead to the generation of excess phase noise in an interferometer when the input fiber is subject to electromagnetic noise. We have found that by using depolarized source light, the dependency of the interferometer visibility and phase on the birefringence properties of the input fiber is eliminated at the cost of a generally reduced interference visibility.

Theory: The dependence of visibility on input polarization have been presented in full in previous publications [2,3] . In order to understand the dependence of the interferometer visibility and phase shift on the input SOP, we refer to Figure 1 which shows a schematic of a Mach-Zehnder interferometer and the Poincare Sphere representation of the SOPs of the light at the input to the first coupler (C_i), and in the signal (C_s) and reference (C_r) arms of the interferometer at the point of recombination in the output coupler. Assuming equal power in the two output optical fields, the output fringe visibility is simply given by the cosine of the half angle subtended between C_s and C_r on the sphere [4], i.e.

$$\mathcal{V} = \cos\eta\,. \tag{1}$$

It can be shown [2,3] that the polarization dependence of an interferometer can be described by a single differential elliptical retarder $\mathcal{R}_{r-s}(\Omega_{r-s})$ (Figure 1.c) which depends on the birefringence properties of the two fiber arms. For an input SOP aligned with either of the eigenvectors of $\mathcal{R}_{r-s}$, the SOPs of light derived from the signal and reference arms ***must*** be coincident. Consequently, the two input SOPs aligned with the eigenvectors of $\mathcal{R}_{r-s}$ behave as eigenmodes of the interferometer for which optimum visibility can be achieved. For an arbitrary input SOP, C_i, the visibility and output phase shift can be calculated by resolving the input state onto the two eigenmodes, and summing the output interference intensities (scalar summation of orthogonal polarization components). If the input SOP subtends an angle θ to the eigenmode vector $\mathcal{R}_{r-s}$ then the output interference comprises two components given by

$$I_a = \frac{1}{2} I_o[\cos^2(\theta/2)\{1 + \cos(\phi + \Omega_{r-s}/2)\}] \tag{2a}$$

Springer Proceedings in Physics, Vol. 44 **Optical Fiber Sensors**
Editors: H.J. Arditty · J.P. Dakin · R.Th. Kersten

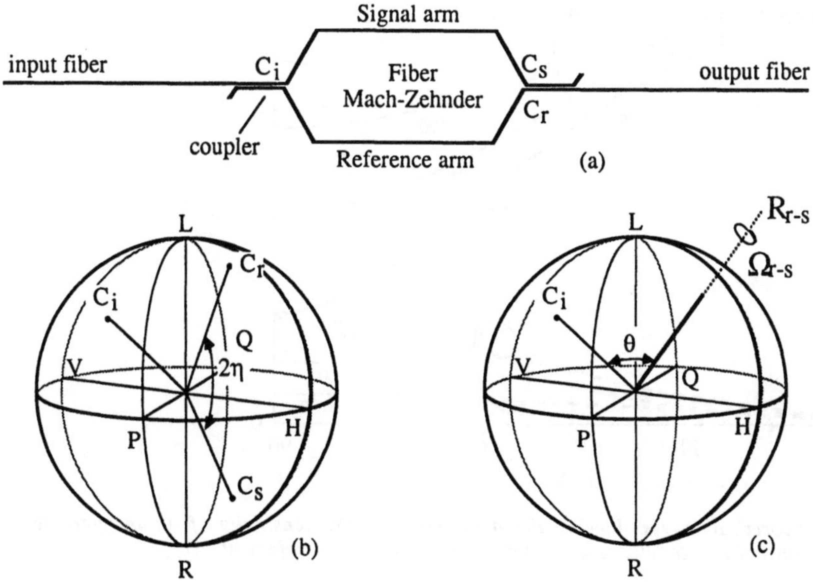

FIGURE 1. (a) Mach-Zehnder interferometer and SOPs at input and output couplers, (b) Poincaré sphere representation of SOPs, and (c) Differential eigenvector representation of the polarization properties of a two-beam fiber interferometer.

$$I_b = \frac{1}{2} I_o[\sin^2(\theta/2)\{1 + \cos(\phi - \Omega_{r\text{-}s}/2)\}] \tag{2b}$$

where $\Omega_{r\text{-}s}$ is the rotation magnitude of $\mathfrak{R}_{r\text{-}s}$. The total output interference intensity is thus $I = I_a + I_b$, which after alegbraic manipulation can be expressed as

$$I = \frac{1}{2} I_o[1 + \mathcal{V} \cos(\phi - \gamma)], \tag{3}$$

where, the visibility, $\mathcal{V}$, and phase offset, γ, can be expressed in terms of θ and $\Omega_{r\text{-}s}$ according to

$$\mathcal{V} = \{1 - \sin^2\theta.\ \sin^2(\Omega_{r\text{-}s}/2)\}^{1/2}, \text{ and} \tag{4.a}$$

$$\tan\gamma = \tan(\Omega_{r\text{-}s}/2)\cos\theta. \tag{4.b}$$

The phase offset term is thus dependent on the position of the input SOP relative to the eigenvector $\mathfrak{R}_{r\text{-}s}$ and the rotational magnitude $\Omega_{r\text{-}s.}$ Random fluctuation in the SOP reaching the interferometer give rise to changes in θ, thus leading to excess phase noise. Such fluctuations can arise due to mechanical perturbation of the input fiber, or as in the case of interest here, when the input fiber is subject to an electromagnetically noisy environment. The phase noise, $\Delta\gamma$, resulting from a polarization-noise fluctuation, $\Delta\theta$, (assuming $\Delta\theta << 1$) is by Eqn. 4.b

$$\Delta\gamma = (\partial\gamma/\delta\theta).\Delta\theta = \frac{\tan(\Omega_{r\text{-}s}/2).\sin\theta}{(1 + \tan^2(\Omega_{r\text{-}s}/2).\cos^2\theta)}\ .\Delta\theta. \tag{5}$$

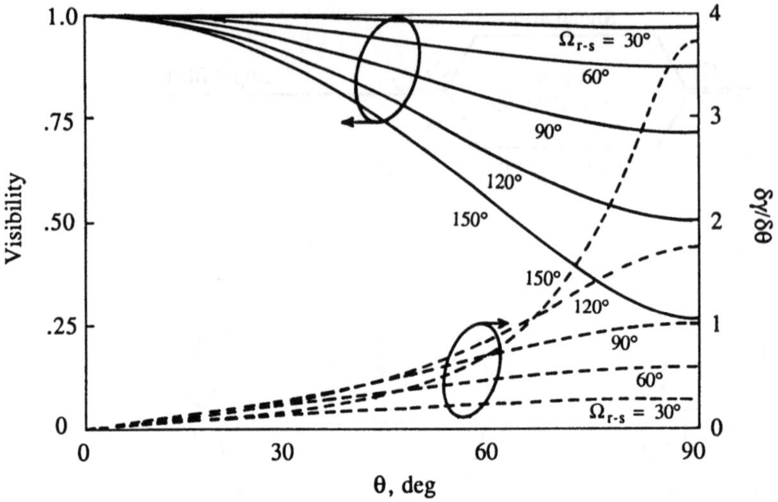

FIGURE 2. Theoretical curves showing the dependence of the sensitivity of an interferometer on input polarization noise, $\delta\gamma/\delta\theta$, as a function of θ, for various values of Ω_{r-s}

Equation 4.a shows that if $\Omega_{r-s} < \pi$ (modulo 2π), then $\mathcal{V} > 0$ for all input SOPs, i.e. the visibility varies between the limits $\mathcal{V}_{max} = 1$, and $\mathcal{V}_{min} = \cos(\Omega_{r-s}/2)$, which occur when $\theta = 0$ and $\theta = \pi/2$ respectively. For a fixed value of Ω_{r-s} the input-polarization-induced phase noise $(\delta\gamma/\delta\theta)$ is minimum when $\theta \sim 0$ (i.e. around maximum visibility) and is maximized when $\theta \sim \pi/2$ (i.e. close to minimum visibility). Figure 2 shows theoretical curves of $(\delta\gamma/\delta\theta)$ and visibility vs. θ for various values of Ω_{r-s}. As seen from this figure, in cases where the visibility can fade close to zero due to variation in the input polarization (i.e. for Ω_{r-s} approaching π), the input-polarization noise to phase-noise conversion factor, $(\delta\gamma/\delta\theta)$, can be relatively large.

Unpolarized source: If we assume the input light is completely unpolarized, it can be resolved into two mutually incoherent orthogonal polarization components of equal intensity. As any aribtrary pair of orthogonal components can be chosen to represent the unpolarized input light, they can be chosen to match the eigenmodes of the differential operator $\mathfrak{R}_{r-s}$. In this case, the interferometer output is thus comprised of two interference terms I^u_1, and I^u_2 (each of optimum visibility), generated by the two components of the unpolarized input light which have a relative phase shift of Ω_{r-s} between them, i.e.

$$I^u = I^u_1 + I^u_2 = \frac{1}{4} I_o\{[1 + \cos(\phi + \Omega_{r-s}/2)] + [1 + \cos(\phi - \Omega_{r-s}/2)]\}$$

$$= \frac{1}{2} I_o[1 + \cos(\Omega_{r-s}/2)\cos(\phi)], \qquad (6)$$

where ϕ is the *mean* phase difference between the signal and reference arms, and I_o is the input intensity of the unpolarized light. The output visibility of a fiber interferometer illuminated using unpolarized light is thus $\mathcal{V}_u = \cos(\Omega_{r-s}/2)$, which is independent of the polarization properties of the input fiber lead, but depends on the rotational magnitude, Ω_{r-s}, of the differential operator $\mathfrak{R}_{r-s}$. The phase shift of the interference output is also independent of the polarization properties of the input lead, and thus input-polarization induced phase noise can be eliminated by depolarization of the input light at the price of a reduced visibility.

Experimental: We have previously demonstrated excess phase noise due to polarization noise induced by mechanically perturbing the input fiber to an interferometer. Here, we report

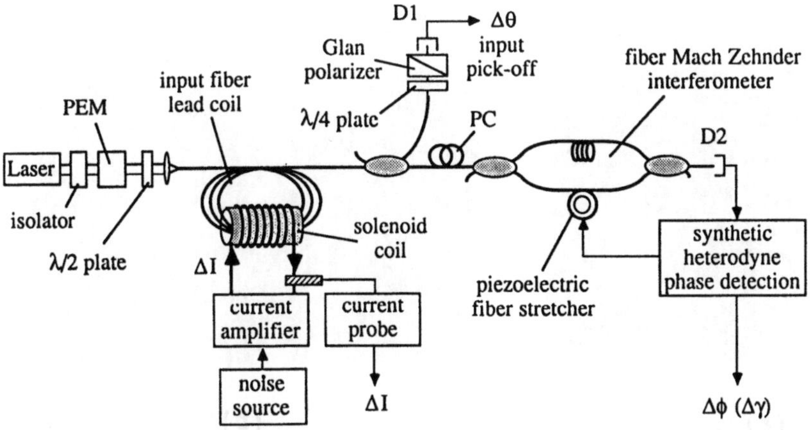

FIGURE 3. Experimental set-up used to investigate excess phase noise due to magneto-optically induced polarization noise in the input to an interferometer

the magnitude of the effect arising due to magneto-optically induced polarization rotation which may be experienced when an input fiber to a sensor is subject to an electromagnetically noisy environment. The experimental set-up is shown in Figure 3. In order to magnetically induce polarization changes in the input fiber, 8 turns of fiber in the input lead were passed through a solenoid coil, which could be excited by a current noise source. The resulting input polarization noise produced was monitored by polarization analysis of light tapped off just before the interferometer using a coupler. The light available at this output was analysed using a λ/4 plate and Glan polarizer, with both elements being rotated to the position which produced maximum noise modulation in the detected intensity passed by the polarizer. Synthetic-heterodyne [5] phase shift detection was applied to the interferometer output to demodulate the interferometric signal. In order to test the effect of depolarizing the input light we included a photo-elastic modulator (PEM) in the input optics. When used, this was driven at frequency of 84 kHz with peak polarization-mode phase delay of 2.4 rad. to effectively scramble the input polarization [6].

Results: By adjustment of the input λ/2 plate and polarization controller PC, the interferometer was found to exhibit a minimum visibility of 0.17 with variation in the input polarization. Using Eqn. 4.a, the magnitude of $\Omega_{r\text{-}s}$ for the interferometer was thus determined to be ~ 160 ° (2.8 rad). Measurements were made with the input polarization adjusted to give close to minimum output visibility, which according to Eqn. 5 and Figure 2 maximizes the sensitivity of the interferometer phase to input SOP fluctuations. Figure 4.a. shows the interferometer output noise spectrum over the range 0 to 250 Hz observed without input polarization noise (lower trace) and with a noise current of ~ 50 mA rms/√Hz applied to the coil (upper trace); the magneto-optically induced excess noise increases the noise floor at 200 Hz by ~ 35 dB - from ~ 60 µrad/√Hz to ~ 4 mrad/√Hz. By analysis of the polarization modulation at the input pick-off, we measured a polarization rotation noise of ~ 0.5 mrad/√Hz. On the Poincare sphere, this translates into a maximum change in θ of 1 mrad rms/√Hz. The experimentally observed $\partial\gamma/\delta\theta$ is thus ~ 4, in agreement with the results of Figure 2, which predicts $\partial\gamma/\delta\theta \leq 3.75$ for $\Omega_{r\text{-}s} \sim 150°$. The excess interferometer phase noise and polarization noise monitored at the input pick-off were well correlated, as shown by Figures 4.b and 4.c.

The spectral noise density of the magnetic field generated in the coil was calculated to be ~ 1.5 Gauss/√Hz over the 0 to 250 Hz bandwidth. This relatively large field was applied over an effective fiber length of 2.5 m in our experiment. However, this results shows that for a 200 m input fiber subject to a noise field of 10 gamma/√Hz (10^{-4} Gauss/√Hz) excess phase noise levels of ~ 20 µrad/√Hz could be produced in an interferometer, which could easily limit phase detection sensitivity of an optimized system.

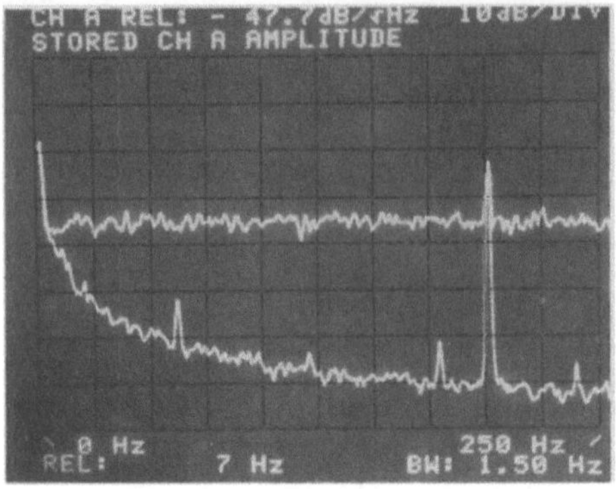

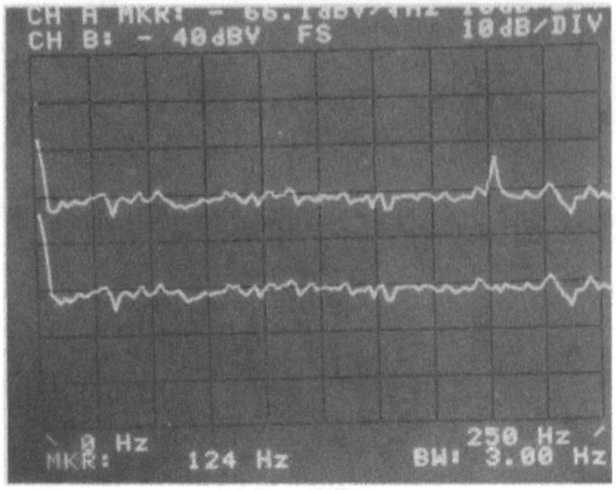

(a) **(b)**

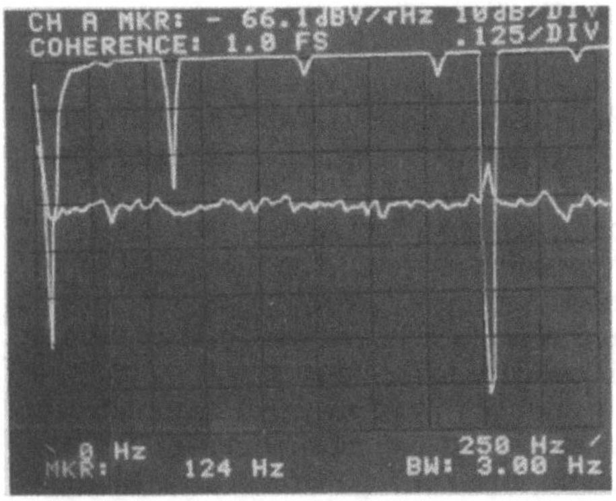

(c)

FIGURE 4. (a) Lower trace - interferometer phase shift noise floor 0 to 250 Hz without input polarization noise (note: 20 mrad test signal at 200 Hz). Upper trace - 4 mrad/√Hz excess noise level in interferometer due to application of noise current to input solenoid coil . (b) Comparison of the noise spectrum observed at the input pick-off to the interferometer to the excess phase noise at the interferometer output. (polarization rotation noise ~ 0.5 mrad/√Hz). (c) Correlation between the noise spectra of (b).

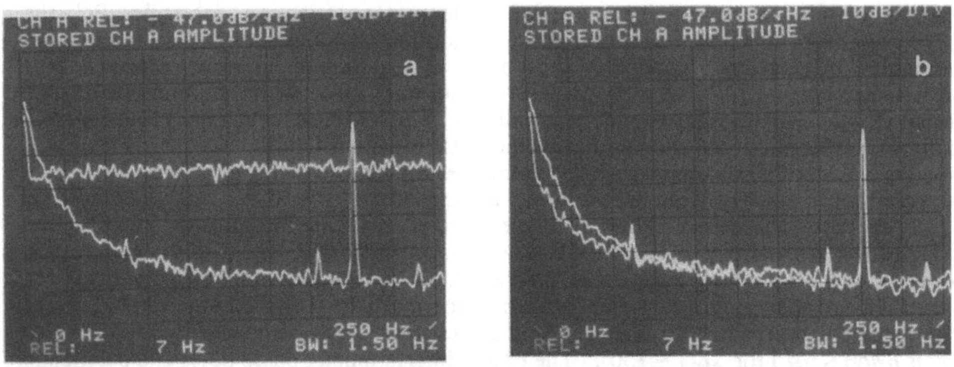

FIGURE 5. (a) Upper trace - interferometer phase shift noise floor 0 to 250 Hz with input polarization noise, lower trace noise reduction with PEM on. (b) Comparison of the interferometer phase noise spectrum observed without input polarization noise and the PEM off (slightly lower trace), and with input polarization noise and the PEM on. Compensation for excess noise to within 2 dB is observed at frequencies > 50 Hz.

Figure 5 shows results which demonstrate the use of polarization scrambling to overcome this problem. The upper trace in Figure 5.a shows the interferometer floor with magnetically induced input polarization noise (as Figure 4.a), and the lower trace shows the effect of applying the appropriate modulation level to the PEM required to effectively scramble the input polarization. As seen by Figure 5.b, which shows a comparison of the interferometer output noise floor without input- polarization noise and the PEM off, and with input-polarization noise and the PEM on, the excess phase noise is nearly completely eliminated by input polarization scrambling.

Conclusions: We have shown that, contrary to the widely held opinion that interferometric fiber sensors are 'lead-insensitive', polarization fluctuations in the input fiber to an interferometric sensor can lead to a changes in the phase shift of the interference output signal. We demonstrated that magneto-optically induced polarization fluctuations in the input fiber of an interferometric sensor can result in the generation of excess phase noise at the output - an effect which may have implications in the design of interferometric sensors for applications where the input fiber is subjected to high levels of EMI. The use of polarization scrambling to eliminate this noise source has been verified.

Acknowledgements: This work was supported by the Office of Naval Technology program on electro-optics.

References:

1. A. D. Kersey, M. J. Marrone and A. Dandridge, Optics Lett., 13, p.847, 1988
2. A. D. Kersey, A. Dandridge, and A. B. Tveten, Optics Lett., 13, p.288, 1988.
3. A. D. Kersey, M. J. Marrone and A. Dandridge, J. Lightwave Technology, 6, p. 1599, 1988.
4. D. W. Stowe et al., IEEE J. Quantum Electron., 18, p. 1644, 1982.
5. A. D. Kersey, A. C. Lewin and D. A. Jackson, Electron. Lett., 20, p. 368, 1984.
6. A. D. Kersey and A. Dandridge, Electron. Lett., 23, p. 634, 1978.

Characterization of a Diode Laser-Pumped Nd : YAG Ring Laser for Fiber Sensor Applications

A.D. Kersey, K.J. Williams, A. Dandridge, and J.F. Weller

Naval Research Laboratory, Optical Techniques Branch, Code 6570, Washington, DC20375, USA

Abstract: We report the measurement of the phase noise, intensity noise and modulation characteristics of a diode laser-pumped solid-state Nd:YAG 1.3 μm laser based on a non-planar ring cavity. A low frequency phase noise approximately three orders of magnitude lower than that typically observed with semiconductor diode lasers is reported. The use of this source in a four-sensor time-division multiplexed array is described.

Introduction: Interferometric fiber sensors have been extensively researched for a wide range of measurands including acoustic, magnetic and electric fields, and acceleration [1]. Although initial work on interferometric sensors used gas laser sources, the need for a more ruggedized source for use with prototype sensors which could be field tested, has led to the wide spread use of semiconductor diode lasers such systems. These lasers provide a compact, reliable source of optical radiation with coherence lengths of several meters (single longitudinal mode). Diode lasers also proved valuable in sensing as a result of the dependence of the optical emission frequency on injection current; this allows the remote passive interrogation of interferometers using laser-FM based techniques such as phase generated carrier [2] and synthetic-heterodyne [3] demodulation. Diode lasers, however, exhibit significant low frequency phase noise or frequency jitter [4] which limits the phase shift detectability in most present-day diode laser-based interferometric sensor systems.

In this paper we report the measurement of the phase noise, intensity noise and modulation characteristics of a commercially available diode laser-pumped solid-state Nd:YAG laser based on a non-planar ring cavity § [5], operating at 1319 nm. Recent studies of this device have shown linewidths < 10 kHz [6], corresponding to coherence lengths > 10 km. We also report the results of an experimental four-sensor time-division multiplexed array based on this source.

Source characterization: Figure 1 shows the experimental arrangement used to measure the low frequency phase noise of this device. In order to provide an interferometric system with a very high resolution to optical frequency, we used an all-fiber Mach-Zehnder interferometer configuration with an imbalance between the arms of 80 m of fiber (equivalent to ~ 120 m optical path difference, OPD). The 'free spectral range' of this interferometer is thus approximately 2.5 MHz, corresponding to a frequency sensitivity of 2.5 μrad/Hz. The interferometer was well shielded from environmental noise sources, such as acoustic noise, by encapsulating it in a housing specially designed for use in the NRL acoustic sensor program. Experiments were conducted with ~ 3 mW of the 15 mW laser output power coupled into the input fiber, although due to the Gaussian output beam of the laser significantly higher launching efficiencies can be achieved. Balanced photodetectors were used at the interferometer outputs to monitor the interference signal and provide rejection of intensity noise. As expected due to the long coherence length of the source, the interferometer exhibited a high interference fringe visibility of > 0.9. Phase noise measurements were made with the interferometer output held in

§ - Lightwave Electronics model 120

Springer Proceedings in Physics, Vol. 44 **Optical Fiber Sensors**
Editors: H.J. Arditty · J.P. Dakin · R.Th. Kersten

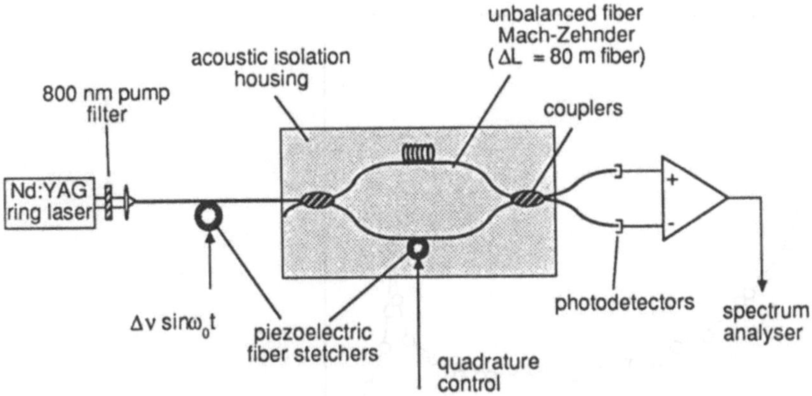

FIGURE 1. Experimental system used to characterize the phase noise of the laser diode pumped Nd:YAG ring laser

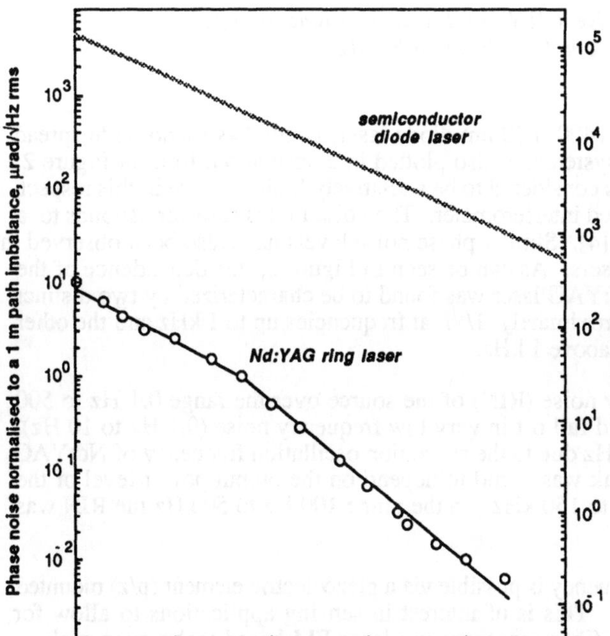

FIGURE 2. Phase noise of the laser diode-pumped Nd:YAG ring laser normalised to a 1 m path imbalance (fiber). Also shown for comparison is the normalised phase noise of an Hitachi HLP 1400 830 nm semiconductor diode laser

quadrature via control of the voltage applied to a small piezoelectric fiber stretcher incorporated into one of the MZ fiber arms. Figure 2 shows the results of these measurements over the frequency range 10 Hz to 250 kHz normalized to a path imbalance of 1 m. The measured phase noise at a signal frequency of 1 kHz was ~ 40 µrad/√Hz, corresponding to a normalized noise of 0.5 µrad/√Hz-m, and a laser frequency jitter of 16 Hz/√Hz. This result is three orders of magnitude lower than typically observed with semiconductor diode lasers; for comparison

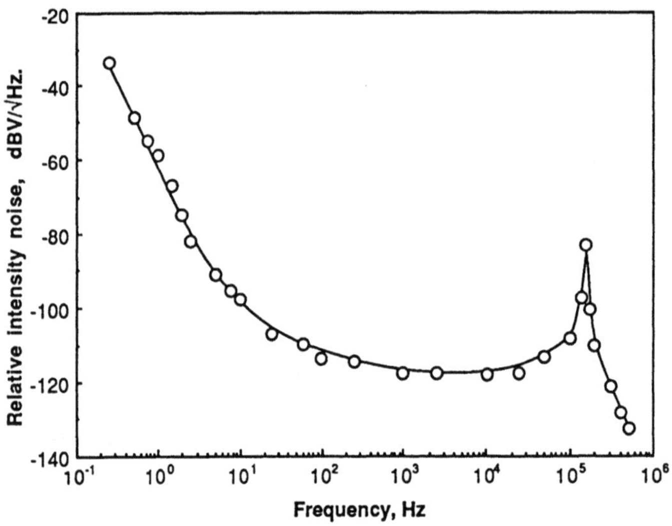

FIGURE 3. Relative intensity noise (RIN) of the laser diode pumped Nd:YAG ring laser from 0.1 Hz to 500 kHz.

the phase noise of an Hitachi HLP 1400 830 nm diode laser, which has found wide spread use in many interferometric sensor systems, is also plotted in a normalised form in Figure 2. The noise of this diode laser, which is considered to be a relatively 'quiet' source in this respect, was measured using a 4 cm unbalanced interferometer. The noise in this case corresponds to a line jitter of ~ 15 kHz/√Hz at 1 kHz [4]. Similar phase noise levels have also been observed in our laboratory using 1.3 μm DFB lasers. As can be seen in Figure 2, the dependence of the phase noise of the diode pumped Nd:YAG laser was found to be characterized by two distinct regions; one falling off at a rate approximately 1/√f at frequencies up to 1 kHz and the other more steeply with a ~1/f dependence above 1 kHz.

Figure 3 shows the relative intensity noise (RIN) of the source over the range 0.1 Hz to 500 kHz. This is characterized by a rapid fall off in very low frequency noise (0.1 Hz to 10 Hz), and a resonant noise peak at ~ 150 kHz due to the relaxation oscillation frequency of Nd:YAG [7]. The frequency of this noise peak was found to depend on the output power level of the laser, varying between ~ 110 kHz to 150 kHz. In the range 100 Hz to 50 kHz the RIN was < -110 dB.

Modulation of the laser emission frequency is possible via a piezoelectric element (p/z) mounted to strain the monolithic ring cavity. This is of interest in sensing applications to allow for passive demodulation of remote all-fiber sensors using laser-FM based techniques such as phase generated carrier [3]. We measured the frequency modulation characteristics of the laser using a 2 m unbalanced interferometer substituted in place of the 80 m unbalanced interferometer in the set up of Figure 1. With the output held in quadrature, the modulation frequency applied to the laser p/z was scanned from 1 kHz to 1 MHz, and the FM induced phase shift in the interferometer was monitored. The results of this experiment are shown in Figure 4. The FM response curve is characterized by a 'flat' region up to ~ 100 kHz, with a normalized FM responsivity $\{\partial\nu/\partial V\}$ of ~ 2.8 MHz/V, beyond which significant resonant behavior is observed. A broad region of high $\partial\nu/\partial V$ (> 50 MHz/V) exists in the frequency window 670 to 900 kHz. Within this window, a voltage of < 2 V p-p was required to drive the 2 m unbalanced interferometer over a complete interference fringe (2π p-p phase shift). Figure 5 shows an example of this modulation capability. The maximum value over which the p/z element could be driven was ± 15 V, corresponding to a frequency deviation of 1.2 GHz p-p at ~ 800 kHz and > 2.5 GHz p-p at the response peak at 720 kHz. These frequency deviations

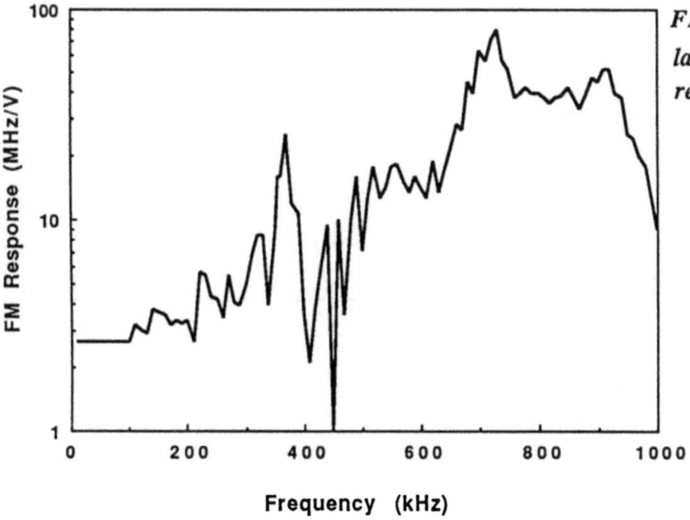

FIGURE 4. Nd:YAG ring laser frequency modulation reponse.

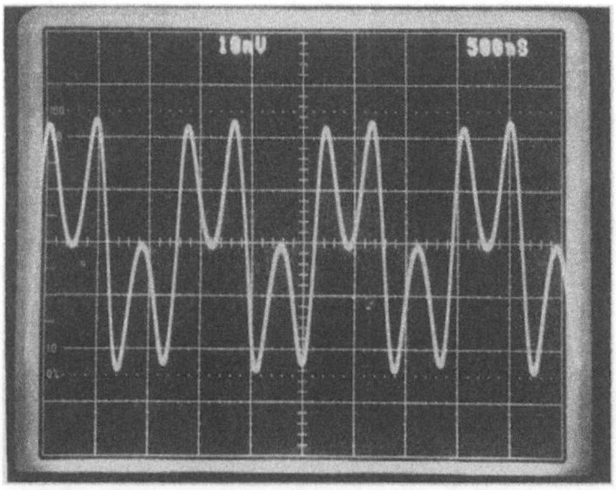

FIGURE 5. Example of the frequency modulation capability of the Nd:YAG ring laser: 2 m unbalanced interferometer driven over 1 complete fringe at 740 kHz.

are much smaller than can be obtained using semiconductor diode lasers, however, due to the low frequency jitter of the Nd:YAG source, path imbalances on the order of a few meters can be used in interferometric sensors to facilitate laser-FM based demodulation without introducing high levels of phase noise. Present systems based on diode lasers employ near-balanced interferometers (2 to 4 cm) to reduce phase noise, while providing sufficient path difference for laser FM based demodulation of the interference output. For sensors using long fiber arms (10 to 100 m), this neccesitates the use of optical techniques to determine the optical paths in order to closely balance the interferometer arms. Furthermore, for an interferometer using ~ 100 m fiber arms, a 0.01 % sensing fiber strain results in a 1 cm error in the path matching; such strains can easily be encountered in sensors due to differential thermal expansions for example. The remarkably low phase noise characteristics and frequency modulation capability of the diode pumped Nd:YAG ring laser alleviates this path matching tolerance, and allows path differences of several meters (< 5 m) to be used while still providing low μrad level phase detection sensitivities and enabling remote passive demodulation of the sensor. To verify this we measured the phase noise of the 2 m unbalanced system. As expected, the phase shift noise floor of the interferometer corresponded to approximately $1 \mu rad/\sqrt{Hz}$ at a signal frequency of 1 kHz, falling off to $\sim 0.3\ \mu rad/\sqrt{Hz}$ at frequencies above 5kHz.

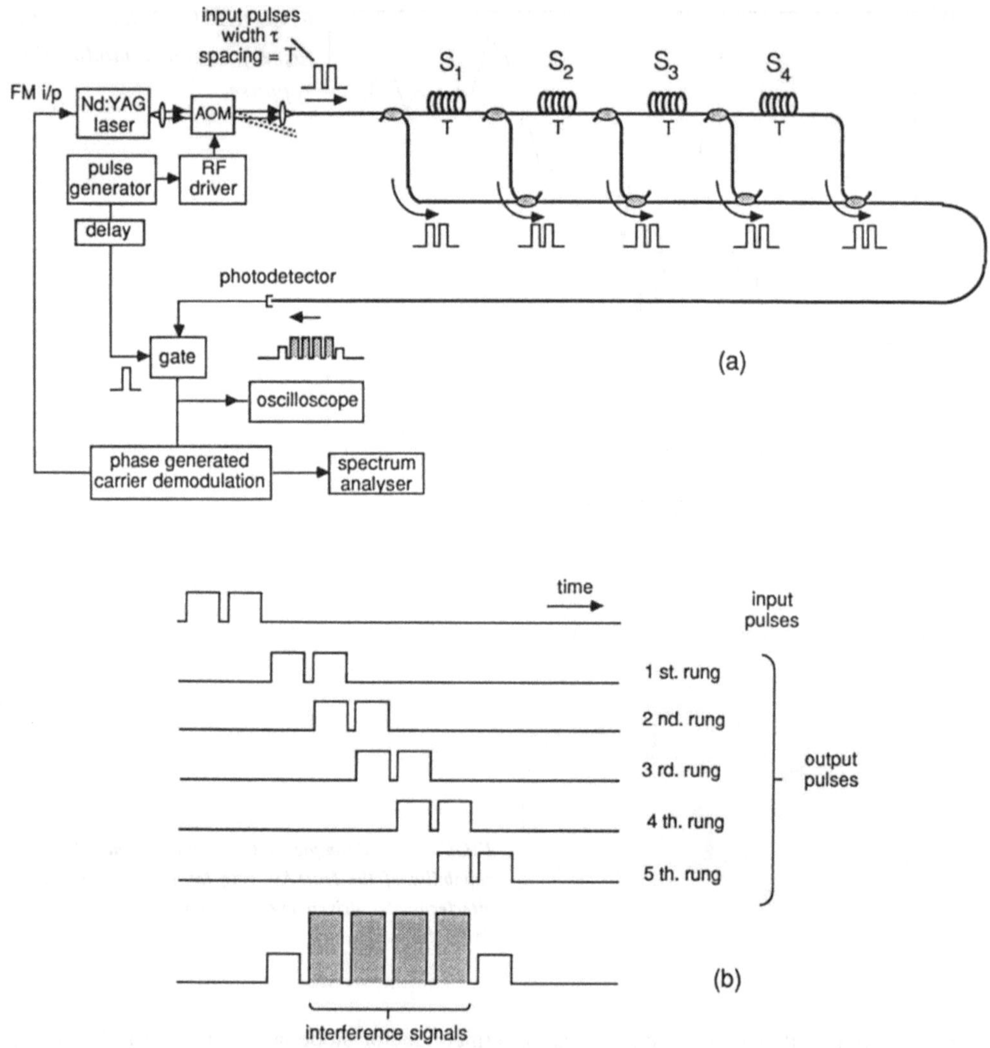

FIGURE 6. Four-sensor time division multiplexed array based on the Nd:YAG ring laser: (a) Array schematic, (b) pulse timing.

Four sensor array: To show the use of this source in a sensor system, we constructed the four-sensor array configuration of Figure 6.a. The array is based on the time-division multiplexing topology described by Brooks et al. [8], but due to the long coherence length of the source no compensating interferometer is required at the output. The array is a ladder-type system with the sensing fiber coils located in the input feed fiber as shown. These sensing coils also serve as delay coils of group delay T. Light from the laser is passed through an acousto-optic modulator (AOM) which is used to pulse the light launched into the input fiber. Two optical pulses of duration τ and separation T are launched into the input. Due to the delay experienced by the light passing through the sensing fiber coils the pulses coupled to the output fiber by each 'rung' of the array are delayed an interval T with respect to the preceding rung. Consequently, the first pulse and second pulses derived from adjacent rungs are time coincident at the output and mix to produce interference signals associated with each sensor element, as shown by the timing diagram of Figure 6.b. Figure 7 shows the output pulse train detected at

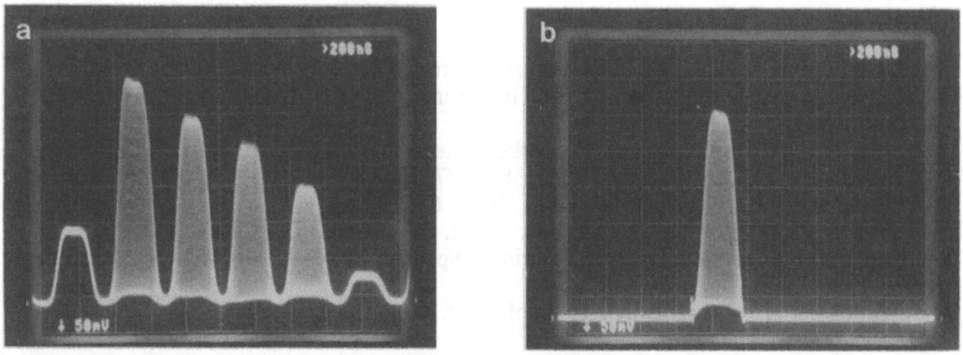

FIGURE 7. (a) Output optical pulse train detected at the photodetector.
Note: central four interference pulses modulated by laser FM.
(b) pulse corresponding to sensor # 2 demultiplexed (gated).

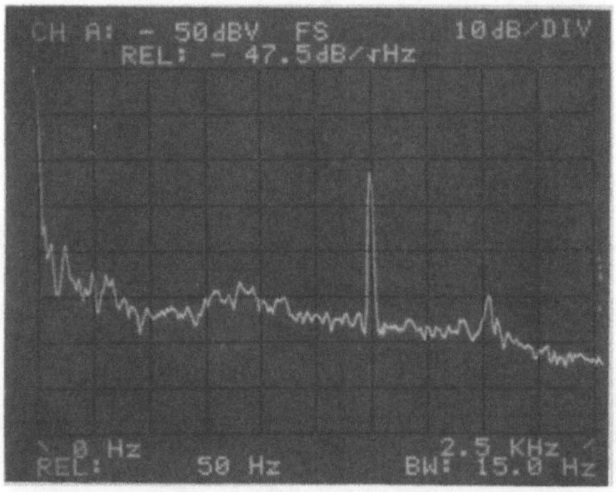

FIGURE 8. Phase detection sensitivity of the TDM array: Sensor # 1 selected: test signal 10 mrad at 1.5 kHz, S/N = 47 dB/√Hz. Noise floor ≈ 45 μrad/√Hz.

the output of the four-sensor array. The sensing fiber lengths were ~ 80 m. Electronic demultiplexing of the outputs and demodulation using the phase generated carrier approach (applied via the laser p/z) yielded a phase detection sensitivity of ~ 45 μrad/√Hz at ~ 1 kHz, in good agreement with the single-sensor measurement of 40 μrad/√Hz discussed earlier.

Conclusions: We have measured the phase noise, intensity noise and frequency modulation characteristics of a diode laser-pumped Nd:YAG 1.3 μm ring laser. The device was found to exhibit a low-frequency phase noise approximately three orders of magnitude lower than that typically observed with semiconductor diode lasers. The use of this source in a four-element time-division multiplexed sensor array has been demonstrated.

Due to its low phase noise and long coherence length, this source should find widespread use in many interferometric fiber sensor applications, in particular, FMCW ranging, coherent optical time domain reflectometry, laser Doppler, and remote vibration analysis.

Acknowledgements: The authors wish to acknowledge the help of A. B. Tveten and A. M. Yurek in part of the experimental work presented in this paper. This research supported by the Office of Naval Technology program on electro-optics.

References:

1. T. G. Giallorenzi et al., IEEE J. Quantum Electron., QE-18, p. 626, 1982.
2. A. Dandridge, A. B. Tveten and T. G. Giallorenzi, IEEE J. Quantum Electron., QE-18, p. 1647, 1982.
3. D. A. Jackson, A. D. Kersey and M. Corke, Electron. Lett., 18, p. 1081, 1982.
4. A. Dandridge et al., Appl. Phys. Lett., 38, p. 77, 1981.
5. T. J. Kane and R. L. Byer, Optics Lett., 10, p. 65, 1985.
6. S. P. Bush, A. Güngör and C. C. Davis, Appl. Phys. Lett., 53, p. 646, 1988.
7. W. Koechner, "Solid state laser engineering", Springer Series on Optical Sciences, Vol. 1, p. 92, 1976.
8. J. L. Brooks et al., J. Lightwave Technol., 5, p.1014, 1987.

1.3 μm Operation of a Coherence Multiplexed Interferometric Sensor

A.M. Yurek, A.B. Tveten, A. Dandridge, and A.D. Kersey

Naval Research Laboratory, Optical Techniques Branch, Code 6570, Washington, DC 20375, USA

Abstract : Operation of a coherence multiplexed fiber optic interferometric sensor is demonstrated at a wavelength of 1.3 μm utilizing a distributed feedback laser (DFB) source. The minimum detectable phase shift is 16 ±1 μrad at 1 kHz. This result represents a 48 dB and 10 dB improvement over the best results reported previously for unmodulated and modulated coherence interrogated systems respectively.

Introduction: Coherence multiplexed fiber systems generally consist of matched sensing and receiving interferometers where the receiving interferometers are located near the electrooptics module and the sensing interferometers are remotely located [1]. This type of approach is often referred to as path matched differential interferometry (PMDI). The optical path differences of each pair of sensing and receiving interferometers are chosen to be the same and significantly greater than the coherence length of the source. The combined interferometer can then be demodulated using a PZT fiber stretcher in the receiving interferometer. Early results for one sensor demonstrated noise levels of approximately 4000 μrad/√Hz [1]. By combining coherence multiplexing with high frequency modulation of the laser diode, it has been shown that the phase induced intensity noise can be upconverted out of the signal band of interest. Interrogation of a single sensor using this technique led to a minimum detectable phase shift of 45 μrad/√Hz [2]. All of the previous experimentation with coherent multiplexed systems have used laser diodes with an operating wavelength of 0.83 μm. The experiment we will describe in this paper used a DFB laser diode operating at 1.3 μm.

Approach: In the previous coherence multiplexed experiments no attempt was made to optimize the system in terms of the source coherence length and the path imbalance of the sensing and compensating interferometers. Even though each interferometer has a path imbalance longer than the coherence length of the source, the coherence of the laser is still finite at these large path imbalances. This leads to high levels of intensity noise, which is generated interferometrically as a result of lasers frequency jitter-induced phase noise [3], in these coherence multiplexed systems. It should be noted that at low frequencies (0.1 Hz to 100 kHz) the lasers frequency jitter is the dominant noise source. The variation of this excess noise as a

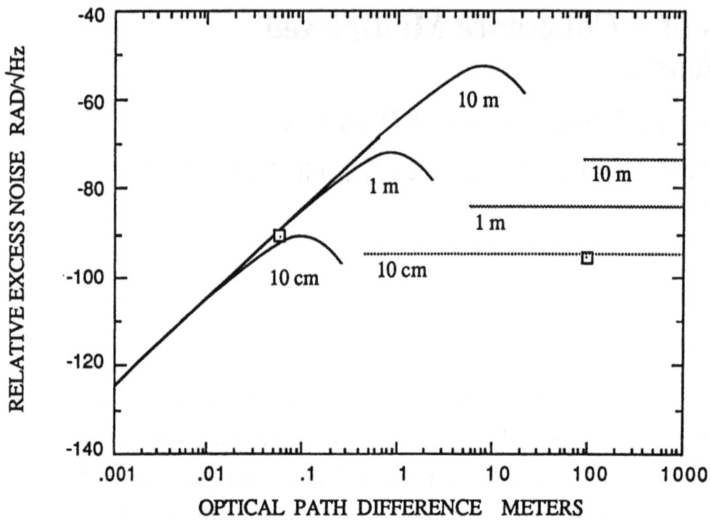

Figure 1. Relative excess intensity noise due to low frequency fluctuations of the source emission frequency (———) and from incoherent mixing (··········) from a single unbalanced Mach-Zehnder interferometer (in quadrature) as a function ofoptical path difference for coherence lengths of 0.1, 1 and 10 meters.

function of the optical path difference (OPD) from a single unbalanced Mach-Zehnder interferometer is shown in Figure 1, this calculation is for a number of different coherence lengths. Clearly it is a requirement of the coherence multiplexed appoach that measurements be made with OPDs beyond the source coherence length. In the previous experiments, however, experimenters typically chose OPDs of a ~20 meters and sources whose coherence length was ~15 meters. As can be seen from Figure 1 this had the effect of unintentionally maximising the noise in the system. In this work we use a path imbalance which is substantially larger than the coherence length of the 1.3 μm source, reducing the excess noise by two orders of magnitude, without requiring high frequency current modulation of the laser. An alternative approach would be to use a super-luminescent diode (SLD), but in this case unacceptable tolerances in matching the interferometer pair (~30 μm) are required.

Experimental: A schematic of the experimental arrangement is shown in Figure 2. The laser source is a Fujitsu DFB laser operating at 1.3 μm. A single sensor 'coherence multiplexed' system was used, i.e. a single sensing and single compensating interferometer. The OPD of each interferometer was ~120 meters (i.e. ~80 meters of optical fiber). Both the sensing and compensating interferometer fiber coils were wrapped on metal mandrels. The compensating interferometer includes a PZT fiber stretcher for demodulation using the phase generated carrier technique [4]. The differential path difference was matched to within 1.5 mm by using standard techniques. The combination of the temperature control of the laboratory environment and the material of the mandrels ensured that this matching was maintained during the experiment. The close balancing of the interferometer pairs meant that direct diode laser

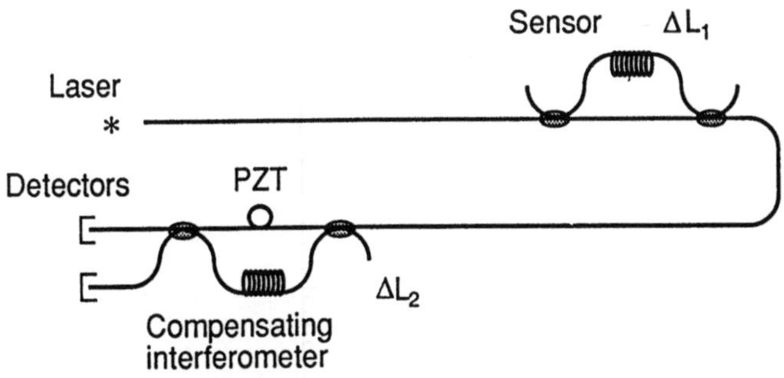

Figure 2. Coherence multiplexed interferometer pair experimental setup.

phase noise due to frequency jitter was reduced to less than 1 μrad/√Hz at 1 kHz. To check the frequency noise properties of the diode laser used in the experiment, a measurement of phase noise from a single 6 cm unbalanced interferometer was made. It was with this measurement that the position of the small OPD portion of the curve in Figure 1 was determined, the point is shown on this figure. The result is comparable with other semiconductor devices operating both at 0.83 μm and 1.3 μm. Measurements from this interferometer indicated that the coherence length of the source was less than 1 meter. Thus the ratio of OPD to source coherence length was over 100. A detector transimpedance amplifier combination was used to detect the output, a balanced detector system was used. Although a high frequency response is not required for this approach, the detector system was designed with a 10 MHz bandwidth to allow some of the diagnostic measurements outlined below.

The minimum detectable phase shift at 1 kHz of the single sensor coherence multiplexed system is shown in Figure 3 as a function of detected optical power. At low optical power the minimum detectable phase shift is limited by amplifier noise in the detection system. At optical power levels greater than 5 μW the minimum detectable phase shift was independent of optical power. At the maximum power on the detector (~20 μW), the system noise was estimated to be ~3 μrad/√Hz at 1 kHz. Typical results at the higher power levels are shown in Figure 4a, a calibration signal of 10 mrad was applied to the PZT fiber stretcher to calibrate the system.

The measurement of the minimum detectable phase shift at 1 kHz was made a number of times and corresponded to 16 ±1 μrad/√Hz. This result is ~48 dB better than previous measurements of unmodulated coherence interrogated sensors. This result is also ~10 dB better than the best result achieved by high frequency current modulation of the diode laser source [2]. In the work of ref [2], the current modulation led to a 38 dB noise reduction which corresponded to the 37 dB theoretically expected, thus the system described in [2] was still limited by the low frequency phase noise term. However, when the high frequency current modulation was applied to the laser in the present system and the minimum detectable

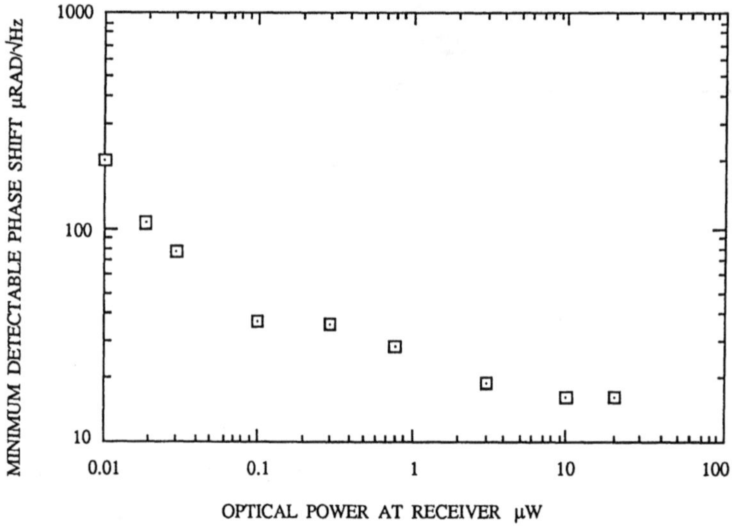

Figure 3. Plot of the minimum detectable phase shift with respect to the detected optical power. Measurements were made at 1 kHz.

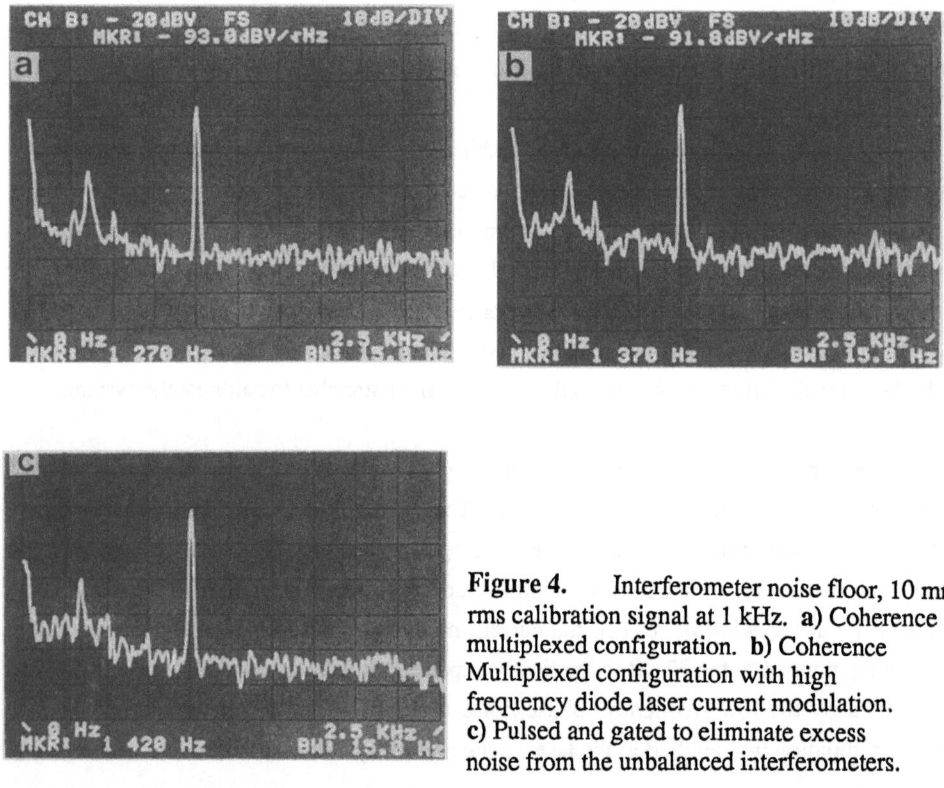

Figure 4. Interferometer noise floor, 10 mrad rms calibration signal at 1 kHz. **a)** Coherence multiplexed configuration. **b)** Coherence Multiplexed configuration with high frequency diode laser current modulation. **c)** Pulsed and gated to eliminate excess noise from the unbalanced interferometers.

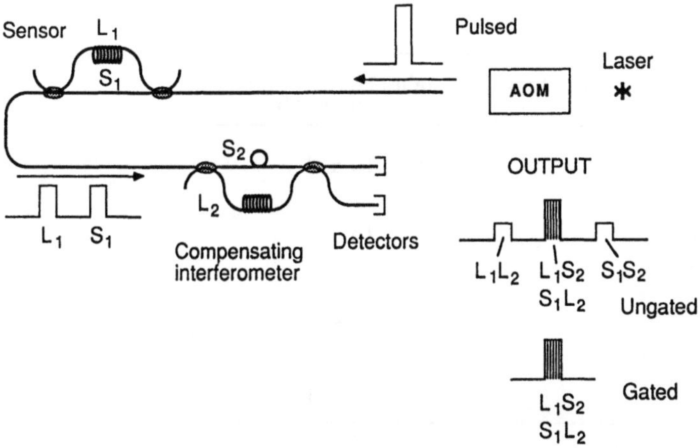

Figure 5. Pulsed interrogation approach to eliminate noise contributions from the unwanted paths which occur in the coherence multiplexed system in the system.

phase shift was determined the result was again 16 ±1 μrad/√Hz (see Figure 4b). Although the frequency modulation characteristics of 1.3 μm lasers are considerably different from 0.83 μm devices, this result implies that the system is no longer limited by the low frequency line jitter induced phase noise from the unbalanced interferometers. To confirm this, we used a pulsed technique to separate, in time, the optical paths with differing transit times [5]. The configuration is shown in Figure 5, a Bragg cell was used to provide the gated output from the cw source. Two methods were used to collect the data firstly by electrically gating out just the interfering term, thus eliminating the intensity pulses travelling through both short arms and long arms of the interferometers. The second approach looked at the total output, which still separates in time the different paths thus eliminating the interference from the unbalanced interferometers, this is shown schematically in Figure 5. Using the electrical gate (Figure 4c) gave a result of 13 ±1 μrad/√Hz, i.e. 2 dB better than the coherence technique (the system noise floor, owing to the reduced time average optical intensity, was 10 ±1 μrad/√Hz). However when the gate was not used, the noise floor rose to 17 ±1 μrad/√Hz, which is basically the same as the coherence technique. The results are summarized in Table 1. It should be noted that the system had a rather low fringe visibility (V=0.25) and owing to only partial subtraction by the balanced detectors, may have led to intensity noise susceptability. However, the estimated intensity noise limit of this system corresponds to a phase detectability of only ~4 μrad/√Hz. Details of the measurements of noise from the single unbalanced interferometers and the origin of the excess noise in terms of the intrinsic linewidth of the laser [6] (as shown in figure 1) will be discussed at the conference.

Conclusion: We have demonstrated a coherence multiplexed system which has a phase detectability comparable with many conventionally interrogated single sensor systems, as well

Table 1 Phase Shift Detectability

Interrogation Approach	Minimum Detectable Phase Shift at 1 kHz µrad/√Hz
PMDI, Coherence Multiplexed	16 ± 1
PMDI, Coherence Multiplexed, with laser modulation	16 ± 1
PMDI, Pulsed and Gated	13 ± 1
PMDI, Pulsed, not Gated	17 ± 1

as state of the art frequency division and time division multiplexed approaches. This approach has the advantage of not requiring the additional complications of pulsing the source and demultiplexing, and is compatible with pigtailed sources. The minimum detectable phase shift is 16 µrad at 1 kHz. This result represents a 48 dB and 10 dB improvement over the best results reported previously for unmodulated and modulated coherence multiplexed systems.

References:

[1] J. L. Brooks, R. H. Wentworth, R. C. Youngquist, M. Tur, B. Y. Kim and H. J. Shaw, "Coherence Multiplexing of Fiber-Optic Interferometric Sensors," *IEEE J. Lightwave Tech.*, **LT-3**, p. 1062, 1985.

[2] A. D. Kersey and A. Dandridge, "Phase Noise Reduction in Coherence Multiplexed Interferometric Fiber Sensors," *Electron. Lett.*, **22**, p. 61, 1986.

[3] A. Dandridge and A. B. Tveten,"Phase noise of Single Mode Diode Lasers in Interferometer Systems," *Appl. Phys. Lett.*, **39**, p. 530, 1981.

[4] A. Dandridge, A. B. Tveten and T. G. Giallorenzi,"Homodyne Demodulation Scheme for Fiber Optic Sensors Using Phase Generated Carrier," *IEEE J. Quantum Electron.*, **QE-18**, p. 1647, 1982.

[5] J. L. Brooks et al, "Time-Domain Addressing of Remote Fiber-Optic Interferometric Sensor Arrays", *IEEE J. Lightwave Tech.*, **LT-5**, p. 1014, 1987.

[6] B. Moslehi, "Analysis of Optical Phase Noise in Fiber-Optic Systems Employing a Laser Source with Arbitrary Coherence Time", *IEEE J.Lightwave Tech.*, **LT-4**, p. 1334, 1986.

Part VII

Systems and Applications

Optical Fiber Environmental Sensors

J.P. Dakin

York Ltd., School Lane, Chandlers Ford, Hampshire, SO5 3DG, UK

This paper describes the potential of optical fiber sensors for remote monitoring of the physical and chemical properties of the surroundings. There is an emphasis on the environmental and health and safety aspects of chemical pollution, with discussion on related factors such as thermal energy wastage. These areas are currently receiving significantly more attention, as a greater realisation of the Earth's limited ability to cope with the greater excesses of Man's activities is becoming apparent. The purpose of this text is to set the scene for a more extensive set of examples to be shown in the lecture.

INTRODUCTION

The limited ability of the Earth to cope with the extremes of Man's pollution has been receiving an increasing amount of political attention in recent years, with greatest media emphasis being on global effects, such as the "greenhouse" effect and the well-publicised "holes" in the ozone layer.

These effects are particularly frightening, if they were to continue unabated to their predicted long-term conclusions. However, there are also many more localised incidents of environmental pollution, for example in home and industrial environments, which currently cause acute problems and may result in cases of severe hazard or death. These are, by their sporadic nature, often difficult to fully assess and quantify. In addition, there are many examples of wastage of valuable resources, such as heat energy loss, chemical spillage etc, which may cause extensive economic loss.

In modern society, the greater capability of technology to determine, more accurately, the level of pollution and wastage is increasing the need for suitable equipment to perform continuous monitoring of the surroundings. The drive for installing sensing equipment is not merely driven by the moral urge to protect Man and the environment, (indeed, certain cynics may claim this is often far from the case) but is becoming of direct economic concern to companies, due to the increasing number and cost of cases of litigation, in response to pollution hazards. In addition, there is now an increasing level of responsible, preventative legislation, aimed at preventing hazardous incidents.

The use of optical fiber sensors in this area has, so far, been extremely limited. However, they do have many potential advantages to offer. Their principal attributes are the following:

Springer Proceedings in Physics, Vol. 44 **Optical Fiber Sensors**
Editors: H.J. Arditty · J.P. Dakin · R.Th. Kersten

1. Low attenuation, allowing remote analysis over many km.

2. Capability to perform distributed measurements again over many km of fiber.

3. Ability to perform a wide variety of measurands via optical fiber.

4. Non-intrusive nature (non-electrical, negligible ignition risk, freedom from lightning damage).

5. Small physical size, even when in armoured form, relative to conventional cables.

6. Highly specific and direct nature of many optical measurements, and high speed of response.

7. Excellent corrosion resistance of optical fibers.

8. Possibility of incorporating self-checking algorithms in optical analysis terminals.

Environmental sensing applications may be conveniently classified, according to the type of area to be monitored. The factors determining the applicability of sensors in these scenarios will now be discussed.

SENSING IN THE HOME

The home environment is where Man spends most of his time, and where he often has a significant exposure to many natural and man-made hazards. However, this is an extremely cost-sensitive area, and therefore the least easy one to employ the first examples of any new technology. Indeed, many of the safety problems in the home tend to be covered by attention to constructural design regulations, rather than by using expensive sensors. Very few monitors, of any kind can be produced at the necessary low costs, the only ones, so far, being smoke and gas sensors of low-cost electrical type. Any optical fiber sensor would require to be of extremely low cost, and probably of a multiplexed nature to find a place in the home. It is unlikely, in the near term,that any fiber sensors will meet the severe cost constraints for home pollution monitoring.

INDUSTRIAL MONITORING WITHIN BUILDINGS

The internal factory environment is the area in which the most extensive use of physical and chemical sensing is made. However, again the cost of individual sensors is generally a major factor, and the constraint of, often-irrational, user conservatism can often be an important consideration. However, modern industry is tending to evolve more rapidly than previously, and is making increasing use of a greater variety of automatic monitoring and control systems. These lead to greatly reduced manufacturing costs and superior quality control.

In the environmental areas, the twin-drive factors, viz. increasing preventative safety legislation and punative damages for accidental leakages, are producing a strong incentive for new monitoring equipment for hazardous materials. In addition, the control of wastage of both material and energy are of economic importance.

Perhaps the two main factors which will fuel the need for fiber sensors for these applications are their high level of intrinsic safety in flammable areas and the valuable information that can be derived using remote, optical spectroscopic analysis. The methods become more cost effective if the opto-electronics terminal units can be centrally located, in a control area, with the passive sensing heads being sequentially polled using a low-loss optical switch. A typical layout is shown in Fig 1. Clearly, the switch must be capable of good repeatability characteristics, as otherwise, some form of alternative referencing method (for example two-wavelength referencing) will be necessary. York has already conducted networking experiments with a commercial switch ("Switchmaster"), of their own design, and achieved a reproducibility of less than 0.1 dB for their multimode version, with feedback-control optimisation.

The two main applications of switch-multiplexed networks, are monitoring of pollution, or analysis of chemical process plant, using fiber harnesses with passive optical sensing heads. Many of the traditional spectrophotometer procedures may be produced in optical fiber remote form,

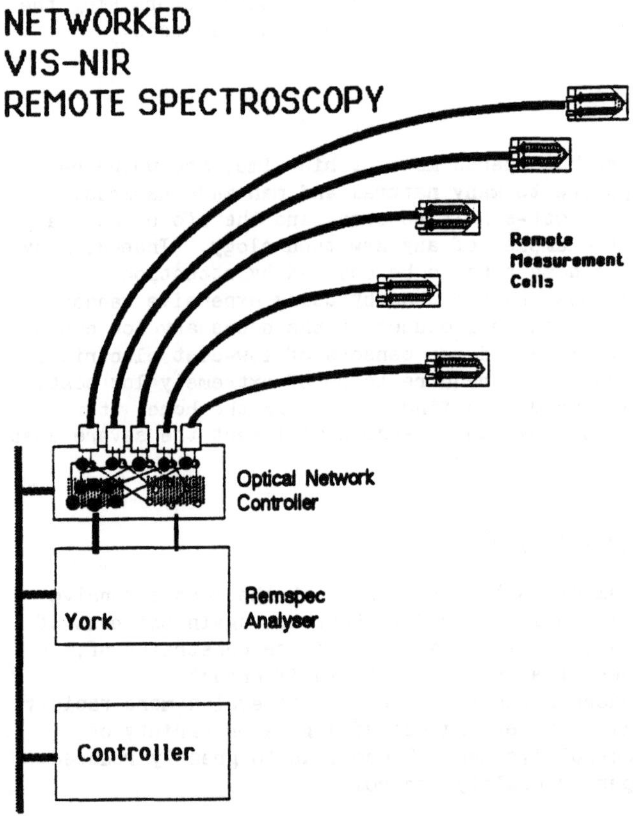

Fig. 1. A schematic illustration of a networked remote spectroscopy system, using a computer controlled switch (York Switchmaster) and fiber-compatible remote spectrometer (York Remspec)

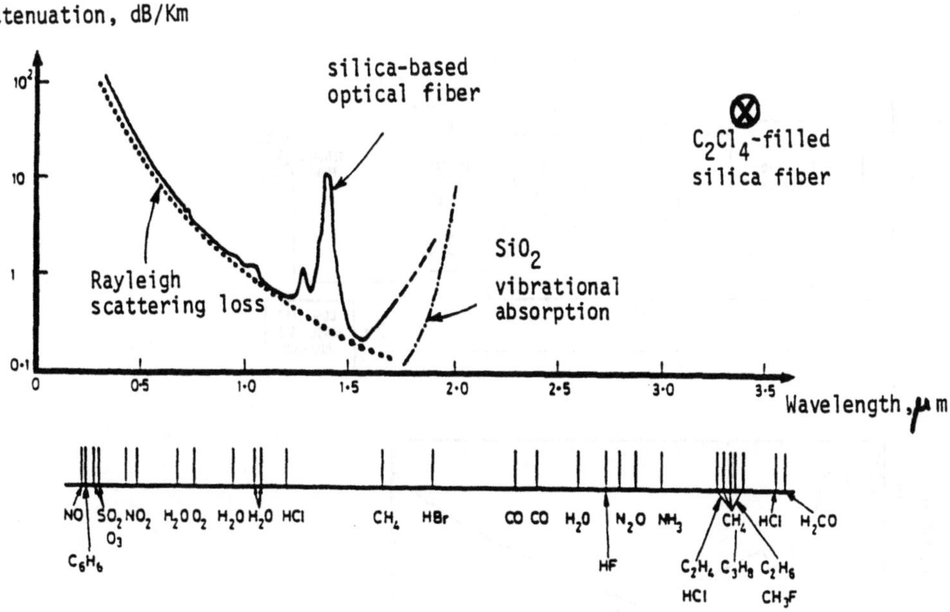

Fig. 2. Attenuation of silica-based fiber (solid line) and the limits imposed by Rayleigh scattering (dotted) and SiO_2 vibrational absorption (dot-dash). Lower scale shows major gas absorption lines of common pollutants. (Diagram is re-drawn from reference 1, by Inaba.)

Note More recently developed fluoride fibers extend the measurable range to approximately 4.0μm in the infra-red region, where absorption lines are much stronger.

removing the need for batch sampling and subsequent analysis. Thus, with such a system, a wide variety of types of environmental monitoring may be performed, provided the media of interest are capable of spectrophotometric anslysis, at the levels required and within the spectral transmission window of the fiber used.

The range of gases that can be monitored using conventional silica optical fibers is shown in Fig 2. This is redrawn from an original paper by Inaba, (Ref 1) who was one of the first to suggest the use of optical fibers for pollution gas monitoring. Since Inaba's early paper, many more techniques have been devised to improve the sensitivity and stability of gas sensing, with the main emphasis being on flammable gases, because of their explosion hazard (Refs 2,3). The optical detection of gases is usually achieved via their electronic or vibrational (or rotational), absorption lines, which are strongest in the ultra-violet and infra-red regions respectively. thus, the availability of special fibers, with enhanced UV or IR transmission, is likely to expand the range of gases which may be viably detected.

Many infra-red fibers (such as fluoride glass types, with transmission out to beyond 4 μm) are already available (Ref 4). However, most commercially-available types appear to have a rather high cost and there are still doubts regarding their mechanical and chemical durability, compared to the well-established silica-based type. Ultra-violet transmitting fibers, with

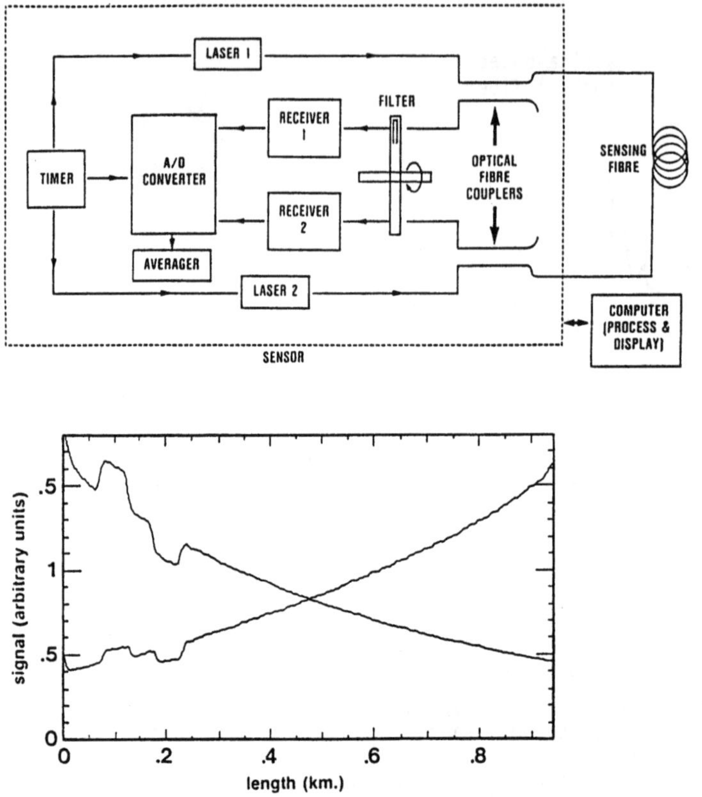

Fig. 3. Schematic diagram of an early version of a commercial distributed temperature sensor system (ref 10). The lower curve shows the measurement of Raman anti-Stokes signals from opposite ends of a fiber, to allow correction for fiber variations. (More recent versions have no moving parts, but allow multiplexing of measurement fibers.)

either high-OH-content silica, or with liquid-filled cores have been available for some time. The use of distributed sensors for monitoring industrial environments is likely to occur in two main areas. The first is that of distributed temperature sensing, which can form the basis of a wide variety of sensors for process monitoring and for estimation of thermal efficiency (or conversely the degree of energy wastage). In addition there are sensors capable of distributed sensing of polluting chemicals. The latter pollutants may be gases, capable of modifying the evanescent field propagation in a reactive fiber cladding (Refs 3,5), or may be radioactive materials, which may more directly modify the absorption loss of the fiber core material, by means of gamma-ray or neutron interactions (Ref 6). In most distributed sensors, some form of Rayleigh scattering (Ref 7,8) or Raman-scattering (Ref 9,10) optical-time- domain reflectometer will be utilised. The first distributed sensor to be commercially available is the Raman distributed temperature sensor (Fig 3). (Diagram taken from an early publication of the basic method.) This system is being considered for a wide variety of thermal monitoring applications in industrial environments. The instrument has significant potential for monitoring (and hence enhancing) the thermal efficiency of industrial plant.

OPEN-AIR AND SPACE SENSING

It is in open-air sensing that the low loss of optical fibers may be used to best effect, to achieve long-range measurements over wide areas. In the spectral region between 1.0 μm and 1.7 μm, the attenuation of silica fibers lies mostly below 1.0 dB/km and may drop to as low as 0.2 dB/km in the 1.45 - 1.6 μm region. Thus, it is possible to carry out spectroscopic measurements on materials, which absorb in the low-loss window of the fiber, at distances up to 10 km, or so, if it were desired.

The main applications for open air gas sensors are currently for methane sensing on oil and gas drilling rigs (Ref 11). However, there is potential demand for control of emission from chemical plants and from exhaust chimney stacks. The main differences, when compared to indoor monitoring, are the greater fiber cable distances that may be involved, and the lower concentrations of gas that will usually be present in the dispersed outdoor environment. As with in-factory monitoring, the capability to multiplex a number of passive networks, using a stable, low-loss switch, is likely to be of major advantage. The development of distributed sensors, using reactive claddings, may be an attractive proposition for the simultaneous monitoring of hazardous leaks of higher concentration.

The use of fiber optics for extra-terrestrial sensing of environmental aspects is still a rather speculative concept. However, the detection of ionising radiation, using long fiber probe lines, could be an attractive research tool. One advantage would be the long length of fiber that could be deployed in weightless conditions. It could, for example, be envisaged to use fiber umbilicals, suspended from orbiting spacecraft, to probe the composition of the outer atmospheric layers of planets. The low effective weight, in these conditions, and the high strength/weight ratios of fibers, would mean that the fiber length would be more likely to be determined by fiber attenuation than by risk of fiber breakage.

An aspect of fiber sensing, which has not yet been mentioned, is that of particle sensing. There are many conventional optical particle sensing methods, based on scattering of focussed optical beams, which may find counterparts in fiber-remoted form. The methods, likely to be most applicable to airborne particle sensing, are those in which scattering from individual particles is monitored, as they pass through a narrow focussed beam. The use of such systems for the detection of smoke, fog and smog may be attractive.

MARINE AND WATER SUPPLY PURITY SENSING

The use of fiber optics in the marine environment has many advantages, due to the lack of electrical signals in a conducting-liquid environment and the high immunity to salt-water corrosion. In addition, the small cross-sectional area can reduce the viscous drag of towed cables. The use of fiber probes may potentially detect a wide variety of polluting conditions, or, with the use of deliberate markers (eg. dyes), be used to examine tidal flow patterns (Ref 12).

As before, the use of a multiplexing switch can ensure the most cost-effective use of a relatively sophisticated measurement station, and again, the use of distributed sensors could have attractions, if suitable fiber claddings can be developed.

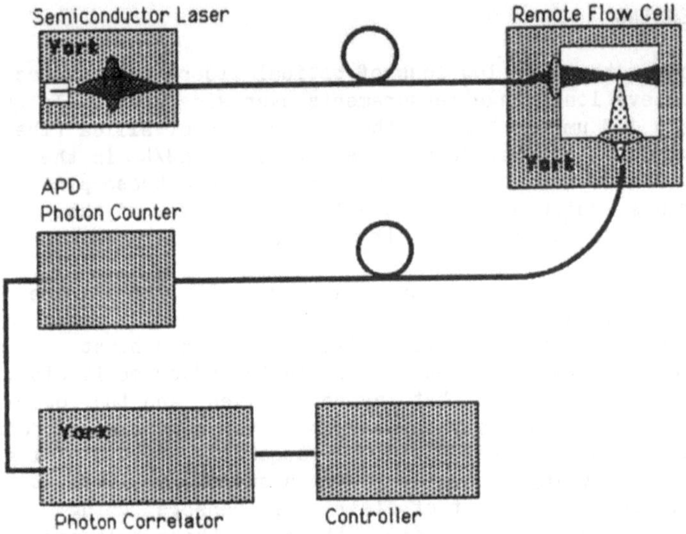

Figure4. **Schematic diagram of a photon correlation spectroscopy apparatus using remote fibre-optic probes.**

One major difference between gaseous and aqueous spectral absorption lines, is that the former are often very narrow, whereas the latter are usually broadened significantly, due to the field interactions arising from the close proximity of solvent molecules. Thus, only species causing a relatively major change in the absorption level will usually be detectable in the case of marine sensing.

However, the aqueous environment is particularly well suited for scattering sensors designed to monitor suspended particulate matter, as the flotation aspects of this denser medium, and the stronger influence of molecular collisions, greatly favour the production of stable turbid suspensions. Thus, aqueous suspensions more readily permit the use of light-scattering techniques, such as photon-correlation spectroscopy, using optical fiber links (Ref 13, Fig.4). This method for particle size analysis involves illumination of a small scattering volume, and the observation of scattered light from the volume. The changes in scattered intensity, due to interference of scattered light from many individual particles, each moving relative to its neighbours, as a result of Brownian motion, allows a measure of the particle size to be calculated. For this method to give accurate results, it is necessary to have a sufficiently high number of particles in the sample volume, to ensure that scattered intensity changes occur primarily due to particle motion, rather than as a result of changes in the number of particles in the scattering volume.

CONCLUSIONS

This paper has attempted to set the scene, by describing the application areas in which fiber optic environmental sensors might be used. The lecture presentation will describe the examples given in more detail, and will present examples of further sensor types, to expand on the treatment above.

In spite of the early stage of development of many optical fiber sensors for this area, there would appear to be significant research potential and, provided the systems are well engineered, they may offer significant advantages over more conventional solutions. The economic success of fiber-based systems is likely to depend on the viability of multiplexed or distributed approaches to measurement problems, and the reliability of the optical solutions, in application areas where the probes may often be subject to surface contamination.

REFERENCES

1. Inaba H et al., "Optical fibre network system for air pollution monitoring over a wide area, by optical absorption method" Electron. Letts. 15 (1979) pp 749-751
2. Hordvik A, Berg A, Thingbo D, "A fibre optic gas detection system" Proc 9th Int. Conf. on Optical Communications,"ECOC 83" Geneva, 1983
3. Dakin J P, "Review of fibre optic gas sensors", Proc SPIE vol 1 011 pp 173-182
4. France P W et al., "Status and prospects of mid I-R fibres" Proc 14th Int. Conf. on Optical Communications, "ECOC 88", Brighton 1988. IEE conf. pub. no. 292 pp 428-432
5. Liebermann R A et al., "Distributed fluorescence oxygen sensor" Proc "OFS 88" Int Conf. on Optical Fibre Sensors, New Orleans 1988, pp 346-348
6. Gaebler W, Braunig D, "Application of optical fibre waveguides in radiation dosimetry" Proc "OFS 83" Int. Conf. on Optical Fibre Sensors, London 1983
7. Dakin J P, "Multiplexed and distributed optical fibre sensors systems", J. Phys E 20 (1987) pp 954-967
8. Hartog A H, "A distributed temperature sensor based on liquid-core optical fibres", J. Lightwave Tech. LT-1 (1983) pp 498-509
9. Dakin J P et al., "Distributed optical fibre Raman temperature sensor, using a semi-conductor light source and detector", Electron. Letts. 21 , (1985) pp 569-570
10. Hartog A H et al., Electron. Letts. 21 (1985) pp 1061-1063
11. Stueflotten S et al., "An infra-red fibre optic gas detection system" Proc. "OFS 84", Int, Conf. on Optical Fibre Sensors, Stuttgart (1984) pp 87-90
12. Dakin J P, King A J, "Limitations of a single optical fibre fluorometer system due to background fluorescence", Proc "OFS 83", 1st Int. Conf. on Optical Fibre Sensors, London, IEE conf. pub. no.221 pp 195-199
13. Chow K M et al., "On line photon correlation spectroscopy using fibre optic probes" J. Phys. E 21 , (1988) pp 1186-1190

Multimode Fiber-Optic Vibrometer

G. Conforti, M. Brenci, A. Mencaglia, A.G. Mignani, and A.M. Scheggi

Istituto di Ricerca sulle Onde Elettromagnetiche, Via Panciatichi, 64, I-50127 Firenze, Italy

A custom-tailored fiber-optic sensor designed for 100 Hz vibration measurements in electrical generators is presented. The relative displacements of an elastic cantilever relative to the body being tested are detected using the optical triangulation technique. PCS optical fibers and GRIN lenses are used. Down-lead independent response is obtained, together with high sensitivity.

Introduction

The measurement of mechanical vibrations plays an important role in industrial machinery. In particular, remote monitoring of the device conditions during operation makes for quick detection of local warning situations. Such monitoring is usually performed suitably housing several small vibration sensors inside the device.

Conventional vibrometers, based on piezoelectric or capacitive working principles, cannot be used advantageously in the presence of high voltages and/or e.m.i.. Optical sensors overcome these drawbacks, although they utilize more complex and expensive technologies. In addition to optical sensors based on interferometric techniques or on laser doppler velocimetry, there exist other optical sensors which utilize transducers connected to the vibrating body, in order to modulate the light transmitted by optical fibers. Typical examples are those based on gratings, microbendings and photoelastic transducers (Refs. 1–7).

This paper deals with the description of a multimode fiber vibrometer especially designed to work inside a turboalternator. For this purpose, it must fit the specific mechanical and electrical requirements of such a device. However, the proposed sensor can be easily applied to different systems.

Requirements

The optical sensor has to be inserted into a slot on the stator of an electrical generator. Hence, severe insulation properties and immunity from e.m.i. are essential. In addition, the electronics must be located far away from the hazardous area.

Fig.1 shows a sketch of the particular portion of the stator where the vibrometer is inserted. The sensor is fixed to a conductive bar inside a ventilation channel between the core laminations of the stator. Geometrical constraints limit the sensor dimensions within 35mm x 12.5mm in the plane perpendicular to the vibration axis. There are no particularly restrictive constraints along the direction of vibration.

Since the main interest is focused on the electrodinamically-induced vibrations, the sensor has been designed for a fixed frequency of 100 Hz (for the European line frequency 50 Hz). A dynamic range up to a peak vibration amplitude of 0.25 mm, a resolution better than 1 µm, and ±3% linearity are the technical requirements.

Springer Proceedings in Physics, Vol. 44 **Optical Fiber Sensors**
Editors: H.J. Arditty · J.P. Dakin · R.Th. Kersten

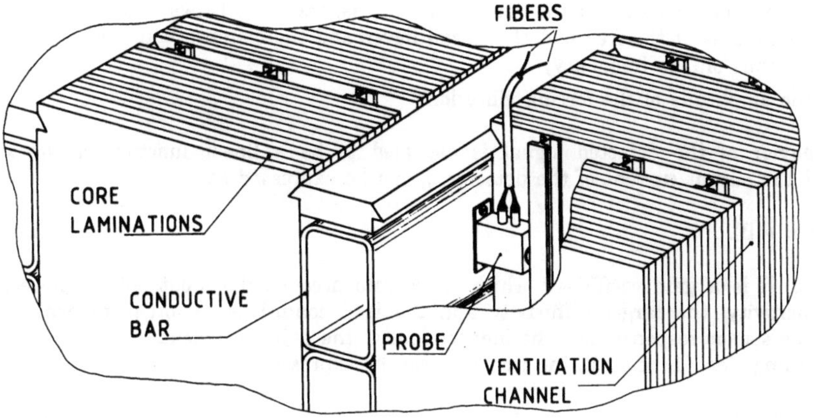

Fig. 1. Sketch of the stator showing the sensor housing.

Working principle

The vibrometer is a multimode optical fiber sensor based on light intensity modulation. The modulation is induced by an elastic cantilever joined to the body being tested and having a diffusive target on its free end.

The working principle is based on the position detection of the cantilever free end by a transmissive fiber optic device which uses the triangulation technique (Ref. 8). The required sensing range must be located a few millimeters from the optics with a linear behavior as wide as possible. The optical head is made of two optical fibers coupled to GRIN rod microlenses in a suitable geometrical arrangement (Fig.2).

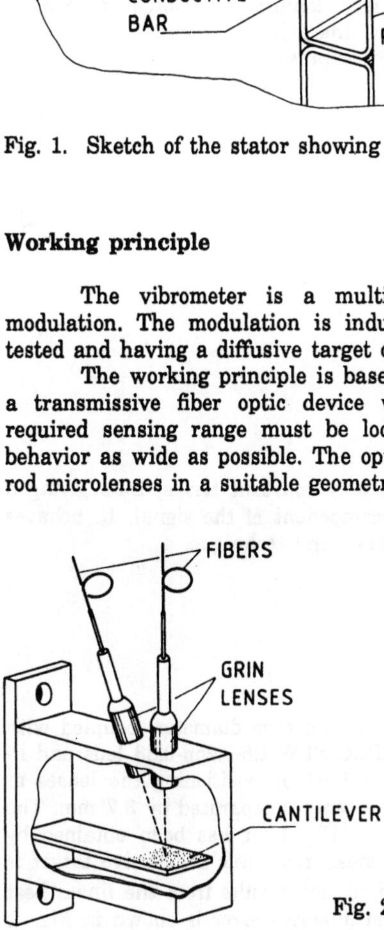

Fig. 2. Sketch of the probe.

When considering a sinusoidal forcing vibration, the current position of the cantilever free end with respect to the optics, x(t), can be expressed as

$$x(t) = X_0 + G(f)\ Z \sin(2\pi ft) \tag{1}$$

where

X_0 the cantilever position at rest;

f the frequency of vibration;

G the mechanical gain of the cantilever, that is, the ratio between the vibration amplitudes of the free end of the cantilever and the clamped end: such a gain depends on the frequency;

Z the amplitude of the forcing vibration.

The behavior of the detected signal is assumed to be a linear function of the x position, so that the output current of the detector, i, can be expressed as

$$i = h\,(A\,x + B). \qquad (2)$$

In Eq. (2) h is a gain coefficient which takes into account the input optical power, the detector sensitivity, the target diffusivity and the lead conditions. In order to achieve good accuracy, the sensor response must be independent of the gain coefficient.

By combining Eqs. (1) and (2), the i current can be expressed as

$$i = I_0 + I_1 \sin(2\pi f t) \qquad (3)$$

where

$$I_0 = h\,(A\,X_0 + B) \quad \text{and} \quad I_1 = h\,A\,G(f)\,Z. \qquad (4)$$

Utilizing Eqs. (4), the vibration amplitude is expressed by

$$Z = k\,I_1/I_0 \qquad (5)$$

where

$$k = (X_0 + B/A)/G(f). \qquad (6)$$

At a given frequency, the k coefficient contains only constant terms, thus giving a sensor response independent of lead conditions. The dc component of the signal, I_0, behaves like a reference channel, making possible the use of a ratiometric technique.

Prototype description

a) Optical probe

The optical head makes use of two PCS fibers (200 µm core diameter) coupled with GRIN rod microlenses by means of optical aligners (SELFOC SLW-1.8-0.25p-0.83 lens and P-OPAL aligner). A suitable geometrical arrangement is obtained by positioning the lenses at an angle of 13° between their axes, and with the fiber-end centers separated by 3.7 mm. The static response curve of such an optical head is shown in Fig. 3; it has been obtained by means of a computer-based system. The output power is measured while a diffusive target is moved by a micropositioning stage (2 µm step). The straight line results from the linear best fit of the experimental curve in the 7.4÷9 mm range. The linearity error is shown in Fig. 4: it is less than ±2% in the whole range, but decreases to ±1% in the reduced 7.5÷8.9 mm range. In order to obtain the lower linearity error for the sensor response, the cantilever position at rest, which is also its average position, is fixed at $X_0 \approx 8.2$ mm.

The elastic cantilever consists of a 100 µm-thick mica slice. At its free end it is loaded with a small weight, together with a diffusive target, so as to obtain the proper gain at 100 Hz. In particular, for vibration measurements up to a 0.25 mm amplitude, giving rise to cantilever displacements in the 7.5÷8.9 mm range, G(@ 100 Hz) = 2.8 has been chosen.

The sensor case is made of polyacetal resin and weighs 12 g; its dimensions are 25 mm height, 31 mm length and 10 mm thickness.

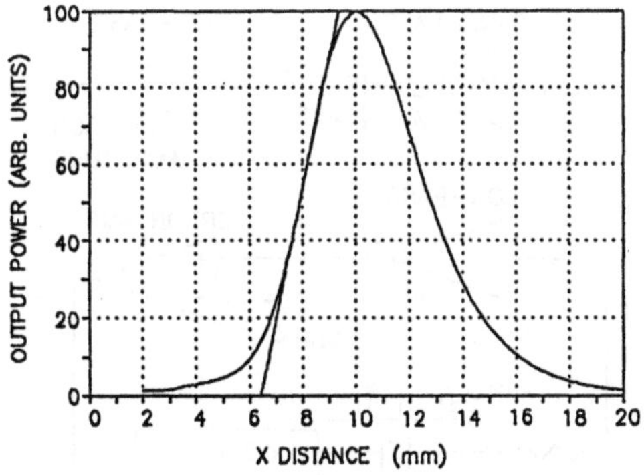

Fig. 3. Static response of the optical head.

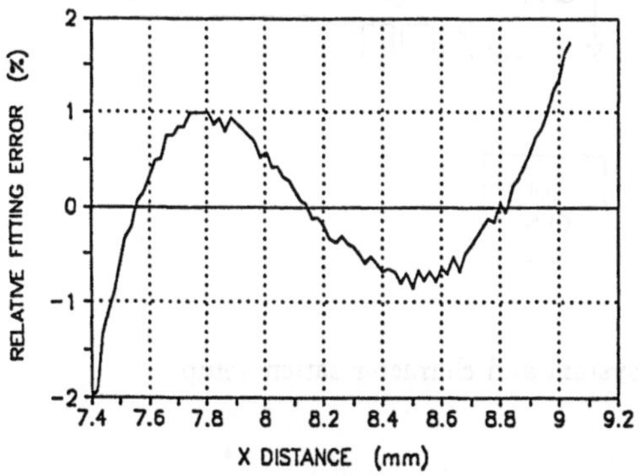

Fig. 4. Linearity error of the static response.

b) Electro-optical system

As shown in Eq. (5), the vibration amplitude Z is proportional to the amplitude of the detected ac component I_1, provided the dc component I_0 is kept constant.

First, the detected signal is amplified; then the ac and dc components are obtained, respectively, by a 1 Hz bandpass filter and by a 10 Hz low-pass (24 dB/octave slope) active filter. By means of a full-wave rectifier and a 4 Hz low-pass filter, having 18 dB/octave slope, a dc output signal proportional to the vibration amplitude is obtained.

In order to keep constant the dc component, a feedback network is utilized (Fig. 5). The LED driver supplies current which is proportional to the voltage across the storage capacitor. This voltage is regulated by positive or negative pulse trains generated by the VCO. The repetitive frequency of the pulses is proportional to the error signal, allowing an high loop-gain together with a stable operation. The V_{ref} voltage can be adjusted in order to obtain the desired dc component.

The light source is a Radiall LED (860 nm wavelength) which couples about 300 µW into the fiber; the average detected light is about 25 nW.

Vibrometer characterization

The experimental setup used for sensor calibration is shown in Fig. 5. The optical sensor is fixed to a Brüel & Kjær shaker table (model 4810); a piezoelectric accelerometer with 2% accuracy is used as reference. The shaker table is driven at 100 Hz, and the vibration amplitude is changed in the 0÷0.25 mm range. Both optical and reference sensor outputs are measured. The ac gain of the optical vibrometer is adjusted to minimize the mean relative quadratic error. A final sensitivity of 10 mV/µm is obtained.

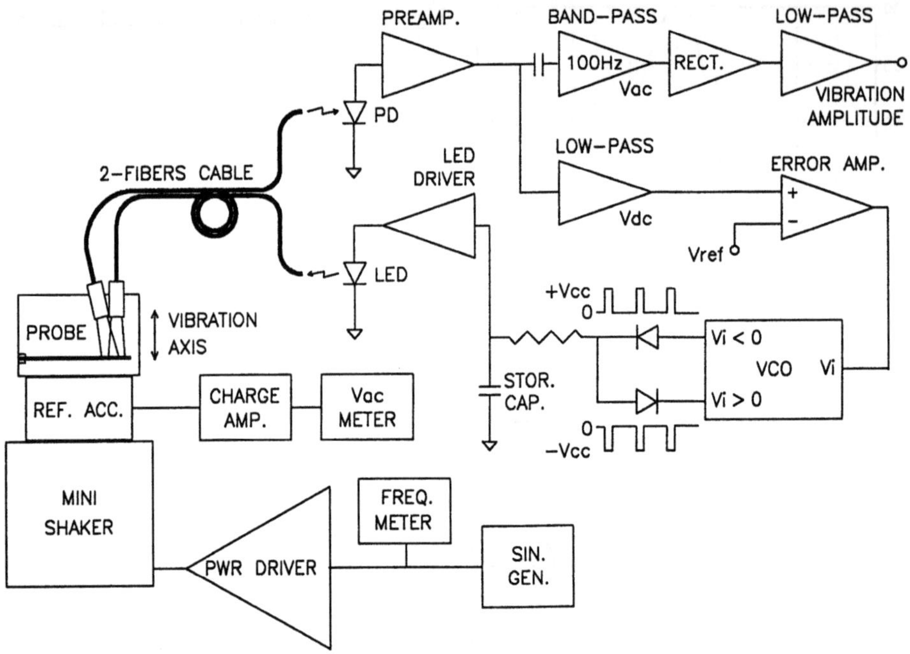

Fig. 5. Block diagram of the electro-optical system and characterization setup.

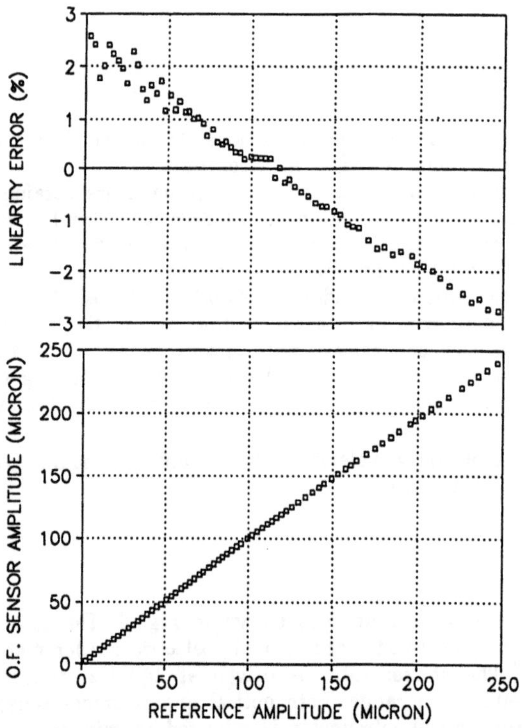

Fig. 6. Vibrometer response (bottom) and linearity error (top).

Fig. 6 displays both fiber-optic vibrometer output (bottom) and the relative linearity error (top) versus the reference output. As shown, the error of the fiber-optic sensor is less than 3%, corresponding to an absolute error of less than 1.5 μm up to a 150 μm vibration amplitude.

It is significant that the relative error is quite linear, so that the effective error could be reduced to the spread around the line of best fit of the experimental points. This could be easily accomplished by a computer-based system, which would provide data acquisition and perform the suitable correction to the linear response. In the present case, where the hardware does not make use of a microcomputer, this goal is difficult to achieve. However, a ±3% maximum error satisfies customer requirements.

Transverse sensitivities with respect to the axes parallel and perpendicular to the cantilever axis have also been measured. Their results are, respectively, less than 0.4 mV/μm and 0.05 mV/μm.

Sensor response is somewhat sensitive to temperature fluctuations. The temperature coefficient has been measured at 0.2 mm vibration amplitude, and it resulted 0.2 %/°C in the 25°÷70°C temperature range.

Conclusions

Figure 7 shows the proposed optical fiber vibrometer, together with the optoelectronic unit. For safe working in the electric plants, 30 m of cabled fibers are used. This prototype has been working for several months on an experimental turboalternator.

The described design criteria allow for modification of the sensor in order to satisfy other requirements. Firstly, the working frequency can be changed to 120 Hz, for the 60 Hz line frequency. Moreover, the use of a cantilever with a resonance frequency closer to the working frequency allows achieving an higher sensitivity, naturally with a reduced amplitude range.

In addition, the probe can be used as an accelerometer at working frequencies which are sufficiently low with respect to the resonance frequency. In fact, it can be proved that, at low frequencies, the G(f) cantilever gain approaches zero as f^2.

The present work has been partially sponsored by S.C.M., via S.Francesco 16, Trieste, Italy, which is now producing the vibration sensor under C.N.R. licence (Italian Patent #9369 A/88, to be extended abroad).

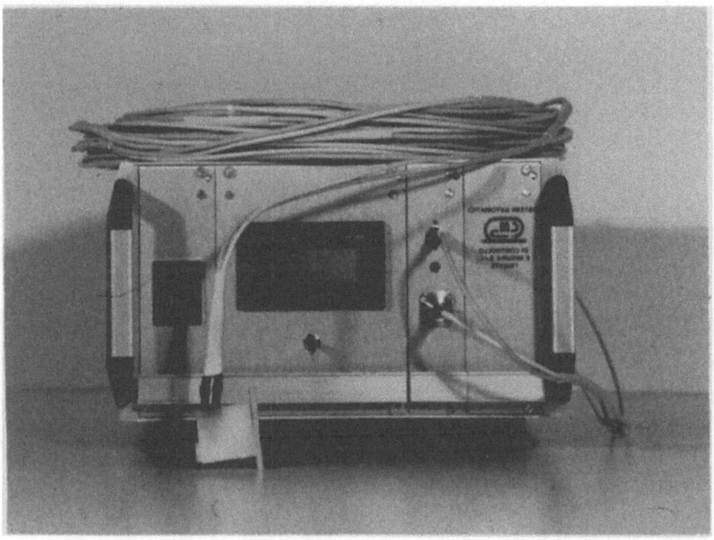

Fig. 7. Vibrometer probe and the electro-optic unit.

References

1. R.I. Laming, M.P. Gold, D.N. Payne and N.A. Halliwell, "Fibre-optic vibration probe", *Electron. Lett.*, **22**, 167, (1986).

2. C. Ovren, M. Adolfsson and B. Hök, "Fiber-optic systems for temperature and vibration measurements in industrial applications", *Opt. & Laser Eng.*, **5**, 155, (1984).

3. Y. Hirose, T. Abe, H. Komai and Y. Tsuzuki, "Optical fibre vibration sensing system", *Trans. Inst. Electron. Inf. & Com. Eng.*, **E70**, 195, (1987).

4. K.Kyuma, S.Tai, K.Hamanaka and M.Nunoshita, "Laser doppler velocimeter with a novel optical fiber probe", *Appl. Opt.*, **20**, 2424, (1981).

5. D.R.Miers, D.Raj and J.W.Berthold, "Design and characterization of fiber-optic accelerometers", *Fiber Optic and Laser Sensors V*, R.P.De Paula and E.Udd ed., Proc. **SPIE 838**, 314, (1987).

6. E.Udd and P.A.Turek, "Single-mode fiber-optic vibration sensor", *Fiber Optic and Laser Sensors III*, E.L.Moore and O.G.Ramer ed., Proc. **SPIE 566**, 135, (1985).

7. K.Yoshinaga, Y.Ida, M.Imataki and T.Takioka,"Fiber-otic vibration sensor using photoelastic substance", *OFS'86 Technical Digest*, edited by Inst. of Electr. and Comm. Eng. of Japan (Tokyo), p.31, (1986).

8. M. Brenci, G. Conforti, A. Mencaglia, A.G. Mignani, A.M. Scheggi,"Fibre-optic position sensor array", *Int. J. Optoelectron.*, **3**, 473, (1988).

An Optical Hot-Fiber Anemometer

A. Samouris [1], *J.P. Davis* [1], *L.C. Bobb* [1], *and D.C. Larson* [2]

[1]Naval Air Development Center, Warminster, PA 18974, USA

[2]Department of Physics and Atmospheric Science, Drexel University, Philadelphia, PA 19104, USA

An optical fiber anemometer has been developed, and used to measure the flow rate of nitrogen gas at flow rates ranging from 0.1 to 2 m/s. The anemometer employs a gold-coated single-mode optical fiber which forms one arm of a Mach-Zehnder interferometer. The device can measure gas flows with greater precision than a conventional hot-wire device.

INTRODUCTION

The flow rates of gases are commonly measured with a hot-wire anemometer which employs a fine wire mounted transverse to the gas flow. The wire is heated with an electrical current and its temperature rise is determined by measuring the electrical resistance of the wire which varies inversely with the flow rate. In the present optical fiber anemometer the fine wire is replaced by a short segment of gold-coated optical fiber forming one arm of an all-fiber Mach-Zehnder interferometer. The fiber is heated by passing a current through the gold coating. The increase in temperature results in an increase in both the length and refractive index of the fiber core, and a corresponding increase in the phase of the light propagating through the sensing arm of the interferometer. The magnitude of this phase change can be used to determine the flow rate. An optical fiber interferometer can measure phase changes with greater precision than resistance changes can be measured with a conventional hot-wire anemometer, and as a result allows more accurate measurements of flow rates.

Optical fibers have been used previously as flow or velocity sensors. Pitt et. al.[1] have discussed several types of optical fiber flowmeters ranging from simple pulsed interruption methods with turbine impeller blades to vortex shedding and correlation techniques. Fiber optic techniques have also been utilized in laser-Doppler anemometry.[2] A catheter-based medical sensor for blood flow measurements has been developed which utilizes a multimode fiber with a temperature sensitive phosphor and infrared absorber attached to the fiber tip.[3] The tip is heated with near infrared radiation through the same optical fiber that optically communicates with the temperature sensor. The change in temperature produced by the infrared radiation is inversely dependent upon the flow rate of the fluid past the sensor. This latter device bears some similarity to our hot-fiber device, but in general previous optical fiber flow sensors are extrinsic sensors where the optical fiber is used to transmit the light to or from the sensor, and not as the sensing element.

We have previously discussed the application of gold-coated single-mode fibers to thermal modulation and to thermal conductivity sensing.[4,5] The thermal modulation application derives from the inherent sensitivity of the optical fibers to temperature when they are in an interferometer setup. In the reference arm of a Mach-Zehnder interferometer, a gold-coated fiber can be heated with a very small voltage to compensate for drift. A comparable phase shift required a much larger voltage applied to a much bulkier PZT stretcher. As a thermal conductivity sensor, the gold-coated fiber is heated in a liquid ambient and the resultant short-time temperature rise is inversely proportional to the thermal conductivity of the medium. The flow sensor described in this paper is very similar to the thermal conductivity sensor, and indeed the hot-fiber configuration can be employed additionally for air pressure and fluid flow measurements.

Editors: H.J. Arditty · J.P. Dakin · R.Th. Kersten

EXPERIMENTAL APPARATUS

An ITT Type T-1601 single-mode optical fiber was used in this study. It has a core diameter of 4 µm and a cladding diameter of 85µm. A 0.1 µm layer of gold was sputtered onto an unjacketed 1 cm length of the silica fiber, and the fiber was inserted into a Mach-Zehnder interferometer which is shown schematically in Fig.1. The interferometer was fabricated from two 2x2 Amphenol 945 single-mode couplers. Light from a single-frequency helium-neon laser was focused onto the core of the input fiber and was split evenly between the two arms of the interferometer at the first coupler. One arm of the interferometer employed a PZT stretcher to maintain the interferometer at quadrature. The other arm of the interferometer contained the flow cell. The light from both arms was combined at the second coupler and collected by photodiodes at the output. The flow cell configuration is shown in the inset of Fig.1. The gold-coated fiber was inserted transversely into a 1 cm diameter tube through which nitrogen was directed. It was connected to the leads with silver paint.

The fiber was heated resistively by applying a repetitive 5-sec square-wave voltage with a power of 0.0976 W and a 50% duty cycle. The interferometer output was recorded on an HP 3561A Dynamic Signal Analyzer. In this manner it was possible to measure the actual phase shift by counting the number of fringes produced upon application and/or removal of the square-wave voltage.

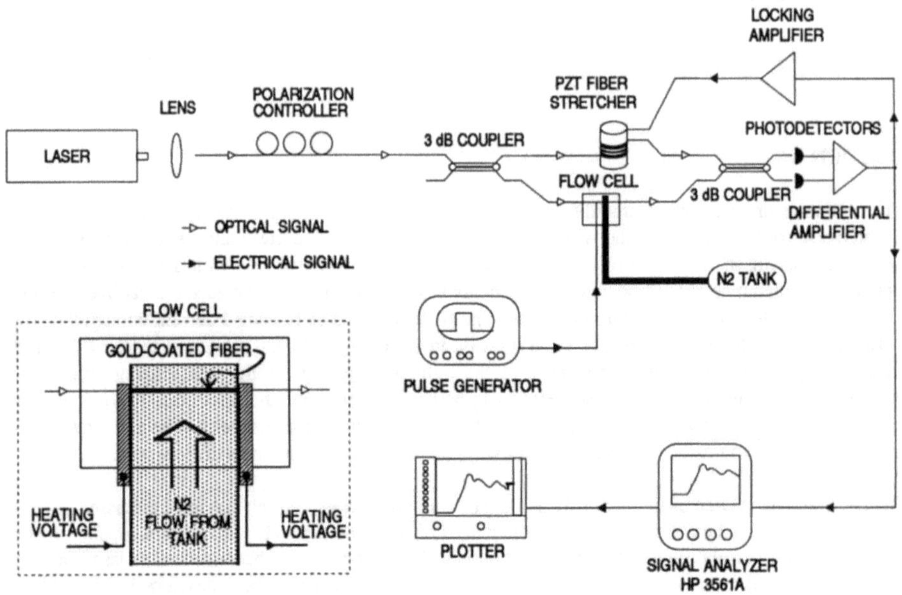

Figure 1. Schematic of the optical hot-fiber anemometer

TEMPERATURE CALIBRATION

In order to assess the usefulness of this sensor to determine the flow rates of various fluids it is useful to know the temperature rise of the gold-coated fiber upon application of the voltage pulses. The temperature sensitivity of the gold-coated fiber was determined both theoretically and experimentally. The phase of a wave propagating in a fiber of length L is given by

$$\phi = 2\pi nL/\lambda \qquad (1)$$

where n is the effective refractive index, which may be approximated by the refractive index of the core, and λ is the wavelength of the light in free space. A change in fiber temperature ΔT results in a phase shift $\Delta\phi$ of the light in the fiber because of the temperature-induced change in the refractive index of the fiber core, the change in fiber length due to thermal expansion, and the photoelastic effect. The sensitivity of the gold-coated section of the fiber may be expressed as

$$\Delta\phi /(L\Delta T) = (2\pi/\lambda)\{(\partial n/\partial T)_\rho + n\,\varepsilon_z/\Delta T - (n^3/2\Delta T)[(P_{11} + P_{12})\,\varepsilon_r + P_{12}\,\varepsilon_z]\} \quad (2)$$

where ρ is the core density, ε_z and ε_r are the axial and radial strains, and P_{11} and P_{12} are the Pockels coefficients. In the analysis, the fiber was considered to consist of four concentric layers, the core, the cladding, the substrate, and the 0.1 μm gold coating. The strains resulting from the temperature change ΔT were calculated using the method of Schuetz et al.[6] The sensitivity was found to be $\Delta\phi/2\pi L\Delta T = 16.3$ fringes/m-K where the first term in Eq. (2) provided the major contribution.

The temperature sensitivity of the gold-coated fiber was also determined experimentally by monitoring the resistance change of the gold film when heated by a steady current. The temperature change was then calculated as

$$\Delta T = \Delta R/R\alpha \quad (3)$$

where ΔR is the resistance change corresponding to a temperature change ΔT, R is the room temperature resistance, and α is the temperature coefficient of resistivity. A value for α was determined experimentally be placing the gold-coated portion of the fiber in a furnace and monitoring the resistance as the temperature was increased. These experimental results were then used to obtain the experimental temperature sensitivity of

$$\Delta\phi/2\pi L\Delta T = 15.9 \text{ fringes/mK} \quad (4)$$

The experimental value is in reasonable agreement with the theoretical value, and is the value used in the thermal conductivity analyses. The greatest experimental uncertainty is the accurate determination of the coated length L of the fiber.

CONVECTION MODEL

The heat transfer from the heated fiber generally occurs by the mechanism of forced convection, although thermal conduction and free convection occur at low flow rates and at short times. A simple forced-convection model which assumes uniform fiber temperature and a temperature-independent convection coefficient is useful in interpreting the data. Upon heating the fiber with a square-wave voltage pulse the energy balance for the coated fiber may be written

$$mc\, dT_h/dt = P_o(1 - \alpha T_h) - hAT_h \quad (5)$$

where m and c are the mass and average heat capacity of the material being heated (primarily silica), T_h is the temperature of the fiber (relative to the ambient temperature) produced by heating, P_o is the initial power applied to the coating, h is the convection coefficient, and A is the surface area of the gold coating. The term $(1 - \alpha T_h)$ accounts for the increased resistance of the gold coating upon heating at constant voltage. The solution to Eq.(5) is

$$T_h = T_\infty[1 - \exp(-t/\tau_h)] \quad (6)$$

where $T_\infty = P_o/(hA + P_o\alpha)$ is the maximum temperature change and $\tau_h = mc/(hA + P_o\alpha)$ is the time constant for heating. Upon removal of the voltage pulse, the fiber cools and the temperature (relative to the ambient temperature) upon cooling is obtained from Eq.(5) with $P_o = 0$

$$T_c = T_\infty \exp(-t/\tau_c) \quad (7)$$

where T_∞ is the same as above and $\tau_c = mc/hA$. The numbers of fringes produced upon heating F_h and cooling F_c are proportional to the temperature changes, and are given by the simple convection model as

$$F_h = F_T[1 - \exp(-t/\tau_h)] \quad \text{and} \quad F_c = F_T[1 - \exp(-t/\tau_c)] \quad (8)$$

where F_T is the total number of fringes corresponding to the maximum temperature change T_∞. Fringe numbers are obtained from temperature changes using the calibration constant from Eq.(4).

EXPERIMENTAL RESULTS

A series of experiments was performed where square-wave voltages were applied across the gold-coated optical fiber immersed in a nitrogen gas flow. The change in the fiber temperature as a function of time was determined by monitoring the number and spacing of the interference fringes produced. A sample recording of fringes generated upon the application of a 5-sec 0.9 V pulse in a nitrogen flow rate of 0.38 m/s is shown in Fig.2. The total number of fringes recorded was $F_T = 14.3$ both upon application and removal of the voltage pulse. Using the calibration constant from Eq.(4), the corresponding maximum fiber temperature change is calculated to be 89.9 C°. The number of fringes produced as a function of time upon both heating and cooling is well represented by Eqs.(8) and in general reasonable exponential fits were obtained at all flow rates.

At each flow rate exponential fits were used to determine values of T_∞, τ_h, and τ_c. In Fig.3, τ_h is plotted as a function of T_∞. A straight line relationship is obtained except at the lowest flow rates which correspond to the maximum values of T_∞. At these flow rates forced convection heat transfer is replaced by free convection with a corresponding change in the relationship between the convection coefficient and the flow rate. The slope of the curve is 1.09×10^{-3} s/°C which compares favorably with the simple convection model value of $\tau_h/T_\infty = mc/P_o = 1.04\times10^{-3}$ s/°C.

In Fig. 4 the total number of fringes F_T is plotted as a function of nitrogen flow rate. The flow rates were measured with a conventional hot-wire anemometer. This calibration shows very good sensitivity at the

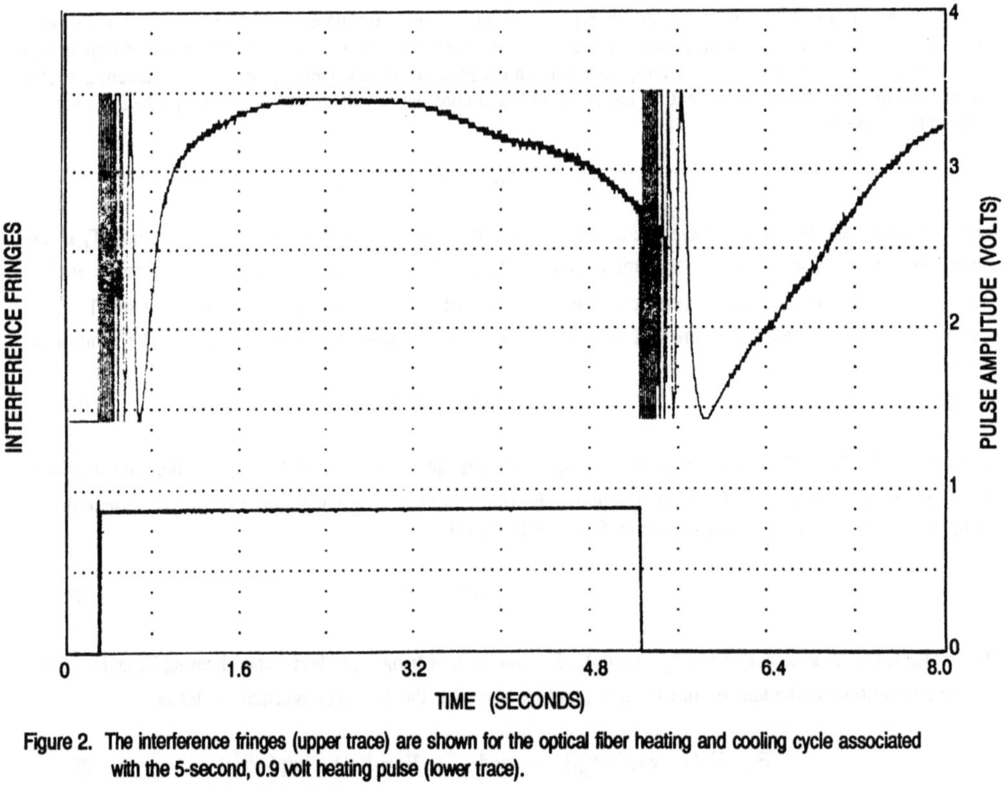

Figure 2. The interference fringes (upper trace) are shown for the optical fiber heating and cooling cycle associated with the 5-second, 0.9 volt heating pulse (lower trace).

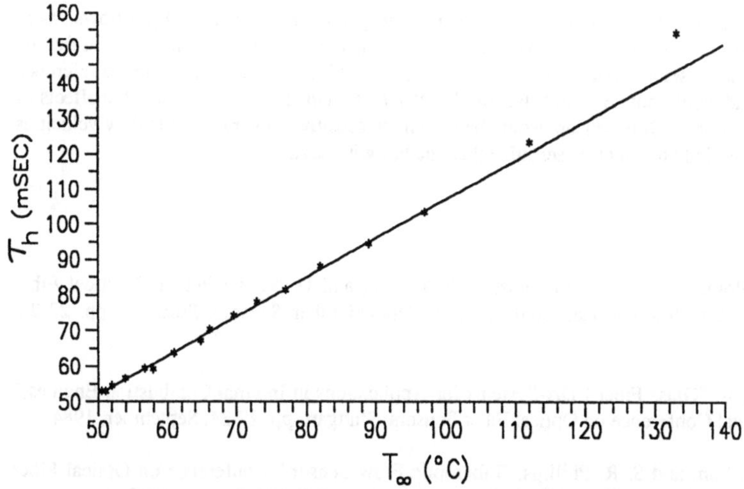

Figure 3. The time constant for heating (τ_h) is plotted as a function of the maximum temperature change (T_∞).
The two highest temperature data points are excluded from the straight line fit.

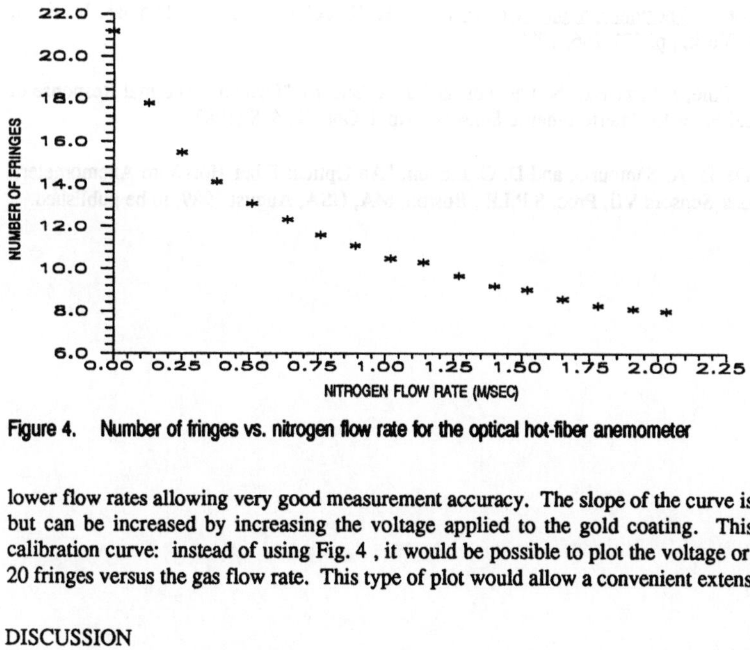

Figure 4. Number of fringes vs. nitrogen flow rate for the optical hot-fiber anemometer

lower flow rates allowing very good measurement accuracy. The slope of the curve is less at higher flow rates, but can be increased by increasing the voltage applied to the gold coating. This suggests an alternative calibration curve: instead of using Fig. 4 , it would be possible to plot the voltage or power required for 15 or 20 fringes versus the gas flow rate. This type of plot would allow a convenient extension to higher flow rates.

DISCUSSION

In a companion publication we have performed a heat transfer analysis of the optical fiber hot-wire anemometer comparing experimental heat transfer coefficients with published values.[7] Using this analysis it is possible to extend the application of the calibration curve to gases other than nitrogen and to temperatures both above and below ambient. In addition the analysis allows a consideration of alternative uses of the 'hot-fiber' device to measure other quantities such as liquid flow rates and air pressure.

In general the 'hot-fiber' device is similar to the conventional hot-wire device. Both devices do not interfere with the gas flow; the hot-fiber sensor has a diameter of 85µm and the hot-wire sensor has a diameter as low as 5 μm. Both devices have very low thermal masses and can respond to rapidly varying gas flows. Both devices require modification in a flow with a variable temperature. The hot-fiber device could be modified by introducing a second gold-coated segment of fiber into the flow stream; this segment would be in the reference arm of the interferometer and would not be heated, but would respond to temperature and cancel the effects of temperature drift in the sensor arm. Both sensors would be extremely sensitive to very low gas flow velocities with the hot-fiber device potentially much more sensitive than the hot-wire device.

REFERENCES

1. G. D. Pitt, A. M. Prabakaran, R. J. Williamson, D. Wilson, and D. N. Batchelder, "Optical Fiber Flowmeters," Proc. 2nd International Conference on Optical Fiber Sensors, Stuttgart, pp. 23-28, September, 1984.

2. F. Durst and H. Krebs, "Glass Fiber LDA-Systems for Applications in Internal Combustion Engines," Proc. 2nd International Conference on Optical Fiber Sensors, Stuttgart, pp. 29-34, September, 1984.

3. M. H. Ikeda, M. H. Sun, and S. R. Phillips, "Fiberoptic Flow Sensor," Conference on Optical Fiber Sensors, New Orleans, LA, USA, Tech. Digest Paper FCC6-1, pp. 438-445, January 1988.

4. B. J. White, J. P. Davis, L. C. Bobb, H. D. Krumboltz, and D. C. Larson, "Optical-Fiber Thermal Modulator," J. Lightwave Tech. **LT-5**, 1169 (1987).

5. B. J. White, J. P. Davis, L. C. Bobb, and D. C. Larson, "Thermal Conductivity Measurements with Optical Fiber Sensors," Thermal Conductivity 20 (D. P. H. Hasselman and J. R. Thomas, Jr., eds.), Plenum Press, New York, pp. 277-286, 1989.

6. L. S. Schuetz, J. H. Cole, J. Jarzynski, N. Lagakos, and J. A. Bucaro, "Dynamic Thermal Response of Single-Mode Optical Fiber for Interferometric Sensors," Appl. Opt. **22,** 478 (1983).

7. L. C. Bobb, J. P. Davis, A. Samouris, and D. C. Larson, "An Optical Fiber Hot-Wire Anemometer," Fiber Optic and Laser Sensors VII, Proc. S.P.I.E., Boston, MA, USA, August 1989, to be published.

Extrinsic Dual-Wavelength Fibre-Optic Based Accelerometer with Common Mode Compensation

A.S. Gerges, T.P. Newson, and D.A. Jackson

Physics Laboratory, University of Kent, Canterbury, Kent CT27NR, UK

We introduce a novel accelerometer, with dynamic range > 10^6 and resolution of $2.5x10^{-7}g$, in which the sensing element is a weighted diaphragm. The acceleration is measured using two miniature hemispherical air-spaced Fabry-Perot interferometers. The interferometer outputs are processed differentially such that the detrimental effects caused by environmental perturbations are strongly minimised.

Introduction

Fibre-optic sensors based on interferometry have been introduced to replace conventional sensing devices for a wide range of measurands [1]. Several fibre-optic accelerometers have been demonstrated, where the single-mode optical fibre itself forms part of the sensing element [2,3,4]. Although these devices offer very high resolution, their performance is critically dependent on the opto-mechanical properties of the fibre.

We have previously demonstrated a type of fibre-optic accelerometer, in which the sensing element was a weighted diaphragm. The displacement of the diaphragm produced by acceleration was measured using a miniature hemispherical air Fabry-Perot interferometer, with the spherical metal mirror attached centrally to the diaphragm [5].

In this work we introduce a new concept for this type of accelerometer such that the deleterious effects of both short and long-term frequency jitter of the optical source as well as environmental effects are strongly reduced. The accelerometer also shows a very small cross-sensitivity to orthogonal components of acceleration.

We also demonstrate how the use of dual wavelength techniques can be used to extend the dynamic range of the accelerometer. Dual wavelength interrogation of interferometers is a well known technique to increase their unambiguous dynamic range, the use of this technique in fibre-optic sensors has been described by several authors [6,7]. An effective wavelength $\lambda_{eff} = \lambda_1 \lambda_2 / (\lambda_2 - \lambda_1)$ is generated, which is longer than either of the wavelengths of the individual sources λ_1 and λ_2 . As $\lambda_2 - \lambda_1$ decreases, the effective wavelength, and hence the unambiguous range of the system increases.

However there are practical considerations which restrict the minimum value of $\lambda_2 - \lambda_1$ which can be used. This is because the error induced in λ_{eff} by wavelength drift, of either source, becomes significant in an interferometer with a finite path-imbalance. We have greatly reduced the problems associated with wavelength drift by the mechanical design of the accelerometer by incorporating dual interferometers of nearly equal optical path difference.

Construction of the accelerometer and transduction mechanism

The construction of the accelerometer head is as shown in figure (1). It basically consists of two identical hemispherical Fabry-Perot interferometers (I_a and I_b) , illuminated via separate monomode optical fibres. Two identical spherical metal mirrors, with radii of curvature of about 8 mm, centrally attached to the diaphragm, one on each side, form the outer mirrors of the interferometers, while the distal ends of the fibre serve as the inner mirrors. The two spherical mirrors act as the mechanical load and form the rigid center of the sensing diaphragm. To reduce the cross-coupling, the mass and the dimensions of the two mirrors are made equal, such that the center of mass of the diaphragm/mirrors assembly lay close to the plane of the diaphragm. The length of the two cavities (l_a and l_b) were adjusted such that the condition of maximum fringe visibility was obtained in each interferometer, this occurs when l_a=l_b [8].

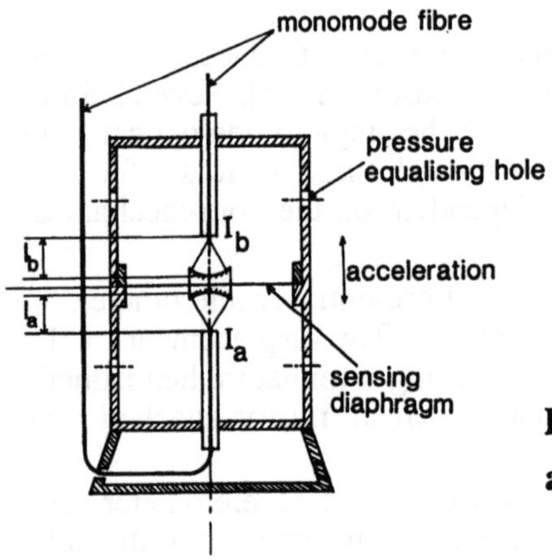

Figure (1) The construction of the accelerometer head.

Within the elastic limit of the diaphragm material, the displacement of the center of the diaphragm, Y_o , is proportional to the applied axial acceleration (r) such that:

$$Y_o = A_s \frac{Mr\, a^2}{16\,\pi\, D}$$

where M is the mass of the diaphragm assembly and a is its radius. D is called the flexural rigidity of the diaphragm, and A_s is a numerical coefficient which depends on its solidity ratio [9].

The optical phase changes induced in interferometer I_a are;

$$\phi_{a1} = \frac{4\pi}{\lambda_1} (l_a + Y_o)$$

$$\phi_{a2} = \frac{4\pi}{\lambda_2}(l_a + Y_o)$$

and in interferometer I_b are;

$$\phi_{b1} = \frac{4\pi}{\lambda_1}(l_b - Y_o)$$

$$\phi_{b2} = \frac{4\pi}{\lambda_2}(l_b - Y_o)$$

As the axial acceleration changes, the corresponding phase changes, induced in each interferometer are equal in magnitude but opposite in sign, such that by measuring the relative phases between the two interferometers (ϕ_{ab1} or ϕ_{ab2}) the sensitivity of the accelerometer is effectively doubled, where :

$$\phi_{ab1} = \frac{8\pi}{\lambda_1} Y_o$$

$$\phi_{ab2} = \frac{8\pi}{\lambda_2} Y_o$$

The difference between these relative phases ($\Delta\phi$) is :

$$\Delta\phi = \phi_{ab1} - \phi_{ab2} = \frac{8\pi}{\lambda_{eff}} Y_o$$

Self compensation of common-mode errors is an another advantage of this configuration apart from doubling the acceleration sensitivity. If the light wavelength drifts by ΔF, the relative phase change between the two interferometers will be $\frac{4\pi\Delta F}{C}(l_a - l_b)$, which vanishes as $l_a = l_b$ (ignoring second order effects due to the movement of the diaphragm $Y_o \ll l_a$ or l_b). ΔF may represent ΔF_1 or ΔF_2 or ΔF_{eff} . It is clear that the system perturbation due to wavelength (frequency) drifts will be reduced by the ratio $(l_a - l_b)/l_a$. It is also clear that any errors caused by temperature changes are greatly reduced, as both interferometers suffer a common mode perturbation, caused by the expansion of the housing forming the body of the accelerometer, indeed this thermally induced error in the differential phase change vanishes if the dimensions of the interferometers are precisely equal.

The resolution of the system is determined by the accuracy with which the phase change of the output carrier of one interferometer relative to that of the other can be determined. The minimum detectable phase change between the two carriers is limited by the signal to noise ratio of each. The phase resolution (ϕ_{min}) is given by:

$$\phi_{min} = \sqrt{(n/s)_1^2 + (n/s)_2^2}$$

where $(s/n)_1$ and $(s/n)_2$ are the signal to noise ratios of the first and second carrier respectively.

Experimental characterisation of the system

The optical arrangement used to demonstrate the accelerometer and also to characterise its performance is shown in figure (2). The outputs from two laser diodes were fed into separate ports of a monomode-fibre directional coupler, where the two wavelengths were mixed and transferred from the two output ports of the coupler to illuminate the two hemispherical Fabry-Perot interferometers. The laser diodes used were a Hitachi HL7801E, emitting at a wavelength of 790.7 nm and a Hitachi HLP-1400, emitting at a wavelength of 855.4 nm. The two laser diodes were thermally isolated, such that a long-term frequency stability of about 1 GHz is obtainable for each. A pseudo-heterodyne signal processing scheme was used [1], where a sinusoidal modulation current, at frequency $f \approx 2KHz$, was applied to each laser simultaneously. The modulation amplitude was adjusted for each laser to avoid carrier amplitude modulations [10]. The optical outputs of the two interferometers were collimated and separated using a ruled echellete diffrac-

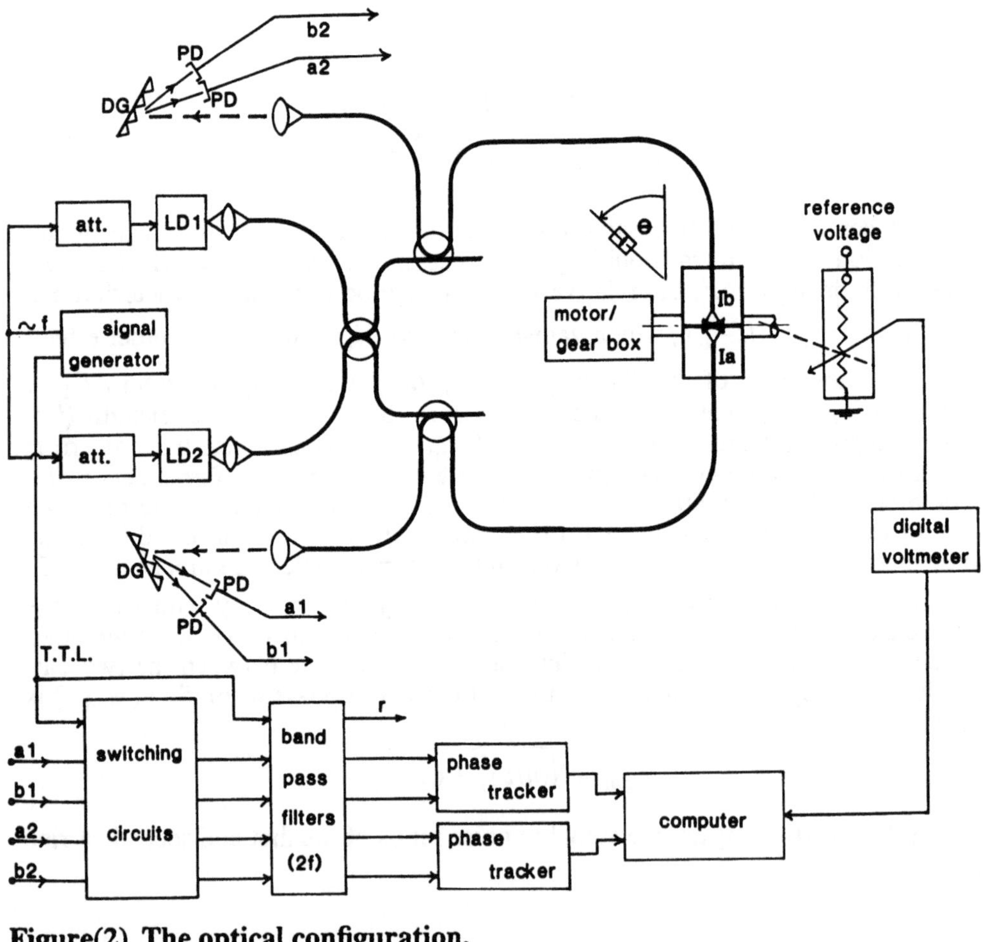

Figure(2) The optical configuration.
LD:laser diode, DC:directional coupler, PD:photodiode, DG:diffraction grating, att.:attenuator.

tion grating, with 1200 lines/mm and a blaze angle of 75° . Four outputs, two corresponding to each interferometer, were detected and gated by synchronous 50-50 duty cycle square pulses and filtered at frequency of 2f to generate four phase modulated carriers. A reference carrier, at a frequency of 2f, was produced by band-pass filtering the synchronous switching pulses as shown in figure (2). This technique of wavelength separation potentially has better performances than the switching technique, described by Kersey et al [6], as the shot noise in each photo detector is reduced by a factor of $\sqrt{2}$ and laser switching noise is completely avoided.

In order to test the quasi-static performance of the accelerometer, it was mechanically coupled to a motor/gear box mechanism such that it could be rotated very slowly through an angle θ about a horizontal axis. A linear potentiometer, mechanically attached to the accelerometer, was used to generate a voltage proportional to θ , this voltage was measured with a digital voltmeter and then transferred to a computer. Digital phase trackers were used to sequentially measure the relative phase changes between the output of one interferometer with respect to the other (ϕ_{ab1} and ϕ_{ab2}), and to transfer their values to the computer, where the differential phase ($\Delta\phi$) was calculated. The sensitivity of the accelerometer was measured by determining the induced phase ϕ_{ar} etc. as it was slowly rotated such that its axis changed from the horizontal ($\theta=90^o$) to the vertical ($\theta=0^o$) positions and back again. The time for each 90° rotation was 5.2 minutes and the corresponding variation in the axial acceleration was r = g Cos(θ), where g is the gravitational acceleration ($g \simeq 9.81 ms^{-2}$). This process was repeated many times in order that the system's repeatability could be studied.

The common mode rejection properties of the accelerometer for temperature and source fluctuations were measured by setting it horizontally and;

a) heating it over a $20^o C$ range ($20–40^o C$).

b) varying the injection current of the laser diode such as to vary its output frequency by about $\pm 12 GHz$ around its operating value.

In each case the ratio of (ϕ_{ab}/ϕ_{ar}) was calculated, where ϕ_{ar} is the phase change between the output of interferometer a and the reference carrier.

Finally the cross-coupling sensitivity of the accelerometer was measured by determining the amplitude of the induced phase changes produced by shaking the accelerometer, with the same vibration frequency and amplitude, along and perpendicular to its axis of symmetry.

Performance

Figure (3) illustrates the outputs of the four photo-detectors. The variations in the induced phases in the accelerometer as a function of the normalised axial acceleration (r/g) is shown in figure (4). The relative phase changes, of the output of one interferometer relative to the other (ϕ_{ab1} and ϕ_{ab2}) were 131.3 rad/g and 121.4 rad/g respectively, with repeatability better than $\pm 8.7x10^{-4}$ rad/g. The differential phase change $\Delta\phi$,corresponding to λ_{eff} , was 9.9 rad/g. The mean signal to noise ratio, at the output of the filters, was approximately 75 dBV, within a bandwidth of 59.7 Hz, equivalent to an optical limit of the phase resolution of $3.3x10^{-5} rad/\sqrt{Hz}$, which corresponds to an acceleration resolution of about $2.5x10^{-7} g/\sqrt{Hz}$ at a frequency of 0.067 Hz (neglecting laser long-term instability).

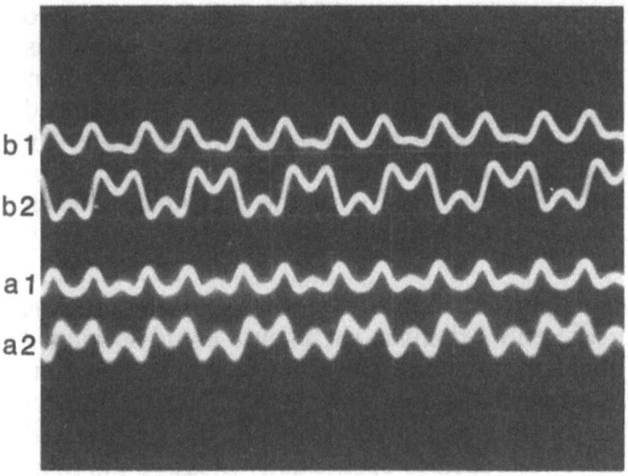

Figure (3) The photo-detector outputs.

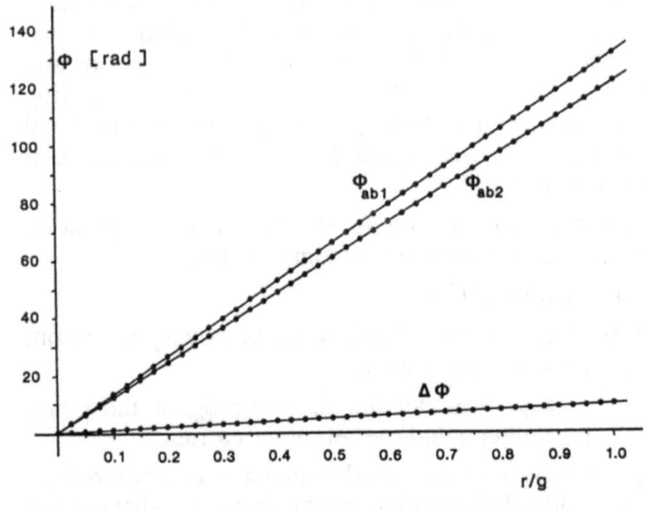

Figure (4) The induced phase changes in the accelerometer as a function of the normalised axial acceleration (r/g).

The common mode rejection of the temperature and wavelength drifts were -27.3 dB and -45.8 dB respectively. The system cross-coupling as measured was better than -36.9 dB. The long-term stability of the system was 34.2 mrad, which corresponds to $2.6x10^{-4}g$.

Discussion

From the results it can be seen that the unambiguous range of the single wavelength output is approximately 0.05 g, whereas that equivalent to λ_{eff} is about 0.64 g. This represents an extension in the unambiguous range of 12.8 times, which could be increased further using two sources with smaller difference in their emitting wavelengths, for example with $\Delta\lambda = 16$ nm a dynamic range of 2.5 g could be obtained whilst the resolution would remain unchanged.

The long-term stability of the system is governed by the stability of the two sources. A technique has been reported that demonstrates a mode hopping-free stable single mode semiconductor laser, with a frequency stability of 28 MHz/24 hours [11]. According to this figure of long-term frequency stability, the corresponding phase stability of one interferometer, in our system, will be about 18.8 mrad/24 hours (path-length imbalance of about 1.6 cm). This value will be reduced to be $9.6x10^{-5}$ rad/24 hours when measuring the relative phase between the two interferometers. This gives our system a long-term measuring stability of $7.3x10^{-7}$ g/24 hours. In our system the differential phase measurement (corresponding to λ_{eff}) needs to be accurate to $2\pi/12.8$. These figures could be achieved by stabilising one of our sources, as described in reference 11, and leaving the other one un-stabilised, other than to provide some thermal isolation from rapid environmental changes.

Summary

A practical optical accelerometer has been described. The resolution and range (acceleration and working frequency) depend on the properties of the diaphragm, thus it may be designed specifically for a given application. Using dual wavelength technique the unambiguous range of the system is increased. A unique feature of this system is that as no conventional optics are used, the accelerometer could be operated over a much wider temperature range than possible with conventional accelerometers.

Acknowledgements

A.S.Gerges would like to acknowledge the support of the government of Egypt.

References

[1] D.A.Jackson " Monomode optical fibre interferometers for precision measurement " J.Phys.E:Sci.Instrum., Vol. 18,1985, P.981.

[2] A.B.Tveten, A.Dandridge, C.M.Davis and T.G.Giallorenzi " Fibre optic accelerometer " Elect.Letters, Vol.16, No.22,1980, P.855.

[3] A.D.Kersey, D.A.Jackson and M.Corke " High sensitivity fibre optic accelerometer " Elect.Letters, Vol.18, No.13, 1982, P.559.

[4] F.Bucholtz, A.D.Kersey and A.Dandridge " DC Fibre optic accelerometer with sub micro g sensitivity " Elect.Letters, Vol.22, No.9, 1986,P.451.

[5] A.S.Gerges, T.P.Newson, J.D.C.Jones and D.A.Jackson," High-sensitivity fibre optic accelerometer ", Optics Lett., Vol.14, No.4, Feb.1989, P.251.

[6] A.D.Kersey and A.Dandridge, " Dual-wavelength approach to interferometric sensing ", SPIE, Vol.798 Fibre optic sensors 2, 1987, P.176.

[7] D.J.Webb, J.D.C.Jones and D.A.Jackson, " Extended range interferometry using a coherence tuned synthesised dual wavelength technique with multimode fibres ", Elect.Lett., Vol.24, No.18, Sep 1988, P.1173.

[8] A.S.Gerges, T.P.Newson, F.Farahi, J.D.C.Jones and D.A.Jackson " A Hemispherical Air Cavity Fibre Fabry-Perot Sensor " Optics Comm. Vol.68, No.3, Oct.1988, P.157.

[9] S.Timoshenko " Theory of plates and shells " McGraw-Hill 1940,Chap.3.

[10] A.C.Lewin, A.D.Kersey and D.A.Jackson," Non-contact surface vibration analysis using a monomode fibre optic interferometer incorporating an open air path ", J.Phys.E.Sci.Instrum., Vol.18, 1985,P.604.

[11] Z.Bingkun, Z.Hanyi, W.Yuanxiang, Z.Jianying, L.Jian and P.Zhenwu, " 24 h frequency stabilisation of an slm external-cavity semiconductor laser " Elect.Letters, Vol 23, No. 5, Feb 1987, P.194.

High Sensitivity Transducer for Fibre-Optic Pressure Sensing Applied to Dynamic Mechanical Testing and Vehicle Detection on Roads

A. Tardy[1], *M. Jurczyszyn*[1], *J.-M. Caussignac*[2], *G. Morel*[2], *and G. Briant*[3]

[1]Laboratoires de Marcoussis, Centre de Recherches de la C.G.E., Route de Nozay, F-91460 Marcoussis, France

[2]Laboratoire Central des Ponts & Chaussées, 58 boulevard Lefebvre, F-75732 Paris Cedex 15, France

[3]Laboratoire Régional de l'Ouest Parisien, 12, rue Teisserenc-de-Bort, F-78190 Trappes, France

This paper describes the design and testing of a fibre-optic pressure sensor based on photoelastic effect. The transduction mechanism is characterized by a ruggedized, elastic steel-glass structure. This transducer leads to high sensitivity as simple fringe counting methods allow interesting applications in the fields of dynamic mechanical testing and vehicle load evaluation.

Sensing elements embedded in laboratory models made of concrete have been implemented for testing structure. The proposed sensing cable has been shown to operate satisfactory for vehicle detection in traffic analysis.

INTRODUCTION

The effect of birefringence induced in a single mode fibre by a lateral force has been applied to measure static force[1]. We propose a transduction mechanism for pressure sensing which leads to a linear,ruggedized gauge more convenient for practical applications than a simple fibre pressed between parallel plates. In the present paper, the sensor design and performances are described. We report main results of two applications:

_ sensing element has been embedded in a concrete test piece and surprizing hardening stress has been found,

_ sensing cable has been install on road to define sensor response when a vehicle passes over the sensor

PRINCIPLE AND DESCRIPTION

Sensor design :

The sensing structure consists of an uncoated low birefringence single mode fibre squeezed between metallic ribbons. The ribbon edges are jointed together by crimping technique. A schematic diagram of the sensor can be seen in Fig.1.

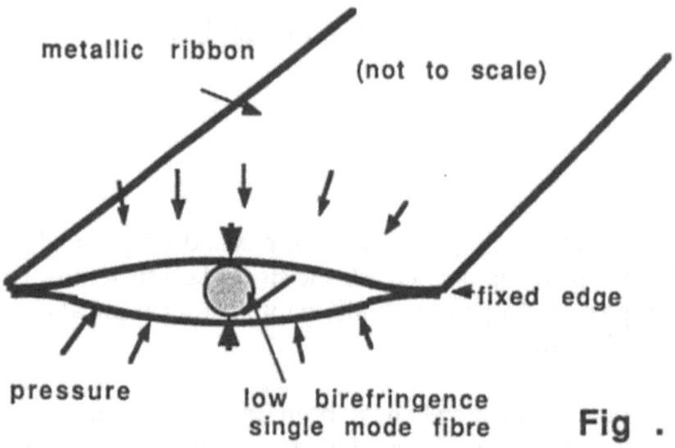

Fig . 1 Sensing structure

Without internal or external pressure, the optical fibre is loaded by a compressive force value of which depends on the ribbon deflection. Each length unit of the ribbons can be considered as a beam, fixed at both ends and bent by a force applied to the center of the ribbons. The deflection is equal to the fibre radius r. The applied force or pre-load per unit fibre length is:

$F/L = 16.E.r\,(h/l)^3.(1-\mu^2)^{-1}$ where E is the Young's modulus of the ribbon material, μ is the Poisson's ratio, h is the thickness and l is the width between stiffened edges of the ribbon respectively.

The birefringence induced by a transverse force is given by[2] $B = 4\,CF/(\pi rL)$.

So the phase difference between the two polarization modes (parallel and perpendicular to the force) is $\Delta\Phi/L=128\,E\,C\,(h/l)^3\,\lambda^{-1}(1-\mu^2)^{-1}$ where C is the stress-optic coefficient of silica glass ($C=3.3\times10^{-12}\,Pa^{-1}$).

For steel, $E=2\times10^5$MPa, $\mu=0.3$ and with the dimensions l=10mm,h=80µm and 2r=125µm, the resulting static phase difference is 56 $rad.m^{-1}$ at $\lambda=0.85$µm.

In the case of an outside pressure P, each unit length element of the ribbon is considered as a beam supporting an uniform load P.l.

_ If P is less than $P_c = 96\,E\,(r/l).(h/l)^3\,(1-\mu^2)^{-1}$ where Pc is a critical value corresponding to the beginning of the width (l) variation.
The additional force applied to the fibre is F/L = 0.5 P.l. The corresponding phase difference is $\Delta\Phi/L = 4\,C\,P\,l\,/(\lambda\,r)$.
_ If P is more than Pc, the ribbon edges come into contact and the free length between fixed ends decreases. The opposite force applied to the fibre is :
$F/L = 2\,[E\,r/(1-\mu^2)]^{1/4}\,.h^{3/4}\,.P^{3/4}$. The corresponding phase delay is:
$\Delta\Phi = 16\,C\,\lambda^{-1}\,[E/(1-\mu^2)]^{1/4}.\,(h/r)^{3/4}.\,P^{3/4}$.

Sensitivity :
With the same figures as the previous numerical application, Pc is equal to 6.75×10^{-2} MPa. In the field of low pressure (P< Pc), the sensitivity is $1/L.d(\Delta\Phi)/dP = 2.5$ rad.$(kPa)^{-1}.m^{-1}$. Compared with a simple bare fibre in a Mach-Zehnder interferometer, the sensitivity of the gauge is higher by approximately the factor 85. In the field of high pressure (P > Pc), the sensitivity becomes $1/L.\,d(\Delta\Phi)/dP = 1.16$ rad.$(kPa)^{-1}.m^{-1}$ at P= 1 MPa.
Sensitivity keeps a high value. So simple fringe counting has been used to determine the change in birefringence. Expressed in term of fringe number, the phase difference is $N/L = 44\,P^{3/4}.\,m^{-1}$. (P expressed in 0.1 MPa or bar).

APPLICATIONS

Concrete structure testing :
14 cm long sensing elements have been embedded in concrete test pieces without special care. Before sensor introduction in the concrete boarding, the ribbons are covered with epoxy resin and sand (1 mm layer). These new sensors have been implemented in a simple compressive test. Uniform stress distribution, $\sigma = F/S$, is assumed where S is the test-piece section and F is the servo-control force of hydraulic-jack. Figure 2 shows a view of the concrete test-piece. Experiments were realised using polarimetric apparatus. First remarks have been a sensitivity loss and a fringe visibility reduction at the output end of the fibre. In spite of employing a laser diode, the reduced degree of polarization indicates relatively large induced birefringence even under low load. A high

FIG.2 Concrete test piece

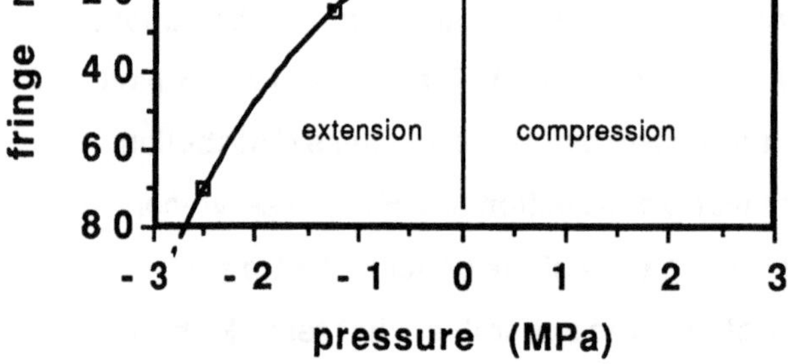

Fig. 3 Experimental response from sensor embedded in concrete

value of the hardening stress is supposed due to shrinking phenomenon of concrete. Evaluation of this contribution is achieved in the following way. Sensor desensitization is accomplished by loading the test-piece in a perpendicular plane to that of ribbons. Taking into account Poisson'ratio, 2.5 MPa are necessary and 70 fringes shift are observed. This unusual high value is due probably to a punching effect by aggregate. From different measurements, the observed sensor response is shown in figure 3 where the stress axis origin corresponds to the hardening pre-load. The fringe numbers versus pressure are in good agreement with the theoretical values.

Vehicle detection applied to traffic :

Installation of 55 cm long sensing elements on road have been implemented. To prevent sensor damage, cable elements are laid into grooves made in the roadway. Laying process consists of: a layer of epoxy resin is poured at the bottom of the groove, sensor cable is laid, then the groove is filled in with elastomer.
Different groove depths have been made.
A reflective coating on the end face of the sensor fibre allows polarization effect to be interrogated by a single fibre with Y coupler.
Typical responses when a car passes over the sensor are shown in figure 4.
Signature characteristicc are :
- a first fringe shift corresponding to the rise load,
- a continuous signal indicating a quasi-static pressure,
- a second fringe shift corresponding to the escape of vehicle from the sensor.

Relevant data about vehicle identification and classification can be fringe number accurately related to the weight in motion and signal duration function of instantaneous speed. Experiments have been implemented with different types of vehicle (car and heavy lorry).
First main results show :
- a sensor response not dependent of groove depth,
- the signal duration follows approximately the speed law,
- the fringe numbers increase with the vehicle load but its dispersion is related to dymamic effect (vehicle suspension and others...).
- twin wheel leads to a specific signature.

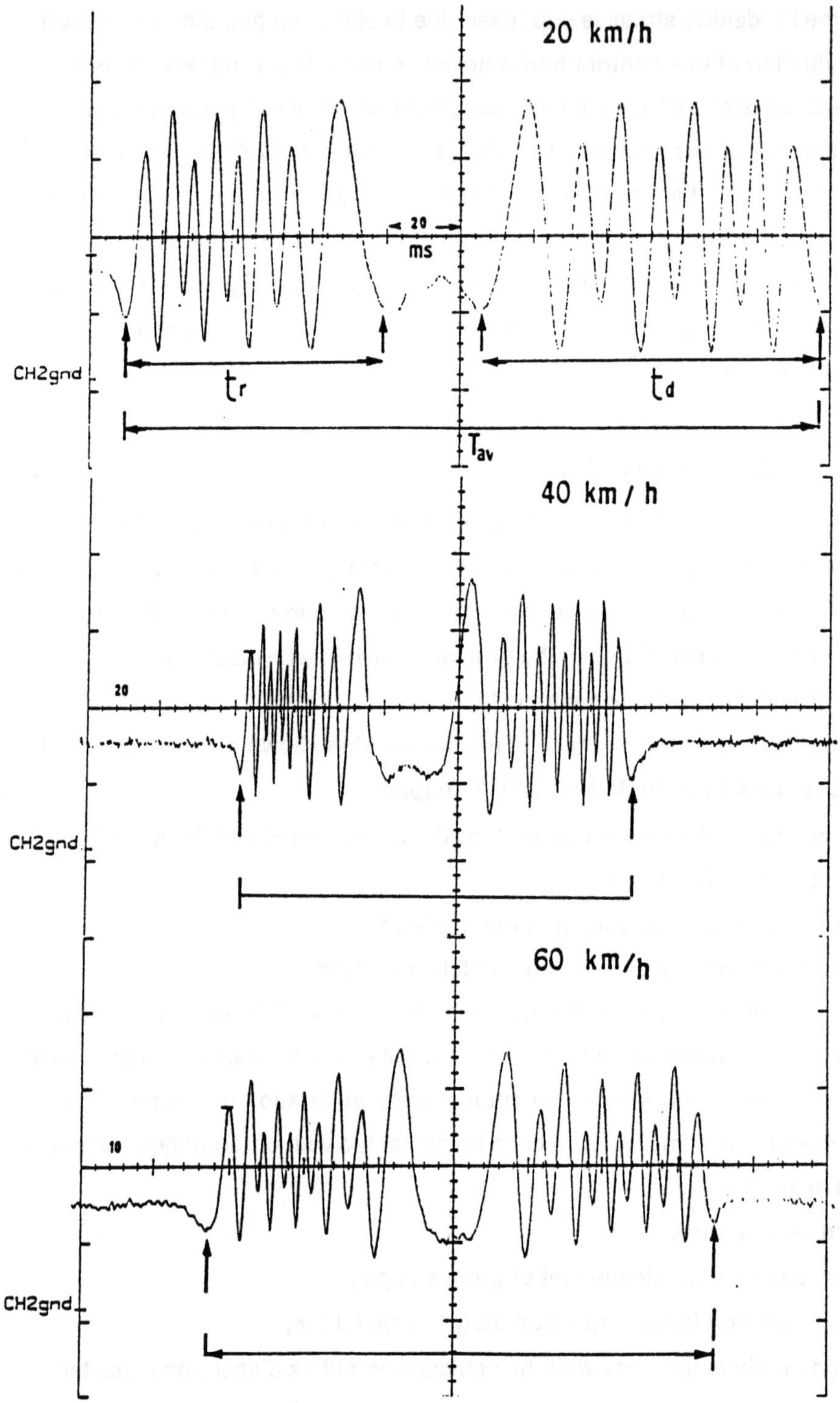

Fig. 4 Optical signatures for vehicle traffic detection
Renault 5 - fore- wheel

Other data depend on the direction of the groove with respect to traffic direction. We are expecting that successively detecting each half-axle, specific characteristics can be obtained for a full vehicle identification.

CONCLUSION

A simple, robust optical pressure sensor has been built. Its high sensitivity and full use in the field of low pressure have not been operated yet. Practical implementations have shown that this sensor is attractive for dynamic material testing and for vehicle detection on roads.

ACKNOWLEDGEMENT

Parts of this work have been supported by le Ministère de la Recherche et le Ministère des Transports (Direction de la Sécurité Routière DSCR)respectively.

REFERENCES

(1)A.Bertholds and R.Dandliker,Applied Optics,Vol.25 N°3,340-343 (1986)

(2)Y.Namihira,M.Kido and Y.Mushiaka,Trans.IECE Japon.60-C,107 (1977)

A Continuous Liquid Level Sensor Using Fluorescent Fibre

J.W. Snow[1], *P.D. Colbourne*[1], *S.H. Woodside*[1], *and B.E. Paton*[2]

[1]Focal Technologies Inc., 40 Thornhill Drive, Unit 7,
Dartmouth, Nova Scotia, Canada
[2]Physics Department, Dalhousie University, Halifax, Nova Scotia, Canada

A fibre optic continuous liquid level sensor which exploits the properties of fluorescent fibre is described. Sensor heads up to 1.2 metres in length which are lead insensitive have been built and tested. While point liquid level sensors have been demonstrated, this is the first time that a rugged continuous liquid level sensor has been built to specifically address commercial packaging and liquid fouling.

Introduction

A continual requirement in the aircraft industry is for improved fuel quantity sensors which answer the demands for increased accuracy and reliability and possess superior safety performance. The dielectric properties of the optical fibre are desirable from the point of view of safety with respect to spark hazard and lightning strikes. High resolution, accuracy and long term stability are also desirable. Present day aircraft fuel level sensors are for the most part capacitance type sensors which lose accuracy when the fuel becomes laden with water and the dielectric constant changes significantly. While there exist many liquid level point sensors, few are continuous. One of the early continuous liquid level sensors [1] used bending losses in loops of ever increasing diameter as the sensor. The difficulty in maintaining a clean interface surface and the potential for damage to the fragile fibre loops make an operational sensor of this type difficult to achieve. We describe here a continuous fibre optic liquid level sensor for the aircraft industry which addresses the stated requirements and in addition addresses many of the packaging problems associated with bringing fibre optic sensors to the market place.

Sensor Configuration

The liquid level sensor probe consists of a large diameter (13mm) acrylic rod waveguide with a smaller diameter (1mm) fluorescent fibre embedded off axis, figure 1. The source or pump optical radiation at 570 nm is injected from a bundle of small diameter plastic fibres at the upper end and placed at an angle midway between the critical angles for an air-acrylic and a liquid-acrylic interface. This arrangement and the numerical aperture of the pump fibres ensures that when air is present, the pump

Springer Proceedings in Physics, Vol. 44 **Optical Fiber Sensors**
Editors: H.J. Arditty · J.P. Dakin · R.Th. Kersten

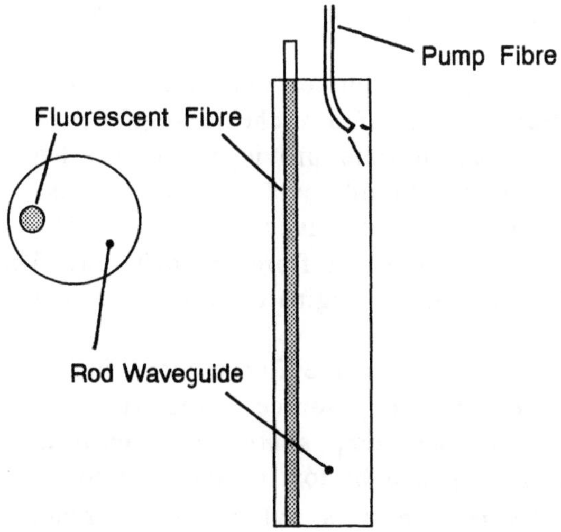

Figure 1 : Sensor configuration for continuous liquid level measurement

radiation spirals down the rod waveguide primarily in large angle skew modes of the rod waveguide. Whenever pump light intersects the fluorescent fibre, part of the energy is down converted to the fluorescent signal centered at a wavelength of 620 nm. Total internal reflection at the air-acrylic interface results in a maximum fluorescent signal when no liquid is present. In the presence of a liquid level outside the rod waveguide, refraction of the pump signal in the liquid results in an attenuation of the pump signal and gives a fluorescent signal which varies inversely with the liquid level.

Fluorescent Fibre as a Detector

The core of the detector fibre is impregnated with an organic fluorescent material [2] which when irradiated with short wavelength radiation, re-emits light at a longer wavelength. Normally, light passing transversely or nearly transversely is not coupled into an optical fibre. However, in the fluorescent fibre, this is not the case. Light at large angles is readily absorbed and the fluorescent sites in the core re-emit light spherically. Light emitted within a cone specified by the numerical aperature of the fluorescent fibre will remain guided within the fibre's core and will propagate to both ends of the fibre. The bidirectionality of the detector fibre is exploited in the lead fibre insensitivity scheme.

Lead Fibre Insensitivity

Intensity based sensors are sensitive to changes in the source intensity, detector sensitivity, optical fibre attenuation, couplers and connectors as well as changes in the sensing signal. It is often impossible to tell the difference between a transient loss in the connecting fibres and a change in the measurand. Even small variations in the lead fibre attenuation can not be tolerated in high performance sensing systems.

Special techniques such as wavelength division referencing [3] can be used to minimize lead fibre sensitivity.

We describe here a technique using two sources and two detectors which are time multiplexed, a modified version of a scheme suggested by Beheim and Anthan [4]. This method is appropriate provided one has two signals which have different intensity versus liquid level curves. This is precisely the case of the fluorescent detector rod waveguide, where the signals at the top and bottom end of the fluorescent fibre are different. The asymmetry is present since the pump radiation originates only from one end.

The second source injects light directly into the fluorescent fibre to provide a reference level at a time when the pump source is off. A measurement is made by first turning on the pump source and measuring the fluorescent signals P_t and P_b at the top and bottom of the detector fibre, respectively. Then repeating the meaurements, when the reference source is on and pump signal is off, yielding reference signals P_{ot} and P_{ob}. The quantity

$$Y = (P_b P_{ot})/(P_t P_{ob})$$

is then calculated, which is independent of attenuation variations in the four lead fibres. Figure 2 shows a typical sensor response including the lead-insensitive signals, for a single cycle of the liquid level. Due to the high NA of the pump fibres, there exists a dc background signal which is

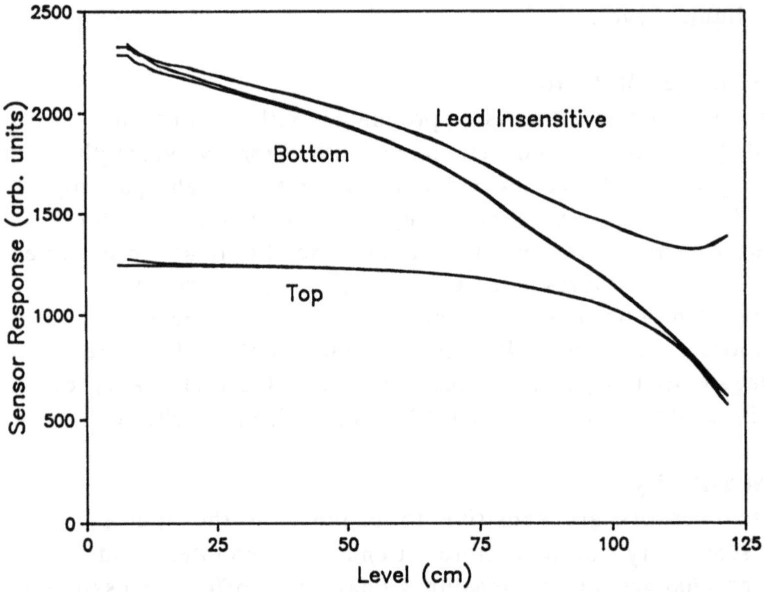

Figure 2 : Sensor response for top and bottom signals from fluorescent fibre and the lead insensitive signal, Y during a single cycle of the liquid level.

insensitive to the liquid level and causes inflection of the lead-insensitive signal near the upper fill level. This effectively reduces the useful length of the sensor. Although this scheme requires time multiplexing and some calculation, the availability of low cost microcomputer and peripheral I/O boards makes implementation straightforward.

Sensor Performance

Rod waveguide sensors were exercised in the test and calibration facility, figure 3. Five discrete liquid level sensors, a pressure transducer and a sight tube were used for calibration and comparision of the rod waveguide sensor with conventional liquid level sensors. Water or other liquids could be repeatedly cycled in the test chamber. A microcomputer provided the control circuitry for time multiplexing the LEDs and digitizing the received signals.

Approximately 0.08 mW of pump radiation is provided by four LED drivers. In air the rod waveguide has an intrinsic attenuation of about 5 db/m. The coupling efficiency of light by the fluorescent sites is only 1 percent. Intrinsic attenuation of the fluorescent fibre at the emitted wavelength is about 6 db/m. These losses together with the sensitivity of the detectors limit the practical length of the sensor to about 2 meters. Some improvement will come with the use of near visible laser sources and longer wavelength fluorescent dyes.

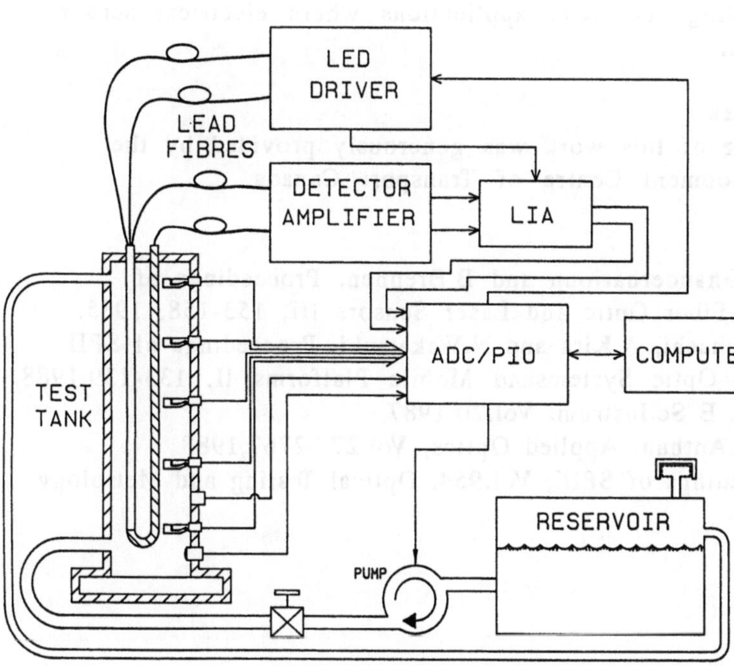

Figure 3: Test and calibration facility for exercising liquid level sensors

Over the range of 100 cm, the rod waveguide sensor displayed a level sensitivity of better then 5 millimetres. Short term drift was better than 1% and long term drift was limited by the temporal stability of the electronics. Initial wetting, lingering liquid droplets or condensation on the waveguide can degrade this performance. This is also a problem with most continuous sensors and can be avoided by careful choice of materials.

To demonstrate the effectiveness of the lead insensitivity scheme, the parameter **Y** was monitored at a static liquid level as the pump source was varied over 6 db. No change was observed in the measured signal.

Rather more significant is the potential problem of fouling through the adhesion of foreign material or other degradation of optical surfaces in contact with the liquid to be sensed. Potential sources of fouling can be biological (fungal), lacquer deposits and chemical clouding in the case of jet fuel or biofouling in the case of water [5]. The effects of various materials and biocides are currently under test and will be reported.

Summary

A novel fibre optic sensor for the continuous measurement of liquid level has been described. A prototype displayed a liquid level sensitivity of better than 5 millimetres over 100 centimeters and proved insensitive to fluctuations in lead fibre attenuation. Sensor lengths up to 200 centimeters should be readily obtainable.

The sensor is rugged and suitable for many liquid level measurements, including fuel tank applications where electrical sensors pose a safety hazard.

Acknowledgements

Partial funding of this work was generously provided by the Transportation Development Centre of Transport Canada.

References

[1] M.Belkerdid, N.Ghanderharioun and B.Brennen, Proceedings of SPIE Vol.566 Fiber Optic and Laser Sensors III, 153-158, 1985.

[2] H.Sawada, E.Takahashi, A.Kira and N.Wakatsuki, Proceedings of SPIE Vol.989 Fiber Optic Systemsand Mobile Platforms II, 133-139,1988.

[3] J.P.Dakin, J.Phys. E Sc.Instrum. Vol.20,1987

[4] G.Beheim and D.Anthan, Applied Optics, Vol.27, 2767,1988

[5] J.W.Snow, Proceedings of SPIE, Vol.954, Optical Testing and Metrology II, 1988.

Prototype Industrial Multi-parameter F.O. Sensor Using White Light Interferometry

D. Trouchet, B. Laloux, and P. Graindorge

Photonetics, 52 Avenue de l'Europe, F-78160 Marly-le-Roi, France

A prototype multi-parameter fiber-optic sensor is presented. The use of white light interferometry, combined with a compact design make this device particularly well suited to industrial applications. The main features include measurement stability, lead insensitivity and also a user-friendly interface for diagnostics, calibration and reconfiguration purposes.

1. Introduction

Many fiber-optic sensors which have been proposed in the past years rely on light intensity modulation to achieve transduction of the physical parameter to be measured to a light beam. This has led to quite a few successful devices. Nevertheless, the drawbacks of this technique are well-known, and mostly related to the non-constant loss of optical-fiber systems when placed in every-day industrial environments.

Hence, classical intensity-modulation F.O. sensors behave quite well in relatively well controlled environments, and when fiber lengths are not too large, but have a poor long term stability.

To overcome this drawback, various "compensation" schemes have been proposed, which usually establish a "reference" light path, which is unperturbed by the physical parameter to be measured, whereas the "measurement" path interacts with this parameter. A proper ratio of the intensities of the corresponding light beams provides a loss-independent measurement. Here again, the main difficulty is to make sure that both beams are affected by exactly the same perturbation in exactly the same amount. This is usually rather difficult to achieve in practice.

2. White light interferometry

Because of this, various principles have been proposed, based upon a modulation of the phase or wavelength of the light beam by the physical parameter. Most of these rely on interferometry to achieve the transduction. The scheme described here is known as white light interferometry. Various publications have demonstrated the principle [1-6]. We describe here a compact unit with no moving parts.

2.1. Principle

In an interferometer used with a monochromatic light source, such as a laser, the absolute path unbalance of the interferometer cannot be known, but its relative evolution can be measured.

If the the same interferometer is used with different wavelengths, or with a broad spectrum source, this uncertainty disappears, and the absolute position of the interferometer can be determined without knowing its past history.

When the source is a broad spectrum light source (i.e. the coherence length is shorter than the interferometer unbalance) , the device is known as a "white light interferometer". The spectrum of the light emerging from the interferometer is modulated by a series of troughs and peaks which correspond to the variation with wavelength of the phase difference between the two beams in the interferometer. The spacing of these "fringes" in the light spectrum is related to the unbalance of the interferometer.

2.2. Spectral analysis demodulation

One of the ways to retrieve the information conveyed by the returning light beam is to analyse its spectrum. As this spectrum is amplitude-modulated by the interferometer, the determination of this modulation gives the information on the unbalance of the measurement interferometer. The spectrum of the light beam is first obtained with a spectroscopic-type arrangement ; then the frequency (in the spectral domain) of the modulation is measured, which in turn gives the value of the path unbalance in the interferometer.

It is interesting to note that the measurement of the modulation frequency is in fact equivalent to achieving a Fourier transform of this modulation, and measuring the position of the resulting peak. The displacement of this peak is then proportional to the path change in the interferometer. Another way of understanding this is to note that the modulation of the light's spectrum is the result of a Fourier transform of the measurement interferometer response. Measuring the modulation frequency of the incident light's spectrum is equivalent to achieving a reverse Fourier transform, which yields the initial value of the interferometer unbalance.

This explains why this method is particularly reliable : in amplitude-modulation sensors, every perturbation to the light path is analyzed as a change of the physical parameter to be measured ; in phase-modulation sensors, the information is conveyed by the phase of the light, but has to be converted into amplitude modulation at one point of the optical path ; from this point on, the whole device is also sensitive to perturbations, whereas in white light interferometers, the information is conveyed by a modulation frequency up to the detector where demodulation is achieved. This principle thus exhibits all the advantages of frequency modulation over amplitude modulation devices.

2.3. Readout interferometer demodulation

Another way to demodulate the information carried by the light beam is to use a second interferometer whose unbalance can be adjusted to match that of the measurement interferometer. The readout interferometer is scanned, and the output monitored, until the transmitted intensity exhibits a contrast maximum. When this occurs, the path differences of both interferometers are identical, and reading the unbalance of the readout interferometer provides the value for the measurement interferometer.

It can be easily seen that the only difference between both methods is the place where the reverse Fourier transform is achieved. In the spectrum analysis scheme, the Fourier transform is made on the electrical signal which is output by the spectrometer. Conversely, in this second scheme the Fourier transform is made optically by the scanning readout interferometer. The difference is in fact the same as that between classical grating spectrometers, and Fourier transform spectrometers, which achieve the Fourier transform of the output of a scanning interferometer to obtain the spectrum of the incident light beam.

3. Sensor design

Our purpose was to design a white-light interferometry based sensor for applications in industrial environments. Consequently, it had to be compact, self-contained, and capable of transmitting easy-to-use information to the main process. In the selected architecture, the system is composed of two main components. The first one is the opto-electronic module, which includes the source, receiving optics and electronics, and the system controller. The input characteristics of the receiving optics are fixed, and determine the type of transducer that can be analyzed by this module. The second part of the system is the set of transducers and fiber leads. Different types of transducers can be used on the same opto-electronic module as long as their optical characteristics are matched to those of the module. The type of transducer is indicated by the user to the controller, which then selects the appropriate conversion table.

3.1. Opto-electronic module description

The block diagram of the sensor is represented in fig 1. The broad spectrum light from a LED source is launched into a multimode fiber, transmitted through a fiber cable to the transducer where it is encoded and reflected. The return beam is directed to the receiving optics by a fiber coupler located in the module. The light is demodulated by a bulk optic system and detected by a CCD detector linear array. The signal generated by this array is processed by the single-board system controller, composed of a 68 010 microprocessor and all necessary peripherals. The value of the path unbalance is computed and converted into the proper units by means of a look-up table. The data (measurement type, value and unit, error messages) are displayed on a LCD alphanumeric display, and can be sent either to the process computer through a RS 232 interface, or to analog receivers through 4-20 mA or 0-10 V outputs.

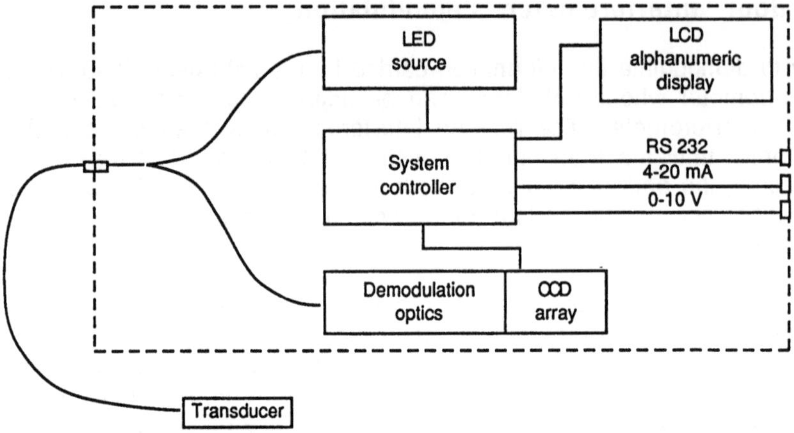

Fig 1 : Sensor block diagram

The choice of a CCD array as a detector enables the use of compact demodulation optics with no moving parts. This, added to the small size of the solid state source, and of the microprocessor controller, results in a compact (110 x 100 x 190 mm) and versatile system. It is packaged in a standard 19" rack compatible module.

The opto-electronics module architecture allows the implementation of powerful user-interface features such as automatic defect detection and diagnostic routines, the result of which are displayed, or transmitted on the RS 232 interface. Multiple sensor calibration tables can be stored, enabling the user to easily switch from one sensor to another without recalibration. If an extra calibration is necessary, the calibration table can be downloaded from an external computer, and stored in non-volatile memory.

All these features make this sensor realistic for industrial applications, easy to use, fail-safe, and easy to reconfigure from one measurement type to another.

4. Transducers ; performances

4.1. Types of transducers

As white light interferometry enables the measurement of the absolute path length in an interferometer, various physical parameters can be measured, according to the way they interact with the interferometer. The change of the physical parameter must induce a change in the optical length of the interferometer. This can be accomplished by two different means : either by changing the physical length of the cavity, or by modifying the index of refraction of the material inside the cavity. The type of parameters that can be measured include displacement (displacement of one mirror with respect to the other), pressure (pressure is applied on a flexible membrane supporting a mirror), index of refraction (by placing the liquid inside the interferometric cavity) and temperature (by using a material with a strong index temperature dependence).

Each of these transducers has to be designed in order for the characteristics of the interferometer to be matched to the input characteristics of the "receiver". When this is the case, either of those transducers can be linked to the receiver which then measures its absolute path length. All that has to be done to retrieve the parameter value, is to convert the cavity length in parameter units, either by computation from the transducer design parameters, or by an experimental calibration. The advantage of this latter method is to suppress at the same time the defects in the response curve which could be introduced for example by the non-linearities of the receiving optics. Once the response curve for a given transducer is known, it can be stored in the module's non-volatile memory.

4.2. Example of a pressure sensor

The pressure measurement cell, shown in fig 2, was realized as a first type of transducer for this system. It is composed of a collimation optics, and a Fizeau interferometer packaged into a case which is mechanically compatible with existing industrial transducers.

The Fizeau interferometer is made of a fixed semi-transparent substrate, and a mobile semi-reflective membrane. A spacer ensures constant distance at the outer diameter between this membrane and the substrate. The membrane stiffness is chosen so that its deflection at the center, when pressure is applied, is in the range which can be measured by the receiving optics of the system. This cell was designed so that its dimensions are compatible with existing, thin film on ceramic, resistive pressure cells (manufactured by JPB in France). This enables one to take advantage of existing hardware, and possibly of some of the manufacturing process, in order to reduce production costs, and increase reliability by using proven technologies.

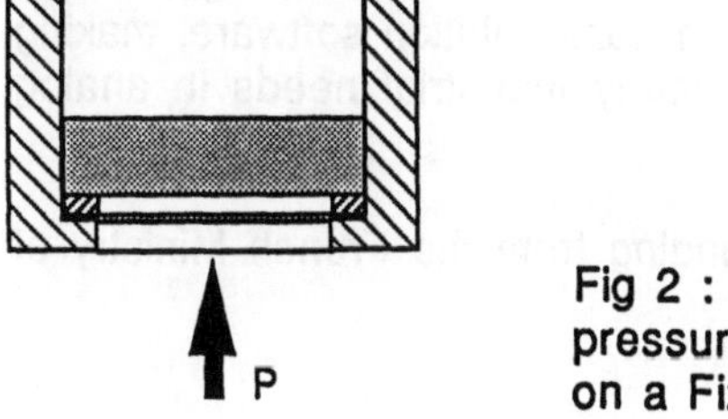

Fig 2 : Schematic view of a pressure transducer based on a Fizeau interferometer.

4.3. Experimental results

Tests are presently being conducted with this transducer for two different pressure ranges : a 0-100 mbar range and a 0-2 bar range. A first estimate of the present resolution of the system is 500 points over the full range. A resolution of 10 000 points can probably be obtained by increasing the efficiency of the demodulation algorithm. A calibration curve is shown below (fig 3), which shows a slight non-linearity in the response of the sensor. This non-linearity is compensated for in the look-up table.

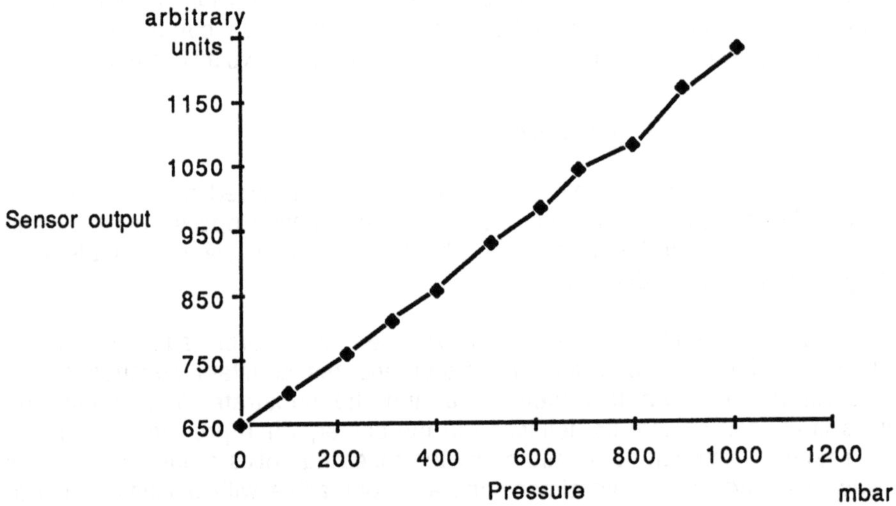

Fig 3 : Calibration curve

5. Conclusion

The feasibility of a white-light interferometer sensor for industrial applications is demonstrated. The combination of solid state components, of an optical subsystem with no moving parts, and a microprocessor based controller, leads to a particularly compact, reliable and user-friendly system (see the photograph of the complete system in fig 4). Different types of sensors can be used with the same demodulation hardware to measure various parameters, or different ranges of a single parameter. The present resolution, estimated to 500 points over the measurement range, should be extended to 10 000 points with small changes in demodulation software, making this system a high resolution system, suited for many industrial needs in analog measurement.

This development was conducted under partial funding from the French Ministry of Research and Technology.

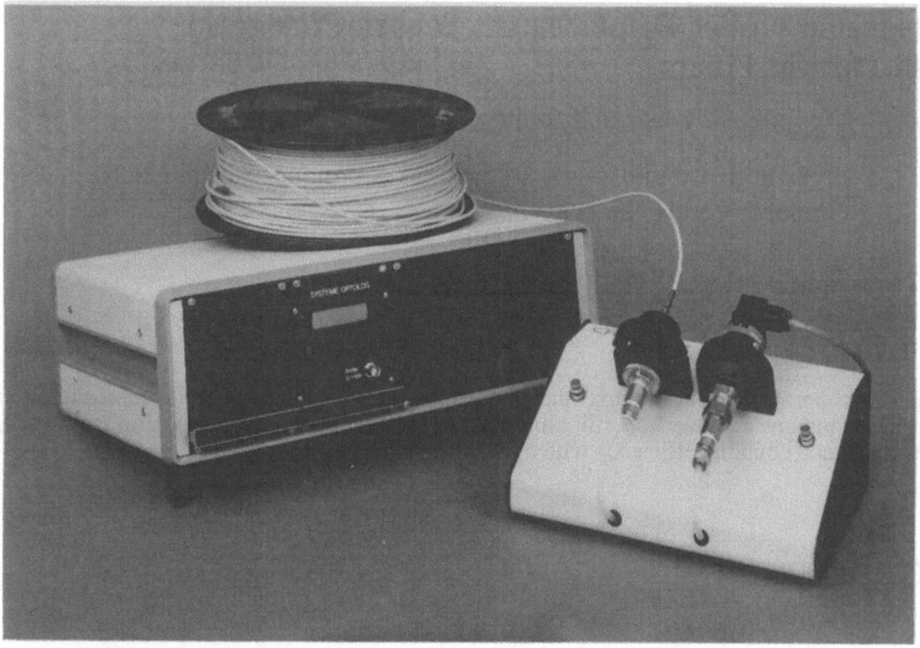

Fig 4 : Photograph of the system

References

[1] C. Delisle, P. Cielo. "Application de la modulation spectrale à la transmission de l'information". Can. J. Phys., **1047, (1975)**.

[2] P. Dabkiewicz, R. Ulrich. "Spectral encoding for fiber optic industrial sensors". EFOC/LAN 85, 212-214, (1985).

[3] T. Bosselmann, R. Ulrich, H.J. Arditty. "Multimode fiber coupled interferometric position sensor". Proc Opto **85**, 168-169.

[4] P. Dabkiewicz, R. Ulrich. "Fiber-optic angular sensor with interleaved channel spectra". Optics Letters, **11**, 543-545, (1986).

[5] M. T. Velluet, P. Graindorge, H. J. Arditty. "Fiber-optic pressure sensor using white light interferometry". S.P.I.E.,**838**, 78-83, (1987).

[6] C. Mariller, M. Lequime. "Fiber-Optic white light birefringent temperature sensor". S.P.I.E., **798**, (1987).

Integrated Optics Displacement Sensor Connected with Optical Fibers

C. Erbeia, S. Valette, J.P. Jadot, P. Gidon, and S. Renard

LETI, a Division of Commissariat à l'Energie Atomique, 85 X, F-38041 Grenoble Cedex, France

In this paper we present the first studies on an Integrated Optics vibration sensor realized on a new stucture achieved on silicon substrate by chemical vapor deposition of silica, allowing a good coupling efficiency with optical fibers.

INTRODUCTION

Silicon based Integrated Optics has become more and more attractive in a wide range of application, such as optical signal processing, optical telecommunications, optical fiber optics sensors.

A Michelson Interferometer used as a vibration sensor has been realized in our laboratory with the IOS1 Technology (for Integrated Optics on Silicon Substrate n°1) using a now standard structure : Si / SiO_2 / Si_3N_4 / SiO_2.[1],[2]. A schematic diagram of this sensor is given in Figure 1.

In this device it can be seen that laser diode and photodetectors are butt coupled to the circuit. An Input / Output connection with optical fibers should give many advantages especially in spatial division of optical and electrical processing and for achieving very small sensor head.

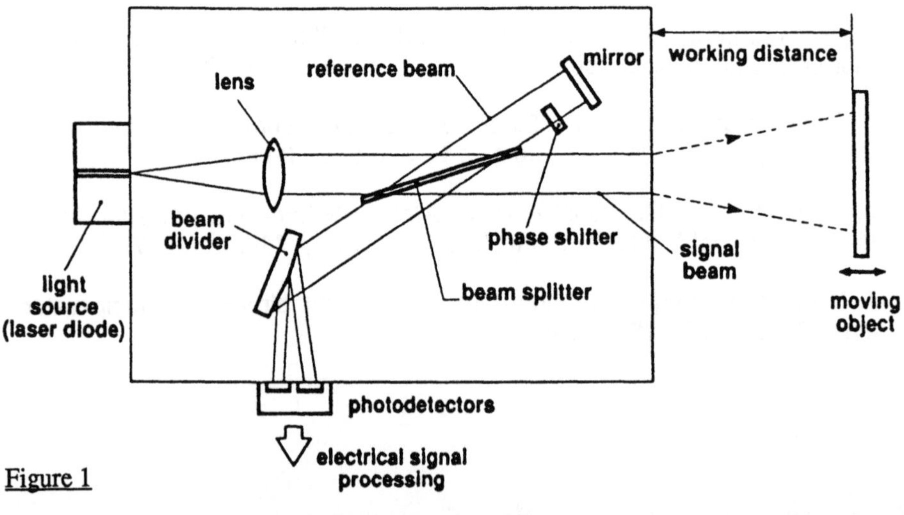

Figure 1

Springer Proceedings in Physics, Vol. 44 **Optical Fiber Sensors**
Editors: H.J. Arditty · J.P. Dakin · R.Th. Kersten

Figure 2

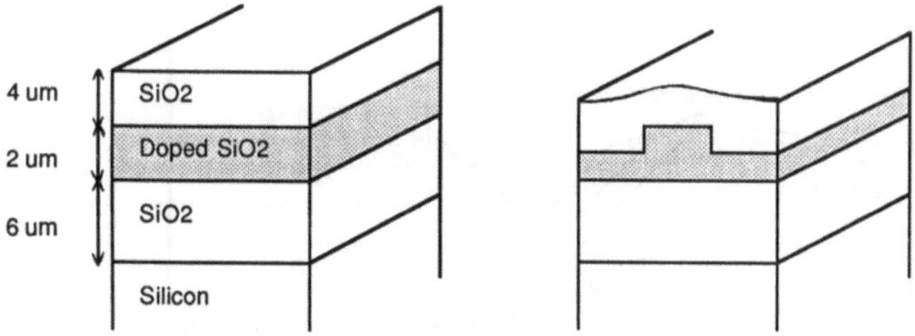

However the main drawback of the IOS1 technology comes from weak coupling efficiency when trying to connect circuit to classical optical fibers.

To solve this problem a new Integrated Optics Technology IOS2 (for Integrated Optics on Silicon Substrate n° 2), using a structure : Si / SiO_2 / Doped SiO_2 / SiO_2 , has been developed at LETI for about 4 years.[1] , [3] , [4]. In this technology all the layers are obtained by PECVD (Plasma Enhanced Chemical Vapor Deposition) which allows to achieve very thick silica layers presenting good optical properties. see figure 2 .

- The first layer constitues the buffer layer and has a thickness of 6-8 µm
- The second layer is the core of the waveguide obtained by the same technique adding phosphorus in the reactive gases in order to increase the refractive index in the range 8.10^{-3} /12.10^{-3}. Its thickness is about 2 µm .
- The third layer is the cladding layer with the same composition as the first layer, and has a typical thickness of 4 µm .

In this structure, the coupling is particularly efficient thanks to input and output channel waveguides, the refractive index and thickness of which are almost identical to those of optical fibers.The most important requirement is obviously to obtain both good optical waveguides and a good control of the waveguide structure especially in terms of refractive index difference between core and surrounding media. The refractive index homogeneity on wafer is +/- 10^{-3}

The main problem associated with IOS2 structure is related to its low refractive index difference between waveguide core and surrounding media. It prevents the achievement of optical components such as integrated lenses. However,catadioptric optics can bring a valuable solution to this drawback. [1] .

PRESENTATION OF THE DEVICE

As the IOS1 sensor, it basically consists of a Michelson Interferometer able to detect the displacement (sign and absolute value) of a reflecting object by counting the fringes of both interference patterns in quadrature. As already mentioned the input light is bounded from a single mode fiber and the output intensity is also carried up to the detectors (twin photodiode) by using optical fiber (singlemode or multimode). In order to limit the diffraction angle due to waveguide and to increase the dynamic range, a spherical lens is placed near the output plane.

An indicative diagram of the device is given in Figure 3. Its fabrication involves five masking steps and their respective alignment during the technological realization is a critical point.

Figure 3

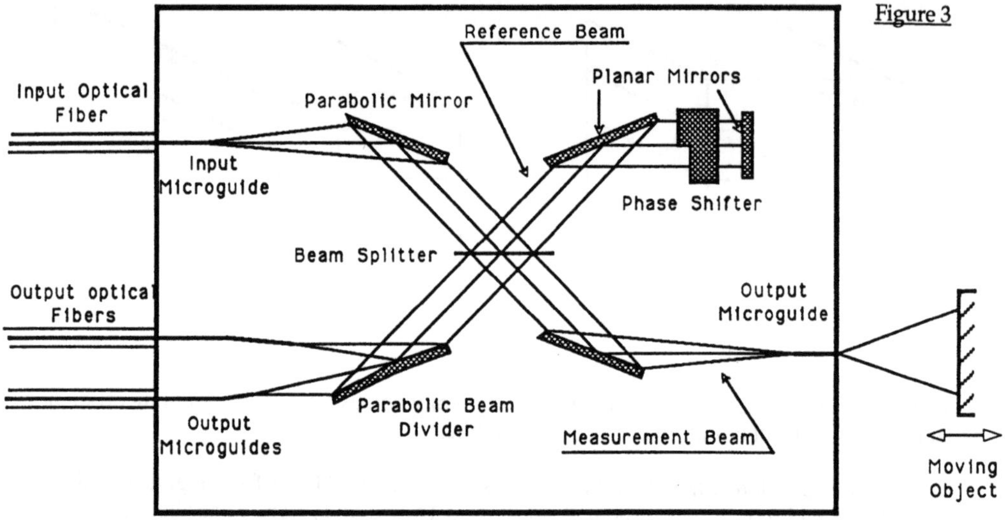

- Microguides are realized by a partial etching of the guiding layer, and then by depositing a silica overlayer, see Figure 2.

So we can obtain large microguides (4 or 5 microns) which allow an experimental coupling efficiency with a singlemode polarisation preserving fiber (York VSOP HiBi) of about 60%.
Obviously microguide's extremity coincides with the focus point of parabolic mirrors. Microguides and mirrors being on two different masks levels, their respective alignment will be a drastic parameter.

- Parabolic and planar mirrors are realized by full etching of the waveguide structure down to the silica substrate. In order to obtain maximum resolution the parabolic mirror has a greater aperture than microguide. This Numerical Aperture has been calculated from the Fourier Transform of the microguide field amplitude, and corresponds to a diffracted beam having an half width angle of about 4 degrees.

Both signals in quadrature are obtained by dividing measurement and reference arms, and by implanting a phase shifter on the last one. In that way, it will be possible to detect the value of the displacement by counting the number of interference fringes and to detect the changes in moving direction.

- Beam splitter is achieved by a local etching of overlayer and guide. First experimental tests have been done on 4 micron width beam splitters working with an incidence of 45 degrees. Best results are obtained for a 6 micron deep etch including the overlayer (i-e a total etch of the guiding layer, see Figure 4). The ratio between reflected and transmitted intensity shows fluctuations in the range .3 /.7 to .6 /.4 for different tested components.

This variation can be explained by the accuracy of 3% on the thickness inhomogeneity on wafer and by the etching depth control on the order of 3% too.

Anyway, this ratio is not a drastic parameter and such fluctuations can be tolerated.

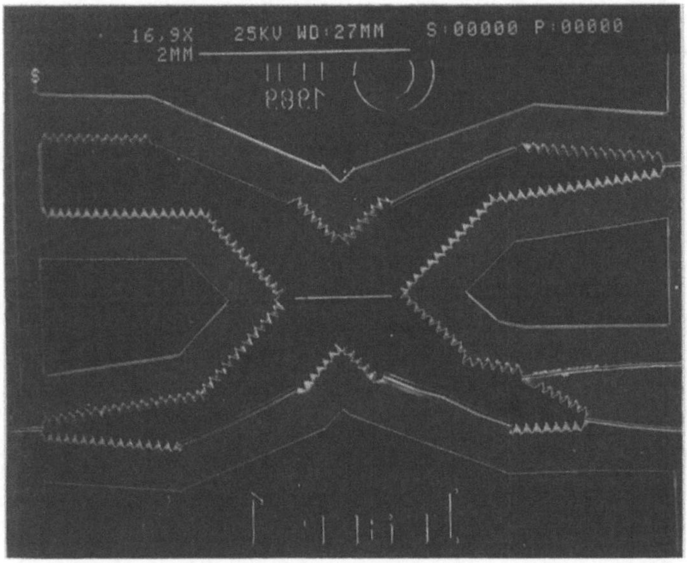

Figure 4

- Phase shifter is achieved by an effective index difference on the mode propagating in the structure, on a length L.

$L = \lambda / 8 / \delta neff$, $\lambda = 0.84\ \mu m$ being the working wavelength.

So, we have to find a solution reducing to the minimum the length variation δL due to all technological fluctuations during a wafer process.
An effective index difference could be reached by:
- total etching of the silica overlayer
- etching guiding layer before depositing silica overlayer

We chose the second solution leading to lesser fluctuations. Another advantage of this solution is that phase shifter will be processed on the same mask level as microguides. So, phase shifter length, performed by a partial etching of 1 micron of the guide, is 37 microns.

EXPERIMENTAL RESULTS

First tests of the sensor, of which photograph is given in Figure 4, without optical fibers connection have been done. Light source is a laser diode Mitsubishi ML2701 and the moving object a piezoelectric transducer working at a 100 Hz frequency. Figure 5 shows a picture of the two systems of interference fringes having $\pi/2$ shift (see Figure 6 in X-Y mode) with a TE input light. Same results have been obtained in TM polarization: the circuit seems to be polarization independent, one reason being the very low effective index difference between TE and TM modes, and the use of catadioptric components.

The feasibility of this new kind of sensor has been demonstrated. Experiments with optical fiber input/output connection will be done and the results presented at the meeting.

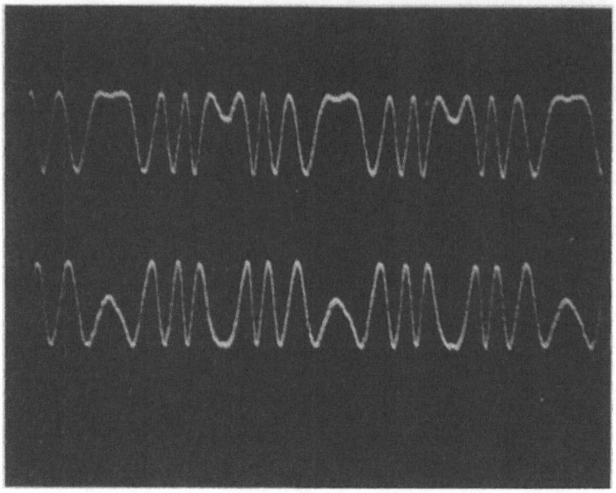

Figure 5

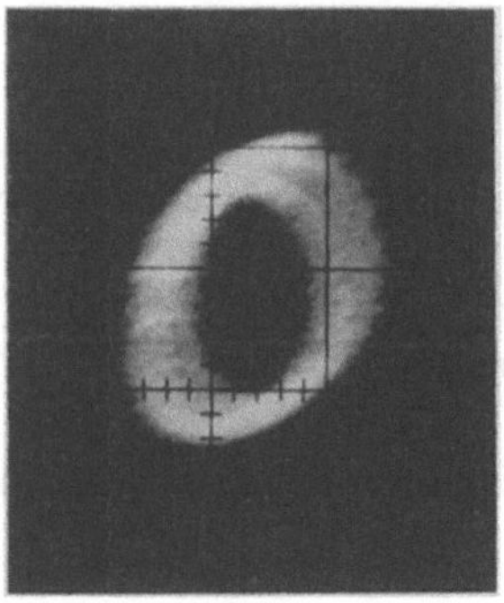

Figure 6

CONCLUSION

The main advantages of the presented device are :

- Miniaturisation of the measurement head comporting only the integrated optics components and which can be located in a volume less than 0.1 cm^3.

- A structure well fitted to the coupling with monomode fibers, and the possibility to achieve accurate positioning and alignment between optical fibers and optical circuit thanks to well controlled grooves realized on silicon substrate.

- No active component in the measurement head permitting its use in hostile environments.

We have also achieved and tested most of the optical components needed in the circuit. The quality of their characteristics is directly related to the quality of the dry etching process which appears as the key technical process in IOS2 structure and more generally in thick multilayer silicon based optics.

The results of the first tests of the sensor connected with optical fibers will maybe be presented at the meeting.

ACKNOWLEDGMENTS

The authors would like to acknowledge Denis H. , Grouillet A.M. , Philippe P. , Brot C. , Fournier A. and Barbier D. , for their assistance .

REFERENCES

[1] Valette S. ,1988, State of the art of integrated optics at LETI for achieving passive optical components. Journal of Modern Optics, Vol 35, n° 6, 993 - 1005.

[2] Lizet J. , Gidon P. , Valette S. , 1987, Integrated optics displacement sensor achieved on silicon substrate.Proceedings of the fourth European Conference on Integrated Optics ECIO Glasgow.

[3] Valette S. , Jadot J.P. , Gidon P. , Renard S. , 1987 ,New integrated optics structure on silicon substrate : application to optical communication and optical interconnects. SPIE Optical Interconnections, Vol 862 , 20 - 26 .

[4] Valette S. , Gidon P. , Jadot J.P. , May 1987, New integrated optical multiplexer demultiplexer realized on silicon substrate.Proceedings of the fourth European Conference on Integrated Optics ECIO Glasgow.

A Sensor for Dimensional Metrology with an Interferometer in Integrated Optics Technology

G. Ulbers

Hommelwerke GmbH, Alte Tuttlinger Str. 20,
D-7730 VS-Schwenningen, Fed. Rep. of Germany

For the purpose of dimensional metrology the application of an interferometer is well known. For a wide application of this method, size and cost of the necessary instruments have to be reduced significantly. Our solution to this demand was the further development of an already existing laboratory example of an interferometer in the technology of integrated optics on silicon at LETI in Grenoble, in cooperation with LETI, to an industrialized product.

The main problem which had to be solved in the development was the stability of the system, concerning the semiconductor laser and the electronic components.

1. Introduction

Optical interferometers provide an effective means of high resolution measurement of displacements combined with a big dynamic range. The simplest configuration of an interferometer is a Michelson interferometer (Fig. 1). For industrial measurements in most cases a He Ne laser is the light source. A conventional interferometer consists further of bulky glass components like mirrors and a beam divider. Those components have to be manufactured very precisely and that makes them expensive, even though the interferometer is realized in microoptics like the prototype Prof. Jäger from the University Ilmenau (GDR) presented at the industrial fair in Hannover this year. Also the He Ne-Laser is not a small and cheap component.

An alternative which doesn't has these disadvantages is the technology of integrated optics on silicon. It has specially the following advantages:
- silicon substrates have a large area (6" diameter), very reasonable price and a good optical quality

Springer Proceedings in Physics, Vol. 44 — Optical Fiber Sensors
Editors: H.J. Arditty · J.P. Dakin · R.Th. Kersten

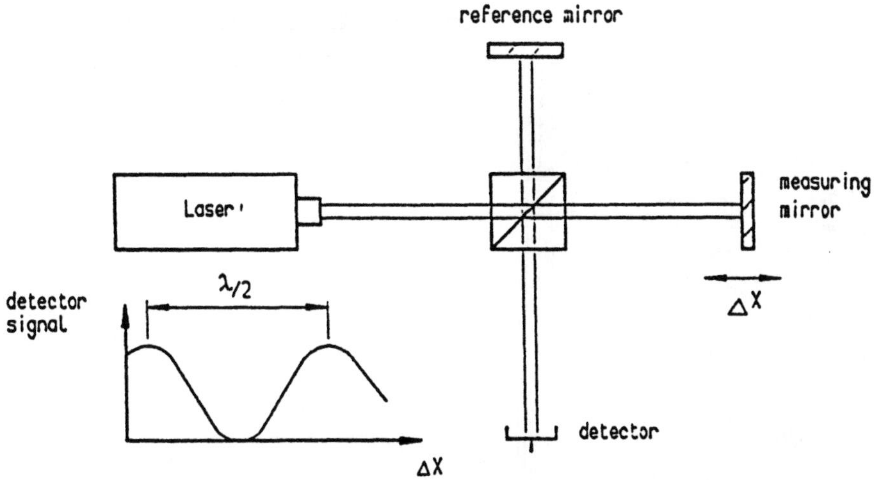

Fig. 1: Principle of a Michelson interferometer

- integrated optical structures can be manufactured with machines in planar technology which have already been introduced into semiconductor technology
- it is possible to integrate photodetectors and electronic circuits for an intelligent sensor
- integration of micromechanical components with the aid of special etching processes may result in sensors for the measurement of force, pressure and acceleration

Therefore the technology of integrated optics on silicon promises to become a cheap alternative with a lot of technical possibilities compared to other possible solutions.

We have developed, in cooperation with LETI in Grenoble and with support of the BMFT in Germany, a displacement sensor with an integrated optics interferometer as a sensor element.

2. The interferometer principle

The realized principle is based on the Michelson interferometer (see Fig. 1). But with this principle, it is not possible to detect the direction of movement of the object to be measured. Therefore a 45° phase shifter is introduced in the reference arm of the interferometer and two output channels instead of one, for a sine and a cosine signal, are created (Fig. 2).

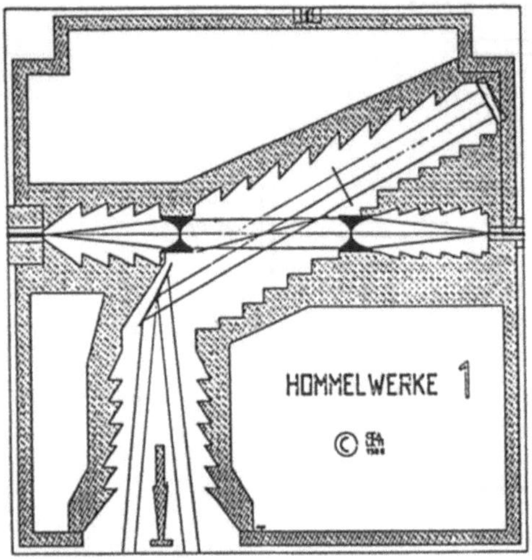

Fig. 2: Interferometer with two output channels

It is necessary to become independent from the tilt of a plane moving mirror, which may cause signal loss or degradation. Therefore we introduced instead of the plane mirror a retroreflector. But also here, looking for a cheap solution, we could not take the corner cube reflector of a conventional interferometer. We developed a special spherical retroreflector with a refractive index of n=2 and a gold coating on the reflecting surface with a diameter of only 2 mm.

In the first step, it was not possible to realize the interferometer with push pull signals, that means with four outputs or with a frequency or phase modulation device. This is a certain drawback considering the signal-to-noise ratio and the drift. This is compensated by the possibility of detecting very fast signals limited only by the frequency response of a fringe counter.

When we move the retroreflector continuously, we get a signal like that shown in Fig. 1 with a periodicity of $\lambda/2$ (λ wavelength of the laser) on a certain DC-level. The DC-level mainly originates from parasitic light in and around the chip.

3. Setup of the interferometer

All interferometer components beside photodetectors and laser are realized on a Si-chip surface of 7.5 x 7.5 mm^2 (Fig.2). The chip is mounted on a carrier (Fig. 3) which also serves as a support for a dual

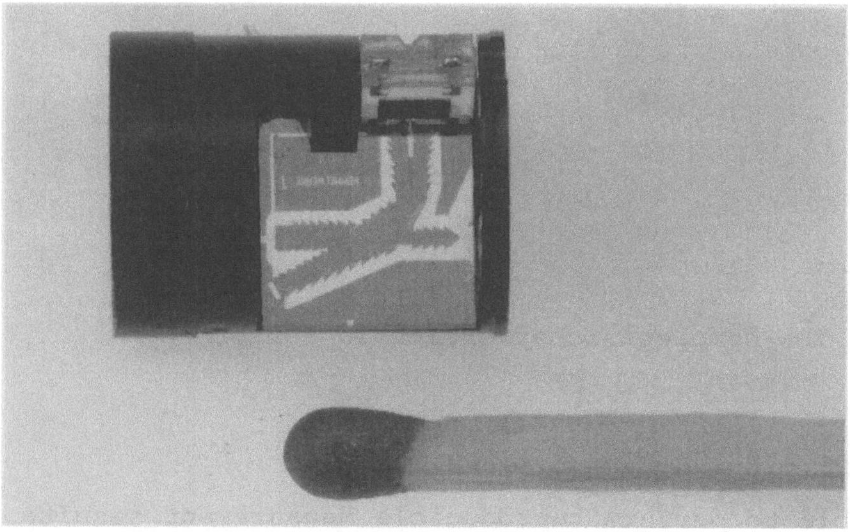

Fig. 3: The interferometer-chip carrier

photodiode and the semiconductor-laser. It is essential for getting a good performance to have a parallel measuring beam. This is achieved by a selfoc lens located directly at the output of the chip in the measurement arm. A nonparallel beam results in very low reflected light intensity and reduces the signal-to-noise ratio very much.

Because the wavelength of the laser is our meter, the laser has to be stabilized by current and by temperature. For temperature control it is sufficient to connect only a small Peltier element with the laser. Because of the small size of the sensor element, the temperature resistance is small which allows a precise temperature control.

We have low level signals in the region of some µA, therefore we need a signal preprocessing close to the interferometer chip. We realized this by a little SMD-board which preamplifies the detector signals and handles the temperature control.

All components are inserted in a tube of 12 mm diameter and a length of 60 mm, connected with a plug socket and sealed against environmental influences (Fig. 4).

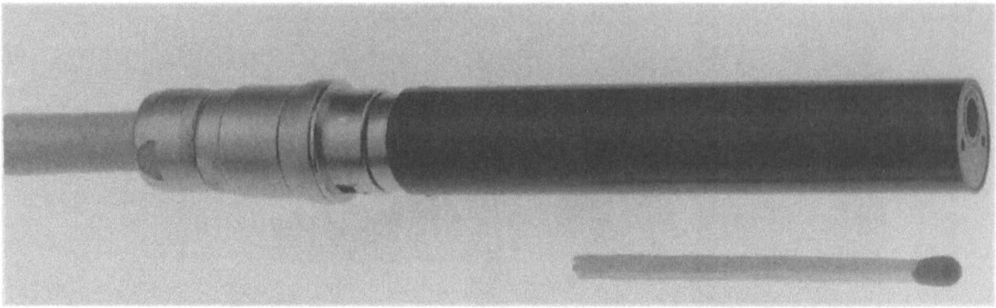

Fig. 4: The complete interferometric sensor

4. Signal processing

To be able to perform reproducible measurement results with a high resolution and accuracy, we need first a stable wavelength of the semiconductor laser. A stability of 10^{-6} of the wavelength is sufficient for many applications. This stability is achieved by stabilizing the laser current to about 1 µA with an accumulator buffered power supply and the temperature to about 5 mK with the already mentioned Peltier element. A good temperature sensor with a quick response and at the right place, is the temperature dependent inner diode resistance of the laser /1/. The performance is much better than any other temperature measurement methods which have been used up to now. The only problem is, that you must be able to measure a voltage drop of 5 µV for a wavelength stability of 10^{-6}. We stabilize the laser around room temperature (18°to 20°C) to minimize the necessary current for the Peltier element.

For a good resolution of the interferomter, we need beside a reasonable signal-to-noise ratio the ability to measure fractions of interference fringes. This is done by an 8 bit AD-converter which divides sin by cos which results in 256 values in one signal period of 2π .In practice, we get with a signal-to-noise ratio of around 40 dB about 40 to 50 distinguishable parts in one period on a measurement distance of 100 mm. That means with a wavelength of 780 nm and one signal period in $\lambda/2$ a resolution ≤10 nm. The fringe fractions are counted with a high speed counter. The result multiplied by $\lambda/2$ is then the displacement of the retroreflector in µm.

When we work with a nonmodulated signal, we have to eliminate the DC-level of the signal. This is simply done by measuring and subtracting. But this works only when the DC-level is constant. A sufficiently constant DC-level is achieved by the already mentioned means for stabilizing the semiconductor laser. Here low frequency noise limits the resolution of the interferometer.

Fig. 5 shows a block diagram of the whole system. For very precise measurements around a distance of 100 mm a device is added for compensating the influence of the refractive index of the air, e.g. a pressure and temperature sensor or a second microinterferometer with a fixed reflector.

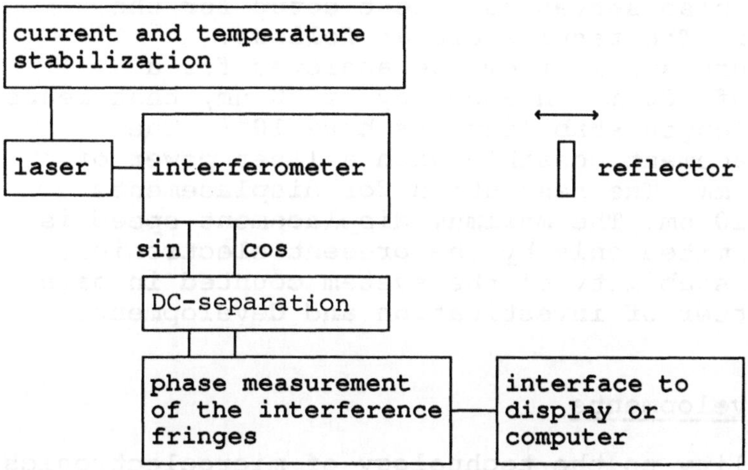

Fig. 5: Block diagram of the complete setup

5. Performance of the interferometer

The interferometer was investigated for measurement ranges between 1 µm and 500 mm. For this range a monomode GaAlAs double-heterostructure laser diode with an output power of 3 mW proved as sufficient. With a wavelength of 780 nm we achieved a wavelength stability of 10^{-6} with a slow tolerable change of the environmental temperature of ±10 K.
To determine the exact wavelength of the laser diode, the interferometer was calibrated with a conventional Hewlett Packard Interferometer with a wavelength stability of 10^{-8}. This can be done on a very short distance in a temperature stable environment, so that the influence of the refraction index of the air is

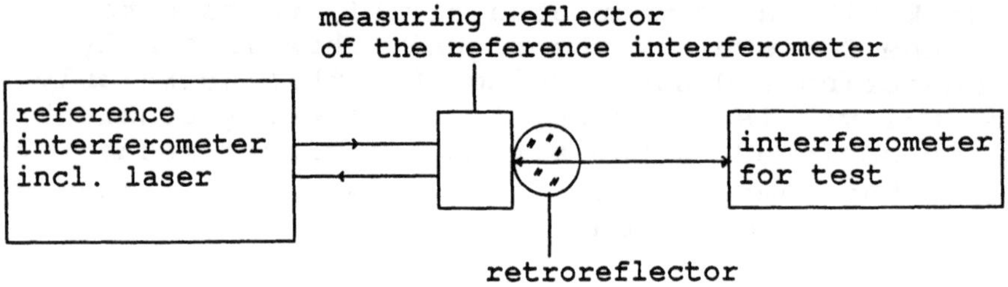

Fig. 6: Setup for calibration and test

negligible. Both interferometers measure a displacement of the same movable axis. On the movable object to be measured, we glued a spherical retro-reflector with a diameter of 2 mm. The calibration setup is shown in Fig. 6, which also serves as a test setup for the interferometer. The results are as follows:
In a temperature stable room, we achieved for a displacement of 100 mm an accuracy of 10 nm, that means that the wavelength stability was here 10^{-7}. The maximum displacement possible with a laser power of 3 mW was 1000 mm. The resolution for displacements < 10 mm is < 10 nm. The maximum displacement speed is now 0.3 m/s limited only by the present electronics. The long term stability of the system counted in days is still a matter of investigation and development.

6. Further developments

The compatibility to the technology of microelectronics has already been proved/2/. It is possible to reduce further the size of the whole system; the electronics like photodetectors and amplifiers will be integrated on the chip together with the interferometer.

To improve the resolution further and with that the signal-to-noise-ratio, it is necessary to introduce a frequency modulated signal on the chip. This can be done e.g. by a Bragg cell on the chip/3/.

There is a demand for absolute displacement measurement. Therefore, it is necessary to introduce a reference point. This can be done by a separate inductive or capacitive sensor or optically after the principle of a compact disc reading head, which can be also made in integrated optics on the same chip.

References

/1/ K.Uehara, K.Katakura: "New Method of Frequency Stabilization of Semiconductor Lasers" Japanese Journ. of Appl. Phys. 27,2 (1988)

/2/ S.Valette, J.P.Jadot, P.Gidon: "Integrated optics and microelectronics toward an integration on a single silicon chip" SPIE Vol.864 Advanced Optoelectr. Techn. (1987)

/3/ P.Mottier, S.Valette, J.P.Jadot: "Broad band Bragg deflector for optical waveguides on silicon substrates" Optics and Laser Technology 18, 2 (1986)

Part VIII

Current and Magnetic Field Sensors

Recent Advances in Faraday Effect Sensors

G.W. Day

National Institute of Standards and Technology, Boulder, CO 80303, USA

This paper reviews recent developments in the application of the Faraday effect to electric current and magnetic field sensing.

The advantages of using the Faraday effect for magnetic field or electric current measurements are well known. Sensors can be constructed entirely of dielectric materials, an important consideration when operating at high voltage or in the presence of electromagnetic interference. Faraday effect sensors are faster than other types of optical sensors for magnetic fields. When diamagnetic materials are used, the fundamental speed limitation is the transit time of light through the sensing element. Either bulk materials or single mode optical fiber can be used as the sensing element, providing great freedom in configuration. And, because they generally use polarimetric, rather than interferometric techniques, they typically are more stable than alternatives. Their principal disadvantage is generally thought to be a lack of sensitivity, but dramatic improvements have been achieved and further gains can be foreseen. The principles and development of Faraday effect sensors have recently been reviewed [1]. In this paper, more recent results will be emphasized.

Current sensors

The principal interest in Faraday effect sensors to date has been for the measurement of electric current. While the current flowing in a conductor can be inferred from a measurement of the magnetic field a known distance from the conductor, a more satisfactory approach is to measure the integral of the magnetic field along a closed (or nearly closed) path around the conductor. It is in this sense that we distinguish a 'current sensor' from a 'magnetic field sensor.'

Although many of the early proposals for Faraday effect current sensors used optical fiber to provide the optical path around the conductor, the polarization properties of the fiber were a severe limitation, and fiber gave way to bulk materials, usually glass, as the sensing element. Sensors based on bulk materials have the further advantages that materials not available in fiber form can be used, and special configurations, such as those in which the sensor can be

Springer Proceedings in Physics, Vol. 44 **Optical Fiber Sensors**
Editors: H.J. Arditty · J.P. Dakin · R.Th. Kersten

slid on the conductor without breaking the optical path can be considered. Current sensors based on bulk materials are now available and have been or are being field tested in several locations. One of the most extensive field tests, completed recently by the Tennessee Valley Authority, demonstrated that accuracy sufficient for metering could be maintained continuously over an 18 month time period [2].

The use of fiber as the sensing medium can be much less costly, however, and fiber also facilitates the use of multiple passes around the conductor to improve sensitivity. As improvements in fiber and fiber processing occur, it seems likely that fiber will become the preferred medium.

The primary problem with fiber is linear birefringence, which is either inherently present in the fiber or induced by the stress associated with bending the fiber around the conductor. Fibers with low inherent birefringence are now readily available, but bend-induced birefringence remains a problem. One method of overcoming the problem is to use a highly circularly birefringent fiber. Circular birefringence can be achieved by twisting[3], by using a helical guiding path[4], or by using a highly linearly birefringent fiber in which the axes of birefringence are rotated along the fiber. The latter of these approaches, though it yields an elliptically, rather than circularly birefringent fiber, has been the most successful in producing small sensing coils[5].

An alternative approach is to remove the bend birefringence by annealing[6]. Coils are wound into the desired form, heated to 800 to 900 °C for approximately 24 hours, and then cooled slowly. When cool, the coils retain their shape without restraint and their linear retardance is nearly eliminated. Using this technique, coils as small as 5 mm diameter with 10 to 20 turns have been produced, and coils of 1 to 3 cm with 100 to 200 turns can be manufactured routinely[7].

Some residual linear retardance remains in these coils, typically 20 to 50 deg at the operating wavelength. At room temperature and above, the residual retardance generally varies with temperature as $(1/\delta)d\delta/dT \sim 10^{-3}/°C$.

Modelling indicates that if the residual birefringence is uniformly distributed through the coil, (and not in the fiber leads) the sensitivity, S, of the sensor (defined as the rotation of the plane of polarization per unit current for small rotations) varies with residual retardance as

$$S \sim (\sin \delta)/\delta. \tag{1}$$

The change in sensitivity with retardance is

$$(1/S)dS/d\delta = \text{ctn}\ \delta - 1/\delta, \tag{2}$$

or for small δ

$$\approx -\delta/3. \tag{3}$$

This, and the typical change in residual retardance with temperature cited above, leads to the estimate that for a coil with 1 rad of residual retardance, the change in sensitivity with temperature will be between 10^{-3} and 10^{-4} /°C.

Recently, separate measurements of the temperature dependence of the Verdet constant of silica have indicated that (1/V)dV/dT can be as small as 0.01%/°C [8], so the effect of the residual birefringence will typically be the limiting factor for coils with a residual retardance greater than 10 or 20 deg. Experiments to confirm these estimates are in progress.

Speed of response. When many turns of fiber are used, the speed of response can be limited by the transit time of light through the coil. The frequency response function is given, in this case, by[9,10]

$$R(\omega) = \mu VN(\sin \omega\tau/2)/(\omega\tau/2) \tag{4}$$

where μ is the magnetic permeability of the fiber, V is the Verdet constant, ω is the radian modulation frequency, N is the number of turns of fiber around the conductor, and τ is the transit time for light through the fiber.

From equation(4), the -3dB bandwidth can be shown to be[10]

$$BW \approx c/(\pi^2 nNR\sqrt{2}) \tag{5}$$

where c is the speed of light in vacuum, n is the effective refractive index of the fiber, and R is the radius of the coil.

Recently, equation (4) has been confirmed to 12 MHz in a 100 turn. 10.7 cm coil having a -3dB bandwidth of 2.7 MHz [11] and higher frequency testing is in progress.

Interferometric designs. Numerous interferometric current sensor configurations have been investigated as alternatives to the more usual polarimetric configurations [12]. Of these, only multiple pass interferometers such as the Fabry-Perot interferometer and the ring resonator, and the Sagnac interferometer seem to have fundamental advantages. Multiple pass interferometers with a finesse greater than one can provide a greater sensitivity than polarimetric configurations [13]. They have the disadvantage of requiring a high coherence source. The Sagnac interferometer [14] is attractive in that it requires no polarizing components and is immune to changes in the input polarization state. A typical design based on a 2x2 fiber coupler has the disadvantage that its response function has zero slope for small currents. It is well known, however, that this limitation can be overcome by using a 3x3 coupler, and such a sensor has recently been demonstrated [15].

Magnetic field sensors

Recently, interest in very small, non-electrical, magnetic field sensors has emerged. The use of the Faraday effect in small elements of bulk material can provide attractive solutions for many of these applications. A sensor based on an 11 mm pathlength in ZnSe has achieved a noise equivalent field of about 200 nT/√Hz at frequencies to greater than 100 MHz [16]. A sensor based on a 5 mm pathlength in YIG has yielded a noise equivalent field of less than 10 nT/√Hz at low frequencies [17]. While these sensitivities are several orders of magnitude worse than the best results with magneto-strictive sensors, the Faraday devices can apparently meet the requirements of many applications with much simpler, and therefore less expensive devices.

References

1. G. W. Day and A. H. Rose, "Faraday effect sensors: The state of the Art," SPIE **985**, 138-150 (1988)
2. (a) P. M. Johnston, Westinghouse ABB, Raleigh, NC
(b) C. Carter and J. Stites, "A magneto-optic current transducer," InTech **34**, 41-45 (June, 1987)
3. S. C. Rashleigh and R. Ulrich, "Magneto-optic current sensing with birefringent fibers," Appl. Phys. Lett. 17, 52-56 (1978)
4. M. P. Varnham, R. D. Birch, and D. N. Payne, "Helical core circularly birefringent fibers," 5th Int. Conf. Int. Opt. and Opt. Fiber Commun. and 11th ECOC, Venice, 1985
5. L. Li, J. R. Qian, and D. N. Payne, "Miniature optical fiber current sensors," Int. J. Optical Sensors, 2, 25-33 (1987)
6. G. W. Day and S. M. Etzel, "Annealing of bend-induced birefringence in fiber current sensors, "5th Int Conf Int. Opt. and Opt. Fiber Commun. and 11th ECOC, Venice, 1985.
7. D. Tang and G. W. Day, "Progress in the development of miniature optical fiber current sensors, "Proc. IEEE LEOS Annual Meeting, Santa Clara, 1988
8. P. Williams, et al., National Institute of Standards and Technology, Boulder, CO, to be published
9. Amnon Yariv, *Quantum Electronics*, John Wiley, 1975, p. 350
10. A. D. Kersey, F. Bucholtz, and A. Dandridge, 'Sensitivity-bandwidth limitations in optical-fibre Faraday-rotation current sensors," Int. J. Optoelectronics 3, 323-332 (1988) [*Note that eq. 9 and the corresponding label in fig. 2 of this reference are incorrect; the correct expression is given as eq. 5, above*]
11. R. W. Cernosek, Sandia National Laboratory, Albuquerque, NM, and G. W. Day, National Institute of Standards and Technology, unpublished

12. R. P. Tatum, M. Berwick, P. Akhavan-Leilabady, J. D. C. Jones, and D. A. Jackson, "Applications of the Faraday effect using monomode optical fiber," SPIE 734, 178-192 (1987)
13. A. D. Kersey, "Fiber optic current sensor based on Faraday rotation in a resonant fiber ring," Proc. IEEE LEOS Annual Meeting, Santa Clara, 1988
14. H. J. Arditty, Y. Bourbin, M. Mapuchon, and C. Puech, "Current sensor using state-of-the-art fiber-optic interferometric techniques," Tech. Digest, Third Int. Conf. on Int. Optics and Opt. Fiber Commun., San Francisco, 1981
15. L. R. Veeser, Los Alamos National Laboratory, and G. W. Day, National Institute of Standards and Technology, to be published
16. G. W. Day, K. S. Lee, A. H. Rose, L. R. Veeser, B. J. Papatheofanis, and H. K. Whitesel, "Optical fiber sensors for electromagnetic quantities," Proc. 34th Int. Instrumentation Symposium, Instrument Society of America, Albuquerque, NM 1988
17. M. N. Deeter, A. H. Rose, and G. W. Day, "Characteristics of polarimetric magnetic field sensors based on Yttrium Iron Garnet," submitted for publication

Heterodyne Detection of Magnetic Fields from 0.1 Hz to 10 MHz in a Magnetostrictive Fiber Sensor

D.M. Dagenais, F. Bucholtz, and K.P. Koo* [1]

Naval Research Laboratory, Optical Sciences Division, Code 6570, Washington, DC 20375, USA

*Permanent address: SFA Inc., 1401 McCormick Drive, Landover, MD 20785, USA

Using a heterodyne technique we have detected ac magnetic fields from 0.1Hz to over 10 MHz in a magnetostrictive sensor. The response is linear and independent of the signal frequency up to a few MHz with no loss in resolution compared to direct ac detection .

I - INTRODUCTION

The strong frequency variation in the response of a metallic glass cylinder typically limits the operating window for direct ac detection to less than 50 kHz. Rather than using that window for direct ac measurements(1,2) a heterodyning technique such as the one suggested by Okamura for a Lorentz force sensor(3) can be implemented. When the carrier (Local Oscillator) tracks and mixes with the signal to be detected in the nonlinear magnetostrictive material the strain at the fixed difference frequency will mechanically couple into the fiber producing an interferometer output proportional to the high frequency signal to be detected. This technique eliminates the problem of frequency response variation and yields resolutions similar to those obtained with direct ac detection(2).

We report for the first time the response of a magnetic fiber sensor over 8 decades of frequency. The response is independent of frequency up to 1 MHz and depends linearly on the amplitude of the signal to be detected. In preliminary measurements we also detected signals up to 1 GHz, however the frequency response becomes more complex above 10 MHz and will be reported elsewhere.

II- HETERODYNING TECHNIQUE

In a metallic glass transducer, two high-frequency magnetic fields mix linearly due to the quadratic response of the magnetostrictive material obtained from the simple Coherent Rotation Model(4),

Editors: H.J. Arditty · J.P. Dakin · R.Th. Kersten

$$e = C\,H^2 \qquad (1)$$

where e is the material's longitudinal strain due to a total applied field H and C is the magnetostrictive coupling parameter. When the two applied fields are $h_1 \cos 2\pi f_1 t$ and $h_2 \cos 2\pi f_2 t$, the strain at the difference frequency $f = f_1 - f_2$ is

$$e(f) = C(f)\, h_1\, h_2 \cos 2\pi f t. \qquad (2)$$

If the frequency f lies within the window corresponding to the mechanical response of the transducer, the strain produces an interferometer phase shift of amplitude

$$\Phi(f) = k\, C(f)\, L\, h_1\, h_2 \qquad (3)$$

where k is the effective wave number $(2\pi n \zeta L/\lambda)$ in the fiber core with refractive index n at laser wavelength λ and strain-optic correction factor ζ. $\zeta = [1 - n^2\{(1-\mu)p_{12} - \mu p_{11}\}/2\ ]$ where μ is the Poisson ratio and p_{ij} are components of the strain-optic tensor ($\zeta \approx 0.78$ in silica glass fibers). L is the length of the sensing fiber.

As can be seen from Eqn.(3), when the difference frequency f and the amplitude of the local oscillator h_1 are fixed and the frequency f_2 is swept, the signal $\Phi(f)$ is linearly proportional to the amplitude of the signal to be detected (h_2). Furthermore, the signal at f is independent of the DC magnetic field . It should be noted that Eqn.(3) is only valid if the magnetic fields are not attenuated by eddy currents or if the mixing is not quenched by large DC fields [5].

By selecting f to fall on a mechanical resonance of the transducer, the coupling is optimized and maximum signal-to-noise ratio is obtained .

III- EXPERIMENTAL SET-UP AND RESULTS

The experimental set-up shown in Fig.1 is a typical fiber-optic Mach-Zehnder interferometer. The semiconductor laser is a 0.83 µm low-noise Hitachi HLP 1400 and the light is launched into the interferometer through an isolator and half-wave plate (for fringe visibility optimisation). The single mode fiber has 31m of length wrapped around a transversely annealed Metglas 2605 S2 cylinder (5cm long by 4.4cm diameter). The interferometer is balanced to less than 1mm and is kept at quadrature by active phase compensation using the PZT cylinder. The ac magnetic fields are obtained from a single loop emitter centered around the transducer and the DC field is obtained by driving a solenoid concentric with the transducer. The two ac fields are obtained by driving the magnetic loop with the combined output of an HP 3326A two-channel synthesizer. The outputs are kept at

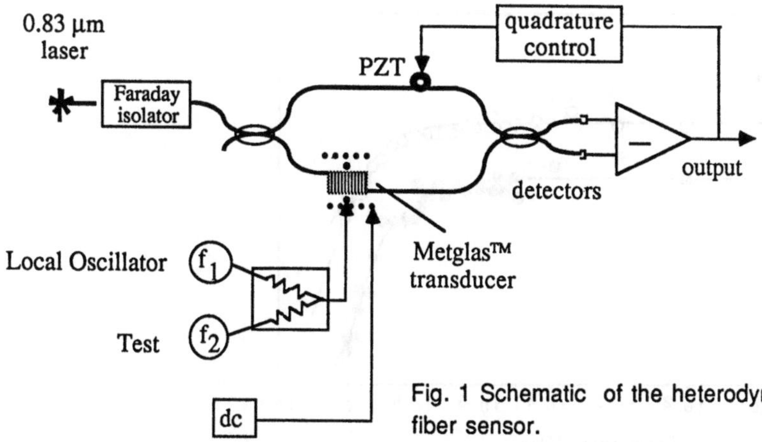

Fig. 1 Schematic of the heterodyne magnetostrictive fiber sensor.

a fixed frequency difference and exhibited frequency stabilities of less than 1 Hz for the duration of the measurements. The measured value of the interferometer phase shift was reproducible to within ±1.5 dB.

We investigated the direct frequency response of the transducer from 10 kHz to 100 kHz. As seen in Fig.2, direct ac detection is limited to a 20 kHz window centered about 25 kHz, and the sensitivity varies with frequency with main resonances at 25 kHz and 31 kHz. The response at 31kHz is approximately down by a factor of 2 from that at 25 kHz. With the heterodyning technique we detected fields ranging from 0.1 Hz to 13 MHz (the limit of the synthesizer) with the difference frequency set at one of the two resonance frequencies.

The amplitude of the ac magnetic fields was first calibrated using a small pick-up coil located at the center of the emitter loop. When the pick-up coil shows an output voltage V, the magnetic field h at the frequency f is computed by

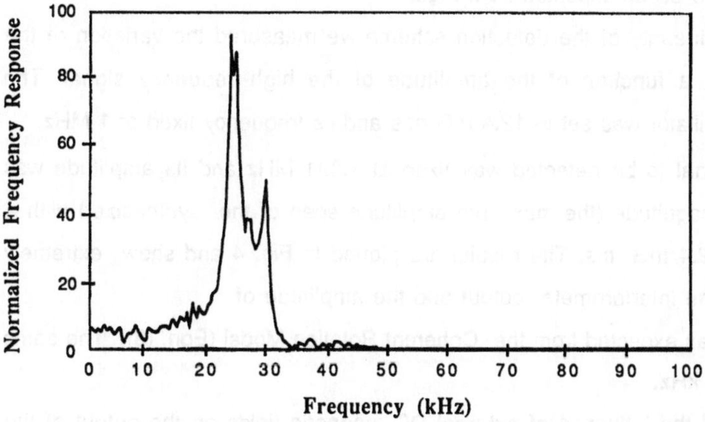

Fig.2- Direct frequency response (linear scale) of the Metglas transducer.

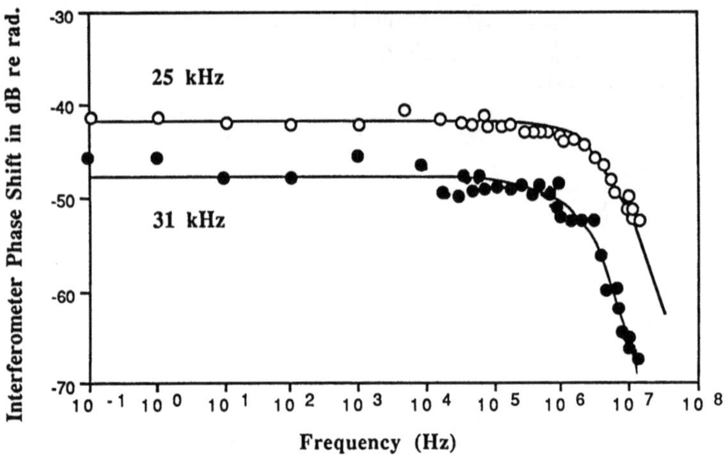

Fig. 3- AC response of the heterodyne magnetic fiber sensor. Field amplitudes are 12.4 mG rms for both test and LO signals.

$$h = V / 2\pi f A \qquad (4)$$

where A is the area of the loop (radius 4mm). We verified that the calibration remained linear in frequency from 1 kHz to 13 MHz. Care was taken to check for and eliminate any direct EM pick-up at the interferometer output and electronics.

By applying two magnetic fields to the transducer and setting the difference frequency equal to one of the resonance frequencies, we obtained the data shown in Fig. 3. According to Eqn.(3), the output of the interferometer for fixed field amplitudes is constant and proportional to the magnetostrictive coupling parameter C(f). Figure 3 shows a constant response of the transducer from 0.1 Hz to 1 MHz and a 20 dB/decade roll-off above 1 MHz which we believe is due to eddy current effects. The ratio of the response at 25 kHz to that at 31kHz is approximately 6 dB as expected from Fig.2.

In order to test the linearity of the detection scheme we measured the variation of the interferometer output as a function of the amplitude of the high-frequency signal. The amplitude of the local oscillator was set to 12.4 mG rms and its frequency fixed at 1 MHz. The frequency of the signal to be detected was fixed at 1.031 MHz and its amplitude was varied over 4 orders of magnitude (the maximum amplitude span of the synthesizer) with a maximum amplitude of 12.4 mG rms. The results are plotted in Fig. 4 and show extremely good linearity between the interferometer output and the amplitude of the signal to be detected as expected from the Coherent Rotation Model (Eqn. (3)). The same linearity was found at 25 kHz.

We also measured the influence of external DC magnetic fields on the output of the interferometer. A variable DC magnetic field was superimposed on a set of fixed high

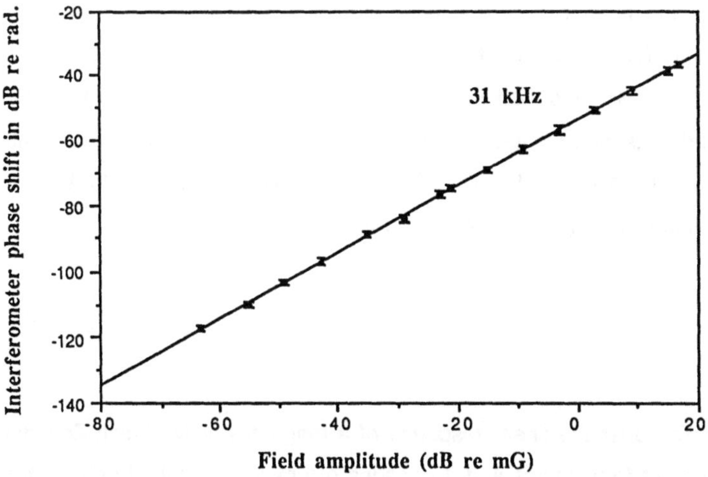

Fig.4- Response of the heterodyne sensor as a function of the signal amplitude. The LO was set at 12.4 mG rms and 1 MHz. The detection bandwith was 18.7 mHz.

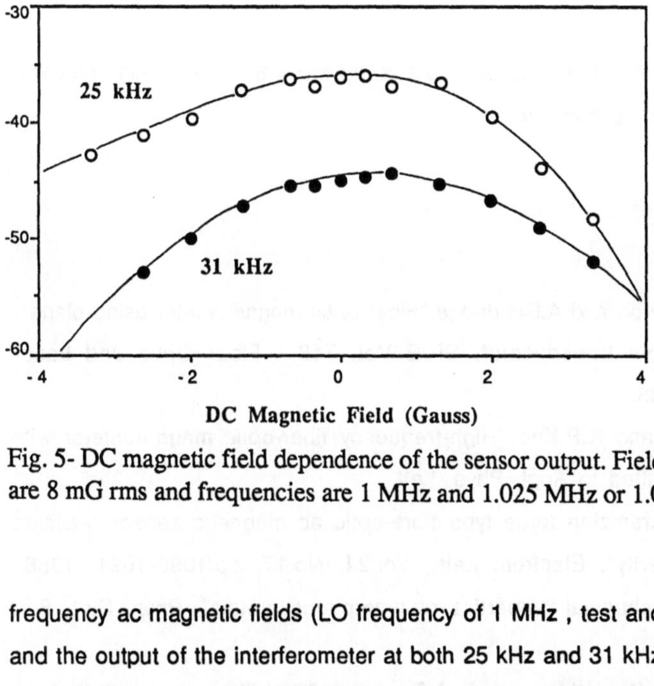

Fig. 5- DC magnetic field dependence of the sensor output. Field amplitudes are 8 mG rms and frequencies are 1 MHz and 1.025 MHz or 1.031 MHz.

frequency ac magnetic fields (LO frequency of 1 MHz , test and LO amplitudes of 7 mG rms) and the output of the interferometer at both 25 kHz and 31 kHz was measured. Fig. 5 shows the variation in the output when the DC magnetic field was varied from -3.3G to +3.3G. As seen from the figure, the response is maximum and flat in the [-0.5G,+1.5G] range corresponding to small DC fields and the quadratic operating regime of Eqn.(1). The response starts rolling off for larger fields as the DC bias begins to saturate the magnetostriction[5]. The 0.5G offset corresponds to the DC ambient magnetic field in the laboratory.

In order to determine the resolution of the heterodyning technique, we measured the power spectrum of the interferometer output in a 10 Hz span around 25 kHz. When the local oscillator was set at 1 MHz and 12.4 mG rms, the noise floor (corresponding to the interferometer noise floor) was approximately 4 μrad/$\sqrt{}$Hz and the minimum detectable field was measured to be 45×10^{-8} G/$\sqrt{}$Hz. The C parameter was measured to be $3 \times 10^{-6}/Oe^2$, the same value obtained from direct ac measurements.

IV CONCLUSION

We demonstrated in this paper the linear response of a magnetostrictive Mach-Zehnder fiber sensor over 8 decades of frequencies and 4 decades of amplitude, limited only by the interferometer floor noise, using a heterodyning technique and making use of the mechanical resonance of the transducer. This is the first demonstration of a linear magnetic sensor with flat response from virtually DC to the MHz range.

We acknowledge the help of J.A. McVicker in preparing the transducer. This work was supported by the Office of Naval Technology.

REFERENCES

1- F.Bucholtz, A.M.Yurek, K.P.Koo and A.Dandridge "Fiber optic magnetometer using planar and cylindrical magnetostrictive transducers", SPIE Vol. 718 - Fiber Optic and Laser Sensors IV, 1987, pp.128-133.

2- F.Bucholtz, D.M.Dagenais and K.P.Koo "High-frequency fiber-optic magnetometer with 500 fT/$\sqrt{}$Hz resolution", submitted to Appl. Phys. Lett.

3- H.Okamura and Y.Fujii "Lorentzian force type fibre-optic ac magnetic sensor realising frequency-independent sensitivity", Electron. Lett., Vol.24, No.17, pp.1090-1091, 1988.

4- J.D.Livingston "Magnetomechanical properties of amorphous metals", Phys. Stat. Sol. Vol.70,pp.591-596,1982.

5- K.P.Koo, F.Bucholtz, A.Dandridge and A.Tvetven "Stability of a fiber-optic magnetometer", IEEE Trans. Magn., Vol. MAG-22, No.3, pp.141-144, 1986.

Input Polarization Coding in Fibre Current Sensors

Z.B. Ren and Ph. Robert

Laboratoire de Métrologie, Swiss Federal Institute of Technology
of Lausanne, CH-1015 Lausanne, Switzerland

A Faraday fibre current sensor using a time multiplexing of two different states of polarization at the input of the fibre is presented. Thus the Faraday rotation measurement becomes independent of the linear birefringence in the fibre. A polarization multiplexing system using a chopper has been built. The experimental results and the way to account for imperfections of the optical components are also discussed.

Introduction

The use of Faraday fibre current sensor has frequently been reported in high current measurement [1]-[4]. But its performance is often limited by the residual linear birefringence in the fibre[5][6], especially the bend-induced birefringence, if the sensor is based on the use of sensing fibre surrounding the electrical conductor. The current measurement via Faraday effect is based on the rotation of the plane of polarization of a linearly polarized light traversing the fibre. This rotation is induced by the magnetic field component along the direction of light propagation. The effect of the linear birefringence is to cause the linear polarization which is acted upon by the Faraday effect to become elliptical. This can dramatically reduce the sensitivity of the device. In the worst case, the Faraday rotation is completely quenched by the linear birefringence and the current measurement becomes impossible. The linear birefringence superposes vectorially with the Faraday rotation in the fibre and it is impossible to separate them at the output by inserting a birefringence compensator.

But by multiplexing in time two different states of polarization at the input of the fibre, the two birefringences can be distinguished from each other. Accordingly, the Faraday rotation measurement becomes linear birefringence independent. This will be dealt with in this paper.

Theory

In the conventional configuration of Faraday fibre current sensor[1], a linearly polarized light is launched into the fibre and the output polarization is analyzed by a Wollaston prism. The propagation of the light through the sensing fibre exhibiting both linear and circular birefringence can be described by the Jones calculus[1][7]

$$\mathbf{E}_{out} = \begin{pmatrix} E_x \\ E_y \end{pmatrix}_{out} = M_w M_f \mathbf{E}_{in} = \begin{pmatrix} \cos\varphi & -\sin\varphi \\ \sin\varphi & \cos\varphi \end{pmatrix} \begin{pmatrix} A & -B \\ B & A^* \end{pmatrix} \begin{pmatrix} E_x \\ E_y \end{pmatrix}_{in} \qquad (1)$$

where M_W, M_f represent the Jones matrices of the Wollaston prism and the fibre, respectively. Furthermore

$$A=\cos(\alpha/2) + j\sin(\alpha/2)\cos\chi \qquad (2)$$

$$B=\sin(\alpha/2)\sin\chi \qquad (3)$$

$$\alpha = [\,\delta^2 + (2F)^2\,]^{1/2} \qquad (4)$$

$$\tan\chi = (2F)/\delta \qquad (5)$$

where F is the Faraday rotation and δ is the total linear birefringence.

When a linearly polarized light oriented along one of the birefringence axes of the fibre is launched, the Wollaston prism being oriented at 45° with respect to this axis, the intensities I_1 and I_2 of the two orthogonal polarizations separated by the Wollaston prism are given by

$$I_1 = 1/2\,(1 + \sin\chi\,\sin\alpha) \qquad (6)$$

$$I_2 = 1/2\,(1 - \sin\chi\,\sin\alpha) \qquad (7)$$

The transfer function T_1 of the sensor system is then

$$T_1 = \frac{I_1 - I_2}{I_1 + I_2} = 2F\,\frac{\sin\left(\sqrt{\delta^2 + (2F)^2}\right)}{\sqrt{\delta^2 + (2F)^2}} \qquad (8)$$

Equation (8) shows that the sensor has the maximum possible sensitivity to Faraday rotation given by $T_1 = \sin 2F$ in absence of linear birefringence or when $2F \gg \delta$. But in the case where $\delta \gg (2F)$, equ.(8) comes to $T_1 = (2F)(\sin\delta/\delta)$ and the sensor sensitivity strongly depends on the linear birefringence. In the worst case, the value of $\sin\delta/\delta$ is close to zero and the measurement of the Faraday rotation by (8) becomes very critical.

In this conventional configuration of the sensor, only one signal T_1 which is a function of two independent variables δ and F is used. Consequently, the Faraday rotation can be obtained only if the linear birefringence is exactly known.The linear birefringence effect can be limited by twisting the fibre which increases the circular birefringence. But the required twist-induced birefringence is often so great that the fibre reaches its fracture limit. Furthermore this birefringence is temperature dependent. The presented solution is based on the measurement of two signals containing comlementary data about F and δ. A short signal processing allows to take into account the effect of δ on the measured value of F.

The superposition of the Faraday rotation with the linear birefringence leads the fibre to become elliptically birefringent. This birefringence acts also on a circularly polarized input light. When such a light is launched into the fibre, its state of polarization is modified by this birefringence along the path of propagation. In this case, if the Faraday rotation is absent, the sensor output will yield the maximum possible sensitivity to the linear birefringence[8]. This suggests the idea to use linearly and circularly polarized light alternatively in order to separate the contributions due to F and δ.

For a circularly polarized input light, the sensor output becomes

$$T_2 = \frac{I_1 - I_2}{I_1 + I_2} = \delta \frac{\sin\left(\sqrt{\delta^2 + (2F)^2}\right)}{\sqrt{\delta^2 + (2F)^2}} \qquad (9)$$

Combining the equations (8) and (9), the Faraday rotation can be directly obtained from the two output signals T_1 and T_2 by

$$2F = \frac{\arcsin\left(T_1^2 + T_2^2\right)^{1/2}}{\left[1 + \left(T_2/T_1\right)^2\right]^{1/2}} \qquad (10)$$

Experiment

A time multiplexing system of the two different states of input polarization using a chopper has been realized as shown in fig. 1. A He-Ne laser (3mw, 633 nm) is used as light source. The linearly polarized light is obtained by a Glan Thompson polarizer P_1 with its transmission direction parallel to one of the axes of the λ/4 plate. The circularly polarized light is obtained by another polarizer P_2 with its transmission axis at 45^0 with respect to the axes of the λ/4 plate. The λ/4 plate is so oriented that its axes coincide with that of the fibre linear birefringence.

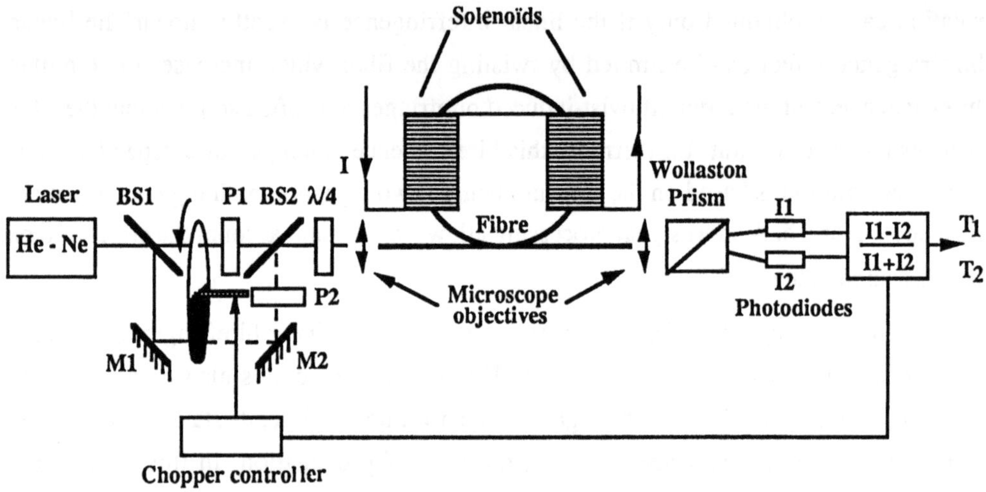

Fig.1 Experimental set-up for the Faraday rotation measurement by time polarization multiplexing.

A York ultra-low birefringence LB_600 fibre with a core diameter of 3.5 μm and cladding diameter of 115 μm is used as sensing fibre. The Verdet constant V of this fibre has been measured, and found to be 2.65×10^{-4} deg/A [9]. Nine turns of this fibre have been wound in a 22 cm diameter groove carved in a PVC plate. An equivalent current of 0-8 KA across the fibre loops is applied by two 800-turns solenoïds in series.

The measurement is actually controlled by a micro-computer. The electronics which yields T_1 and T_2 is synchronized with the chopper controller. The maximum multiplexing frequency is 4 kHz. A DC current of 0-5 A is applied to the solenoïds. A typical measured Faraday rotation against the equivalent current across the fibre loops is represented in fig. 2.

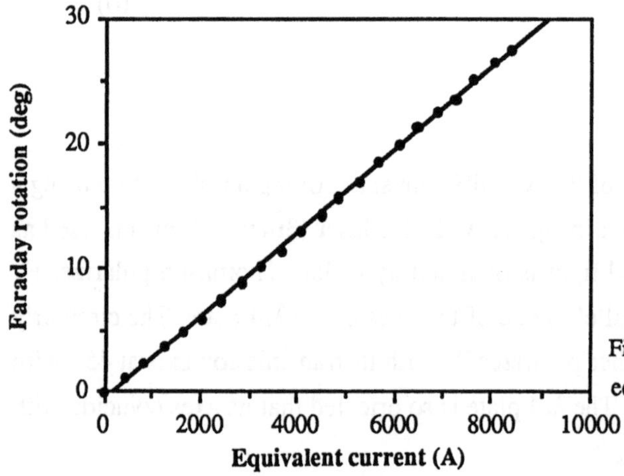

Fig.2 Measured Faraday rotation against equivalent current in fibre loops.

Analysis

The above theory is strictly valid only if the input light is perfectly linearly respectively circularly polarized. The experimental set-up shown in Fig.1 can only grant a perfectly linearly polarized light but no circularly polarized light. This is due to two facts. First, the $\lambda/4$ plate is not perfect. The retardation tolerance Δ of a $\lambda/4$ plate produces an ellipticity [8] given by

$$e = \sqrt{\frac{1 - \sin\left[(2\pi/\lambda)\,\Delta\right]}{1 + \sin\left[(2\pi/\lambda)\,\Delta\right]}} \tag{11}$$

The $\lambda/4$ plates available on the market exhibit values of Δ ranging from $\lambda/20$ to $\lambda/500$, corresponding to ellipticities between 0.73 to 0.99. In the present case $e = 0.97$. Second, the beam splitter BS2 is sensitive to polarization and a total practical value of ellipticity $e = 0.87$ has been measured. When the axis of the polarization ellipse is oriented at 45° with respect to the birefringence axis of the fibre, equ.(9) becomes :

$$T_2' = \frac{1}{1+e^2}\left[\left(1-e^2\right)\cos\left(\sqrt{\delta^2 + (2F)^2}\right) + 2e\,\delta\,\frac{\sin\left(\sqrt{\delta^2 + (2F)^2}\right)}{\left(\sqrt{\delta^2 + (2F)^2}\right)}\right] \tag{12}$$

It is no longer possible to express F in close form. However, the calculus of F by (10) using T_2' instead of T_2 leads to a modified value of F named F': apparent Verdet constant. The Verdet constant as measured in Fig.2 corresponds to F'. It is equal to 3.33×10^{-4} deg/A in agreement with the theoretical calculus according to (12) in which the values of e and δ are the values measured on the set-up. The effect of the ellipticity of the input light can be simulated using (8), (10), and (12). The result appears in Fig.3.

This method is interesting in two ways: it leads to a quasi-linear transfer function of the sensor in spite of high ellipticity of the "circularly" polarized input light and above all, it grants the functioning of the sensor whatever the linear birefringence of the fibre may be. This is not the case of the conventional configuration in which a factor $(\sin\delta)/\delta$ is always present.

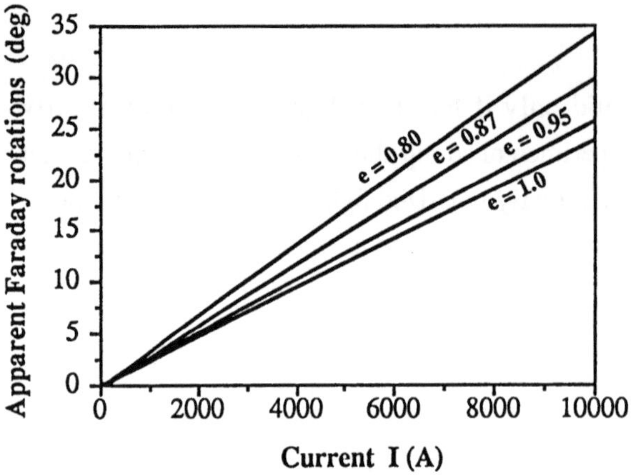

Fig.3 Simulated Faraday rotation measurement for a sensor with diameter of the fibre loop of 0.22 m and fibre turn number of 9. The ellipticity of the circularly polarized input light is taken as parameter. The apparent Verdet constant obtained by dF/dI for the case of e=0.87 is found to be 3.30×10^{-4} deg/A.

Conclusions

A time polarization multiplexing technique applied to the Faraday fibre current sensor is presented. By this method the Faraday rotation can be easily obtained, in the presence of linear birefringence, by means of two output signals. The actual multiplexing system consists of a chopper, two polarizers, and a $\lambda/4$ plate. The accuracy of the proposed sensor system depends essentially on the quality of the input circularly polarized light. The error introduced by the optical components used is constant and can be compensated numerically or analogically.

This research is supported by the Swiss National Science Foundation.

References

1. A. M. Smith, " Polarization and magnetooptic properties of single-mode optical fibre," Appl. Opt., Vol. 17, No 52, 1978.
2. A. Papp and H. Harms, " Magnetooptical current transformer," Appl. Opt., Vol.19, No.22,1980.
3. George I. Chandler, P. R. Forman, F. C. Jahoda, and K. A. Klare, " Fibre- optic heterodyne phase-shift measurement of plasma current," Appl.Opt., Vol. 25, 1770 1986.
4. H. S. Lassing, A. A. M. Oomens, R. Woltjer, P. C. T. van der Laan, and G. G. Wolzak, "Development of a magneto-optic current sensor for high, pulsed currents," Rev. Sci. Instrum. 57(5), 851, 1986.
5. S. C. Rashleigh and R. Ulrich, " magneto-optic current sensing with birefringent fibres," Appl. Phys. Lett. , 34(11), 1979
6. A. J. Rogers, " A vibration-insensitive optical-fibre current sensor," J. Optical Sensors, Vol. 1, pp. 361-370, 1986.
7. R. C. Jones, " A new Calculus for the treatment of optical systems. I," J. Opt. Soc. Am., 31, 488, 1941.
8. Z. B. Ren, Ph. Robert, and P.-A. Paratte, "Linear birefringence measurement in single-mode optical fibre with circularly polarized input light," J. Phys. E: Sci. Instrum., Vol. 18, pp. 895-862, 1985.
9. Z. B. Ren, Y. Wang and Ph. Robert, "Faraday rotation and its temperature dependence measurements in low-birefringence fibres", J. Lightwave Technology, Vol.7, 1989.

Magneto-optic Current Sensor Using a Helical Fiber Fabry-Perot Resonator

F. Maystre and A. Bertholds

Institute of Microtechnology, University of Neuchâtel,
CH-2000 Neuchâtel, Switzerland

The concept of a magneto-optic current sensor which employs a fiber-optic Fabry-Perot resonator shaped into a single-turn helix is presented. The helical shape permits construction of a compact device which has the required polarization properties. Compared to a single-pass sensor, an increase in sensitivity is obtained in proportion to the finesse of the resonator.

Introduction

When an optical fiber is plunged into a properly oriented magnetic field, the Faraday effect on the fiber material produces a non-reciprocal circular birefringence in the core of the fiber. Using this effect, AC and DC electric currents can be conveniently measured by placing a loop of fiber around the conductor [1,2]. Because the Verdet constant is very small in diamagnetic materials (e.g. $V \cong 3\times10^{-6}$ rad/A at λ=780 nm for silica), optical fiber current monitors usually use fiber coils wrapped around the electrical conductor to increase the total optical rotation. In this paper, a new sensor is presented which uses a fiber-optic Fabry-Perot resonator shaped into a single-turn helix (Fig.1). The increase in sensitivity is proportional to the finesse of the resonator, and the single-turn helix produces a short resonator which forms a complete loop and which has the required polarization properties. For the measurement, the optical frequency of a semiconductor laser is locked to a resonance peak of the Fabry-Perot and the phase difference between the two eigenpolarizations of the resonator is then measured using heterodyne polarimetry [3].

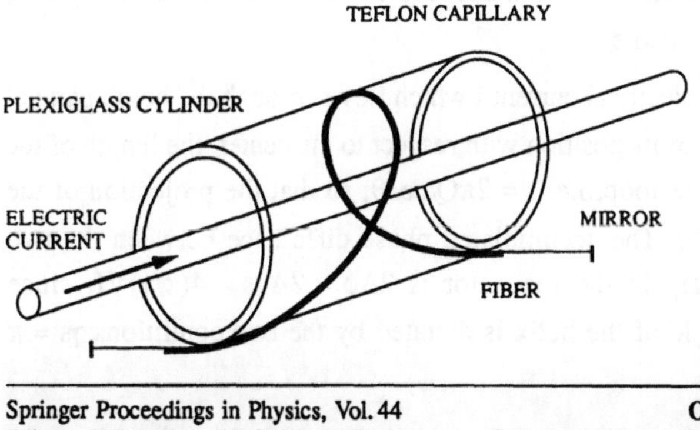

Fig.1. Fiber-optic Faraday current sensor consisting of a Fabry-Perot resonator shaped into a single-turn helix. The cylinder has a diameter of 28 mm and a length of 52 mm.

Editors: H.J. Arditty · J.P. Dakin · R.Th. Kersten

Sensor configuration

When an optical waveguide describes a non planar curve in space, the natural frame to describe the propagation of polarized light has been established by Tang [4], who shows that the polarization does not follow the geometrical torsion of that curve. For a fiber of low intrinsic birefringence wound on a cylinder into a helical shape, the combination of the bend induced birefringence, which follows the geometrical torsion, and the rotation of the polarization, which follows Tang's frame, gives the same effect as a spun fiber with intrinsic birefringence (i.e. a fiber drawn from a rotating preform) [5,6]. Detailed calculations show that in Tang's frame a piece of right-handed helical fiber of arc length s is represented in a circular polarization basis by the Jones matrix

$$\mathbb{M}_c = \mathbb{N}_c\,\mathbb{R}_c = \begin{pmatrix} (\cos\gamma s + i(\tau/2\gamma)\sin\gamma s) & -i(\beta/2\gamma)\sin\gamma s \\ -i(\beta/2\gamma)\sin\gamma s & (\cos\gamma s - i(\tau/2\gamma)\sin\gamma s) \end{pmatrix} \begin{pmatrix} e^{-i\tau s} & 0 \\ 0 & e^{i\tau s} \end{pmatrix}, \qquad (1)$$

where $\gamma = (\tau^2 + (\beta/2)^2)^{1/2}$. The geometric torsion τ of the helix is given by the pitch angle θ and the radius Q of the helix through the relation $\tau = \sin\theta\cos\theta/Q$. The bend-induced birefringence β is proportional to the square of the radius of curvature $\kappa = \cos^2\theta/Q$ of the helix and reads $\beta = (\pi n^3/2\lambda)(p_{11}-p_{12})(1+\nu)(r\cos^2\theta/Q)^2$, where λ is the wavelength and r the fiber radius. The material parameters for silica fibers are $n = 1.45$, $(p_{12}-p_{11}) = 0.14$, $\nu = 0.16$ [7].

In the presence of a magnetic field, the Faraday effect produces an additional circular birefringence which is non-reciprocal and which modifies the matrix $\mathbb{N}_c$. In this matrix, τ must then be replaced by $\tau - VH_s$, where H_s is the component of the magnetic field along the direction of propagation of the light wave and is assumed to have a constant value along the fiber.

For a Fabry-Perot resonator made from such a fiber, the eigenpolarizations are given by the eigenvectors of the matrix for one total round trip and their corresponding phases are obtained from the eigenvalues of that matrix. For H = 0, the eigenpolarizations of the resonator become degenerate when the condition $\gamma s = \pi$ is satisfied. In this particular case, the phase difference after a round trip is exactly zero and the eigenpolarizations of the single-pass matrix in Eq.(1) are circular. Provided that $(\beta/2)^2 << \tau^2$, the eigenpolarizations remain nearly circular for $H \neq 0$ and the eigenvalues for the round-trip matrix become $e^{\pm i\Delta\gamma s}$, with $\Delta\gamma \cong -2(\tau/\gamma)VH_s$ (the approximation holds for $VH_s << \gamma$ and τ).

The sensor is designed to measure an electrical current I which flows through the helix, parallel to its axis. In order to be independent of its position with respect to the center, the length of the helix is chosen to make one complete loop, i.e. $s = 2\pi Q/\cos\theta$, so that the projection of the helix becomes a closed circle (Fig.1). The accumulated phase difference between the two eigenpolarizations for one round trip in the resonator is $2\Delta\phi = 2\Delta\gamma s = 4(\tau/\gamma)VI$, since $H_s = I\cos\theta/2\pi Q = I/s$. The pitch angle of the helix is dictated by the two conditions $\gamma s = \pi$

and $s = 2\pi Q/\cos\theta$. For silica fibers of 80 µm outer diameter, a wavelength of 780 nm and helix diameter $Q > 10$ mm, the pitch angle θ is close to 30°.

At the output of the Fabry-Perot resonator, the phase difference between the two transmitted eigenpolarizations can be written as

$$\Delta\psi = \psi_1 - \psi_2 = 2 \tan^{-1}[\rho \tan(\Delta\phi/2)] + \delta , \quad (2)$$

with $\rho = (1 + 4F^2/\pi^2)^{1/2}$ [3] and $\delta = 2\pi(1 - \tau/\gamma)$, which accounts for the single-pass phase shift between the transmitted eigenpolarizations, according to Eq.(1). Around the resonances ($\Delta\phi = m\pi$) $\Delta\psi$ depends linearly on $\Delta\phi$ and the slope, or sensitivity, is readily seen to be $\Delta\psi/\Delta\phi = \rho$. The sensitivity is therefore increased by a factor ρ with respect to a single-pass sensor, e.g. $\rho = 7$ for $F = 10$, if one works close to the resonance.

Measuring system

The detection scheme for the Fabry-Perot sensor consists of two parts. The first part is a feedback loop, which locks the laser frequency to a resonance peak where the sensitivity is maximum [8]. The second part accurately measures the phase difference $\Delta\psi$ (period) . Because of the large amplitude variations near resonance, it is necessary to measure the phase difference between the two eigenpolarizations independently of their amplitudes, e.g. by heterodyne phase detection [3]. External perturbations affecting the length of the fiber are common-mode and do not change the measured value of induced circular birefringence. Thus, the sensor is insensitive to temperature changes of the Fabry-Perot cavity. However, the dynamic range of a Fabry-Perot sensor is limited to the resonance width, which is about $\Delta\psi = \pm 0.8$ rad (± 45°).

Experiment

Electric current measurements are made with a Fabry-Perot consisting of a 17 cm long single-mode fiber having low intrinsic birefringence (York LB-800, 80 µm outer diameter). The fiber is bent into a single-turn helix having a radius of $Q = 14.3$ mm and a pitch angle of $\theta = 30°$. The fiber has its coating removed and is maintained in proper position by placing it inside a teflon capillary, previously bent in the desired configuration and glued on a plexiglass cylinder. The geometric rotation rate provoked by the helix is $\tau = 30.2$ rad/m and the bend induced linear birefringence is $\beta = 4.4$ rad/m (i.e. $\beta/2\tau = 0.073$ and $\tau/\gamma = 0.997$).

The resonator is formed by a gold mirror of 58% reflectivity and a dielectric mirror of 97% reflectivity glued on the well polished endfaces of the 17 cm long fiber [9]. The free spectral range of the resonator is 600 MHz and the finesse is measured to be $F = 6$, which is somewhat

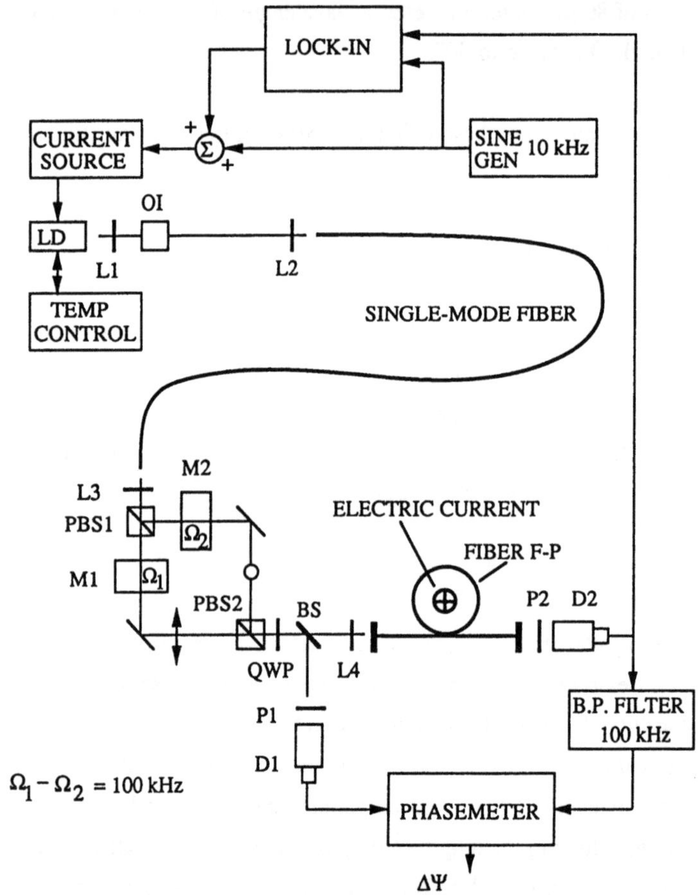

Fig.2. Experimental setup using a feedback loop to lock the optical frequency of the laser to the Fabry-Perot resonance and heterodyne polarimetry to measure the phase difference between the two eigenpolarizations at the output of the resonator. LD: laser diode, L: lenses, OI: optical isolator, PBS: polarizing beam splitters, BS: beam splitters, M: acousto-optic modulators, P: polarizers, QWP: quarter wave plate, D: detectors.

lower than the expected value. The laser diode is optically isolated from the Fabry-Perot resonator and emits a single line of 20 MHz width (FWHM) at 780 nm, which is about 5 times narrower than the Fabry-Perot resonance peaks. The current to be measured flows through the loop formed by the fiber resonator along a straight copper wire parallel to the axis of the helix (Fig.1). Experimental results are shown in Fig.3. Open circles represent experimental data points and the straight lines correspond to the theoretical curves for a resonator with a finesse F = 6.2 and for a single pass through the same fiber coil (i.e. F = 0). AC currents at 43 Hz from 80 A rms down to 15 A rms are measured with a detection bandwidth of 1.45 Hz. The

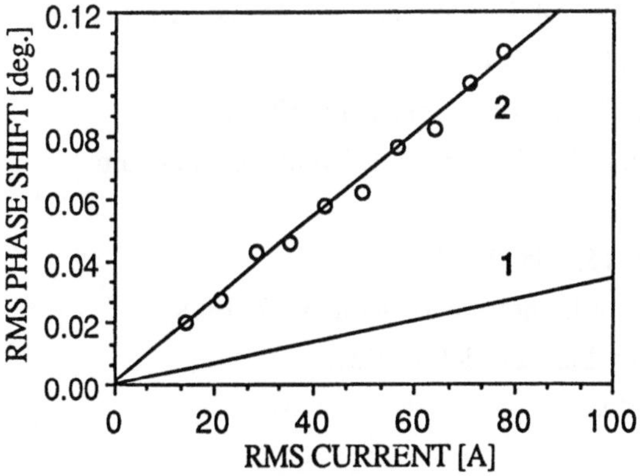

Fig.3. Faraday induced optical phase shift in the fiber optical sensor. 1) response for a single pass through one loop of fiber, 2) response for a Fabry-Perot with a finesse of F = 6.2, compared with the experimental results (open circles).

minimum measurable current corresponds to the noise floor produced by the amplifier of the output detector. This value is rather high because the large number of optical elements between the laser and the Fabry-Perot add up to a low peak transmission of the resonator, so that the optical output power is only about 40 nW. The experimental points follow closely the theoretical curve for a Fabry-Perot resonator of finesse F = 6.2 and the slope is $\Delta\psi/I = 1.4 \times 10^{-3}$ deg/A, which is 4 times larger than the sensitivity of a single-pass sensor.

Conclusions

A helical fiber Fabry-Perot resonator with circular eigenpolarizations in transmission has been constructed. It consists of a low intrinsic birefringence fiber wound into a single-turn helix of about 30 mm diameter and 30° pitch angle. The bend induced birefringence is quenched by the rotation of the polarization due to the geometrical torsion of the helical path. It has been verified experimentally that a resonator with a finesse F = 6 has a sensitivity of 1.4×10^{-3} deg/A, which is 4 times larger than for a single-pass configuration, as predicted by theoretical analysis.

The authors wish to thank R. Dändliker, G. Frosio and P. Gannagé for helpful discussions.

References

[1] A.M. Smith, Appl. Opt. **17**, 52 (1978).

[2] A. Papp and H. Harms, Appl. Opt. **16**, 1315 (1977).

[3] F. Maystre, P. Gannagé and R. Dändliker, OFS, 1988 Technical Digest Series, Vol.2, (Opt. Soc. Am. Washington, D.C. 1988) pp. 424-432.

[4] C. H. Tang, IEEE Trans. Microwave Theory Tech., **MTT-18**, 69 (1970).

[5] D. N. Payne, A. J. Barlow and J. J. Ramskov Hansen, IEEE Journ. of Quantum Electron., **QE-18**, 477 (1982).

[6] J. N. Ross, Opt. and Quant. Electron. **16**, 455 (1984).

[7] A. Bertholds and R. Dändliker, Journ. of Lightwave Technol. **6**, 17 (1988).

[8] J-L. Picqué and S. Roizen, Appl. Phys. Lett. **27**, 340 (1975).

[9] J. Stone, Electron. Lett. **21**, 504 (1985).

A Polarization-Based Fibre Optical Sensor System Using a YIG Optical Waveguide for Magnetic Field Sensing

H. Sohlström[1], *U. Holm*[1], *and K. Svantesson*[2]

[1]Royal Institute of Technology, Instrumentation Laboratory,
P.O. Box 70013, S-10044 Stockholm, Sweden

[2]Swedish Institute of Microelectronics, P.O. Box 1084,
S-16421 Kista, Sweden

A sensor system utilizing a polarization maintaining fibre for the optical signal transmission and a planar optical waveguide for magnetic field measurements is presented. The system is based on polarization modulation originating from the TE to TM mode conversion in a magneto-optical thin film of Gd,Ga substituted YIG (Yttrium Iron Garnet).

Introduction

In a fibre optical sensor system based upon amplitude modulation one generally wants a separate reference channel. By using this reference one may in principle correct for signal transmission perturbations, assuming that such influences are common to both the signal and the reference channel.

In this paper we describe a measurement system where one polarization maintaining fibre alone provides the two transmission channels. The two eigenmodes of the fibre propagate practically independently of each other and are to a great extent equally affected by environmental conditions. Furthermore the sum of the light intensities of the two modes serves as a normalization quantity when evaluating the signal from the sensor element.

Sensing Principle

The magnetic field sensing is based upon the Faraday effect in a garnet thin film, YIG (Yttrium Iron Garnet), epitaxially grown on a GGG (Gadolinium Gallium Garnet) substrate. The waveguiding sensing element can, in contrast to the bulk YIG crystal sensors previously described,[1] efficiently be coupled to a polarization maintaining fibre.

In a waveguide, the Faraday effect can be described as a circular birefringence that is dependent on the magnetic field in the light propagation direction. This birefringence causes a coupling between the TE and TM modes of the same order. The transfer of power follows the expression

$$F = \left[1+\left(\frac{\Delta\beta}{2\theta}\right)^2\right]^{-1} \sin^2\left\{\left[\theta^2+\left(\frac{\Delta\beta}{2}\right)^2\right]^{\frac{1}{2}} \cdot z\right\} \tag{1}$$

where F is the fraction of power in the mode not excited at the input, $\Delta\beta$ is the linear birefringence, θ is the circular birefringence and z is the travelled distance.

If $\Delta\beta$ is zero the modes propagate synchronously and a complete transfer of power is possible. This is in effect a rotation of the polarization direction just as in the bulk case.

In the systems presented here, light is launched into one of the polarization modes of the planar waveguide. During the propagation there is a transfer of power into the other mode due to the Faraday effect. This gives a power distribution between the modes which is magnetic field dependent. The light is then coupled into the return fibre, which is oriented in such a way that each of the modes of the planar guide is coupled into one of the corresponding fibre polarization modes. At the output end of the fibre, the light from the two fibre modes is separated and detected. The detector outputs P_1 and P_2 are then combined to give the intensity independent signal S which is a measure of the magnetic field:

$$S = \frac{P_2}{P_1 + P_2} \qquad (2)$$

Below we present two sensor configurations, one transmissive and one reflective. In the reflective configuration the same fibre is used to carry the light both to and from the sensing element.

To be able to understand the behaviour of the sensor one has to consider the magnetic properties of the garnet materials used in this work. Ferrimagnetic materials such as YIG exhibit a spontaneous magnetization in the form of a magnetic domain structure. In the absence of an external field these domains are oriented in such a way that the total magnetization of the material vanishes. If an external magnetic field is applied, the domains will change their distribution, size and shape. The circular birefringence, which is proportional to the magnetization parallel to the light propagation, will in general not have a useful relation to the applied field.

A sufficiently large external field will align all domains into a single domain. A varying field together with such a bias field gives a uniform change in the orientation of the magnetization and thus in the Faraday rotation. This is the approach used here[2]. A similar result can be achieved without a bias field with materials that have a strong magnetic anisotropy.

System Realization

TRANSMISSIVE SENSOR

The transmissive sensor configuration is shown in fig 1. A polarization maintaining fibre was used to send polarized light to the sensing element. Since the fibre core diameter and the YIG film thickness were approximately the same it was possible to couple a large fraction of the light from the fibre into the zero order TE mode of the film, which allowed 16 modes. For a correctly positioned fibre we could observe less than 3 % of the total power in mode 2, less than 1 % in mode 4 and no significant amount in any other mode.

At the opposite edge of the film, the light from the TE and TM film modes was coupled into the two polarization modes of a second polarization maintaining fibre. As the coupling between the

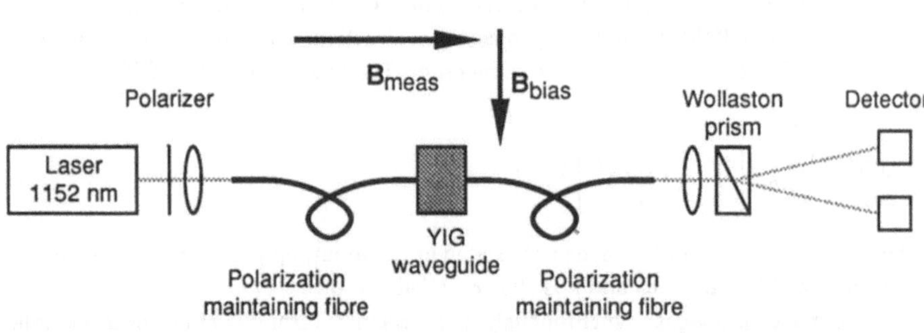

Fig 1. The transmission sensor set-up. The waveguide is 4 mm long and 6.7 μm thick.

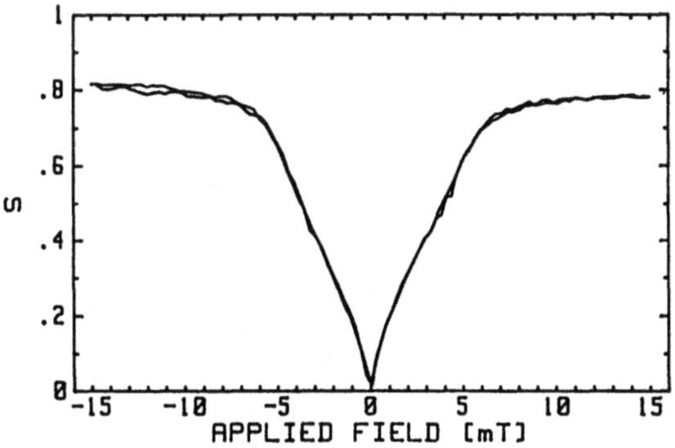

Fig 2. System output S versus B_{meas} with a sensor according to fig 1 and a bias field of 3 mT.

modes in the fibre is negligible, their power ratio remaind practically unchanged throughout the fibre length. The light from the two fibre modes was geometrically separated by a Wollaston prism and detected by two germanium photodiodes.

As the light was not laterally confined in the waveguide, only a rather small fraction was coupled into the second fibre. Results from measurements with the transmissive sensor are shown in fig 2.

REFLECTIVE SENSOR

To be able to collect a large fraction of the light some kind of focusing optics had to be introduced. Also, from a practical point of view, it would be advantageous if the fibres were attached on the same side of the sensing element, or even better, if only one fibre could be used.

These objectives were met by the one-fibre system in fig 3. The same polarization maintaining fibre was used to carry the light to and from the sensing element, which is shown in greater detail in fig 4. The rear edge of the sensing element, which was polished into a semi-circular shape and gold coated, focused the light back into the fibre. Due to the nonreciprocity of the Faraday effect, the effective interaction distance will be twice the radius.

Power that is reflected from other points in the system than the rear edge of the waveguide may cause interference problems. To reduce some of the critical reflections, the input end of the

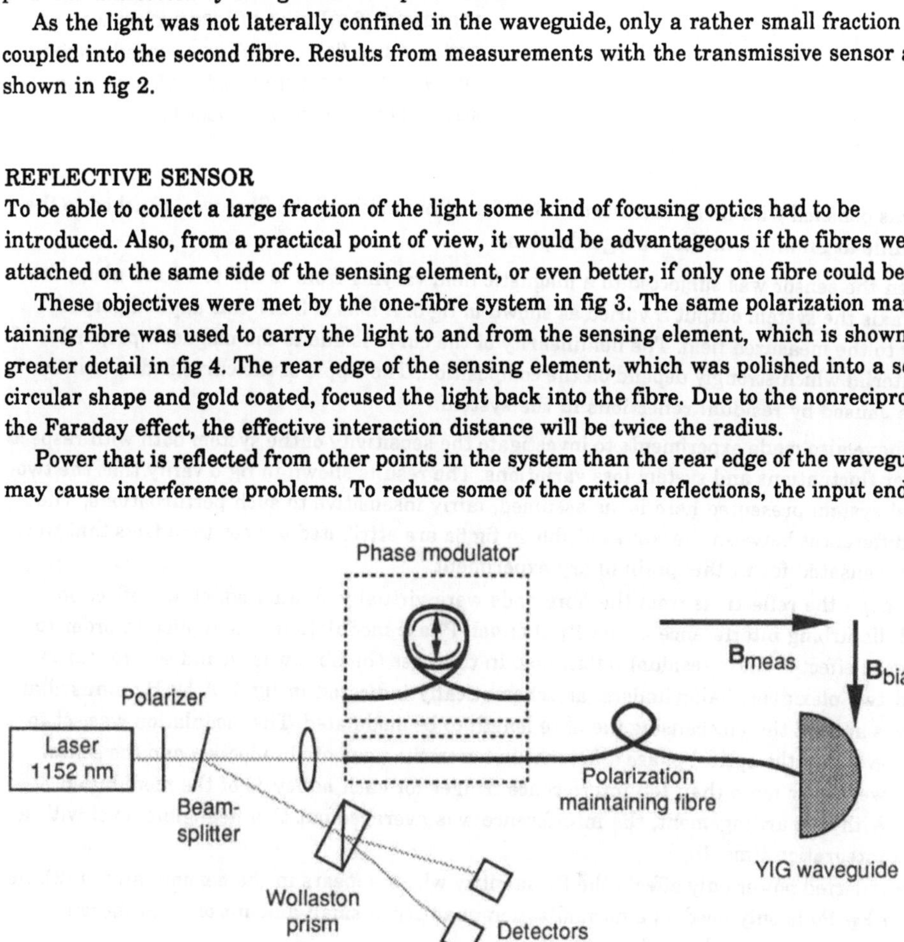

Fig 3. The reflective sensor set-up. For the results presented below, 2 m of Furukawa polarization maintaining fibre for 1.3 µm was used. The use of phase modulation is described in the text.

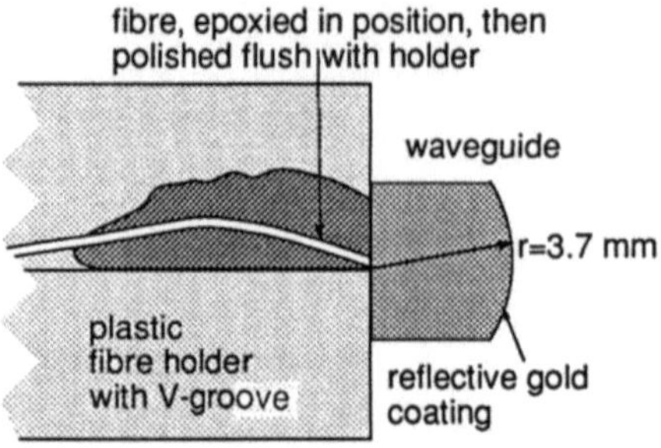

Fig 4. Experimental reflective sensor. Waveguide as in fig 1.

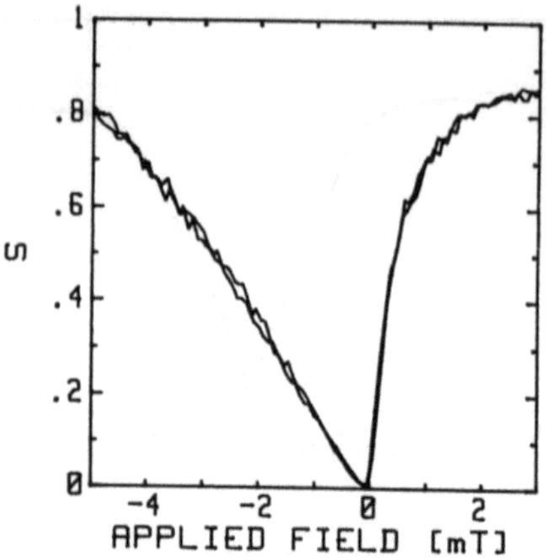

Fig 5. System output when the applied field was varied from -6 mT to +4 mT and back in two cycles. A bias field of 3 mT was applied perpendicular to the measured field. 1.15 µm HeNe laser light and no phase modulation.

fibre was polished with its surface normal 6° off the fibre axis and the fibre was attached to the waveguide at an angle in the waveguide plane.

When the sensor was subjected to a magnetic field varying from -6 mT to +4 mT along the sensor axis the system output S varied as shown in fig 5. A bias of 3 mT was applied perpendicular to the measured field. The nonlinearity of the curve is due to the magnetic properties of the material which strongly depend on the composition. The ripple of the curve is due to interference caused by residual reflections in the system.

We have also made experiments to investigate the sensitivity of the system both with respect to power fluctuations and system loss variations. The results shown in fig 6 verify that the two-channel system presented here is, as assumed, fairly insensitive to such perturbations. The small differences between the curves visible in fig 6a are attributed to detector offsets that were not compensated for in this preliminary experiment.

Although the reflections from the fibre ends were virtually eliminated, other reflections caused disturbing interference in the P_1 channel. Phase modulation was applied in order to reduce the effect of these residual reflections. In this case the fibre was wound several turns around two plexiglass halfcylinders, as schematically indicated in fig 3. A 50 Hz sinusodial force was applied thereby causing the fibre length to be modulated. The modulation was set to such a level that the optical phase difference between the unwanted reflection and the sensor signal swept over more than ten interference fringes for each halfcycle of the modulation signal. With this arrangement, the interference was averaged out to a negligible level with a 100 ms integration time, fig 7.

The reflected power only affects the P_1 quantity which appears in the denominator of (2). As the sum P_1+P_2 is only used as a normalization quantity, a small amount of non-coherent additive power may be tolerated.

Another way to reduce the interference disturbance is to use a light source with a much shorter coherence length than that of the gas laser used in the above measurements. Experiments with a conventional laser diode will be carried out later in order to verify this statement.

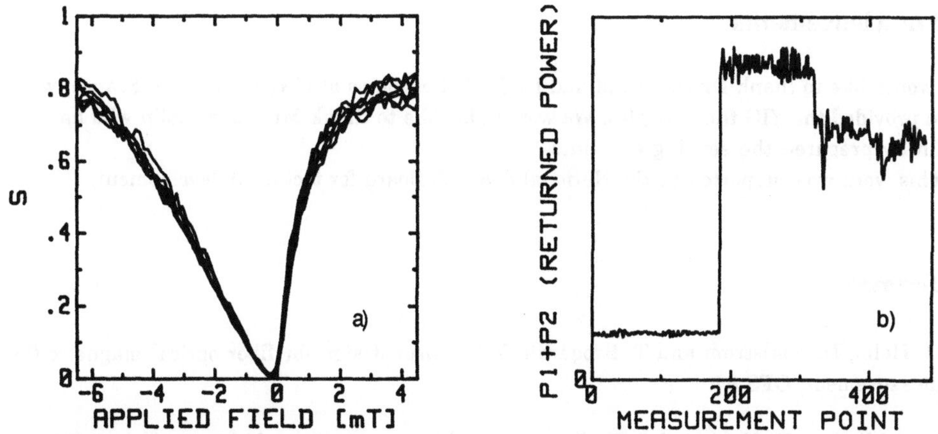

Fig 6. a) System output S during a measurement sequence of -6 mT to + 4 mT and back 3 cycles with a bias field of 3 mT. During the first cycle, the laser power was reduced. During the second cycle, the system was undisturbed and during the last cycle the fibre was perturbed by winding it around a 10 mm diameter rod.

b) P_1+P_2 (≈returned power) during the measurement sequence.

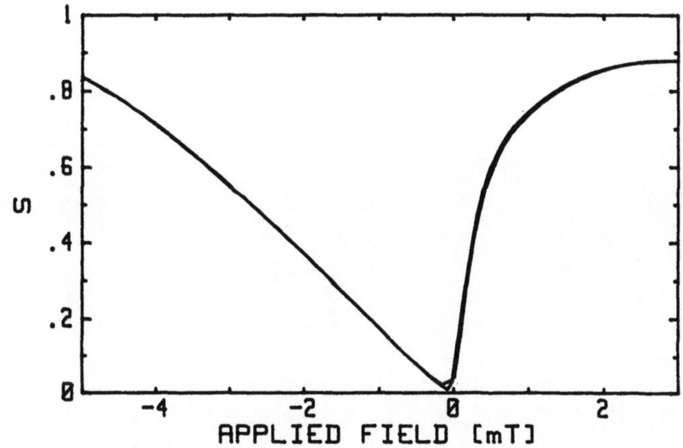

Fig 7. Result from measurement similar to the one in fig 5, but with fibre length modulation.

Conclusions

We have presented a sensor system utilizing a polarization maintaining fibre and a planar optical waveguide for magnetic field measurements. The system is based on polarization modulation originating from the TE to TM conversion in a magneto-optical thin film of Gd,Ga substituted YIG. Both a transmissive and a reflective sensor systems have been demonstrated. We have shown that the use of the two polarization modes of a polarization maintaining fibre as measurement and reference channel gives a system which is highly insensitive to power and loss fluctuations. We have also demonstrated that phase modulation further reduces the noise in the system.

Acknowledgements

We would like to thank Dr H. LeGall and Dr J. M. Desvignes at CNRS, Meudon, France, who have provided the YIG film samples. We would also like to thank Mr Bengt Molin who has skillfully prepared the sensing elements.

This work was supported by the National Swedish Board for technical development.

References

1 U. Holm, H. Sohlström and T. Brogårdh: YIG-sensor design for fiber optical magnetic field measurement, OFS 84.

2 H. Sohlström, U. Holm and K. G. Svantesson: Characterization of Magnetooptical Thin Films for Sensor Use, ECO 2, Paris 1989, SPIE Proc, 1126.

Fidelity/Sensitivity Evaluation of Fiber Optic Magnetic Gradiometer

R.D. Rempt

Boeing Aerospace and Electronics, Mailstop 87-60, P.O. Box 3999, Seattle, WA98124, USA

Retrieval of magnetic gradient signal from a Mach-Zehnder fiber interferometer is presented. Common mode rejection of the field is discussed and data given. Calibration procedures are summarized with experimental data for a gradient noise level of $.045\gamma/\mathrm{ft}\sqrt{\mathrm{Hz}}$ rms at 1.0 Hz. These are the most sensitive fiber optic magnetic gradient data reported.

Introduction

Over the past several years, Mach-Zehnder interferometers employed as fiber optic sensors have enjoyed a good deal of attention. This paper concerns itself with a magnetic gradiometer, in which each of the arms of the Mach-Zehnder interferometer contains a magnetic transducer.[1,2] Since the interferometer output tracks the phase difference between the two arms, when properly nulled and balanced, its output is then proportional to the magnetic field gradient, with the field itself being common mode rejected.

Gradient Signal Recovery

The coupling of the magnetic field to each arm of the transducer is accomplished via a non-linear magnetostrictive process. The strain imparted to a length L of fiber which is bonded to the magnetostrictive material is given by CH^2, where C is a constant, and H is the total field strength experienced by the transducer.

A convenient transducer geometry is a cylindrical magnetostrictive foil layer around which are wrapped many turns of fiber which is affixed with a flexible adhesive. When the field strength increases along the transducer's cylindrical axis, the cylinder strains axially,which is accompanied by a corresponding radial compression. This compresses the fiber, and thereby the optical path of that interferometer arm. When the field strength decreases, there is a corresponding extension of the optical path. The transducers used were 10 cm long and 4.53 cm in diameter with 381 turns of 633 nm single mode fiber for a total fiber length of approximately 50 m each.

The sensor configuration is shown in figure 1. Each transducer is immersed in a DC and AC field provided by its bias solenoid. The path difference between the two interferometer arms is then given by

$$\Delta L = L_1 K_1 H_1^2 - L_2 K_2 H_2^2 \tag{1}$$

Where the K's are material constants and the L's are the lengths of the fiber in each arm which are bonded to the magnetostrictive material. As pointed out by Koo & Sigel,[1] H_1 may be represented as the sum of several fields:

$$H_1 = H_1^{dc} + H_{1E} + H_{ext}^{dc} + H_1^{ac} \tag{2}$$

Editors: H.J. Arditty · J.P. Dakin · R.Th. Kersten

Figure 1. Gradiometer Block Diagram

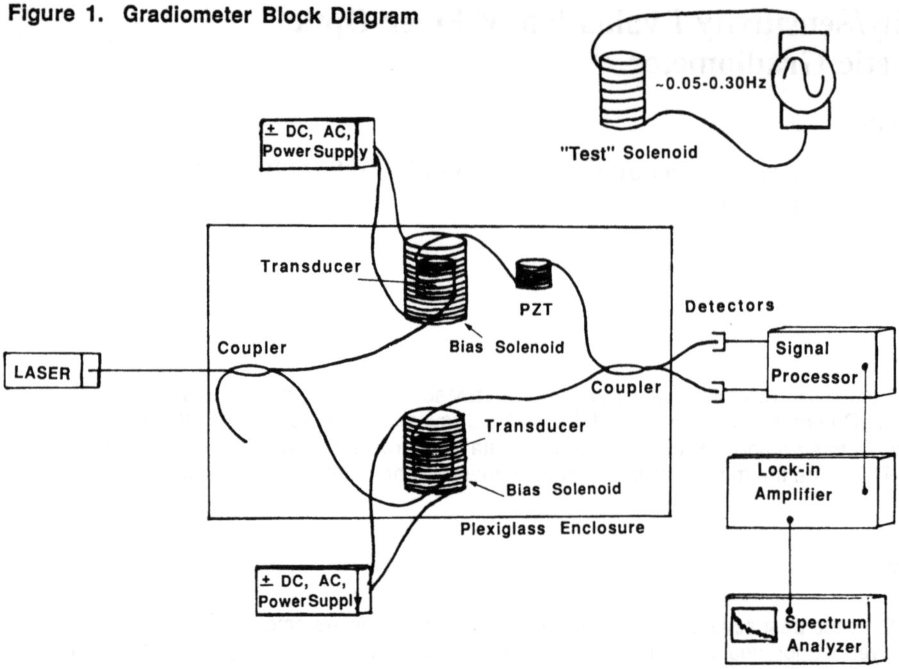

where H_{1E} is the earth's field at transducer 1, H_1^{dc} the applied dc field from the solenoid, H_1^{ac} the applied ac field from the solenoid, and H_{ext}^{dc} the external field of interest.

At transducer 2, H_2 is as follows:

$$H_2 = H_2^{dc} + H_{2E} + H_{ext}^{dc} + \Delta H_{ext}^{dc} + H_2^{ac} \tag{3}$$

where similar notation applies, with the additional expression of the external field of interest being incremented by ΔH_{ext}^{dc} which represents the gradient.

The applied AC "dither" fields are at a suitably selected frequency ω, which is provided by a single oscillator to both solenoids. Squaring H_1 and H_2, inserting into equation (1), and collecting only terms corresponding to the frequency ω yields:

$$\Delta L|_\omega = 2L_1 K_1 H_1^{ac} (H_1^{dc} + H_{1E}) - 2L_2 K_2 H_2^{ac} (H_2^{dc} + H_{2E})$$
$$+ 2(L_1 K_1 H_1^{ac} - L_2 K_2 H_2^{ac}) H_{ext}^{dc} - 2L_2 K_2 H_2^{ac} \Delta H_{ext}^{dc}. \tag{4}$$

As pointed out by Koo and Sigel, the following conditions may be set experimentally:

$$H_1^{dc} + H_{1E} = 0,$$
$$H_2^{dc} + H_{2E} = 0. \tag{5}$$

This is done by adjusting the applied DC field in each solenoid to minimize the interferometer output at ω.

Also, the dither amplitudes of one or both of the solenoids may be adjusted such that

$$L_1 K_1 H_1^{ac} - L_2 K_2 H_2^{ac} = 0. \tag{6}$$

When conditions (5) and (6)are satisfied, then equation (4) simplifies to

$$\Delta L|_{\omega} = - 2L_2 K_2 H_2^{ac} \Delta H_{ext}^{dc}. \quad (7)$$

So the interferometer output at the dither frequency is proportional to the gradient when nulling conditions (5) and balance condition (6) are established.

Common Mode Rejection

Experimentally, it is observed that conditions (5) may be set, and remain over periods of hours, and sometimes several days. The balance (6), if not precise, causes some signal containing the field H_{ext}^{dc} to be present at the interferometer output.

The output signal then has the form:

$$\Delta L|_{\omega} = \frac{dBz}{dx} \Delta x + K B_z \quad (8)$$

for a field in the z direction and transducer separation along x.

Most desirable applications of magnetic gradiometers detect dipole fields, in which the gradient varies as the inverse fourth power of the distance and the field as the inverse third power. For this reason, the presence of even a small amount of field signal can "swamp" the gradient signal.

Figure 2 shows 3 sets of plots of equation (8) for various values of 1/K. The plots are generated for a sensor detecting a dipole at a distance large compared to the dipole dimension. In Figure 2 a, even for 1/K=20, the shape of the plot is totally field dominated. The gradient starts to show in Figure 2b at a 1/K value of 40, and in figure 2c, the signal is almost "pure" gradient when 1/K is about 1000. This indicates that more than 60 dB of common mode rejection is required to accurately recover the gradient signal. The symmetry of the gradient signal may also be used to estimate the field rejection.

Experimental data demonstrate the above analysis as shown in Figure 3. These data show the output of the Lock-in amplifier for 3 consecutive "passes" of a bar magnet being carried past the sensor. The bar magnet was hand held, with its axis parallel to the sensor axis, located several feet away. The bar magnet was carried past the sensor along a path parallel to the line between the two transducers. The first"pass" has a shape which is somewhat "gradient-like" in appearance, but the lower peak is smaller in magnitude than the upper peak, indicating a slight imbalance. A slight adjustment in the dither amplitude of one of the bias solenoids (H_1^{ac} or H_2^{ac}) was then made. The next shows a much more symmetric case, with approximately 60 dB of common mode rejection. The third pass was done without further adjustment to demonstrate repeatability. The slight humps between the passes are due to the bar magnet quickly being returned to the starting position.

Balancing Technique

In practice, it is convenient to perform the DC nulling for each transducer at a frequency distinguishable from the common dither frequency used to detect the gradient. This technique has been described elsewhere[3], and will not be repeated here.

Sensitivity Evaluation

One of the attractive advantages with a common mode rejection sensor is that the necessity of making absolute measurements is circumvented. In order to calibrate the gradiometer a "test" coil was placed several feet from the sensor as shown in Figure 1.

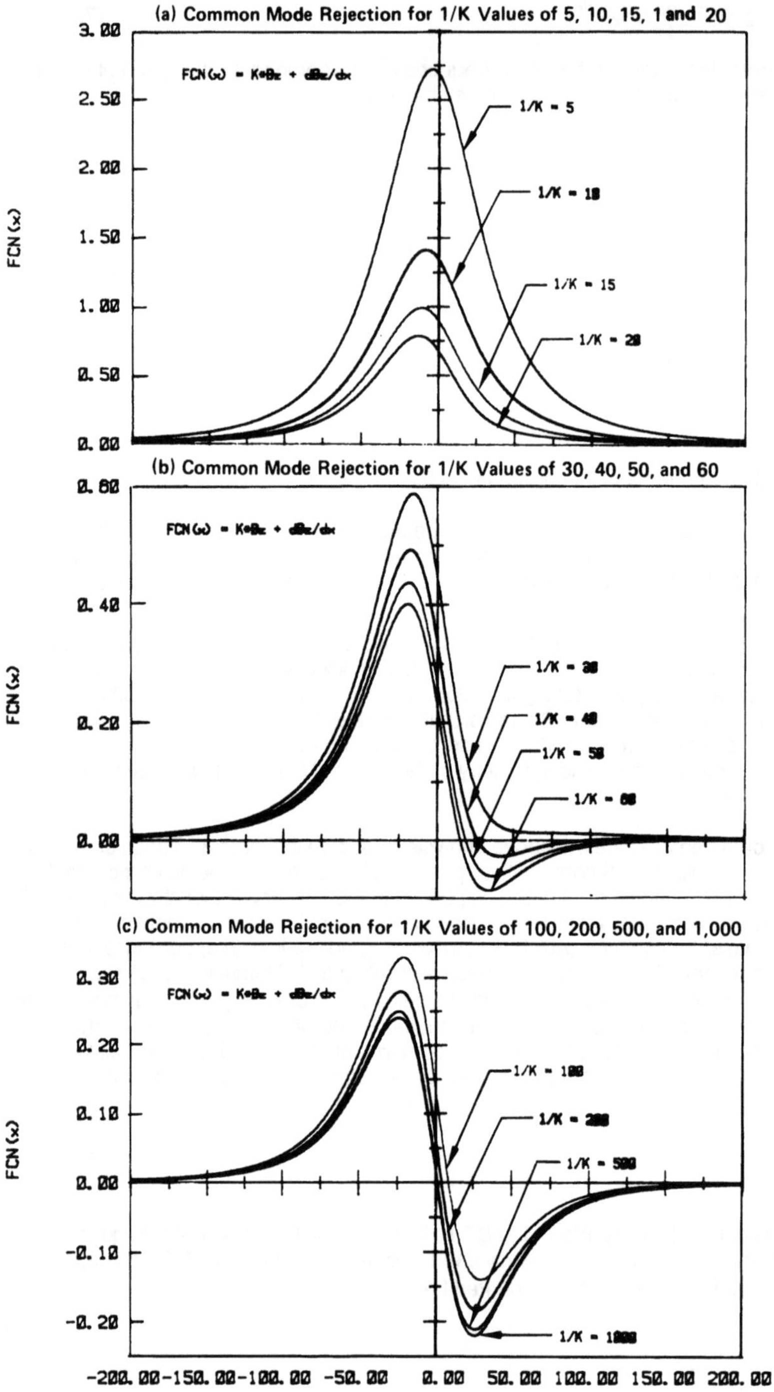

Figure 2. Common Mode Rejection

Figure 3. Lock-In Amplifier Output for 3 Passes of Bar Magnet

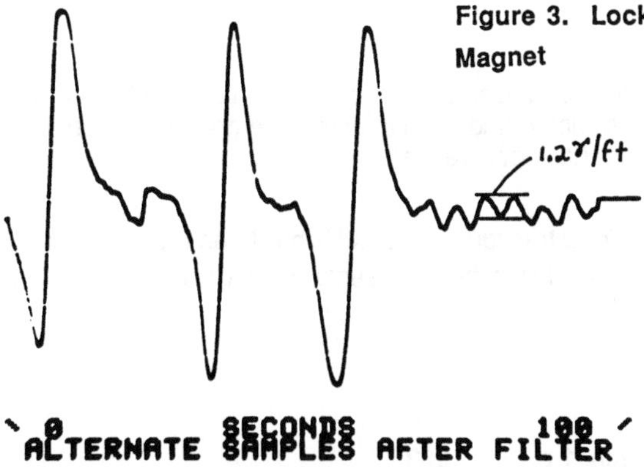

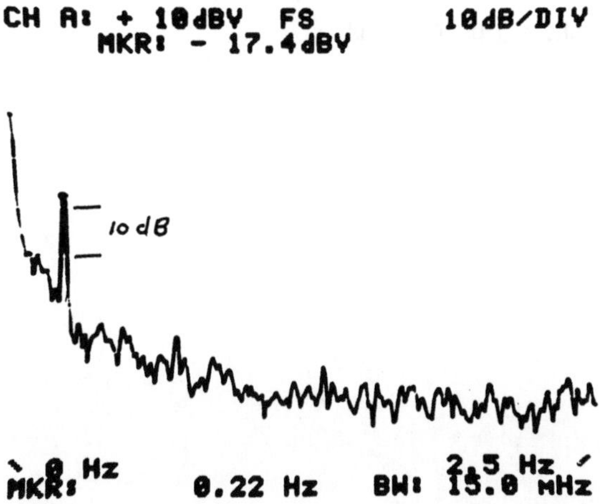

Figure 4. Noise Spectrum for Figure 3

The axis of the test coil is parallel to that of the transducer axes. It's location was picked so that it was closer to one of the transducers than to the other, therefore producing a different field at each transducer, constituting a field gradient between the two transducers..

A fluxgate magnetometer was placed carefully at the position of each of the transducers and the field recorded as a function of test coil current with a sinusoidal signal at approximately 1/2 Hz. This low frequency allowed discrimination from other noise in the laboratory, and was far enough from DC to make the measurements easy and repeatabe. The two linear plots of field versus test coil current are then subtracted to determine the calibrated gradient signal versus test coil current. Even if there is some offset in the fluxgate magnetometer, the difference readings are still believable, since the offset subtracts out when the difference is taken. The settings of the fluxgate magnetometer were not changed throughout the process, and the linear plots are linearly extrapolated from the 10-30γ peak to peak region to make the measurements down to the ~1γ region for the calibrated gradient.

At the right of the lock-in amplifier output shown in Figure 3, there is a sinusoidal signal. This was turned on at 0.22Hz, after the third "pass", and is provided by the test coil, and calibrated to be 1.2γ/ft. peak to peak by the procedure described above. The noise spectrum for this signal is shown in Figure 4. The signal at .22 Hz is approximately ~28dB above the noise, and at 1.0 Hz approximately ~38dB above the noise. This sets the noise level at .045γ/ft $\sqrt{Hz}$ rms at 1.0 Hz. For the 15mHz bandwidth used to make the measurements, this translates into a gradient noise level of 5.5m γ/ft rms at 1.0Hz.

Summary

Signal recovery and sensitivity evaluation for a fiber optic magnetic gradiometer have been discussed. Common mode rejection may be estimated and used to determine gradient signal fidelity. Confirmation of gradient signal "shapes" has been verified.

Noise level for a gradiometer with cylindrical transducers at 1.0Hz has been measured at 45mγ/ft $\sqrt{Hz}$ rms which is 1.5 x10^{-8} G/cm $\sqrt{Hz}$ rms. This is the most sensitive magnetic field gradient detected by a fiber sensor that has been reported.

References

1. K.P Koo & G.H. Sigel, J. Lightwave Tech. 1 p. 509, 1983.

2. C.M. Dube, S. Thordarson, & H.K. Wanser, Proc SPIE 883 p. 17, 1987.

3. R.D. Rempt, Photonics Technology Letters, 1 p. 105, 1989.

All-Fiber Faraday-Rotation Current Sensor with Remote Laser-FM Based Heterodyne Detection

A.D. Kersey and M.A. Davis

US Naval Research Laboratory, Optical Techniques Branch, Code 6574, Washington, DC 20375, USA

Abstract: We describe an all-fiber Faraday-rotation current sensor which utilizes an optical frequency-induced polarization modulation approach to the detection of the magneto-optical polarization rotation. Remote passive current sensing over insensitive fiber leads utilizing heterodyne-type signal processing is demonstrated.

Introduction: The measurement of current using the Faraday effect in optical fibers has generated considerable research interest over the past decade. Research has focused both on system aspects, such as techniques to monitor the magneto-optically induced rotation using essentially homodyne- [1-3] and heterodyne-based [4] detection, and the development of speciality fibers, such as helical core [5], spun high-birefingence fiber [6], doped fibers with enhanced Verdet constants [7], and annealed fiber coils [8] to reduce intrinsic and bend induced birefringence. In this paper we report the design and operation of an all-fiber Faraday-rotation current sensor which provides remote passive sensing over insensitive fiber leads utilizing a heterodyne approach to the detection of polarization rotation.

Principle and experimental: Figure 1 shows a schematic of the overall sensor design. Light from a laser source is coupled into a fiber polarizer, the axes of which are aligned with the polarization modes of a high-birefringence (HB) fiber input lead. At the sensing head, the input light is coupled into an 'optical-frequency-to-polarization' (OFP) modulator; this comprises a section of HB fiber (length L) followed by a fiber $\lambda/4$ wave plate. The input light, which is guided in one mode of the input HB fiber, is coupled equally into the two polarization modes of the OFP HB-fiber section through a 45° splice. Due to the effective optical path difference between the polarization modes of this fiber, modulation of the laser emission frequency (by means of the injection current) gives rise to changes in the state of polarization (SOP) of the light at the output of the OFP section. The principle of operation of this section is as follows. As the two polarization modes of the OFP HB-fiber are equally excited by the input optical field, the SOP at the point immediately before the fiber $\lambda/4$ plate is confined to a great circle PLQR on the Poincare sphere which passes through the poles (left and right circular states) and linear states at $\pm 45°$ to the polarization axes (denoted X and Y) in the fiber, as shown in Figure 2.a. Passing this light through the $\lambda/4$ waveplate oriented with its axes at $\pm 45°$ to the modes of the OFP HB-fiber rotates the great circle PLQR on the Poincare sphere by 90° into the equatorial plane XPYQ, as shown in Figure 2.b. The light after the $\lambda/4$ plate is thus linearly polarized at an azimuth which depends on the net polarization phase delay in the OFP HB-fiber section. Consequently, with this arrangement, a sweep in optical emission frequency of the laser results in a sweep in the plane of polarization of the linearly polarized light at this point. By analysis of the optical system the azimuth of the output plane of polarization (relative to the X mode of the OFP fiber) can be expressed as

$$\psi = \pi B L \nu / c \tag{1}$$

where, B is the modal birefingence of the fiber, ν is the laser emission frequency, and c is the velocity of light. A modulation in the laser frequency, $\Delta\nu$, thus results in a rotational modulation in the plane of polarization $\Delta\psi$ given by

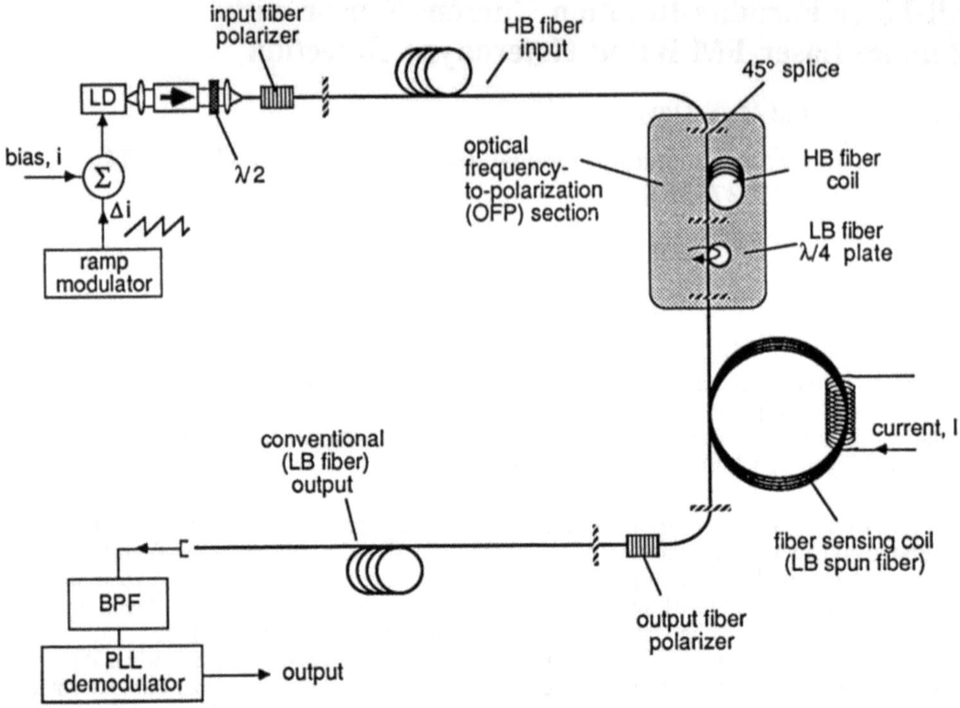

FIGURE 1. Schematic of the all-fiber Faraday-rotation current sensor. ----- fiber splice, HB - high birefringence, LB - low birefringence

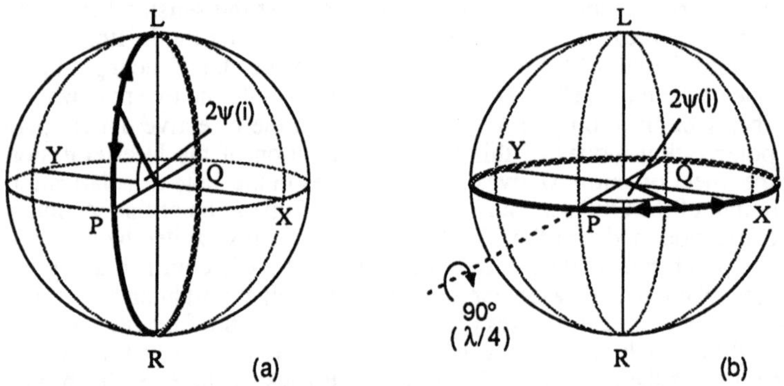

FIGURE 2. Poincaré sphere representation of the SOPs, (a) after the HB-fiber section of the OFP, and (b) after the λ/4 waveplate. Note in (b) the output SOP is confined to the equatorial plane PYQX, i.e all linear states.

$$\Delta\psi = \{\delta\psi/\delta\nu\}\,\Delta\nu = \pi\,B\,L\,\Delta\nu\,/\,c. \qquad (2)$$

For typical values of $B \sim 4.10^{-4}$ (~ 2 mm beat length at 830 nm), the frequency-to-polarization factor $\{\delta\psi/\delta\nu\}$ = 12°/GHz (.21 rad/GHz) for a 50 m length of fiber.

As shown in the schematic of Figure 1, light at the output of the OFP section is coupled into a low-birefringence current sensing coil, the output of which is coupled directly to a second fiber polarizer. The light passed by this output polarizer is guided by a conventional fiber link to the detection electronics. In the experimental system, the input and output fiber links were each 100 m in length, and a 70 m coil of high-birefringence fiber was used as the OFP section. The sensing coil consisted of 187 turns of ultra-low birefringence fiber wound on a 32 cm diameter former. This fiber was twisted a a low rate of 2 twists/turn to counteract the effects bend-induced linear birefringence introduced by coiling the fiber. The current was passed through a coil of 974 turns of 20 AWG wire wound in a a torodial configuration around the fiber coil. The net Faraday rotation in this arrangement is given by

$$\phi_f = V N_f N_w I, \tag{3}$$

where V is the Verdet constant of silica, I is the current in the conductor coil of N_w turns, and N_f is the number of fiber turns in the sensing coil. At the output of the sensing coil, the azimuth of the output polarization can be written as

$$\psi_0 = \psi(\nu) + \Delta\psi(\Delta\nu) + \phi_0 + \phi_f(I), \tag{4}$$

where, $\psi(\nu)$ and $\Delta\psi(\Delta\nu)$ are due to the laser bias point (i) and modulation (Δi), ϕ_0 is a rotational bias in the sensing coil due to circular birefringence introduced due to twisting of the fiber, and $\phi_f(I)$ is the current dependent Faraday rotation. After the output fiber polarizer, the intensity of light coupled into the output fiber is given by

$$\mathfrak{J}_0 = \mathfrak{J}_i \cos^2(\psi_0 - \alpha), \tag{5}$$

where α is an angular offset which accounts for the arbitrary orientation of the output polarizer axes, and $\mathfrak{J}_i$ is the intensity at the input to the system (after the input polarizer). Eqn. 5 can be expressed as

$$\mathfrak{J}_0 = \mathfrak{J}_i \frac{1}{2}\{1 - \cos(2\Delta\psi + 2\phi_f(I) - \zeta)\}, \tag{6}$$

where, $\zeta = 2\alpha - 2\phi_0 - 2\psi(\nu)$ is the net phase offset term.

The heterodyne-type technique used here to detect Faraday rotation $2\phi_f(I)$ is identical to that previously described for use with interferometric [9] and polarimetric sensors [10]. Modulating the laser emission frequency with a sawtooth-waveform of deviation sufficient to induce a rotational modulation $\Delta\psi = \pi$ rad. p-p., ramps the output transfer function described by Eqn. 6 over a complete cycle. Bandpass filtering of the detected output at the fundamental of the ramp frequency, ω_r, produces a output which can be written as

$$S = k\,\mathfrak{J}_i \cos(\omega_r t + 2\phi_f(I) - \zeta), \tag{7}$$

where k is an electronic gain factor. This describes a carrier signal of frequency ω_r, phase-modulated by the magneto-optically induced rotation $2\phi_f(I)$. In our experimental system a phase-locked loop was used to demodulate the output PM carrier.

It should be noted that the carrier described by Eqn. 7 is also phase modulated by the offset term ζ. Drift in this is possible due to laser frequency instability, temperature induced changes in the birefringence of the OFP HB-fiber section and drift in the rotational offset ϕ_0 in the sensing fiber coil, and is indistinguishable from D.C. Faraday rotation phase shifts induced by D.C. currents in the sensor. However, the detection of A.C. currents, which is usually of interest in applications such as the power generation and distribution industry, is not compromised by this effect.

Results: Before constructing the overall sensor system, the fiber sensing coil was calibrated. The result of this measurement is shown in Figure 3. A measured Faraday rotation of 0.51 rad/A (29°/A) was determined for the arrangement which is within 7 % of that calculated using a fiber Verdet constant of $2.6 \cdot 10^{-6}$ rad/A. The second stage of the experiment was to construct and test the OFP element, which comprised a 70 m length of high-birefringence fiber of 2.0 mm beat length wound in a 10 cm dia. coil and a short section of low-birefringence fiber with a single 3.2 cm dia. loop to form a fiber $\lambda/4$ plate. Figure 4.a shows the set-up used to test the operation of the OFP. The fiber waveplate was mounted such that the plane of the fiber loop could be rotated with respect to the polarization axes in the OFP HB-fiber. Linear polarized light from the input fiber was coupled into the OFP fiber via a 45° fusion splice. The orientation of the $\lambda/4$ plate was adjusted until it was found that a linear output SOP was generated at the output, the azimuth of which depended linearly on the laser injection current (emission frequency). This was checked by varying the laser drive level over several mA and rotating a Glan analyzer to determine the plane of output polarization and the degree of extinction. Typically, extinction ratios ≥ 30:1 were achieved over a wide current change, indicating that the output polarization of the OFP was maintained close to a linear SOP. Modulating the laser current by a ~ 5 mA p-p current ramp (laser frequency ramp ~ 15 GHz p-p) was found to be sufficient to induce a polarization-rotation modulation $\Delta\psi = \pi$ rad. p-p in at the output of the OFP. The resulting form of the intensity passed the analyzer is as shown in Figure 4. b. This signal was passed through a bandpass filter centered at the ramp modulation frequency of 10 kHz to produce a carrier signal

$$S = k\,\mathfrak{I}_i \cos(\omega_r t + 2\psi(\nu) - 2\beta) \tag{8}$$

where, β is the relative angular orientation between the polarization axes in the HB-fiber section of the OFP and the analyzer axis. Figure 4.c. shows the carrier signal produced for three settings of the analyzer. Here, the analyzer was rotated from its arbitrary setting by approximately ± 30°; the resulting phase shift of the carrier can be seen to be approximately ± 60° as expected.

Figure 5.a shows the output carrier of the overall all-fiber sensor system. Here, the upper trace shows the unmodulated carrier (no coil current), whereas the lower trace shows the phase modulated carrier observed with a coil current of 3 A p-p. The phase modulation in this case is seen to be 180° p-p, or 60°/A. As the phase modulation term is twice the Faraday rotation ang (see Eqn. 7), this is in excellent agreement with the calibration curve of Figure 3. The upper trace of figure 5.b shows the demodulated output in response to a 1 A p-p current (lower trace

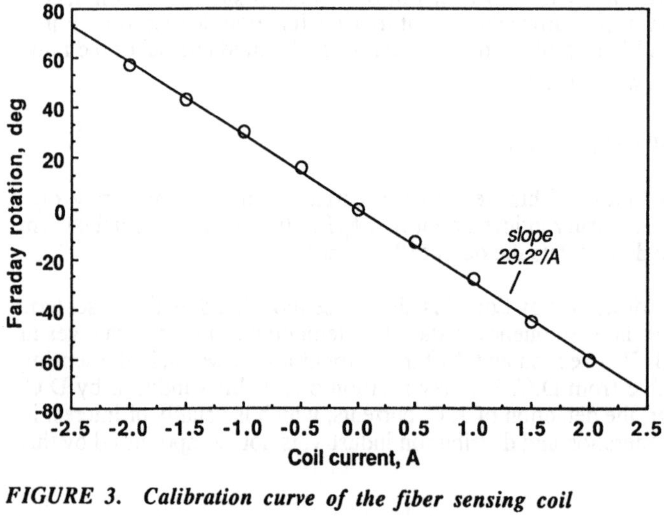

FIGURE 3. Calibration curve of the fiber sensing coil

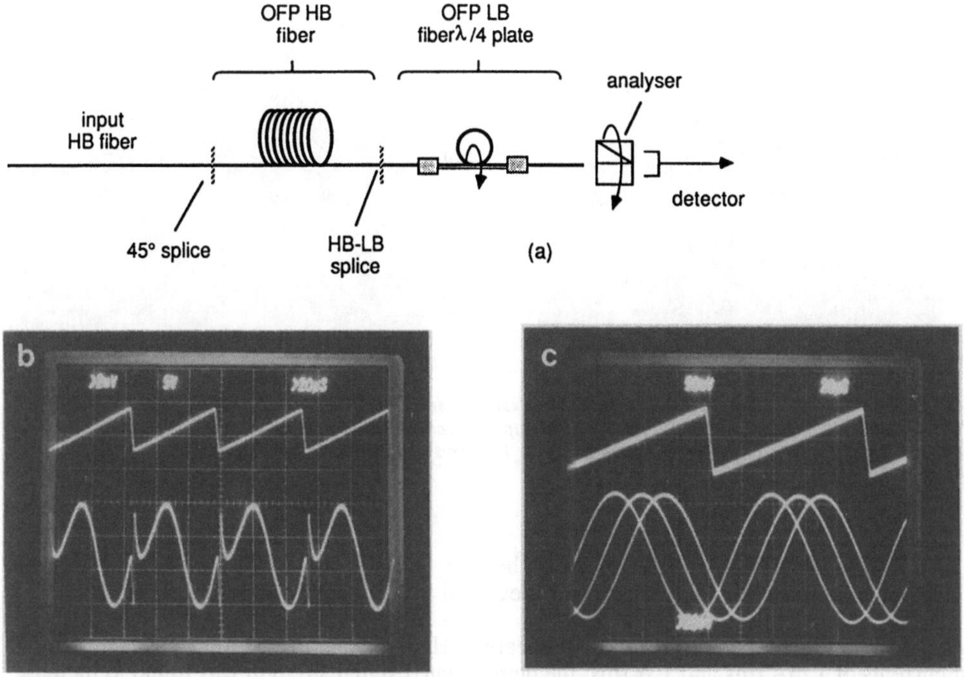

FIGURE 4. (a) Set-up used to test the operation of the OFP section,
(b) detected signal with a current ramp applied to the laser (π p-p modulation in ψ)
(c) BP filtered output carrier showing phase shift due to rotation of analyser axes by approximately $\pm$ 30°

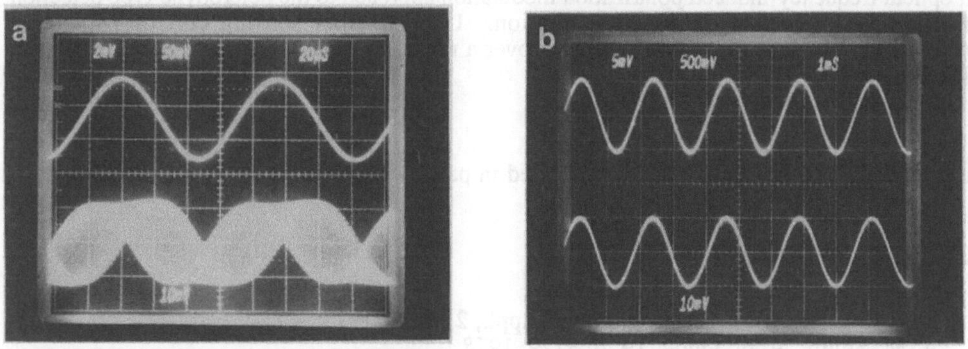

FIGURE 5. (a) Output carrier of the current sensor; upper trace - unmodulated (no coil current), and lower trace - modulated at $\approx \pm$ 90° ($\pm$ 1.5 A at 440 Hz).
(b) Upper trace - demodulated sensor output (PLL output) response to 1 A p-p at 440 Hz, lower trace - conventional current probe output.

shows the output of a conventional current probe). Figure 6.a shows a measure of the minimum detectable current. Here, a test current of 1 mA rms at 440 Hz was passed through the current coil, and the output of the PLL demodulator was monitored on a spectrum analyzer. The S/N of the recovered signal was 30 dB/√Hz, corresponding to a minimum detectable current of 30 μA/√Hz at 440 Hz. This minimum detectable current phase shift corresponded to

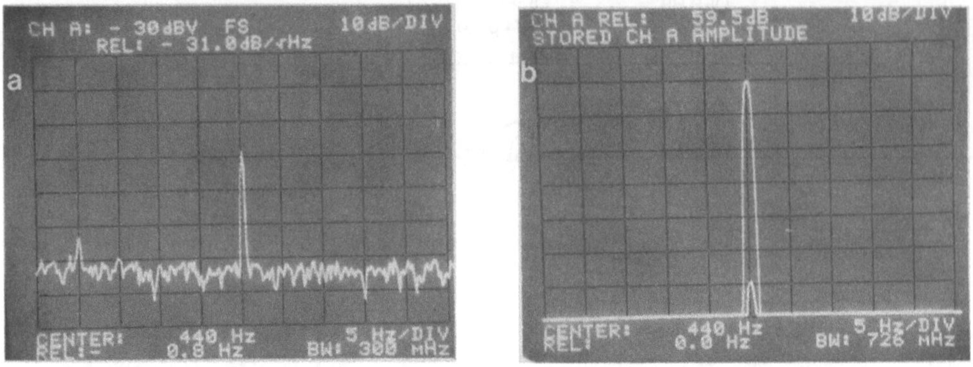

FIGURE 6. (a) Minimum detectable current; test current 1 mA rms at 440 Hz, S/N ≈ 30 dB/√Hz. Noise ≡ 30 μA/√Hz. (b) Output spectra showing sensor linearity: lower trace - 1 mA rms, upper trace - 1 A rms (note: ratio of output signals = 59.5 dB)

~ 30 μrad√/Hz, which was determined to be due to laser phase noise. A lower current equivalent noise floor of ~ 20 μA/√Hz was measured at a frequency of 1 kHz.

Figure 6.b shows a measure of the system linearity. Here the output spectrum was recorded for test currents of 1 mA rms and 1 A rms; the demodulated signal strength was found to increase by 59.5 dB. Operation over a dynamic range of ~ 100 dB (30 μA/√Hz to 3 A rms) was possible, limited only by excessive heating in the coil windings at higher currents. This corresponds to a range in the enclosed current (i.e. $I.N_w$) of ~ 30 mA to 3 kA (equivalent to approximately 5.6 A to 0.56 MA for a single fiber loop (i.e. $N_f = 1$) sensor configuration).

Conclusions: We have described an all-fiber Faraday-rotation current sensor which utilizes an optical frequency-induced polarization modulation approach to the heterodyne-type detection of the magneto-optical polarization rotation. Using a fiber sensing coil of 187 turns, enclosed-current equivalent measurements over a range 30 mA to 3kA have been made with high linearity.

Acknowledgements: This work is supported in part by the Office of Naval Technology.

References:

1. A. J. Rogers, IEE J. Electron. Power Appl., 2, p. 20, 1977.
2. A. M. Smith, Appl. Optic, 19, p. 3729, 1978.
3. S. C. Rasheigh and R. Ulrich, Appl. Phys. Lett., 34, p. 768, 1979.
4. A. D. Kersey and D. A. Jackson, J. Lightwave Technol., 4, p. 640, 1986.
5. M. P. Varnham *et al.*, Digest ECOC/IOOC Conference Venice, p. 135, 1985.
6. L. Li *et al.*, Electron. Lett., 22, p. 1142, 1986.
7. K. Shiraishi *et al.*, Appl. Optics, 24, p. 1896, 1985.
8. G. W. Day and S. M. Etzel, Digest ECOC/IOOC Conference Venice, p. 871, 1985.
9. D. A. Jackson *et al.*, Electron Lett., 18, p. 1081, 1982.
10. A. D. Kersey *et al.*, Proc. OFS'84, Stuttgart, p. 247, 1984.

Fiber-Optic Sensing with Micromechanical Resonant Metallic Glass Structure

H. Wölfelschneider, J. Philip, and R. Kist*

Fraunhofer-Institut für Physikalische Meßtechnik, Heidenhofstr. 8,
D-7800 Freiburg, Fed. Rep. of Germany
*On leave from Cochin University of Science and Technology, Cochin, India

Micromechanical resonators fabricated out of metallic glass ribbons by laser cutting have been optically excited and interrogated via standard multimode fibers. The resonant frequencies of the magnetically soft metglass cantilevers show excursions of the order of 10 to 50 percent as function of magnetic field variations of up to 35 kA/m. High resolution, magnetic field based fiber optic sensors are feasible which are frequency encoded, i. e. downlead insensitive, and can be realized using cheap components and fabrication techniques that are suitable for mass production.

1. Introduction

Fiber-optic sensors for industrial process control and other applications need to be downlead insensitive, i.e. independent of optical power degradation or losses due to source or detector aging, varying coupler losses, fiber bends, etc.. Since intensity - encoded fiber-optic sensors are principally susceptible to such losses they require special efforts such as referencing the intensities of a measuring and a reference channel in order to minimize corresponding errors.

As with conventional sensors frequency encoding of the measurand to be detected seems to be the best approach to reach downlead insensitivity. One concept for frequency encoded fiber - optic sensors makes use of micromechanical resonant structures that are anisotropically etched in silicon /1,2,3/. Work that has been performed recently at our institute with a fiber-optic sensor for the detection of the angular velocity of rotating ferromagnetic toothed wheels led to first investigations of a new class of micromechanical fiber-optic sensors using metallic glass as material for the resonant structure.

Let us first describe the angular velocity sensor which allows - in contrast to inductively operating versions - to measure down to very small angular velocities as well as in electro-magnetically contaminated, dirty and elevated temperature environments. As shown in Fig.1, the hermetically sealed sensor head contains the fiber holder, a little permanent magnet and a very small strip of metallic glass in front of the fiber tip.

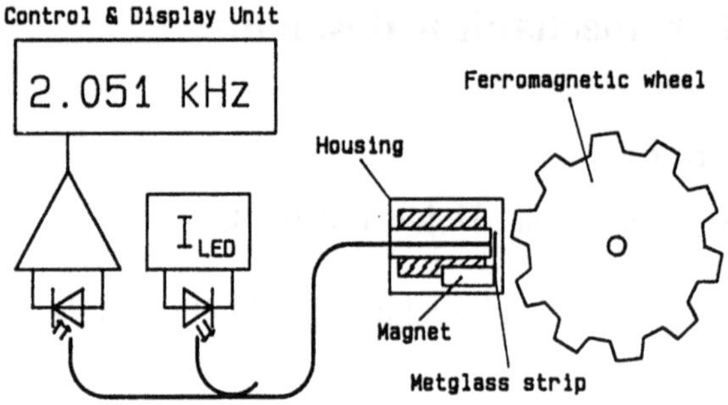

Fig. 1: Principle of the fibre-optic rotation sensor.

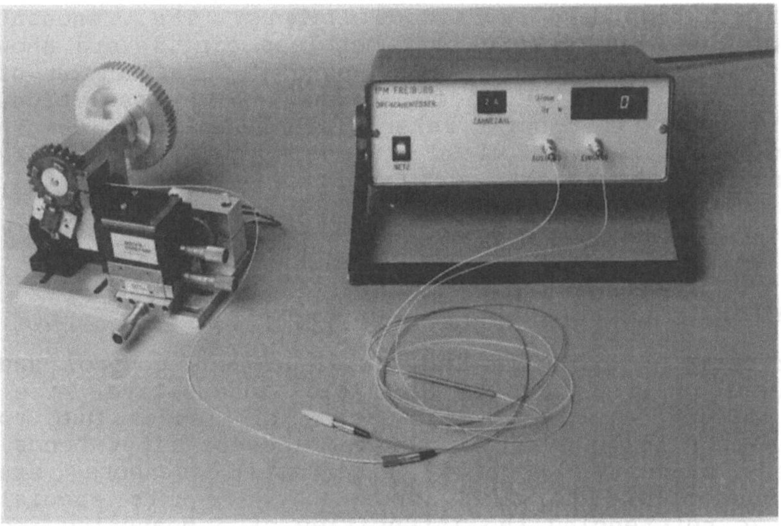

Fig. 2: Photograph of the fibre-optic rotation sensor.

The strip is slightly deformed due to the change of the magnetic flux produced by the passage of the ferromagnetic wheel tooth. This modulates the intensity of the light back reflected by the strip into the fiber, the frequency of this modulation being a measure for the angular velocity given the number of teeth on the wheel. Fig.2 is a photograph of the laboratory mount of the sensor showing the toothed wheel, the sensor head within the positioning unit, the connecting fiber and the electronic box. Fig.3 demonstrates the small decrease of the amplitude of the fiber-optic sensor output pulses (due to a high pass filter in the detecting unit) with decreasing rotation rate as compared with the strong decrease of the signal amplitude of an inductively operating angular velocity sensor mounted closely to the fiber optic sensor head. It turned out by mechanical excitation experiments that the various metglass strips used for the development of the sensor had mechanical resonances between 2.3 and 10 kHz.

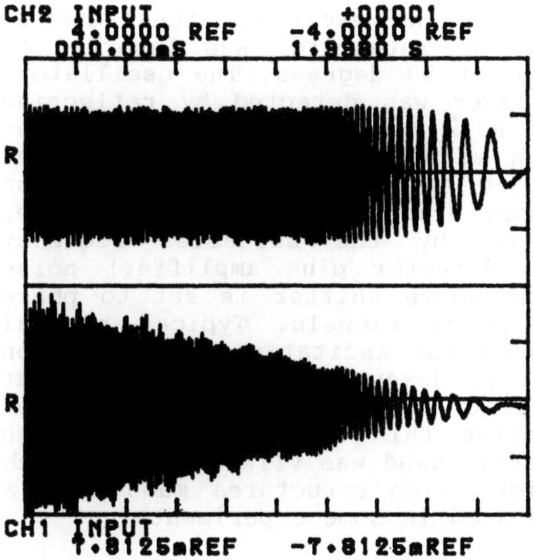

Fig. 3: Traces of the signals picked up by the fibre-optic (upper) and conventional magnetic (lower) sensors.

2.Optically excited metglass structures

In the context of these investigations with the fiber-optic angular velocity sensor the idea was born to optically excite the mechanical resonance of metallic glass structures in order to use this material for micromechanical fiber optic sensing. From the material properties point of view metallic glass is interesting since it is magnetically soft, but mechanically hard /4/, and allows for easy and cheap structuring such as laser cutting.

Fig.4 shows schematically the measuring configuration of the experimental setup. A mechano-opto-electronical closed loop self oscillating system including phase shift and amplification has been realized in order to automatically tune the modulation frequency of the laser diode source (Hitachi HL 7801, λ=790 nm) to the resonance frequency of the system which is determined by the mechanical resonant element excited via a first fiber.

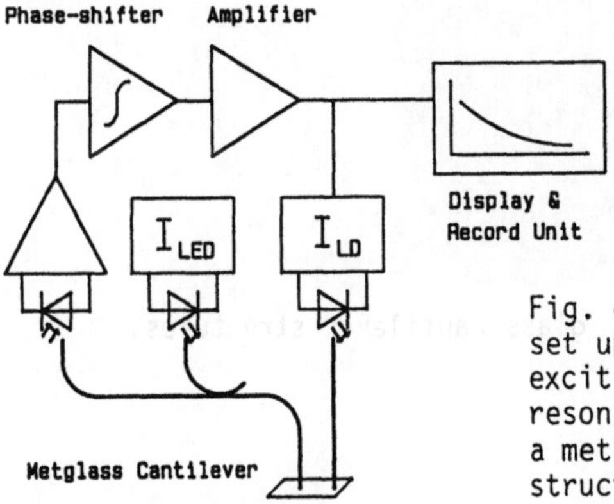

Fig. 4: The experimental set up for optically exciting and picking up resonant oscillations in a metallic glass cantilever structure.

Since the detected signal leads the exciting signal nearly constant phase by 90 degrees, an integrator was inserted to provide a constant phase shift of 90 degrees. The oscillatory motion of the metglass cantilever was detected by reflecting the light of a LED (IRE 160 B, λ = 830 nm) at the metglass surface and detecting it via a second fiber and fiber coupler by a PIN diode. The fibers used for excitation and detection were both of standard 50 µm core diameter gradient index type. Self oscillation is initiated by the suitable frequency component of the (source and detector plus amplifier) noise spectrum and amplified, if the phase shifter is set to phase match the excitation and response signals. Typical optical power levels in the fibers used for excitation and detection are 2 mW and 30 µW, respectively. However, it is expected that much lower levels for excitation will be possible when optimizing the system (absorptive thin films on the metglass etc.). The metallic glass material used was VITROVAC 7505 which comes as a ribbon, 16 µm thick. Also structures made from a ribbon of 25 µm thickness were used in some experiments.

Fig.5 shows a set of 4 pairs of metglass cantilevers, 3 mm long and 16 µm thick, with widths equal to 0.1, 0.2, 0.4, and 0.8 mm. The structure has been cut by means of a Nd-YAG - Laser. For comparison a 250 µm/125 µm outer/inner cladding diameter fiber is shown.

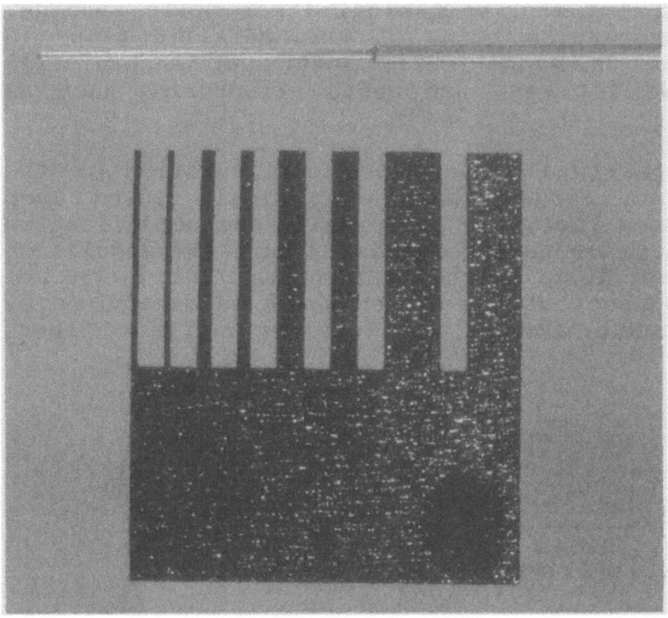

Fig. 5: Photograph of a set of metallic glass cantilever structures.

3.Basic results

The metglass structures turned out to respond easily to the fiber optical excitation, the modulation depth of the received signal being typically between 1 and 10 %. The resonance amplitude of the structure was measured with an interferometrical arrangement, using a laser diode instead of the LED. Typical amplitudes are of the order of 1 to 10 µm.

As shown in Fig.6a for the 0.2 mm wide structure, its resonance at f = 1648 Hz can be detected with a signal to amplitude noise ratio of approximately 70 dB and 40 dB for optical excitation power levels of 2 mW and 0.1 mW, respectively. Fig. 6b shows measured frequency fluctuations for signal gating times of 160, 60, and 10 ms, respectively.

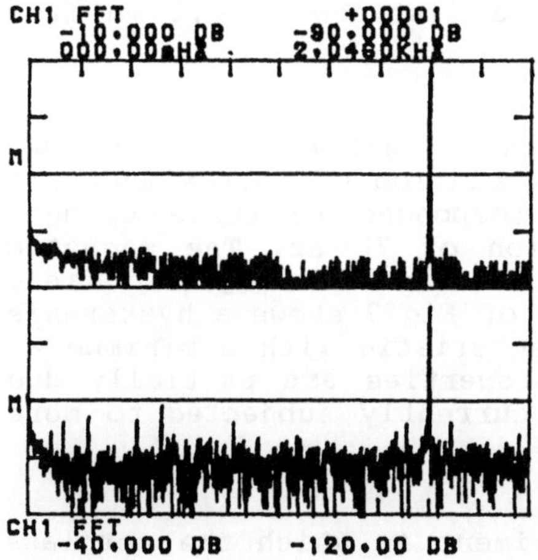

Fig. 6(a): Frequency spectra of a structure at two different excitation powers (upper trace: 2mW; lower trace: 0.1mW).

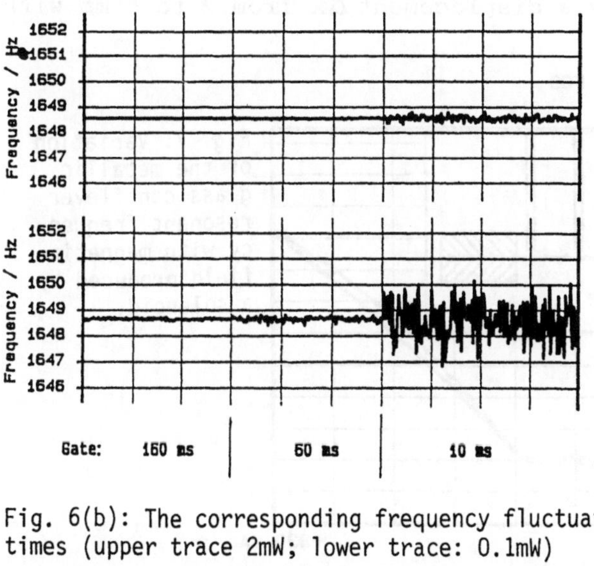

Fig. 6(b): The corresponding frequency fluctuations at different gating times (upper trace 2mW; lower trace: 0.1mW)

4.Results relevant to sensing

Since the material of the resonating structures is magnetically soft, the first set of experiments aiming for sensing applications was devoted to magnetic field based measurements. To this end the metglass cantilever was exposed to a magnetic field produced by a magnetic coil with ferromagnetic core as well as small permanent magnets. In the first case magnetic field or electric current sensors, in the second case position or displacement sensors as well as sensors for parameters that are derived from displacement can be conceived.

Fig.7 shows the change in resonant frequency as function of the magnetic field produced by a magnetic coil operated at varying currents. The experimental arrangement is shown schematically in the inset of the figure. The coil axis is collinear with the metglass cantilever (of thickness 16 µm), the end face of the ferromagnetic core being placed at a distance of 0.2 mm from the cantilever end.

The coil current I was changed from - 330 mA to + 330 mA, corresponding to a magnetic field variation of approximately ± 40 kA/m. The metglass cantilever responded to these changes with a resonance frequency excursion of 70 Hz. The magnetic field was separately measured with a thin film magnetic field probe and a hall element. The curve of Fig.7 shows a hysteresis effect as well as a nonlinear characteristic with a minimum at low magnetic field values. These properties are partially due to the coil core material and are currently subjected to more detailed investigations.

Fig.8 shows the result of an experiment in which the metglass cantilever of Fig.7 was subjected to the field of a serial set of four small permanent magnets (type G3 - 2i, 2 x 3x1 mm^3 each) that were moved by a displacement Δx from 2 to 5 mm with

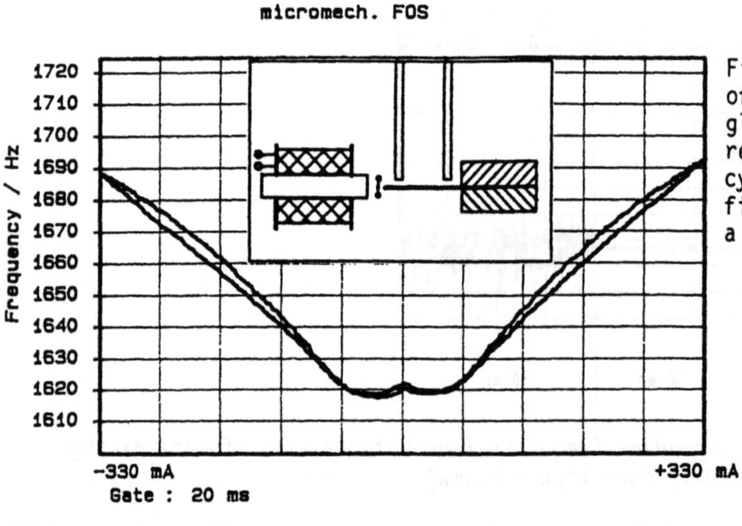

Fig. 7: Variation of the metallic glass cantilever resonant frequency with magnetic field produced by a solenoid.

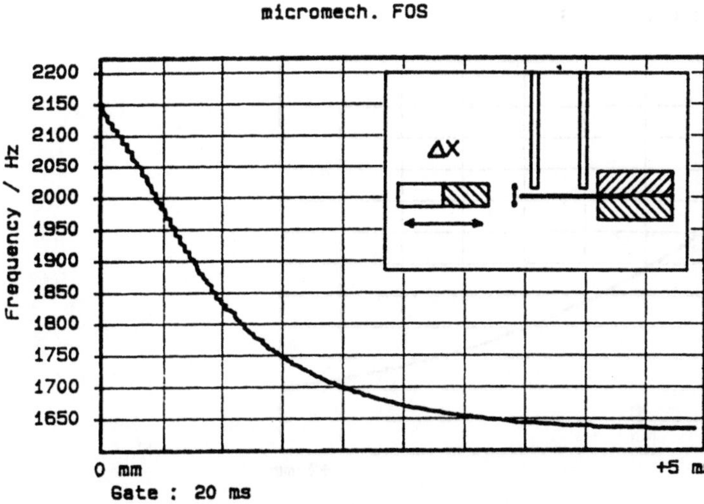

Fig. 8: Frequency variation with displacement of a permanent magnet along the cantilever axis.

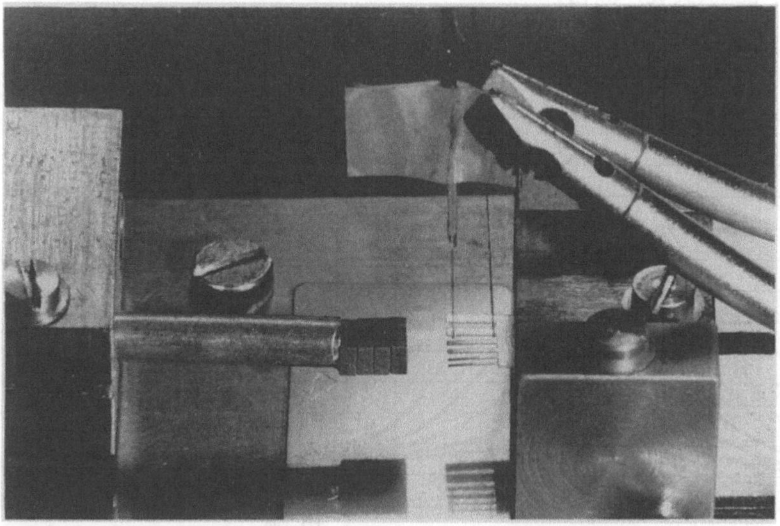

Fig. 9: Photograph of the experimental arrangement showing the structure in the field of a permanent magnet.

respect to the metglass cantilever end as shown in the figure's inset. The photograph of Fig.9 illustrates the arrangement of the experiment further. The characteristics of this position sensor shows a corresponding frequency shift of about 500 Hz, which corresponds to a position resolution of the order of 1 µm if the resonance frequency is determined with an accuracy of 0.1 Hz.

Fig.10 shows a portion of the curve of Fig.8 in the range Δx = 0...2 mm with higher resolution using a conventional differential transformer as inductive displacement sensor. The

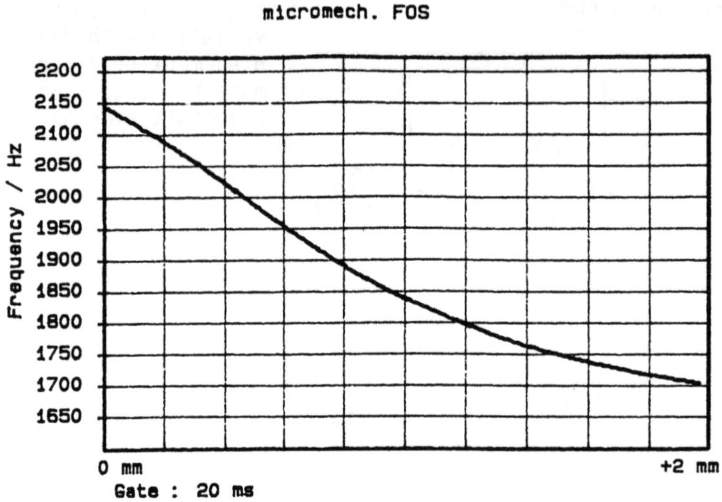

Fig. 10: Variation of the resonant frequency with displacement of a permanent magnet along the cantilever axis.

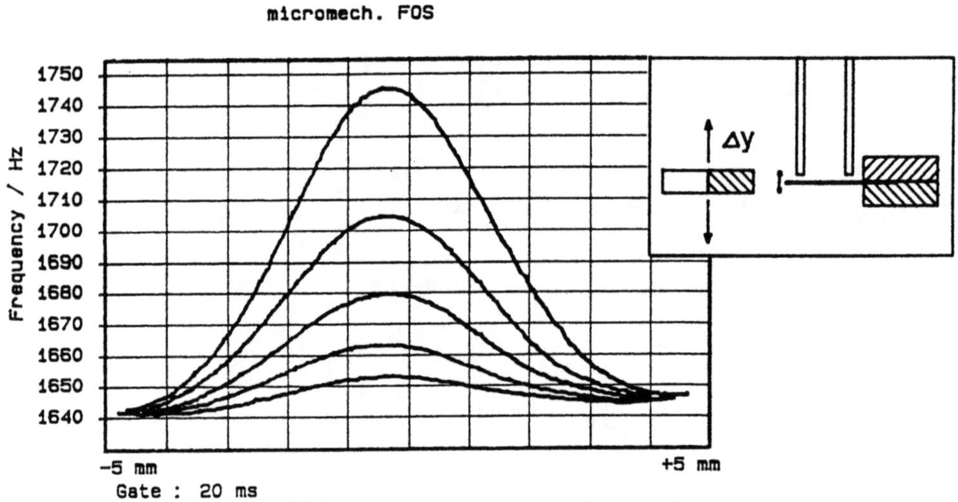

Fig. 11: Frequency variation with displacement of a permanent magnet transverse to the cantilever axis.

noise equivalent frequency increment is lower than 0.1 Hz so that this simple arrangement of a fiber optic metglass resonator sensor provides already a resolution of the order of 0.5 µm.

Fig.11 finally shows the variation of the resonant frequency when the permanent magnets are moved laterally, i.e. in y - direction with respect to the metglass cantilever, for various axial distances x as shown in the inset. Within a range of Δy = ± 4 mm at distances Δx smaller than about 1.8 mm the lateral position can be measured with an accuracy of about 5 µm.

5.Conclusions

We have shown in this paper for the first time that micromechanical resonant structures fabricated from metallic glass material using laser cutting techniques are suitable for optical excitation and interrogation by standard multimode fibers. Possibly wet etching processes or other cheap structuring techniques are also usable to produce the structures. The resonance frequency shows very large excursions of the order of 10 to 50 % as function of magnetic fields applied at the location of the structure. Thus high resolution, magnetic field based fiber optic sensors in which cheap components and techniques suitable for mass production are used can be constructed.

The specific material properties of amorphous metals or metal glasses can be choosen by suitable alloy compositions. For example metglass alloys with large as well as small magnetostrictive coefficients can be produced. In addition the elastic modulii of metglasses show a strong dependence on the magnetic field strength.

Thus a wide variety of frequency encoded, line loss insensitive fiber optic sensors with metallic glass micromechanical resonators seems to be feasible.

6.Acknowledgement

The authors would like to thank A. Wollscheidt for his committed contribution to the realization of the angular velocity sensor, G. Knoll for scientific assistance and valuable discussions, E. Neske for laser cutting the metglass structures, and H.-P. Anderer and A. Fischer for technical assistance. The work has been sponsored by the Bundesministerium für Forschung und Technologie (BMFT), Bonn.

7.References

1. S. Venkatesh, B. Culshaw, "Optically activated vibrations in a micromachined silica structure", Electron. Lett. 21 (1985), pp. 315-317.
2. H. Wölfelschneider, R. Kist, G. Knoll, S. Ramakrishnan, H. Höflin, FhG Institut für Physikalische Meßtechnik; W. Benecke, L. Csepregi, A. Heuberger, H. Seidel, FhG Institut für Mikrostrukturtechnik; "Optically excited and interrogated micromechanical cantilever structure", SPIE, Vol. 798, Fiber Optic Sensors II (The Hague, 1987), pp. 61-66.
3. M. Hildebrand, H. Wölfelschneider, "Micromechanical structures excited by noise-modulated light via optical fibers", SPIE Vol. 1011 Fiber Optic Sensors III (Hamburg, 1988), pp. 95 - 100.
4. F.E. Luborsky, "Amorphous Metallic Alloys". London, Boston, Toronto; Butterworths-Verlag (1983).

Part IX

Poster Papers

Thermal Characteristics of Semiconductor Ring Laser Sources for Integrated Optical Sensor Devices

W. Chen[1], *P.J.R. Laybourn*[1], *A.F. Jezierski*[1], *and P.W. Webb*[2]

[1]Department of Electronics and Electrical Engineering, University of Glasgow, Glasgow, G12 8QQ, UK

[2]Department of Electronics and Electrical Engineering, University of Birmingham, Birmingham, B15 2TT, UK

Semiconductor ring lasers have been fabricated for use as sources in monolithic integrated optical sensor and other circuits. The very compact nature of the devices brings increased thermal problems. A thermal model of the ring laser has been developed and is compared with infrared microscanned images of operating ring lasers.

INTRODUCTION

Monolithic integration of semiconductor optical sensor devices requires the development of integrated laser sources with feedback mechanisms other than cleaved cavities and little influence of the output termination on the lasing action. Miniature ring resonant waveguide lasers, shown in Fig.1, have been investigated at Glasgow [1]. A variety of ring configurations between 10 and 100μm in diameter have been formed by reactive ion etching double heterostructure and quantum well material to a depth of about 7μm. They have attracted considerable attention because high quality mirrors are not needed, the output termination has little influence on oscillation conditions and the small dimensions make the longitudinal mode spacing large so that singlemode operation is favoured. The

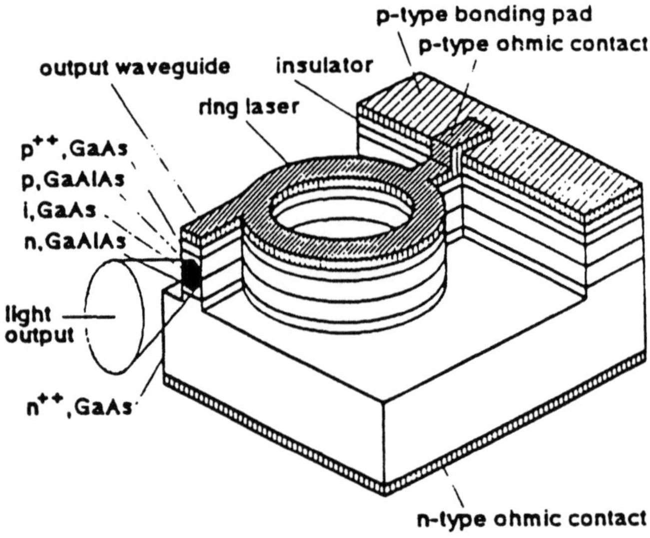

Fig.1 Semiconductor ring laser device

Springer Proceedings in Physics, Vol. 44 **Optical Fiber Sensors**
Editors: H.J. Arditty · J.P. Dakin · R.Th. Kersten

bending loss in a rib waveguide depends inversely on the refractive index difference between the waveguide and outer boundary material [2]. A rib waveguide configuration was used to maximise optical confinement and minimise bending loss at small ring diameters. Owing to the simplicity of fabrication of the structure by planar processing, it can be more easily integrated with an optical fibre sensor signal processing circuit.

So far, devices have lased only with very short current driving pulses. CW operation has not yet been demonstrated. This is partly due to thermal problems. Temperature rises within the ring laser structure are expected to be excessive, as the heat source is more concentrated and the heat generated difficult to dissipate due to the extremely small dimension. The temperature increase may degrade the operating characteristics and life time. Therefore, it is essential to characterise the temperature behaviour of the ring laser with the aim of understanding its thermal characteristics and improving heat dissipation so that CW operation may be possible.

The theoretical analysis of thermal problems in semiconductor laser devices has been demonstrated by many authors, [3] and [4]. But their solutions were based on analytical methods and were obtained only for simple geometries. This work is concerned with characterising the thermal effects of a semiconductor ring laser by numerical simulation methods and infrared imaging techniques. The temperature distribution in the integrated device and its thermal resistance were computed by a finite difference method. An RM−20 infrared image scanner was used to observe its surface temperature distribution. The experiment will show a good assessment of their thermal behaviour which can be used to compare qualitatively with the theoretical results.

COMPUTER MODELLING

(1) Heat flow equation and finite difference formulation

The nonlinear heat flow equation governing heat dissipation is given by

$$K\frac{\partial^2 T}{\partial x^2} + K\frac{\partial^2 T}{\partial y^2} + K\frac{\partial^2 T}{\partial z^2} + q^* = \rho c\frac{\partial T}{\partial t} \quad . \qquad (1)$$

Here q^* is heat generated density (input power density) and ρ the density of material. K and c are the thermal conductivity and thermal capacity per unit volume respectively.

For steady state and a circularly symmetrical body, whose temperature distribution is independent of azimuth, eq.(1) can be obtained in two dimensional cylindrical coordinates, r(radius) and z(axis) as

$$\frac{\partial^2 T}{\partial r^2} + \frac{1}{r}\frac{\partial T}{\partial r} + \frac{\partial^2 T}{\partial z^2} + \frac{q^*}{K} = 0 \quad . \qquad (2)$$

Heat distribution can be obtained in the steady state within the ring laser devices by solving eq.(2) with certain boundary conditions.

The useful method used to solve the equation is the finite difference technique. The device to be analysed is divided into small regions or nodes. Eq.(2) can be substituted by the finite difference formula and the temperature at each node is given by

$$T_{i,j} = \frac{1}{4}[T_{i,j-1}+T_{i,j+1}+(1-\frac{1}{2i})T_{i-1,j}+(1+\frac{1}{2i})T_{i+1,j}+\frac{q^*_{i,j}L^2}{K}] \quad . (3)$$

(2) Thermal model

The ring laser chip is placed on the top of a heat sink surrounded with air. Consideration of only two dimensions is sufficient due to symmetry along the azimuthal direction. Half of the cross section in cylindrical coordinates is shown in Fig.2. The heat generated in the active layer is assumed only to transfer to the heat sink and heat transfer across the chip surface by conduction through air is neglected because the thermal conductivity of the laser chip material is much bigger than that of air. Heat transfer by convection and radiation from any surface is also assumed to be negligible [5]. Therefore, the boundary condition on any side except the base is given by

$$\left.\frac{\partial T}{\partial z}\right|_s = \left.\frac{\partial T}{\partial r}\right|_s = 0 \quad .$$

The finite difference form is

$$T_{n+1} = T_{n-1} \quad . \tag{4}$$

The temperature on the base of the chip is forced by the heatsink down to an adjustable temperature, e.g.20 °C.

For simplicity, the laser chip is divided into two parts, substrate and lasing rib, and their temperature distributions are analysed separately.

(3) Temperature distribution in ring laser substrate

The thermal model for the substrate is shown in Fig.3. Only the junction of the laser is assumed to act as a heat source, on the top of the substrate. The influence of the lasing rib will not be considered here.

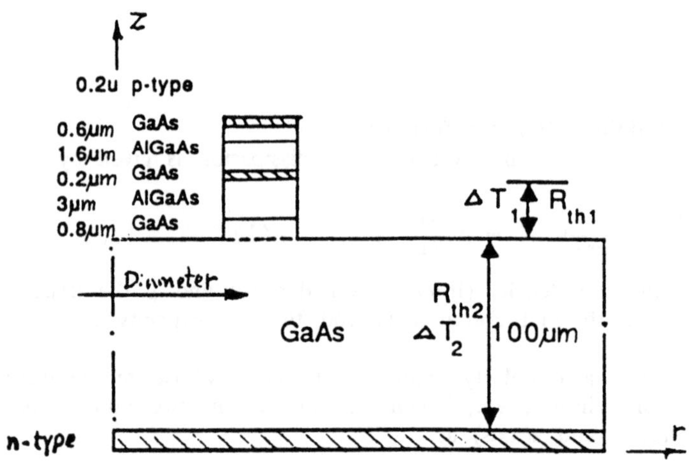

Fig.2 The cross section of ring laser chip along a radius (substrate+rib)

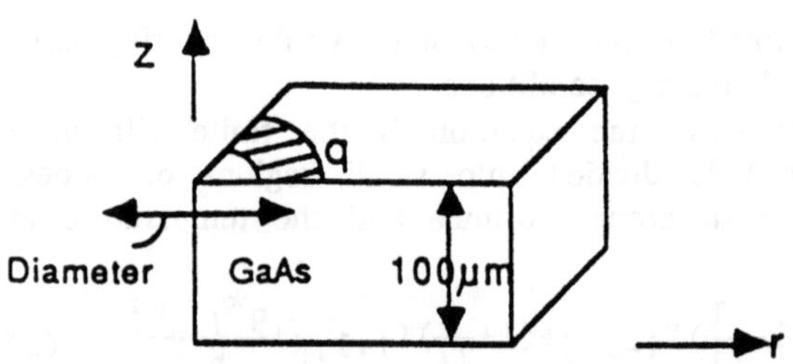

Fig.3 The cross section of ring laser substrate along a radius of the ring

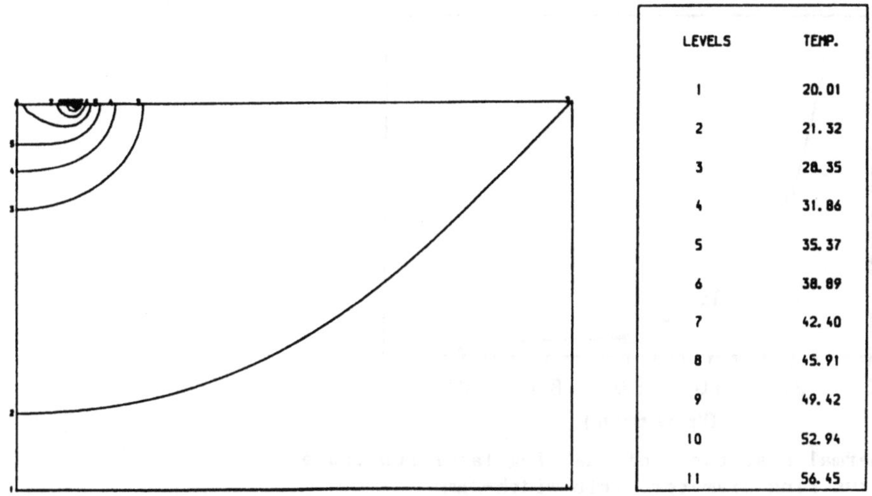

Fig.4 Temperature distribution of ring laser substrate
(diameter=40μm, power=200mW)

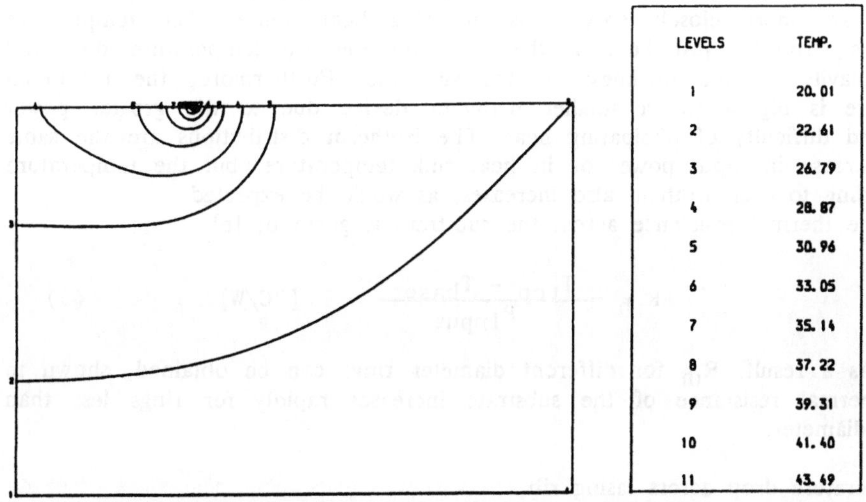

Fig.5 Temperature distribution of ring laser substrate
(diameter=100μm, power=200mW)

The finite difference program used to solve eq.(3) and find the heat distribution within the structure was run on a IBM 3090 computer. The program contains an iteration procedure to calculate the value at each point over and over again until the value at each point in two subsequent iterations changes very little and eventually the contour of the temperature distribution will be plotted out.

Various diameters of the ring laser substrate (D=12,20,40,60,100μm) and for the same input power of 200mW were simulated, (e.g. Fig.4 and 5). The horizontal and vertical directions correspond to radius and height, respectively, in the ring laser chip. Temperature of the bottom of rectangle is forced down to the adjustable temperature of the heat sink, 20^{0}C in these graphs. The curves within the structure are isotherms which correspond to the different temperatures in the table on the right. A small region on the top of rectangle, where the

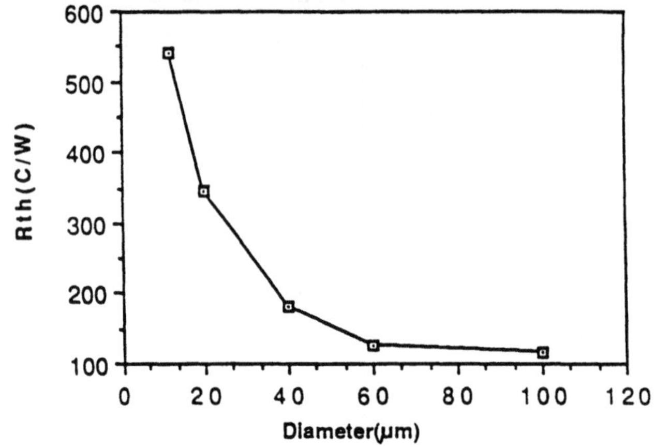

Fig.6 Thermal resistance of the ring laser substrate verus ring diameter. (rib width=2μm)

isotherms are more closely spaced, is the ring heat source. The temperature gradient is strongest just beneath the rib, and the rib temperature does not depend heavily on the thickness of the substrate. Furthermore, the maximum temperature is higher for a smaller diameter device due to the greater power density and difficulty of dissipating heat. The isotherm distributions are the same for an increase in input power or in heat sink temperature, but the temperature corresponding to each isothem also increases, as would be expected.

The thermal resistance across the substrate is given by [6]

$$R_{th} = \frac{T_{top} - T_{base}}{P_{input}} \quad [^{o}C/W] \quad . \qquad (5)$$

As a result, R_{th} for different diameter rings can be obtained, shown in Fig.6. Thermal resistance of the substrate increases rapidly for rings less than 50μm in diameter.

(4) Temperature drop across lasing rib

The thermal influence imposed by the lasing rib has not been considered up to now. However, in some cases, it cannot be ignored and becomes more significant for small rings or narrow ribs. Let us consider the rib model, as shown in Fig.7.

We assume that the heat generated in the active layer flows entirely into the substrate so that the part above the active layer can be neglected. The finite difference technique can be applied in this model. Temperature drop ΔT and

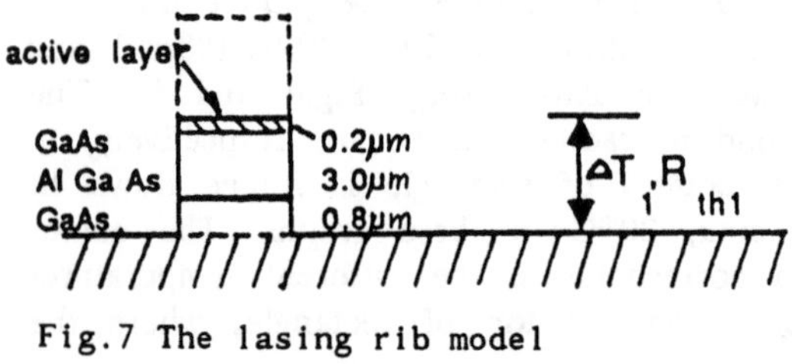

Fig.7 The lasing rib model

Table 1 Comparison of the temperature drop and thermal resistance between lasing and substrate of the ring lasers

diameter(μm) \ width(μm)			2	4	8	10
40 power=42mW	rib	ΔT1(C)	8.79	4.64	2.61	2.22
		R1(C/W)	209.2	110.5	62.14	52.86
	substrate	ΔT2(C)	7.72	7.0	6.73	6.53
		R2(C/W)	183.8	166.7	160.2	155.5
100 power=200mW	rib	ΔT1(C)	16.24	8.29	4.32	3.54
		R1(C/W)	81.2	41.45	21.6	17.7
	substrate	ΔT2(C)	23.5	14.0	12.96	12.6
		R2(C/W)	115.0	70.0	64.8	63.0

thermal resistance R_{th} between the active layer and the top of the substrate can be estimated by using similar methods. Two ring diameters (40 and 100μm) with supplied power of 42mW and 200mW respectively have been considered. In addition, the temperature drop ΔT_1 and the thermal resistance R_{th1} with different rib widths have been calculated and presented in Table 1. In comparison, temperature drop ΔT_2 and thermal resistance R_{th2} between the top and the bottom of the substrate under the same conditions are presented in the same table. For 100μm diameter ring with 200mW power supplied, the temperature drop of the wider rib is smaller than that of the substrate. In this case the rib thermal effect can be ignored for simplicity. However, for narrower ribs, the temperature drop of the rib is larger than that of the substrate. In the 40μm diameter ring, the temperature drop is even more significant.

With reference to their thermal resistances, as shown in Fig.8 and 9, the thermal resistances of both rib (R_{th1}) and substrate (R_{th2}) increase with decreasing widths and R_{th1} increases more quickly than R_{th2}. Fig.9 shows that for wider ribs (more than 7μm), R_{th2} is three times larger than R_{th1}. However,

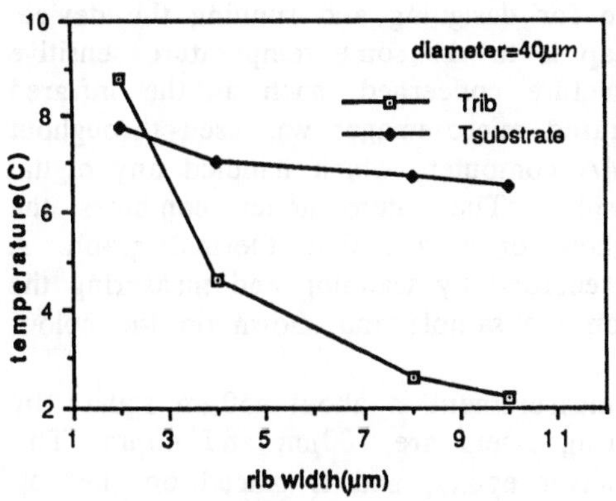

Fig.8 Thermal resistance of lasing rib and substrate with rib width

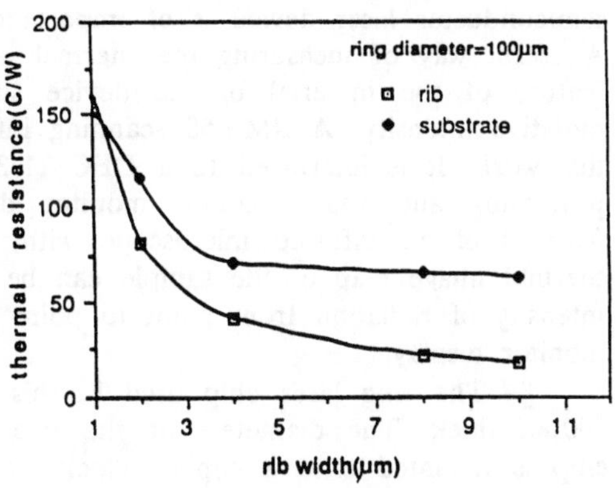

Fig.9 Thermal resistance of lasing rib and substrate with rib width

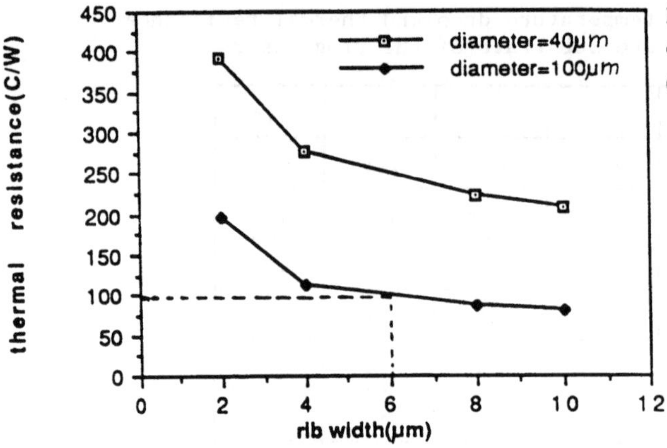

Fig.10 Thermal resistance of whole ring laser chip (substrate+rib)

when the width is less than 4μm, R_{th_1} increases more quickly than R_{th_2} and eventually it is bigger than R_{th_2}. In Fig.8, the 40μm diameter ring laser has relatively large thermal resistances for both the substrate and the rib. The thermal resistance of the rib (R_{th_1}) increases sharply with increasing rib width. When the width is less than about 2.5μm. R_{th_1} is bigger than R_{th_2}.

The results obtained show that the thermal effect of the lasing rib is significant for a small ring or narrow width rib. Furthermore, a shallower and wider lasing rib can be of advantage because the thermal resistance of the rib is proportional to its height.

If the results in sections (3) and (4) are combined, we can find the thermal resistance of the whole ring chip (substrate+ rib), shown in Fig.10. It illustrates that a large ring and wide rib structure can have low thermal resistance and would be better for heat dissipation. It also suggests that for 100μm diameter ring laser, the thermal resistance of only the structures in which the rib width is more than 6.0μm can be less than 100 ^{0}C/W, which may represents a device with acceptable thermal characteristics.

INFRARED IMAGING MEASUREMENT

Direct observation of the temperature profile on the top of the semiconductor laser device is of importance for designing and running the device. A useful way of measuring the thermal map is to use some temperature sensitive feature of the material of the device structure concerned, such as the infrared radiative intensity. A RM−50 scanning infrared micro imager was used throughout this work. It is interfaced to a DEC 11/34A computer, which handled any digital processing and has a colour monitor display. The microscanner combines the features of an infrared microscope with those of a real time thermal graph. A thermal image map of the sample can be recorded by scanning and measuring the intensity of radiation from point to point on the sample and shown on the colour monitor display.

The ring laser chip used in this measurement is about 450μm square by 100μm thick. The diameters of the two ring lasers are 100μm and 40μm. The chip is mounted onto a copper block by silver epoxy, and is placed on the top of the heater in the scanner system. The problem which arises is that low material emissivity makes measurement difficult, especially for this work which has

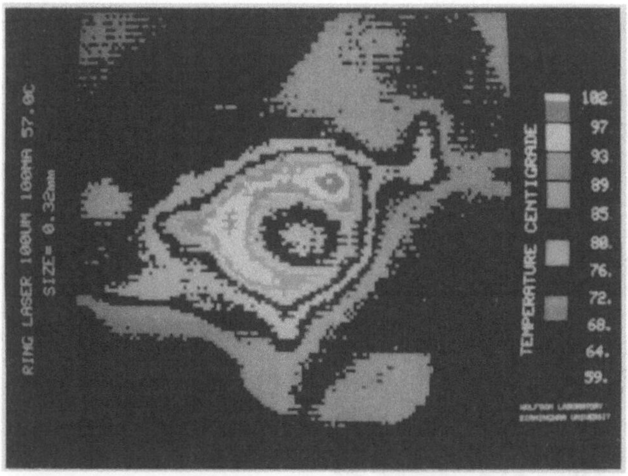

Fig.11 Infrared temperature map (Average temp. on the top of ring is 92°C

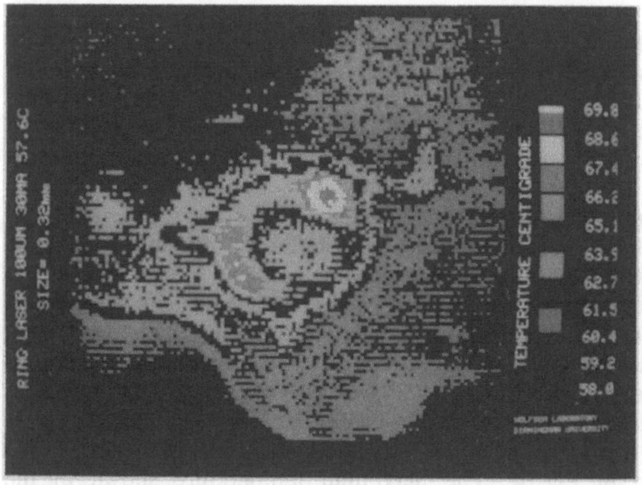

Fig.12 Infrared temperature map

low supply current (10mA–100mA). A solution to the problem has been found by using a coating of known high thermal emissivity material, silicon monoxide, which increases the minimum emissivity up to values exceeding 0.4.

A 100μm diameter ring laser was measured with 100 and 30mA supply current. The thermal maps are presented in Figs.11 and 12. Increase of power supply in the laser device makes it not only emit more light, but also generate more heat.

The thermal maps clearly illustrate the temperature distributions on the top of the sample, hotter in the ring region than outside the ring. Different colours on the maps correspond to different temperatures which are shown on the right hand side. It is clear from the photographs that the temperature on the top of the ring is not uniform. For example, in Fig.11, there is a hot point, 102 °C, at the Y junction and the left half of the ring is hotter than the right half. The average temperature on the ring can be estimated roughly to be about 92 °C. It is not easy to explain the nonuniform temperature distribution because it depends on

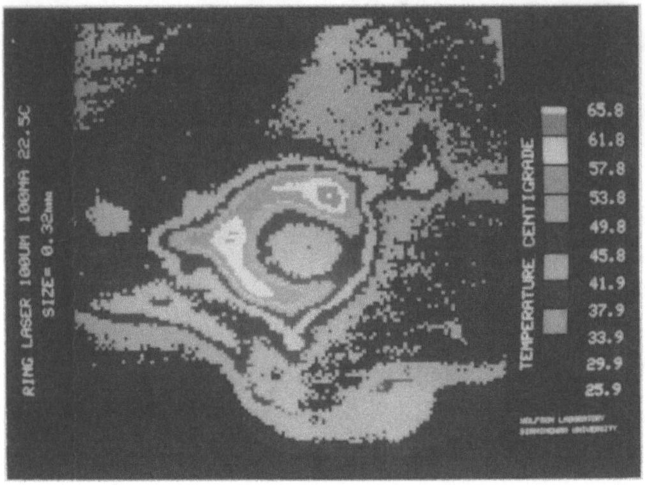

Fig.13 Infrared temperature map (Average temp. on the top is 54.5°C)

many factors, e.g. low resolution of instrument, emissivity, nonuniformity of the material and its surroundings. The reason for the higher temperature in the left half of the ring may be explained by the fact that there is a big drop of epoxy, the conductivity of which is 0.02W/cm°C, which is much smaller than that of GaAs, 0.44W/cm°C, on the top left corner of the chip. As a result, heat is more difficult to dissipate in the left half of the ring. The hottest point is due to Y junction coupling where more current is concentrated.

Fig.11 and Fig.13 show temperature distributions for different base temperatures, $T_S = 57.\,^{\circ}C$ and 22.5 °C, under the same 200mW power supply. It is clear that the results obtained are in general agreement with the theoretical simulation results above, i.e. the temperature profiles remain unchanged, but temperatures in the table shift to lower temperatures. For example, the difference between maximum and base temperatures in Fig.11 is about 45°C, which is nearly equal to the equivalent difference, 43.3°C, in Fig.13.

The thermal image map for a 40µm diameter ring with 51mW (30mA) power supply is demonstrated in Fig.14. The maximum temperature shown in the temperature table is the average temperature of the whole ring, because the ring is too small for different temperatures to be distinguished on the ring.

Table 2 shows the comparison of temperature drop and thermal resistance in theory and experiment. It illustrates that the theoretical results are bigger than the values obtained experimentally. These differences between theoretical model and experiment can be explained by noting that the resolution of the infrared scanner is 10µm at best which is larger than the width of the lasing rib. Therefore, the temperature of the ring shown in the thermal maps is actually the average temperature of the rib and surrounding substrate, see Table 2. Where the rib is widened, for example, at a Y junction, the scanner will indicate a higher temperature.

It has been found that the thermal maps show the temperature distribution so obviously and directly that we can say it is a good method to observe the thermal behaviour of the device. Although it cannot indicate rib temperature accurately due to the low resolution of the apparatus, it does give some idea of the temperature distribution of the device, which is in broad agreement with theoretical expectation.

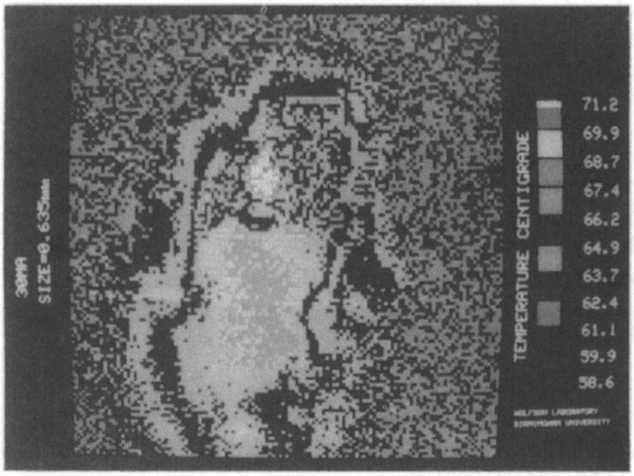

Fig. 14 Infrared temperature map (Diameter - 40 um, power - 42 mw, T_s - 61°C)

Table 2 Comparison of the temperature drop and thermal resistance of the ring lasers in the theory and experiment

	diameter=100μm power=200mW	diameter=40μm power=42mW
theoretical		
substrate	ΔT=23.5 C, R=115.0 C/w	ΔT=7.72 C, R=183.0 C/w
rib+sustrate	ΔT=39.74 C, R=198.7 C/w	ΔT=16.51 C, R=393 C/w
experiment	ΔT=32.5 C, R=162.6C/w	ΔT=9.8 C, R=209 C/w

CONCLUSION

The theoretical thermal modelling of the ring laser device was based on the finite difference technique. Two thermal models have been set up — ring laser substrate and lasing rib models. Temperature distributions in them have been obtained that fit the experimental device fairly well, although a cylindrical symmetry model was used which assumed that all the heat was transferred to the heat sink. For simplicity, no convection or radiation has been taken into account. The results obtained indicate that for smaller rings and narrower ribs, temperatures rise and the increase in the thermal resistance of the laser is more severe. Furthermore, the drop in temperature in the substrate is concentrated beneath the rib. Finally, more attention should be paid to the influence of the thermal behaviour imposed by the lasing rib.

The simulation results suggest that in successful operation of the ring laser the following factors have to be considered:

(1) shallower and wider sandwich rib structure;

(2) larger diameter laser ring (for 100μm diameter, if rib width>6μm, R_{th}<100 °C/W);

(3) lower threshold current;

(4) pulse operation(although only CW has been analysed);

(5) lower heatsink temperature;

(6) sample placed up side down (i.e. top surface of the device attached on the heat sink, allowing the heat generated in the rib to be transferred to the heat sink directly rather than passing through the substrate);

The infrared thermal image scanning technique has been utilized to explore the surface temperature distribution of the ring laser. The thermal maps taken from the RM-10 infrared image scanner show the surface temperature distribution clearly and directly. Although the scanner cannot provide very good quantitative results which could be compared with the theoretical results, mainly due to low resolution of the apparatus, it can still give a fair qualitative assessment of the thermal behaviour of a ring laser.

REFERENCES

[1] JEZIERSKI,A.F AND LAYBOURN,P.J.R: Integrated semiconductor ring laser, IEE Proc.Pt.J, 135, 17-24, (1987)

[2] WALKER,R.G AND WILKINSON,C.D.W: Integrated optical waveguide structures made by silver ion-exchange in glass, Appl.Opt.,22, 1029-1039,(1984)

[3] JOYCE,W.B. AND DIXON,R.W: Thermal resistance of stripe lasers, Appl.Phys, 46, 855-862, (1975)

[4] NAKWASKI,W.: Dynamical thermal properties of stripe geometry laser diodes, IEE Proc., 131, 49-102, (1984)

[5] CHEN,W: Thermal problems in semiconductor ring lasers, M.Sc thesis, University of Glasgow, U.K., G12 8QQ

[6] PAOLI,T.L.: A new technique for measuring the thermal impedance of junction lasers, IEEE J.Q.E., 11, 498-503, (1975).

Optical Fiber Interferometer Sensor with Immunity from Environmental Disturbance Induced in the Lead-Fiber

Der-Tsair Jong and K. Hotate

RCAST, Research Center for Advanced Science and Technology, The University of Tokyo, 4-6-1 Komaba, Meguro-ku, Tokyo 153, Japan

We propose a novel technique to eliminate the environmental disturbance induced in the lead-fiber of the optical fiber interferometric sensor. A heterodyne output and a homodyne output are produced independently to distinguish the signal from the disturbance. The disturbance can successfully be removed, in the experiment, without any active device at the sensor head.

1. Introduction

In this paper, we propose a novel technique to remove the environmental disturbance induced in the lead-fiber of optical fiber sensors. The optical fiber interferometric sensor is expected to have a quite high sensitivity, in which, however, the disturbance induced in the long length lead-fiber limits its performance.

In the sensor system presented in this paper, the coherence characteristics of the light and an optical frequency shifter are used to distinguish the signal from the disturbance. The signal is obtained as an optical heterodyne output, and the disturbance is obtained as an optical homodyne output. Consequently, the disturbance induced in the lead-fiber can sufficiently be eliminated without any active device at the sensor head. The experimental setup has been constructed. The results demonstrate that the disturbance is fairly suppressed.

2. Disturbance induced in the lead-fiber

Fiber sensor is usually categorized into amplitude or phase sensor. Systems using the amplitude modulation are

Editors: H.J. Arditty · J.P. Dakin · R.Th. Kersten

simple and do not require a coherent light. However, they are of low sensitivity and application is accordingly restricted. Phase sensors have been shown to possess extremely high sensitivity. The interferometer must have several schemes to detect accurately the relative optical phase shift between the two arms. First of all, the low frequency random drift induced by the temperature and/or the pressure fluctuations must be compensated. Change in the source power or the loss of the system affects the scale-factor. When the input level is high, the degradation of the linearity takes place.

To overcome these problems, several detection schemes have been studied, such as active homodyne, heterodyne, and quasi-heterodyne manner. They require optical and/or electrical devices such as phase shifter or frequency shifter. By the way, to keep the reliability of the sensor system, it is desirable that the sensor head should be as simple as possible, preferably without electrical connections. The sensor head should not incorporate the phase shifter or the frequency shifter. In the system to realize such requirements, the output signal is mixed with some environmental fluctuation in the lead-fiber which connect with the sensor head. Especially, in the case of a fiber sensor with a long length lead-fiber, the enviromental disturbance from the lead-fiber limits seriously its performance.

Several techniques to remove the enviromental disturbance mixed in the lead-fiber have been proposed[1-5]. However, problems are left yet: excessive optical coherence for a given sensitivity, difficulty in setting and keeping the operating point, difficulty in demodulation, difficulty in keeping the sensor bandwith, and low sensitivity. Another technique by using source frequency control may be applied, in which the reference route is set near the sensing route. In this case, the feedback delay induced by a long length lead-fiber limits the upper frequency of the input signal. Moreover, the reference route should not be set near the sensing field, but set in a stable circumstences.

Considering the situation mentioned above, we propose a novel scheme, in which the enviromental disturbance induced in the lead-fiber can fairly be removed.

We have investigated the manner in which the conbination of the optical delay loop having the optical frequency shifter and the light source with the adjusted coherence[6] achieves a novel function in the fiber sensor. The method proposed in this paper is another application of this manner.

3. Principle and optical setup

Figure 1 shows the schematic diagram of the proposed interferometric fiber sensor. The light source LS1 is selected so that its coherent length is sufficiently shorter than the optical length of the sensor loop and that of the reference loop. Moreover, its coherence length is sufficiently longer than the optical path length difference between the optical path ABD and ACD, and that between the sensor loop and the reference loop.

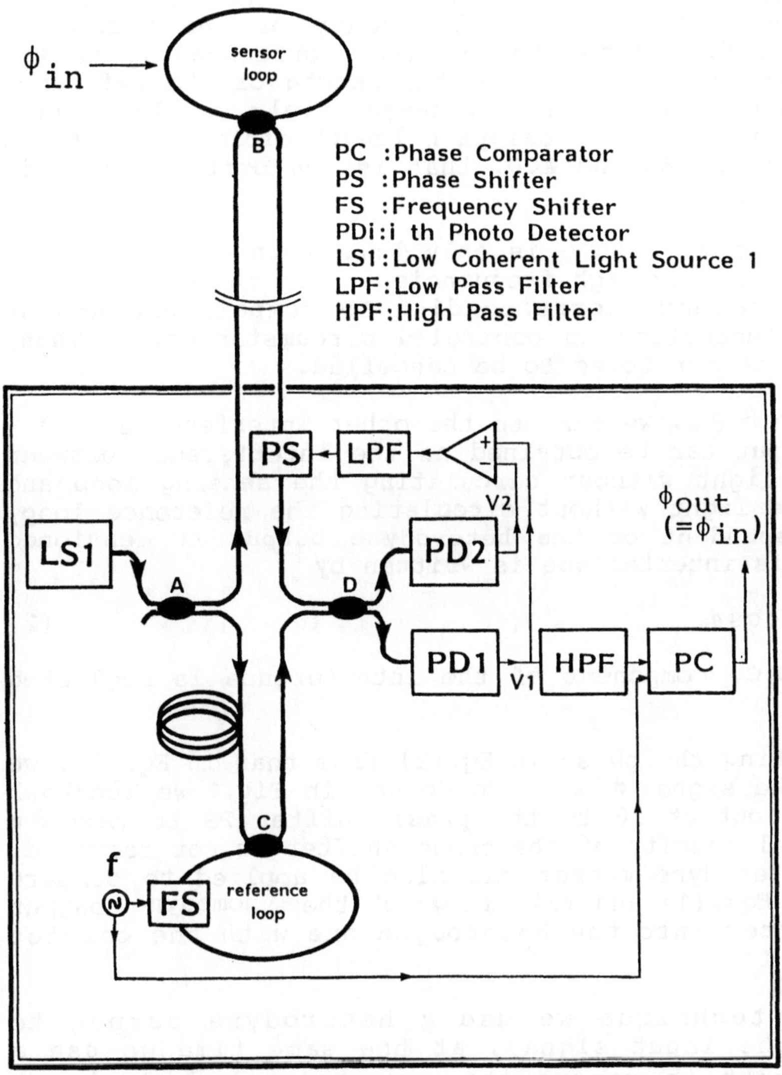

Figure 1. Schematic diagram of the proposed fiber sensor with immunity from the environmental disturbance induced in the lead-fiber.

Consequently, the selective interference can be obtained. The sensing light circulating one turn in the sensor loop and the reference light circulating one turn in the reference loop can interfere to produce the heterodyne output with intermediate frequency f, which is written as

$$I1=(1-T_s)(1-T_r)(t_s t_r)^{1/2}\cos(2\pi f t+\phi_{in}-\phi_r-\theta_0-\delta), \qquad (1)$$

where, the D.C. component of the interference intensity is neglected. The coupling coefficient of the coupler located at point B and point C are denoted as Ts and Tr, respectively. The transmission ratio of the sensor loop and the reference loop are denoted as ts and tr, respectively. The driving frequency of the frequency shifter FS is f, and the phase shift corresponding to the input and that corresponding to the length of the reference loop are denoted as ϕ_{in} and ϕ_r, respectively. In Eq.(1), θ_0 corresponds to the optical path length difference between the optical path ABD and ACD, that is the drift induced in the lead fiber.

What we want to have is only ϕ_{in}. In ordinary case, ϕ_{in} is in relatively high frequencies. On the contrary, ϕ_r is in low frequency corresponding to temperature and/or mechanical fluctuation in controled circumstances. Then, only θ_0 must be monitored to be cancelled.

To monitor θ_0, we can use the other interference. A homodyne output can be obtained as the interference between the sensing light without circulating the sensing loop and the reference light without circulating the reference loop, that is independent of the heterodyne output I1 mentioned aboved. This interference is written by

$$I0=(T_r T_s)^{1/2}\cos\theta_0, \qquad (2)$$

where, the D.C. component of the interference is neglected again.

Subtracting the phase in Eq.(2) from that in Eq.(1), we can obtain the signal ϕ_{in}. To do so, in Fig.1 we feedback the homodyne output I0 to the phase shifter PS to make θ_0 fixed. The linearity of the phase shifter is not required. The quasi-heterodyne manner can also be applied to compare the phase in Eqs.(1) and (2), in which the homodyne output I0 is converted into the heterodyne one with the carrier frequency of f.

In our technique we use a heterodyne output to demodulate the input signal, at the same time we use a homodyne output to remove the enviromental disturbance induced in the lead-fiber.

4. Experimens and results

Considering the signal-to-noise ratio corresponding to the noise induced by the phase-fluctuation of the source[6] and the difficulty in the system fabrication, a light source with several cm or several decimeter coherence length is appropriate in the experimental system. However, such a light source is not avaible as a compact semiconductor light source. Therefore, we adopted the modulation onto a laser diode with a noise current [7] to achieve several decimeter coherence length. Figure 2 shows the spectrum of the light source measured by the delayed self-homodyne method. Figure 2(a) is the original laser diode whose spectrum width is about 5MHz. Figure 2(b) is the same laser diode modulated by a noise current. It has about 100MHz spectrum width, corresponding to several decimeter coherence length.

The sensor loop has 10 meter optical path length, and the lead-fiber has 3 meter optical path length. The driving frequency of the frequency shifter is 40MHz. The optical setup is adjusted through such process: <1>The sensor route is taken off and the light shown in figure 2(a) is launched only into the reference route. Under this condition the reference loop acts as an optical resonator. The lights interfere with one another, then the output of the detector PD1 has frequency components of 40MHz, 80MHz, and so on, as is observed in figure 3(a). <2>The laser diode is modulated by a noise current so that its coherence length is shortened to several decimeter as shown in figure 2(b). By this way, the lights which travel multi-turn in the loop

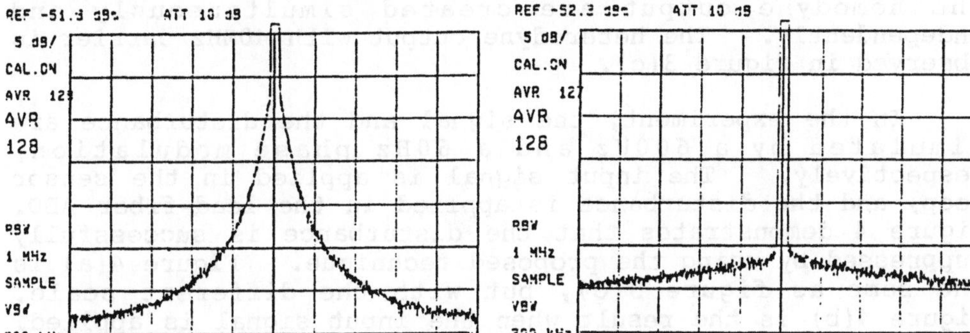

(a) (b)

Figure 2. Spectrum of the light measured by the delayed self-homodyne method. (a) Original laser diode. (b)Laser diode modulated by a noise current.

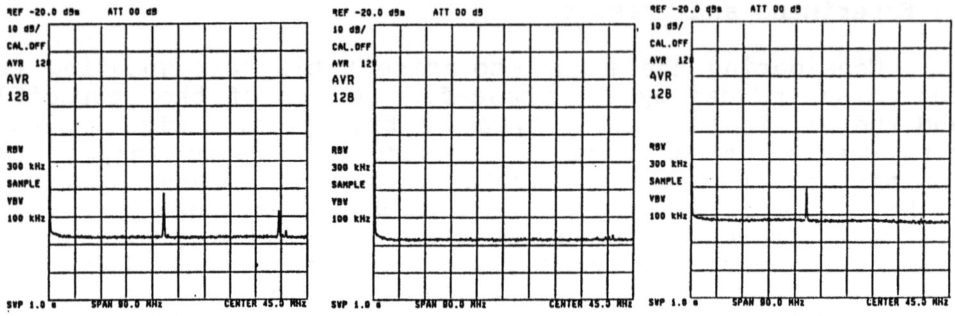

(a) (b) (c)

Figure 3. Output of the sensor system during the adjustment observed on the spectrum analyzer.
(a)LD without the noise-modulation is used, and the sensor route is removed.
(b)LD with the noise-modulation is used, and the sensor route is removed. The multi-turn interference can be successfully reduced.
(c)LD with the noise-modulation is used, and the sensor route is set. Heterodyne and homodyne output are obtained simultaneously and independently

are not able to interfere with one another. The result is shown in figure 3(b). <3>Finally, the sensor route is set. The sensor loop is adjusted as same length as the reference loop, and the optical path ABD is adjusted as same length as the optical path ACD. Therefore, the heterodyne output and the homodyne output are created simultaneously and independently. The heterodyne output with 40MHz carrier is observed in figure 3(c).

In the experiment, the signal and the disturbance are simulated by a 600Hz and a 60Hz phase modulation, respectively. The input signal is applied in the sensor loop, and the disturbance is applied in the lead-fiber ABD. Figure 4 demonstrates that the disturbance is successfully suppressed by using the proposed technique. Figure 4(a) is the same as figure 3(c), but with the different scale. Figure 4(b) is the result when the input signal is applied. It is found that the 40MHz carrier is modulated in the phase by a 600Hz signal. Figure 4(c) is the result when the disturbance is applied. The 40MHz carrier is modulated in phase by a 60Hz. Figure 4(d) shows the signal and the disturbance are applied simultaneously. With the optical feedback to the phase shifter PS, the disturbance is fully suppressed over 25dB, which is shown in figure 4(e).

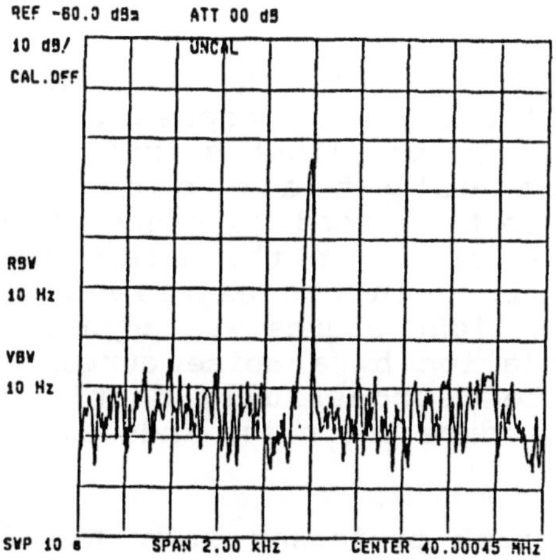

(a)

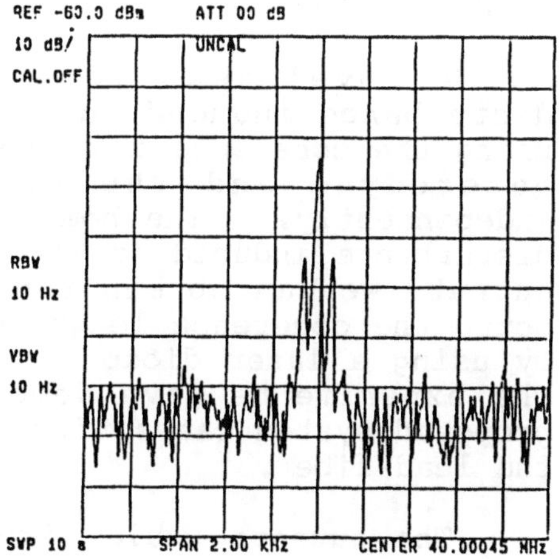

(c)

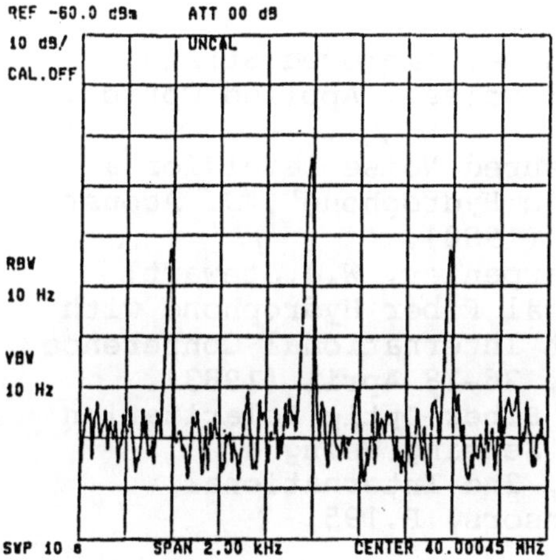

(b)

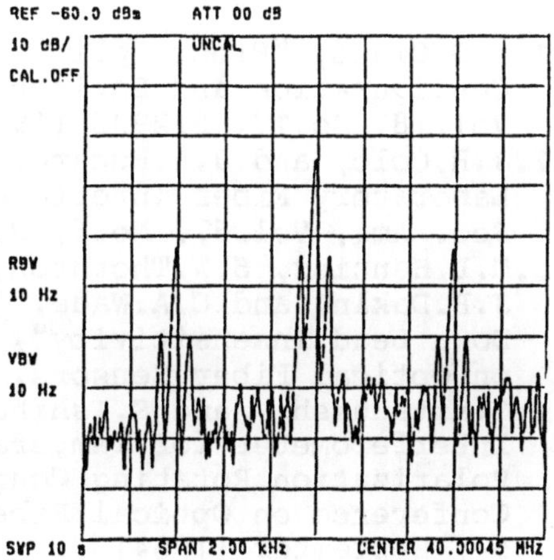

(d)

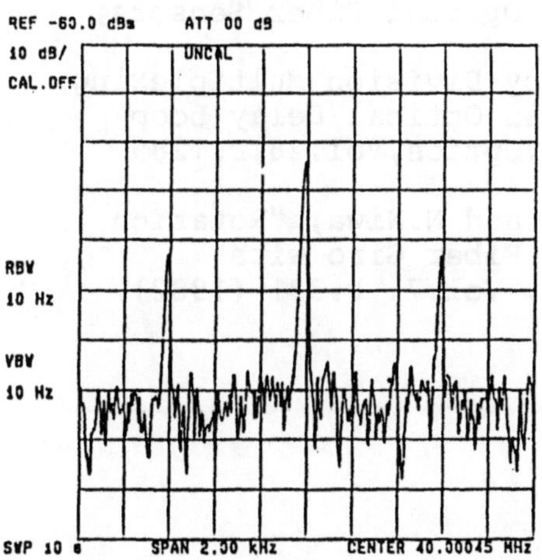

(e)

Figure 4.
Experimental results when the simulation signal and the simulation disturbance are applied. Signal of 600Hz is applied in the sensor head and disturbance of 60Hz is applied in the lead-fiber.
(a)Heterodyne output, same as the carrier in Fig.3(c), but with different scale.
(b)Only the signal is applied.
(c)Only the disturbance is applied.
(d)Signal and Disturbance are applied simultaneously.
(e)Disturbance is fairly suppressed with the feedback to the phase shifter PS.

5. Conclusion

A novel technique to remove the environmental disturbance induced in the lead-fiber was proposed. By using the coherence characteristic of the light source, the heterodyne and the homodyne output are obtained independently. The homodyne output is used to cancel the disturbance induced in the lead-fiber. In the heterodyne output, we can obtain the signal. In the experimental setup the coherence length of the light source was adjusted by using a laser diode with modulation by a noise current. The experiments have demonstrated successfully that the proposed system can sufficiently reduce the noise induced in the lead-fiber.

The authors would like to thank professor T.Okoshi of our Research Center for his encouragement.

References

1. P.G.Cielo, "Fiber Optic Hydrophone: Improved Strain Configuration and Environmental Noise", Applied Optics, Vol.18, No.17, P.2933 (1979).
2. J.H.Cole, and J.A.Bucaro, "Measured Noise Level for a Laboratory Fiber Interferometric Hydrophone", J. Acoust. Soc. Am., Vol.67, No.6, P.2109 (1980).
3. M.L.Henning, S.W.Thornton, R.Carpenter, W.J.Stewart, J.P.Dakin, and C.A.Wade, "Optical Fiber Hydrophone with Down Lead Insensitivity", First International Conference on Optical Fiber Sensors, P.23, 26-28 April (1983).
4. Y.Mitsuhashi, and S.Ishihara, "Fiber-Optic Polarization Interferometer for Temperature Sensing Using a Polarization Rotating Coupler", 2nd International Conference on Optical Fiber Sensors, P.195, 5-7 September (1984).
5. S.W.Thornton, M.L.Henning, and R.J.Langston, "Experimental Performance of an Optical Hydrophone", 2nd International Conference on Optical Fiber Sensors, P.369, 5-7 September (1984).
6. D.T.Jong and K.Hotate, "Frequency Division Multiplexing of Optical Fiber Sensors Using an Optical Delay Loop with a Frequency Shifter", Appl.Optics,Vol.28,P.1289 (1989).
7. K.Hotate, N.Okuma, M.Higashichi and N.Niwa, "Rotation Detection by Optical Heterodyne Fiber Giro with Frequency Output", Optics Lett., Vol.7, P.331 (1982).

A Coiled Fiber Displacement Sensor Immune to Temperature Noise

Y. Imai, M.A. Rodrigues, and K. Iizuka*

University of Toronto, Department of Electrical Engineering, Toronto, Ontario, Canada M5S 1A4

*On leave of absence from Department of Computer Science and Electronics, Kyushu Institute of Technology, Iizuka, Fukuoka 820, Japan

A temperature insensitive fiber sensor based on the bending induced birefringence is proposed. Two types of transducers using a pair of fiber coils spliced with a ninety degree rotation are examined. The minimum sensitivity of 362°/mm (retardation/displacement) is guaranteed in both schemes with 30 turn coils.

1. Introduction

In a fiber optic sensor, the birefringence method [1, 2] uses the interference between the two orthogonal modes in a polarization maintaining fiber. One of its biggest advantages is its high sensitivity and simplicity of the sensor. A disadvantage is its temperature dependence. Birefringence is induced in a bent fiber [3]. This allows us to utilize a fiber coil as a transducer for a displacement measurement.

This paper presents two new methods of a fiber coil transducer capable of removing the temperature dependence of the sensor. The fiber transducer of the proposed methods consists of a pair of coils of a flat PANDA type polarization maintaining fiber. In one type (type A), the two coils spliced with a ninety degree rotation are differentially squeezed. While the phase retardations caused by the differential squeeze are constructively added, the retardations caused by the environmental temperature are destructively added, removing the temperature effect. In the other type (type B), the fibers in the two coils are wound along a different principal axis. Then, the two coils with a 90 degree connection are squeezed in the same manner. This type also enhances the retardations induced by the displacement, followed by the removal of the temperature effects.

2. Principle of Operation

Figure 1 shows the two geometries of a proposed transducer. Type A, in (a), the polarization maintaining fiber is wound into a coil with its slow axis always parallel to the axis of the coil. Two such coils are squeezed differentially. When one is compressed, the other is released. The change $\Delta\phi_1$ in the phase retardation in coil 1 due to the bending that makes the radius of the curvature smaller than the original shape of the coil is [3]

$$\Delta\phi_1 = c_0 r^2 2\pi N[1/R_0 - 1/(R_0 - x/2)], \quad (1)$$

where $c_0 = -7.7\times10^7$ (deg/m), R_0 is the radius of curvature of the original coil, $2r$ is the diameter of the fiber, N is the number of turns of the

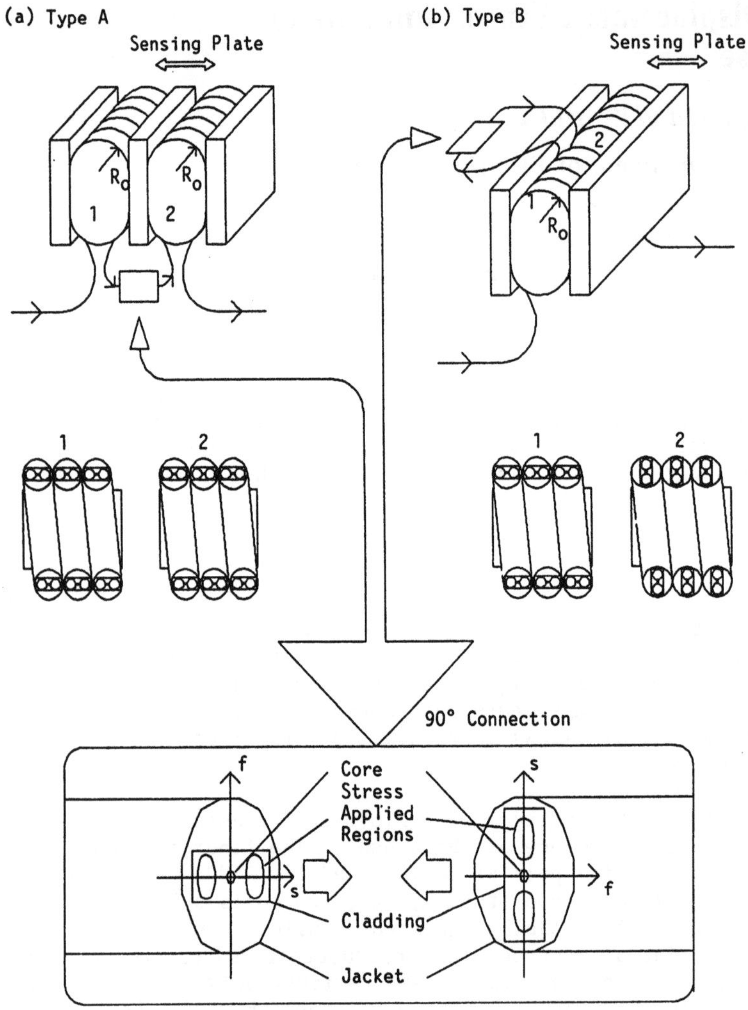

Fig.1 A transducer using a pair of fiber coils

coil, and x is the displacement of the radius of curvature of the coil from the original shape. Similarly, the change $\Delta\phi_2$ in coil 2 is given by replacing $(R_0-x/2)$ by $(R_0+x/2)$ in Eq.(1).

If the slow and fast axes are interchanged, the sign of the phase retardation reverses and the total retardation by two coils become additive. Thus the splicing of the two coils with 90° rotation enhances the two retardations:

$$\Delta\phi_A=|\Delta\phi_1-\Delta\phi_2|. \tag{2}$$

Let us assume that the temperature is uniform over the two coils of an identical fiber length. The retardation changes by the same amount with the same sign in each coil. Thus,the temperature effects are cancelled because of the 90° connection.

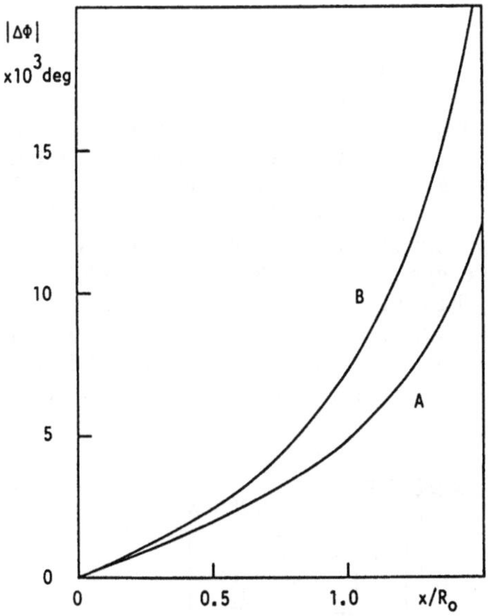

Fig.2 Change in retardation

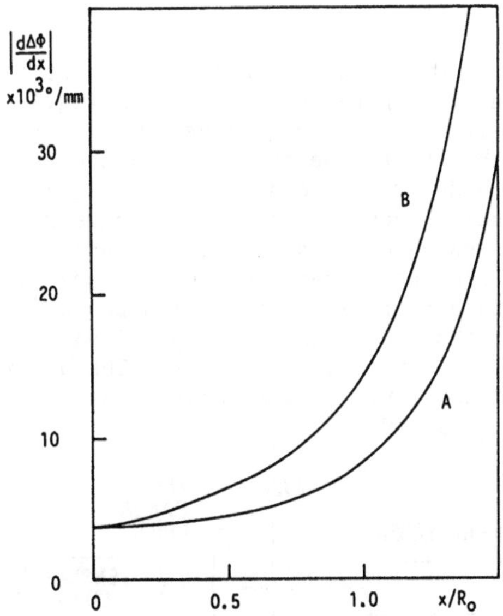

Fig.3 Sensitivity

Type B, in (b), the polarization maintaining fiber is wound into each coil with a different principal axis parallel to the coil axis. Two such coils are squeezed identically. The retardations occurring in the two coils change in the opposite way. Thus the total change in retardation $\Delta\Phi_B$ is given, in consideration of the effect of a 90° connection, by

$$\Delta\Phi_B = |2\Delta\phi_1| . \tag{3}$$

Similar to type A, the temperature effect is cancelled in this type because of the 90° connection. A comparison of a change in retardation between type A and type B is shown in Fig.2. In a calculation, a coil radius and a core diameter are assumed to be 10 µm and 100 µm. The number of turns of the coils was N=30. The retardation in type B changes effectively for the displacement.

The sensitivity, defined by the derivations of Eqs.(2) and (3), in type A and type B is given, respectively, by

$$\frac{d\Delta\Phi_A}{dx} = \left| c_0 r^2 2\pi N \frac{R_0{}^2 + (x/2)^2}{[R_0{}^2 - (x/2)^2]^2} \right| , \tag{4}$$

$$\frac{d\Delta\Phi_B}{dx} = \left| c_0 r^2 2\pi N \frac{1}{(R_0 - x/2)^2} \right| . \tag{5}$$

A comparison of the sensitivity is shown in Fig.3. Parameters used in the calculation are the same as those in Fig.2. The transducer of type B is more sensitive than type A. The minimum sensitivities given at $x/R_0=0$ are identical and are 362° /mm (retardation/displacement) in both types. The transducers become sensitive as the number of turns N in the coil increases and the radius R_0 of the coil decreases.

3. Experiments

Only a preliminary experiment of type A was demonstrated. The experimental arrangement is shown in Fig.4. Linearly polarized light from a He-Ne laser passes through a halfwave plate to adjust the direction of polarization to be 45° to the principal axis of the polarization maintaining fiber. In the transducer scheme shown in Fig.1(a), the transmitted light is affected by both bending-induced loss and the birefringence. The emerging light from the fiber passes through an analyzer at 45° from the principal axis to obtain a polarimetric interference output. A special birefringent fiber (Flat PANDA SM63) with a flat cladding (70x130 μm) was used. This fiber was easily wound keeping the slow axis always parallel to the coil axis, because of the rectangular cross section of the fiber. The original radius of curvature R_0 was 7.5 mm, and the range of displacement was within ±10 mm. The number of turns of the coil was 30.

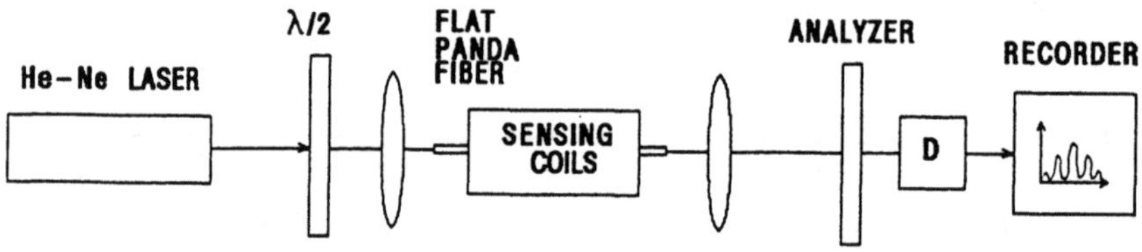

Fig.4 Detection scheme of polarimetric displacement sensor

4. Results

A typical experimental result of the output intensity vs the displacement is shown in Fig.5. The envelope of the curve represents the bending-induced loss that modulates the amplitude of the polarimetric interference curve. The change in retardation vs the displacement is shown in Fig.6 in which the solid line is obtained from Eq.(2) with 2r=70μm. Although the loss effect is not taken into account in Eq.(2), the experimental data are in good agreement with the theory.

In the experiments, significant improvement in the stability of 4°/h (retardation /time) for the temperature was recognized, as is seen from Fig.7.

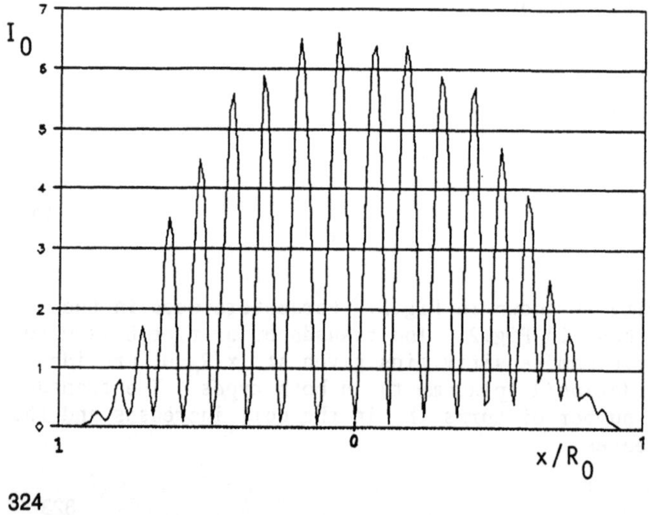

Fig.5 Polarimetric output intensity

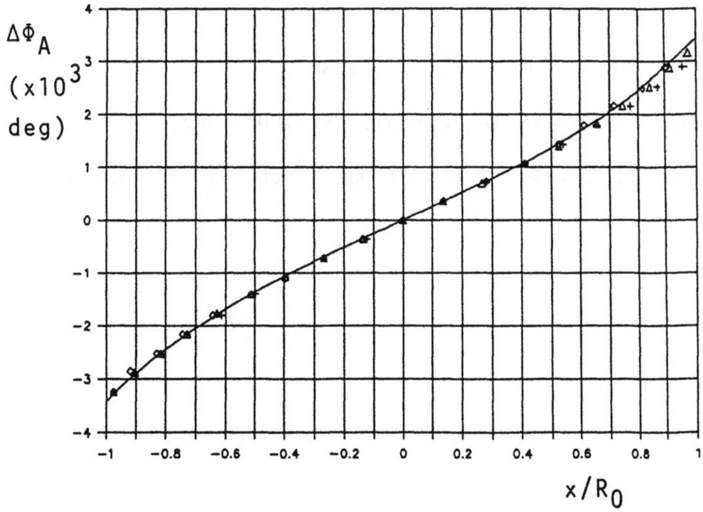

Fig.6 Retardation v.s. displacement

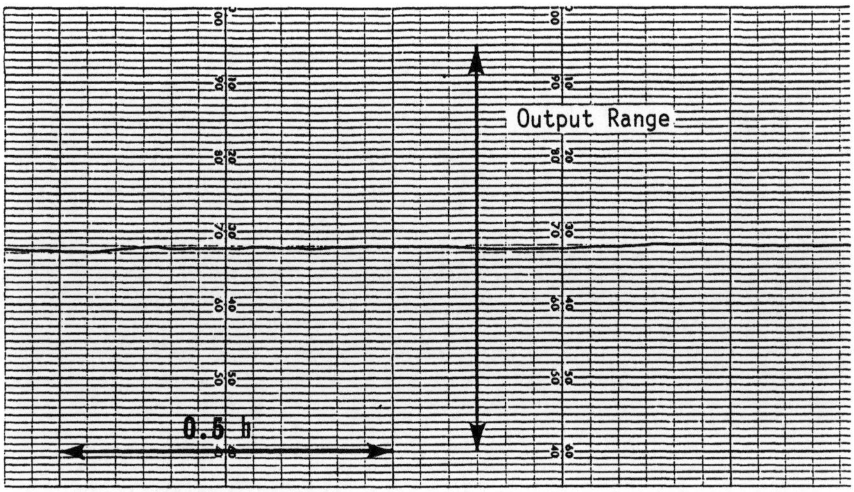

Fig.7 Output stability

5. Conclusions

A coiled fiber displacement sensor was proposed. Two types of a transducer capable of removing a temperature noise were examined. In a preliminary experiment of type A, a minimum sensitivity of 362° /mm and a long term stability of 4° /h were attained.

Acknowledgement

We are grateful to Mr. F. Suzuki of Fujikura Ltd. of Japan, for providing us with a flat PANDA fiber. Mr. Blaha of Avionics Division of Canadian Marconi inspired us to initiate this research. This work was supported by the Ontario Laser and Lightwave Research Centre at the University of Toronto.

References

[1] Y. Imai and Y. Ohtsuka; Appl. Opt., 25, 4444(1986).

[2] Y. Imai and Y. Ohtsuka; J. Lightwave Technol., 5, 1008(1987).

[3] R. Ulrich, S.C. Rashleigh, and W. Eichkoff; Opt. Lett. 5, 272 (1980).

Optical-Fibre Surface-Plasma-Wave Polarisers

M.N. Zervas

Electronic and Electrical Engineering Dept., University College London, Torrington Place, London WC1E7JE, UK

In this paper, theoretical and experimental results concerning the operation of optical-fibre surface- plasma-wave polarisers are presented. Variations in the TM-like polarisation transmission characteristics and polarisation extinction ratio of the device due to variations in the depth of polishing, metal film thickness and optical wavelength are presented. The role and degree of excitation of the various surface-plasma-wave modes are also discussed.

1. Introduction.

All-fibre-optic polarisers are of the most important components in interferometric and polarimetric sensor applications [1,2]. So far, several high quality configurations have been proposed and demonstrated. The basic idea is to allow the light to interact with a polarisation-selective configuration in such a manner that eventually one polarisation is extinguished while the orthogonal one is transmitted unaffected. All-fibre-optic technology has the advantage of extremely low insertion losses and results in mechanically robust devices.

Surface-Plasma-Wave (SPW) fibre-optic polarisers utilize the SPW resonance as the polarisation-selective mechanism and have recently received great attention because of the relative ease of manufacturing and the achieved large polarisation extinction ratios (> 50 dB) accompanied by small insertion losses (< 0.5 dB). Their operation is primarily based upon the evanescent field interaction between the fundamental HE_{11} fibre mode and the surface polariton-like mode supported by a thin metal film which is brought into close proximity with the polished optical fibre. A thin metal film with negative-real-part dielectric constant (surface-active medium) surrounded by two semi-infinite dielectrics is known to support four different surface polariton-like modes, namely the symmetric and antisymmetric, bound or leaky, SPW modes, which are TM-polarised and exhibit different phase velocities and losses depending upon the thickness of the metal, the refractive indices of the dielectrics and the wavelength [3,4].

Two alternative configurations of SPW fibre-optic polarisers with thin-metal-films have been reported so far [5,6]. In the first, a dielectric buffer layer is formed between the polished fibre surface and the thin metal film and the transmission of the TM-like polarisation varies with the buffer-layer refractive index. The thickness of the metal film and the buffer layer are found to affect the transmission characteristics of the device. This configuration offers the

Editors: H.J. Arditty · J.P. Dakin · R.Th. Kersten

flexibility of selectively transmitting the TM-like or TE-like polarisations by choosing correctly the buffer-layer refractive index. In the second configuration, the metal film is deposited directly onto the polished fibre surface and covered by a semi-infinite dielectric. When this configuration is used, the performance of the polariser is enhanced dramatically by depositing an extra thin lossy layer on top of the surface-active thin metal film [7].

In attenuated-total-reflection (ATR) studies of the propagation characteristics of SPWs, it has been shown that the SPW excitation efficiency depends critically upon the dielectric properties of the metal and surrounding materials as well as the prism-coupling-gap height [8,9]. In previous work [6], it has been suggested that the performance of the SPW polariser with the metal film deposited directly onto the polished surface is determined entirely by the metal film thickness. In this paper, experimental evidence is presented advocating that the overall performance of the device is critically dependent upon the metal film thickness as well as the depth of polishing and the optical wavelength.

2. Surface - plasma - wave modes – Theory.

As it has already been mentioned, the polarisation selective structure in the device under consideration is a thin metal film of negative-real-part dielectric constant surrounded by two semi-infinite dielectrics. The dielectric (ε_1) / thin-metal-film (ε_m) / dielectric (ε_3) structure supports four different SPW modes whose propagation characteristics vary with the metal-film-thickness, the dielectric constants ε_1, ε_3 and the wavelength. The dispersion relation the modes supported by the three-media slab structure has the form

$$\tanh(S_2h)(\varepsilon_1\varepsilon_3S_2^2 + \varepsilon_m^2S_1S_3) + [S_2\varepsilon_m(\varepsilon_1S_3+\varepsilon_3S_1)] = 0,$$

where h is the metal-film-thickness, $S_j = \sqrt{(\beta^2 - \varepsilon_j k_o^2)}$ $j = 1,3$, $S_2 = \sqrt{(\beta^2 - \varepsilon_m k_o^2)}$ and $\beta = \beta_R - i\,\beta_I$ the propagation constant of the mode [3]. β_R is related to the phase velocity of the propagating mode and β_I to its propagation losses. It has been shown that only four modes have physical meaning and are experimentally important, namely the symmetric-bound (s_b) and -leaky (s_l) and the antisymmetric-bound (a_b), -leaky (a_l) modes [3].

The dispersion relation has been solved numerically and results are shown in Figure 1. The parameter values used were $\sqrt{\varepsilon_1} = 1.462$, $\varepsilon_m = -29.8 - i\,11.6$ (aluminium), $h = 12$ nm, $\lambda_0 = 633$ nm and ε_3 as variable. In Fig. 1(*a*), the variation of the normalised propagation constant (β_R/k_o), i.e. the effective refractive index, of each of the supported SPW modes is shown. In Fig. 1(*b*), the $\log(\beta_I/k_o)$ of each mode is plotted versus the superstrate refractive index n_3.

The theoretical results show that both the phase velocity and the losses of the various SPW modes vary with the superstrate refractive index. Of particular complexity and importance is the s_b mode. In Fig. 1(*b*), it is shown that this mode changes character from the usual attenuated Fano-type mode ($\beta_I > 0$) to a growing-type mode ($\beta_I < 0$) over certain n_3 range.

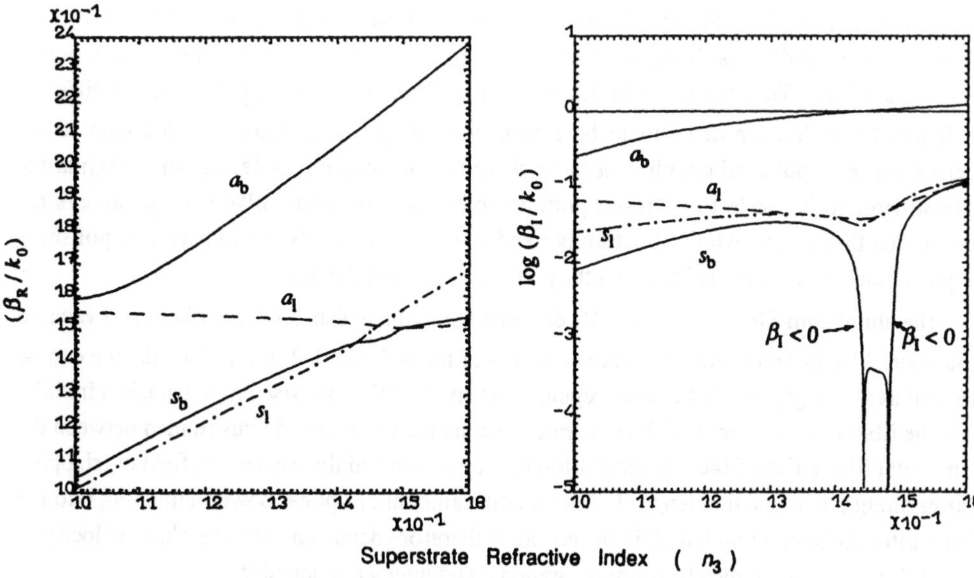

FIGURE 1 : Propagation constant variation versus superstrate refractive index.

The growing-type mode is characterised by an increase of the reactive power, at the expense of real power, associated with it and is considered to be *cut-off* [3,10]. It should be stressed also that the s_l (a_l) mode is always localised at the ε_3-ε_m (ε_1-ε_m) interface and radiated into ε_1 (ε_3) medium, over the entire n_3 range.

3. Device description – Principle of operation.

The device configuration is schematically depicted in Figure 2. To maintain and ensure the state of polarisation at the interaction region, high birefrigence bow-tie fibres (York technology) single-moded at 633 nm were used to build the device. The fibres were aligned with their slow axis perpendicular to the glass substrate block using an elasto-optic technique, and then epoxied into a 25 cm radius curved slot in the substrate. The surface of the fibre was initially ground and then polished to a few microns of the core resulting in an effective

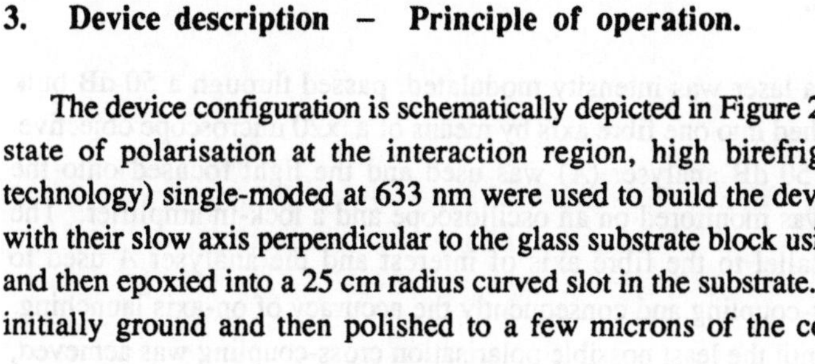

FIGURE 2: Surface-Plasma-Wave Optical-Fibre Polariser configuration.

interaction length L of about 1.5 mm [11]. The half block was regularly subjected to the drop-of-oil test and the polishing continued until around 30 dB of the launched optical power was coupled out. This procedure results in slightly different depths of polishing, for different half blocks, which are difficult to be estimated exactly. This causes no problem to the performance of polished couplers because there is still freedom for lateral and longitudinal adjustments to be made to achieve optimum coupling. As far as SPW fibre polarisers are concerned though, the whole structure is fixed and no adjustments are allowed and polishing depth variations are expected to affect the performance of the device.

The aluminium films were thermally deposited onto the polished fibre surfaces in a vacuum chamber. The distance from the fibre core was of the order of 1-2 μm and the device can be viewed as a strongly coupled two-waveguide system. TM-like polarised light launched initially into the fibre is coupled into SPWs as it enters the interaction area and redistributed between the thin metal film and the fibre. The metal-to-core separation and the evanescent field overlapping varies along the interaction length L. The relative amount of optical power remaining into the fibre, after the interaction length is surpassed, is therefore determined by the phase velocity of the SPW the minimum metal-to-core separation d and interaction length L.

The thin metal film thickness as well as the refractive index of the dielectric superstrate determine the type [3] and the phase velocity [4] of the propagating SPW mode and, therefore, are expected to affect the transmission characteristics of the device. Since the two waveguides are strongly coupled, small variations of the metal-to-core separation d are expected to affect the response of the device considerably. Finally, the device performance is expected to be wavelength dependent.

4. Experimental results.

Light from the output of a laser was intensity modulated, passed through a 50 dB bulk crystal polariser (P) and launched into one fibre axis by means of a x20 microscope objective. At the output end, a similar 50 dB analyser (A) was used and the light focused onto the photodetector whose output was monitored on an oscilloscope and a lock-in amplifier. The polariser P was oriented parallel to the fibre axis of interest and the analyser A used to measure the polarisation cross-coupling and consequently the accuracy of on-axis launching. The procedure was repeated until the least possible polarisation cross-coupling was achieved, typically - 30 dB. Once the fibre-axis of interest was located, the analyser was aligned parallel to it and intensity measurements carried out.

Cargile liquids of various refractive indices were placed on top of the thin aluminium film and the transmission of the device monitored on the lock-in amplifier. The total optical power launched into the system was determined by performing cut-back power measurements and used to normalise the transmission of the device. The experiment was performed for TM-like and TE-like polarisations, alternatively. To separate the effects of the polishing depth d and metal film thickness h on the transmission characteristics of plasmon wave polarisers, two different sets of experiments were carried out.

(a) Constant metal-film thickness: Two half blocks polished to different depths, designated HB1 and HB2, were placed next to each other in the vacuum chamber to ensure identical deposition conditions and thin metal films of exactly the same thicknesses. The film thickness was measured to be 12 nm. The TM-like polarisation transmission of both half blocks was measured for different dielectric superstrates of various refractive indices and the results are summarised in Figure 3(*a*). They clearly show different transmission characteristics which are attributed to the difference in the depth of polishing.

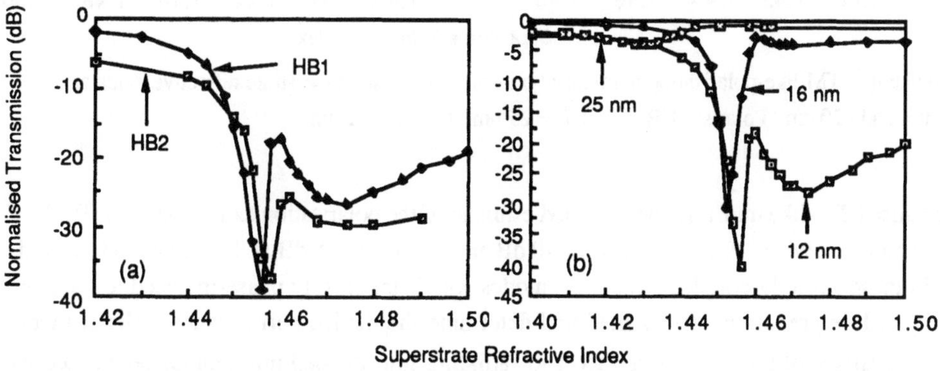

FIGURE 3 :Normalised TM-like-polarisation transmission; (*a*) Constant metal-film thickness, (*b*) Constant polishing depth.

(b) Constant polishing depth: The thin aluminium film was etched away from the polished surface of the half block HB1 and a 20 nm aluminium film was redeposited. The TM- and TE-like polarisation transmissions of the new device were measured and the thin aluminium film was etched away once again and the half block was covered with a 25 nm aluminium film. The experimental results are summarised in Figure 3(*b*). The aluminium etching (90% phosphoric acid, 10% nitric acid) was washed out as soon as the metal film was removed completely leaving the depth of polishing unaffected. Therefore, the marked change in the transmission of the device was attributed exclusively to the metal film thickness variation.

Finally, the implemented polarisers were tested at the wavelength of 820 nm. SPW polarisers constitute strongly coupled structures, as it was clarified in Section 3, and, therefore, their transmission is expected to be wavelength dependent. Typical results, regarding polariser *HB1*, are presented in Figure 4.

5. Discussion.

In Figures 3 and 4, the normalised transmission of the TM-like polarisation is shown alone. The normalised transmission of the TE-like polarisation , i.e. the insertion loss of the device, was in all cases less than 0.4 dB and not plotted for clarity reasons. Therefore, the

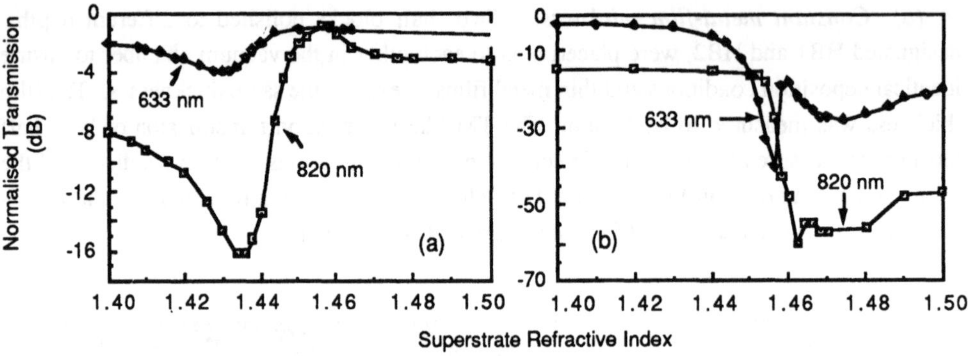

FIGURE 4 : TM-like-polarisation normalised transmission versus superstrate refractive index at 633 nm and 820 nm. Polariser HB1; (*a*) $h = 25$ nm, (*b*) $h = 12$ nm.

normalised TM-like transmission is a good estimate of the polarisation extinction ratio (PER) of the particular device. PERs in excess of 40 dB at 633 nm and 55 dB at 820 nm were achieved.

In these strongly-coupled structures, modes which are closely phase-matched to the fibre mode will be predominantly excited and determine the device performance. The s_l mode, being localised at the ε_3-ε_m interface and radiating into ε_1 medium, *cannot* be excited by evanescent field interaction in the configuration of Fig. 2. The a_l mode, on the other hand, having the reverse field distribution and opposite power flow, is *excitable* and determines the device response in the n_3-range above 1.462, where it is well matched to the fibre mode (n_{eff}≈1.462) (c.f. Fig. 1(*a*)). In this range, the s_b mode is cut-off and, therefore, *not excited.* This is in excellent qualitative agreement with the experimental results, where the observed transmission-variation in the aforementioned n_3-range was due to light leaking into the superstrate.

In the n_3-range 1.42 - 1.46, the s_b mode is above cut-off and exhibits extremely low propagation losses. On the other hand, it is well phase-matched to the fibre mode (n_{eff}≈1.462) and, therefore, *predominantly excited* (c.f. Fig. 1(*a*)). This again is in excellent qualitative agreement with the experimental results, where the observed dip in the transmission was accompanied by no radiated light and, therefore, attributed to bound-mode excitation. So far, the a_b mode plays no significant role in the device performance, for it is considerably phase-mismatched and exhibits large propagation losses (c.f. Fig. 1). This mode becomes important and determines the device response only when n_3=1 (air). In this case, the s_b mode is cut-off while the a_b mode exhibits its least phase-mismatch with the fibre mode and lowest propagation losses.

Practical devices will have fixed metal film thickness and be covered with a certain dielectric material of constant refractive index. Given the limited range of the available dielectric materials with refractive indices 1.3 - 1.5 at 633 nm, from the data in Figures 3 and 4, it is inferred that the correct combination of metal film thickness and polishing depth should be chosen so that the maximum possible polarisation extinction ratio is attained.

As far as the effect on coupling conditions is concerned, it has been shown that the depth of polishing affects considerably the coupling coefficient while leaves unaffected the effective interaction length of the half block.[11,12] On the other hand, when high-birefrigence fibres are used, the depth of polishing has been shown to affect the effective refractive index of the particular polarisation in a rather complicated way.[13,14] Therefore, depth-of-polishing variations affects several parameters simultaneously which, in turn, affect the coupling conditions and the performance of the device.

6. Conclusions.

High quality optical-fibre SPW polarisers have been constructed and tested. Typical PERs in excess of 40 dB at 633 nm and 55 dB at 820 nm were obtained. The insertion losses were typically below 0.4 dB. Polishing-depth and metal-film-thickness variations were shown to affect the device performance. Depending on the superstrate refractive index, the device response was due to predominant excitation of symmetric-bound, antisymmetric-leaky or antisymmetric-bound SPW mode.

ACKNOWLEDGEMENTS : M. N. Zervas was partially supported by the Bodossakis Foundation, Athens, Greece. Helpful discussions with Dr. I.P. Giles, M. Farhadiroushan and S. Markatos are gratefully acknowledged. Special thanks are due to Mr A.A.Stride for the patient thin metal film evaporation and Tony Kerr for the fibre polishing.

REFERENCES.

[1] T.G. Giallorenzi, J.A. Bucaro, A.T. Dandridge, G.H. Sigal, J.H. Cole, S.C. Rashleigh and R.C. Priest, *IEEE J. Quantum Electron.* **QE-18**, 626 (1982).

[2] M. Farhadiroushan and I.P. Giles, *Proc. SPIE,* Vol. **949**, pp. 162-169 (1988).

[3] J. J. Burke, G. I. Stegeman and T. Tamir, *Phys. Rev. B* **33**, pp. 5186-5201 (1986).

[4] M. N. Zervas, S. Markatos and I. P. Giles, *Proc. SPIE,* Vol. **988**, pp. 70-76 (1988).

[5] S. Markatos, M. N. Zervas and I.P. Giles, *Electron. Lett..* **24**, pp. 287-288 (1988).

[6] W. Johnstone, G. Stewart, B. Culshaw and T. Hart, *Electron. Lett.,* **24**, pp. 866-868 (1988).

[7] M.N. Zervas and I.P. Giles, *Electron. Lett.,* **25**, pp. 321-323 (1989).

[8] D. Sarid, R.T. Deck, A.E. Craig, R.K. Hickernell, R.S. Jameson and J.J. Fasano, *Appl. Opt..* **21**, pp.3993-3995 (1982).

[9] D.L. Begley, D.A. Bryan,R.W. Alexander and R.J. Bell, *Appl. Opt..* **16**, pp. 1549-1552 (1977).

[10] M.N. Zervas, *Ph.D. thesis,* University of London, July 1989.

[11] M. J. F. Digonett and H. J. Shaw, *IEEE J. Quantum Electron.* **QE-18**, pp. 746-754, (1982).

[12] R. Tewari and K. Tyagarajan, *J. Lightwave Technol.* **LT-3**, pp. 386-390 (1986).

[13] R. H. Stolen, A. Ashkin, J. E. Bowers, J. M. Dziedzic and W. Pleibel, *J. Lighwave Technol.* **LT-3**, pp. 1125-1129 (1985).

[14] H. C. Lefevre, P. Simonpietri and Ph. Graindorge, *Proc. SPIE,* Vol. **988**, pp. 63-69 (1988).

Wavelength-Flattened Couplers Fabricated from Single-Mode Fibers with Different Core Parameters

H. Hanafusa, Y. Takeuchi, and J. Noda

NTT Opto-Electronics Laboratories, Tokai-mura, Ibaraki-ken 319-11, Japan

Wavelength-flattened couplers are fabricated using fibers with different core parameters and the same cladding diameter. Maximum coupling ratios are determined by the difference in core parameters, and are independent of fabrication condition. The present method is superior in productivity to the conventional one in which fibers with different cladding diameters are used.

I. Introduction

Wavelength-flattened couplers (WFC), with coupling ratios insensitive to wavelength, are significant components in optical communication systems and fiber sensor systems. Such couplers are fabricated by fusing two fibers with different cladding diameters and tapering them. Fibers with different cladding diameters are prepared by being selected from commercial fibers,[1,2] being pre-etched before fusion,[1,2] or similarly being pretapered.[3] However, fibers with different diameters are inconvenient for connection, and couplers fabricated by pre-etching or pretapering are expensive because of additional unproductive process. Arrayed fiber couplers, which are fabricated using two four-fiber ribbons,[4] have big potential for reducing the cost and integrating the package. The conventional method for WFC fabrication cannot be applied to arrayed fiber couplers.

This paper presents WFC fabrication by more productive method. This method is based on using fibers with different core parameters and the same cladding diameter.

II. Fabrication

Parameters of single-mode fibers used are listed in Table 1. These fibers were selected from commercially available fibers for optical communication

Springer Proceedings in Physics, Vol. 44 Optical Fiber Sensors
Editors: H.J. Arditty · J.P. Dakin · R.Th. Kersten

Table 1. Parameters of Single-Mode Fibers

Fiber	Cutoff Wavelength (μm)	Mode-Field Diameter (μm)	Fiber Diameter (μm)
#1	1.12	9.7	124
#2	1.15	9.7	125
#3	1.26	10.0	125

use. Fused-tapered 2×2 fiber couplers were made using an apparatus with a pair of oxygen/propane microtorches and a pair of elongation stages.[5] A laser diode with 1.46 μm wavelength was used to monitor coupling ratio and excess loss during fabrication. In order to examine effects of fusion condition on coupling characteristics, constituent fibers were fused at different temperatures and over different lengths. After fusion, length and width of the fused region were measured with a microscope. Tapering was stopped when the coupling ratio had a maximum value at 1.46 μm.

III. Results and discussion

A. Maximum coupling ratio

Plots of coupling ratio vs. wavelength are shown in Fig. 1. Couplers fabricated using fibers with different cutoff wavelengths λ_c have maximum coupling ratios less than 100%; the maximum coupling ratio is about 90% for couplers made from fibers #3 and #2, and is around 50% for couplers made from fibers #3 and #1. A plot of maximum coupling ratio R_{max} vs. difference in λ_c between constituent fibers is shown in Fig. 2. The value of R_{max} decreases with increasing difference in λ_c. This indicates that any values of R_{max} are obtained by selecting constituent fibers with a difference in λ_c, as theoretically studied by Snyder.[6,7] Some scatter of R_{max} was observed for six couplers made from the same pair of fibers as shown in Fig. 2. The greater the difference in λ_c, the larger the scatter of R_{max}. Origins of the scatter will therefore be discussed.

When a pair of single-mode waveguides with propagation constants β_a and β_b are uniformly coupled over length z, the coupling ratio R is given by[7]

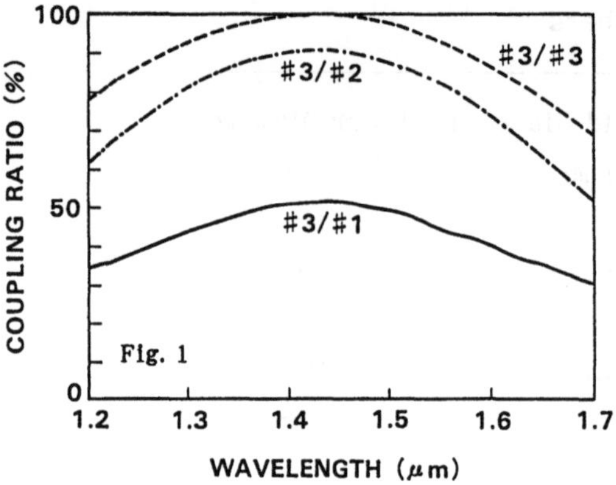

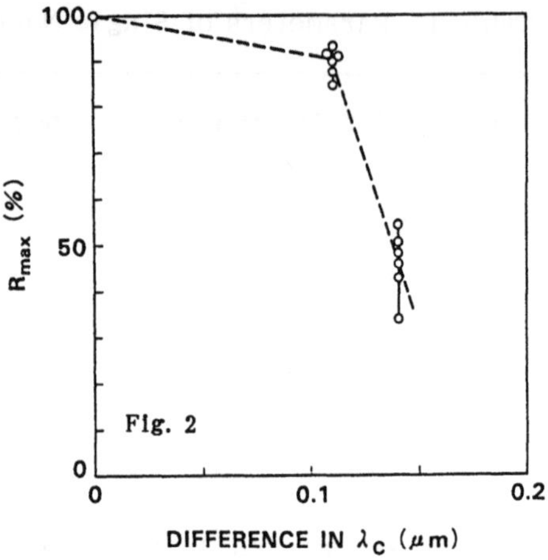

Fig. 1. Coupling ratio vs. wavelength for couplers made from fibers with different cutoff wavelengths.

Fig. 2. Maximum coupling ratio vs. difference in cutoff wavelength between constituent fibers.

$$R = R_{max} \sin^2 (\beta_o z) , \quad (1)$$

$$R_{max} = 4\kappa^2 / \{4\kappa^2 + (\beta_a - \beta_b)^2\} , \quad (2)$$

$$\beta_o = \{4\kappa^2 + (\beta_a - \beta_b)^2\}^{1/2} / 2 , \quad (3)$$

where κ is coupling coefficient. In actual fiber couplers, propagation constant and coupling coefficient are nonuniform, and depend on the shape of the fused-tapered region. Scatter of R_{max} is therefore ascribed to variation of these parameters, caused by some irregularity in the coupler fabrication process.

Effects of fusion condition on coupling characteristics were examined. Plots of R_{max} vs. length of the fused region are shown in Fig.3. The effect of fusion length on R_{max} is unclear because of scatter. Scatter of R_{max} decreases with decreasing fusion length. This indicates that WFCs are fabricated with better yield by being fused over a shorter length.

Plots of R_{max} vs. degree of fusion are shown in Fig. 4. Here, degree of fusion is defined as an aspect ratio b/a of cross section in the fused region as shown in the inset of Fig. 5. The value of R_{max} is insensitive to b/a. On the other hand, it was reported that R_{max} varied with the degree of fusion

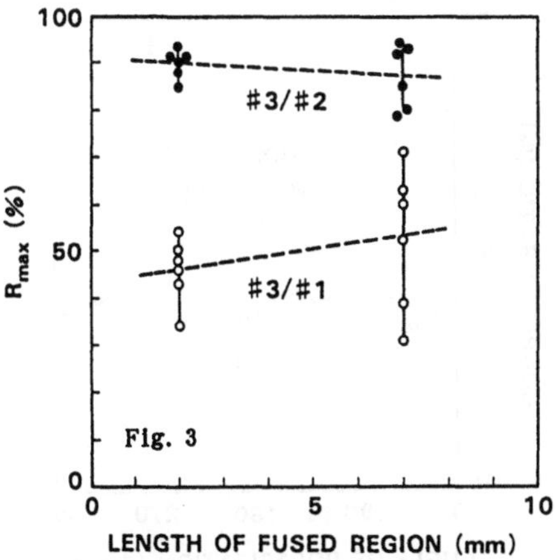

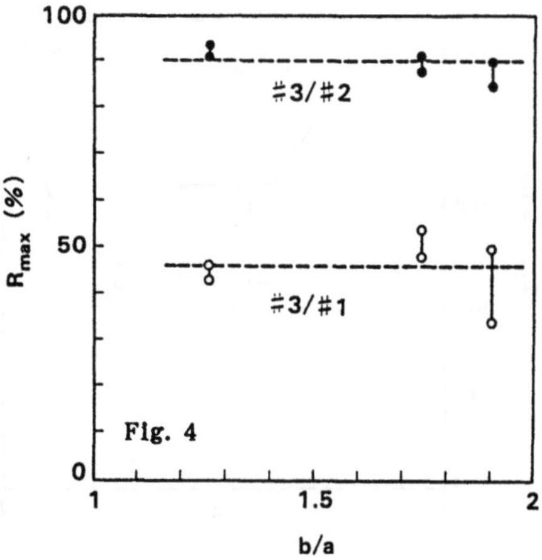

Fig. 3. Maximum coupling ratio vs. length of fused region.

Fig. 4. Maximum coupling ratio vs. aspect ratio of cross section in fused region.

in WFCs made from fibers with different cladding diameters.[8] This suggests that the present method is more productive than the conventional one. In the present method, R_{max} is determined by the difference in core parameters, and is independent of the degree of fusion. This means that it is easy to obtain the same coupling ratio.

B. Flatness of coupling ratio

A plot of flatness of coupling ratio vs. aspect ratio b/a is shown in Fig. 5. Here, flatness of coupling ratio is defined as a wavelength width $\Delta\lambda_{80}$ in which coupling ratios are higher than 80% of R_{max}. The value of $\Delta\lambda_{80}$ increases with increasing b/a; when b/a is 1.9, $\Delta\lambda_{80}$ is broadened to 0.36 μm. For realizing the wavelength-flattened characteristics, it is desirable to lightly fuse constituent fibers.

The excess loss is around 0.1 dB for every couplers. The effect of polarization on coupling ratio is shown in Fig. 6. Variation in the coupling ratio is less than ±0.4%.

IV. Conclusions

Wavelength-flattened couplers were fabricated using single-mode fibers with different core parameters and the same cladding diameter. Maximum coupling ratios R_{max} are determined by the difference in core parameters between constituent fibers, and are independent of fabrication condition.

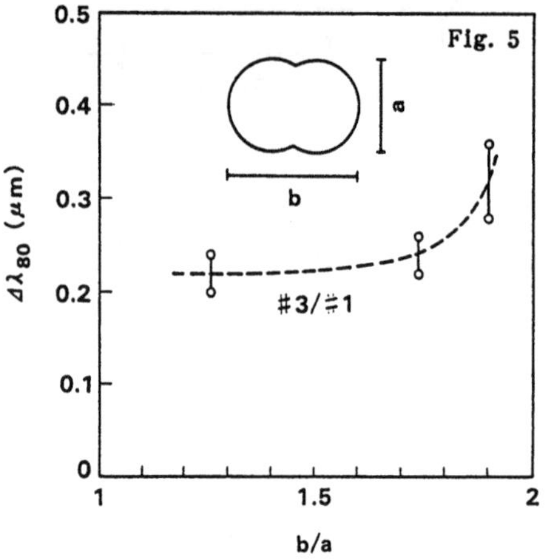

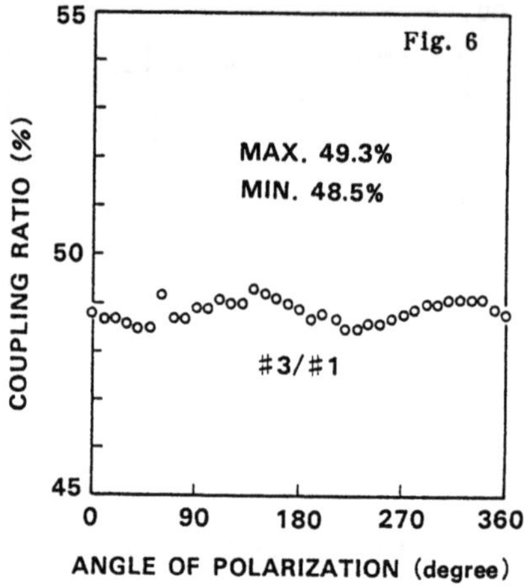

Fig. 5. Wavelength width for 80% R_{max} vs. aspect ratio of cross section in fused region. The inset shows cross section in fused region.

Fig. 6. Effect of polarization on coupling ratio at 1.31 μm wavelength.

When cutoff wavelengths are 1.12 μm and 1.26 μm, R_{max} is about 50%. The wavelength width in which coupling ratios are higher than 80% of R_{max} is 0.36 μm. The excess loss is around 0.1 dB, and the coupling ratio variation by polarization is less than ±0.4%

Acknowledgment

The authors would like to thank Dr. Naoya Uchida for fruitful discussion.

References

1. R. G. Lamont, K. O. Hill, and D. C. Johnson, Opt. Lett. 10, 46 (1985).
2. R. G. Lamont, K. O. Hill, and D. C. Johnson, in Digest of Topical Meeting on Optical Fiber Communication (Optical Society of America, Washington, D. C., 1985), p. 78.
3. D. B. Mortimore, Electron. Lett. 21, 742 (1985).
4. I. Yokohama, K. Chida, H. Hanafusa, and J. Noda, Electron. Lett. 24, 1147 (1988).
5. I. Yokohama, J. Noda, and K. Okamoto, J. Lightwave Technol. 5, 910 (1987).
6. A. W. Snyder, J. Opt. Soc. Amer. 62, 1267 (1972).
7. A. W. Snyder, and A. Ankiewicz, J. Lightwave Technol. 6, 463 (1988).
8. T. A. Birks, and C. D. Hussey, Opt. Lett. 13, 681 (1988).

Temperature Dependence of Magnetic Field Sensors Using $(Cd_{1-x}Mn_x)Te$

N. Mikami[1], *C. Nagao*[1], *T. Sawada*[1], *Y. Furukawa*[2], *and E. Aikawa*[2]

[1]Mitsubishi Electric Corp., 1-1 Tsukaguchi-Honmachi, 8-chome Amagasaki, Hyogo 661, Japan

[2]Chubu Electric Power Co., 20-1 Ohdaka-chu Aza-Kitasekiyama Midori-ku Nagoa, Aichi 456, Japan

The temperature dependence of magnetic field sensors using a LED light source and $(Cd_{1-x}Mn_x)Te$ is discussed by means of effective Verdet constant defined by an average weighted by transmission light spectra. Temperature variation is less than 0.5% over 80°C temperature range, with adjustment of composition to visible LED spectra.

Introduction

Since a giant Faraday rotation has been found[1)] in semimagnetic semiconductor $(Cd_{1-x}Mn_x)Te$ solid solutions with zinc blende structure, these magnetic, optical and magnetooptical properties have been widely investigated[2)-4)]. The materials have been determined as candidates for magnetooptical devices such as magnetic field sensors,[6)] isolators and modulators[5)]. Especially for magnetic field sensors, a large Faraday rotation with low light transmission loss and no fear of losing the S/N ratio by stiff motions of ferromagnetic domain walls bring great advantages to these materials.

The temperature dependence of Faraday rotation was reported to be large[1)] and decrease with temperature increase, as is usual in the case of paramagnetics. At a specific wavelength, Faraday rotation was found to be temperature independent,[6)] because the opposite temperature variation exists at wavelengths near the absorption edge that moves to longer wavelengths with increasing temperature. The specific wavelength varied with manganese concentration x, along with the band gap of semiconductor. Magnetic field sensors were proposed[6)] with use of this material and the light sources of the specific wavelength such as a light emitting diode (LED) with narrow band pass filter or fluorescence of glass excited by a LED. These light sources have low light intensity, which reduced the signal-to-noise ratio of the sensors. As regards using a laser diode as light source, speckle noise has a bad influence on the sensitivity of the sensor's response[10)].

The representation of sensitivity for magnetic field sensors using light sources with comparatively wide spectra and magnetooptical materials whose Verdet constant is strongly dependent on wavelength was considered in another article[8)] by the authors. Sensitivity is expressed by the effective Verdet constant defined by an average of the Verdet constant weighted by transmission light intensity in wavelength.

In the present paper the temperature dependence of a magnetic field sensor using a comparatively wide range light source LED and this promising material is discussed. Temperature variation of the sensor is less than 0.5% over a range from -20°C to 60°C. according to choice of light source and composition of material.

Determinations of characteristics of (Cd,Mn)Te

Single crystals were grown by the Bridgman method. Gradual composition shifts have often occurred in solid solution crystals grown by the Czochralski or Bridgman methods, because of the existence of a two phase region in a phase diagram. Accordingly, the composition of the grown crystal does not coincide with that of the melt generally. The composition of $(Cd_{1-x},Mn_x)Te$ crystals was determined by inductively coupled plasma emission spectrometry, and then compared to the absorption edge shown later.

Optical transmission spectra of $(Cd_{1-x},Mn_x)Te$ were measured in several Mn concentrations and temperatures. The absorption edge was obtained from the spectrum by linearly exatrapolating the steep edge to zero transmission in a 1mm thick sample. This definition is available in practical use. Fig.1 shows the composition dependence of the absorption edge that increases in proportion to composition for x less than 0.5. Absorption edge pinning can be observed in the present experiment in which an absorption coefficient on the order of only $10^{-2}cm^{-1}$ is detectable. Contrary, the fundamental absorption edge was determined to have a linear relation up to 0.5 in Mn concentration by the measurement of the absorption coefficient on the order of $10^{-4}/cm$[4]. The absorption edge varies linearly with temperature. Fig.2 shows the composition dependence of the temperature coefficient of the absorption edge. The small temperature coefficient in the composition range higher than 0.5 is considered to be due to the influence of absorption edge pinning.

The temperature dependence of the Verdet constant was measured by ellipsometry[8] at various wavelengths and Mn concentrations. In the ellipsometry, a coherent light source such as a laser diode, He-Ne or dye laser was used. Although the

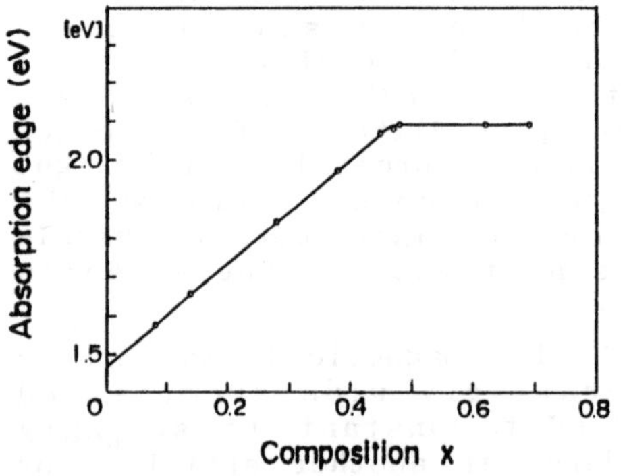

Fig.1 Composition dependence of absorption edge in $(Cd_{1-x}Mn_x)Te$.

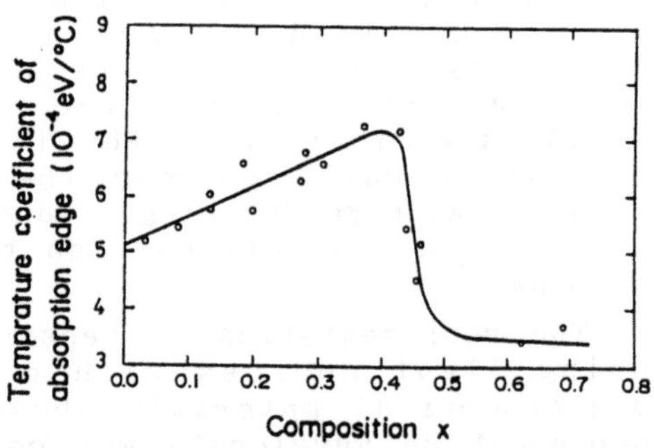

Fig.2 Composition dependence of temperature coefficient in $(Cd_{1-x}Mn_x)Te$.

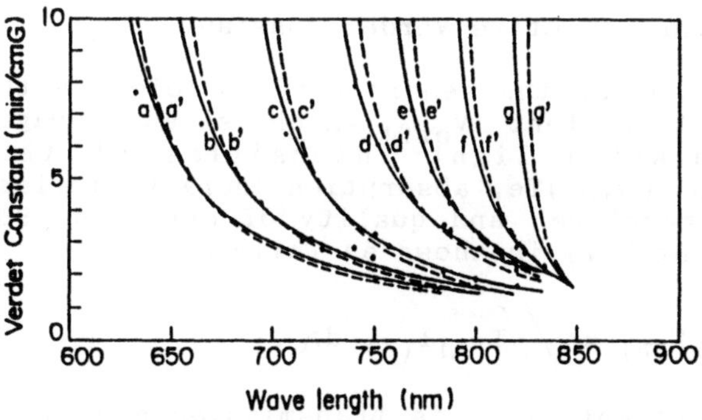

Fig.3 Verdet constant spectra in $(Cd_{1-x}Mn_x)Te$ of various compositions at 20°C (solid line) and 100°C (broken line). The compositions are (a)x=.48, (b)x=.38, (c)x=.29, (d)x=.19, (e)x=.14, (f)x=.09, (g)x=.04.

temperature dependence of the Verdet constant seems to be linear, a little curve is observed in all cases. The deviation from a straight line is about 1% in a range of 80°C. Fig.3 shows the spectra of the Verdet constant in various compositions at 20°C and 100°C. The theoretical value of the Verdet constant was calculated in each case as follows;

$$V = \frac{A}{\lambda^2 - \lambda_0^2},$$

where λ is the wavelength, λ_0 is the wavelength related to the exciton energy which causes Faraday rotation, and A, which is constant, is related to oscillator strength. Fig.4 shows the composition dependence of three photon energies. The first (a) is the exciton energy corresponding to the energy of the wavelength λ_0, the second (b) is the absorption edge, the third (c) corresponds to the photon energy of the specific wavelength at which temperature independent Faraday rotation will be found on the assumption that the Verdet constant changes linearly with temperature.

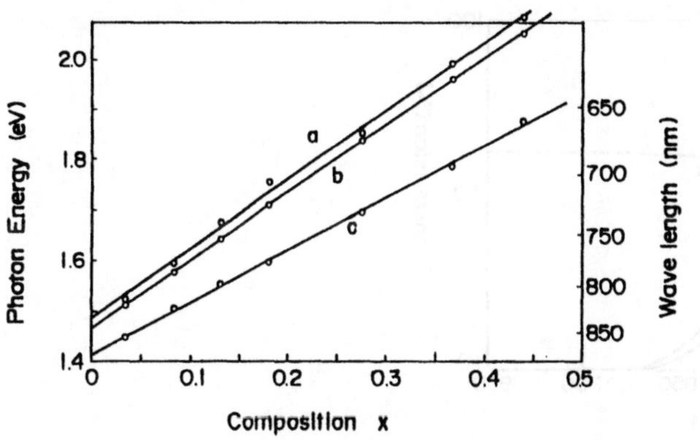

Fig.4 Composition dependence of photon energies. (a)Exciton energy, (b) absorption edge, (c) photon energy related to the specific wavelength of temperature independent Faraday rotation.

Temperature dependence of the effective Verdet constant

The temperature dependence of the sensitivity coincides with that of the effective Verdet constant V_{eff}, which is an average weighted by the transmission light intensities of the wavelengths. Especially in (Cd,Mn)Te, absorption characteristics depend on composition, temperature and quality of crystals. So the effective Verdet constant V_{eff} is shown as follows:

$$V_{eff} = \int V_{(\lambda)} T_{(\lambda)} I_{(\lambda)}\, d\lambda \Big/ \int T_{(\lambda)} I_{(\lambda)}\, d\lambda ,$$

where $V_{(\lambda)}$ is the Verdet constant, $T_{(\lambda)}$ is transmission and $I_{(\lambda)}$ is light intensity.

Available light sources are light emitting diodes in infrared wavelengths between 700nm and 900nm, and in the visible range around 650nm wavelength. Fig.5 shows several optical spectra. The First one (a) is a couple of transmission spectra of $(Cd_{1-x},Mn_x)Te$ for x=0.10 at 20°C and 100°C. The Second (b) is an emission light spectrum of a LED light source whose center wavelength is adjusted to that of temperature independent Faraday rotation, and the half width of the spectrum was about 50nm. The third (c) is a couple of spectra of light from LED transmitted through the material at each temperature, and these spectra are different in distribution and total transmission intensity from each other. In addition to temperature variation of the true Verdet constant, the effective Verdet constant decreases with temperature along with a shift of the spectrum toward longer wavelengths. Fig.6 shows the temperature dependence of the effective Verdet constant of various center wavelengths of the light source. The effective Verdet constant was calculated by transmission spectra in fig.5, the temperature coefficient of the absorption edge, the verdet constant and the LED spectra which was obtained to shift the spectrum in measurement. Neither an increase with temperature nor temperature independence of the effective Verdet constant appears in all cases, though the Verdet constant increases with temperature at shorter wavelengths.

Generally speaking, temperature independence of the effective Verdet constant cannot be observed when a steep absorption edge

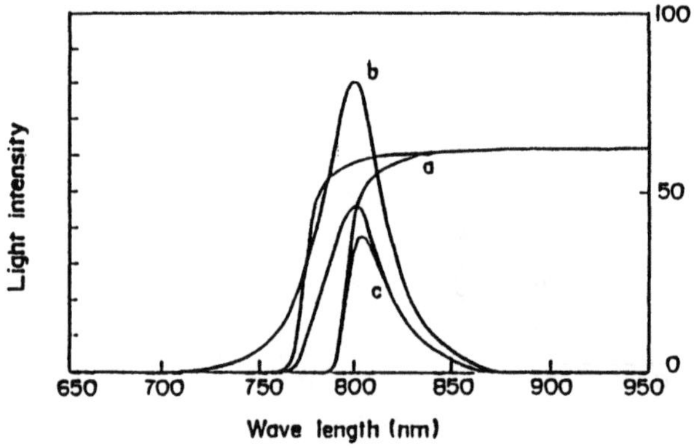

Fig.5 Optical spectra (a) transmission at 20°C and 100°C, (b) LED, (c) the light transmitted through $(Cd_{.90}Mn_{.10})Te$ at each temperature.

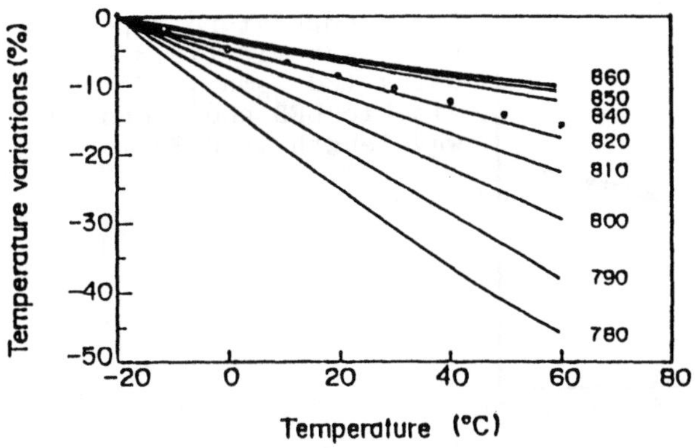

Fig.6 Temperature variation of the effective Verdet constant with $(Cd_{.90}Mn_{.10})Te$ whose center wavelength is shown in the figure.

of the material and a light source spectrum overlap each other, and more than several percent reduction of light intensity with temperature increase exists. As shown in fig.4, the intervals of absorption edge and photon energy related to the specific wavelength of temperature independent Faraday rotation become wider with Mn concentration increase in (Cd,Mn)Te. Therefore a tolerance to the light source spectrum in order to minimize temperature variation will grow at higher concentrations, because the overlap of light source spectra and absorption spectra of materials is considered to be small.

Fig.7 shows temperature variations of the effective Verdet constant for $(Cd_{1-x}Mn_x)Te$ for x=0.48 with various visible LED whose half width of spectra were about 25nm in measurement. Lines are calculated values of the effective Verdet constant and circles are the measured modulation depth of sensors. In the composition range around 0.45, temperature variation of the effective Verdet constant with visible LED and the material is

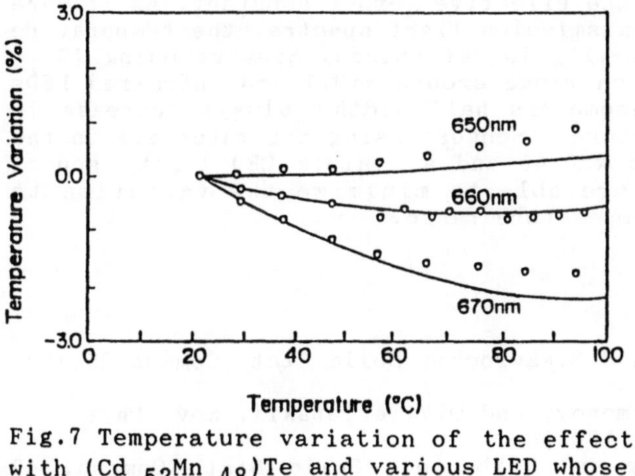

Fig.7 Temperature variation of the effective Verdet constant with $(Cd_{.52}Mn_{.48})Te$ and various LED whose center wavelength is shown in the figure.

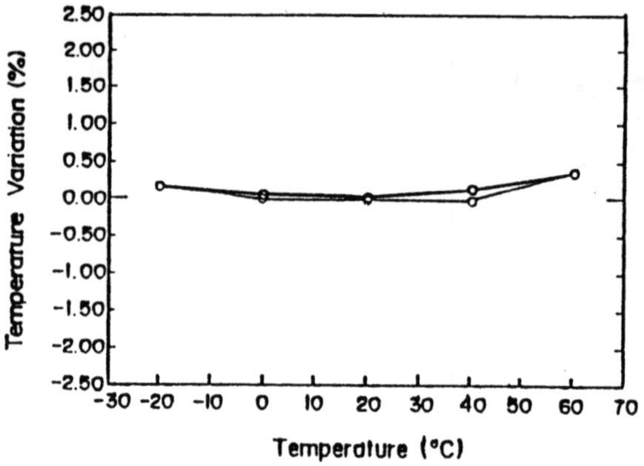

Fig.8 emperature dependence of a magnetic field sensor using $(Cd_{.56}Mn_{.44})Te$ and a visible LED whose center wavelength is 663nm.

minimized within 0.5% with adjustment of LED spectra, for a wide interval of the energies and comparatively narrow half width of light source. Fig.8 shows the temperature dependence of a magnetic field sensor constructed from a polarizer, $(Cd_{.56}Mn_{.44x})Te$, analyzer, whose azimuth is at 45 degree from that of the polarizer, and a LED light source whose center wavelength is adjusted to 663nm. The configuration of the sensors is the same as that described before. Temperature variation was less than 0.5% between -20°C and 60°C.

Conclusion

For optical fiber sensors for the measurement of magnetic field or electrical current, temperature variation and sensitivity are the most important problems, especially for sensors using materials such as (Cd,Mn)Te which have large variations of Verdet constant and absorbance with temperature and wavelength. The present paper gives a method to examine the temperature dependence of sensitivity using such materials.

The Verdet constant and optical transmission of $(Cd_{1-x}Mn_x)Te$ are determined as a function of composition x, wavelength, and temperature. By use of the effective Verdet constant, as it were a weighted average by transmission light spectra, the temperature dependence of the sensitivity is determined. Sensors using $(Cd_{1-x}Mn_x)Te$ in the composition range around x=0.1 and infra-red LEDs with several tens of nanometers half width always decrease in sensitivity with temperature. Sensors using the materials in the composition range around x=0.45 and a visible LED light source adjusted in wavelength are able to minimize the variation to within 0.5% in a 80°C temperature range.

References

1)J.A.Gaj, R.R.Galazka and M.Nawrocki, Solid Stat. Commun.25,193 (1978)

2)S.M.Ryabchenko, Y.G.Semenov, and O.V.Terletskii, Sov. Phys. Solid Stat.27(10)1746 (1985)

3)J.Diouri, J.P.Lascaray and Triboulet, Solid Stat. Commun. 42 (3), 231 (1982)

4)M.E.Amrani, J.P.Lascaray and J.Diouri, Solid Stat. Commun. 45, 351 (1983)
5)A.E.Turner, R.L.Gunshor, and S.Datta, Appl. Opt. 22, 3152(1983)
6)N.Kullendorff and Bertil Hok, Appl.Phys.Lett. 46(11),1016(1985)
7)K.Kyuma,S.Tai,M.Nunoshita,N.Mikami, and Y.Ida, IEEE J.Lightwave Tech. LT-1,93 (1983)
8)N.Mikami, C.Nagao, T.Sawada, Y.furukawa and E.Aikawa, submitted to J. Appl. Phys.
9)N.Mikami, C.Nagao, T.Sawada and K.Sato, Japan.J.Appl.Phys. 26 suppl. 26-2,152(1987)

Splicing of Polarization-Maintaining Fibers

T. Watanabe, K. Osaka, E. Sasaoka, S. Tanaka, and Y. Asano

Sumitomo Electric Industries, Ltd., 1, Taya-cho, Sakae-ku,
Yokohama 244, Japan

Polarization maintaining fibers (identical, nonidentical) were spliced. Both low splice loss and low extinction ratio were established simultaneously. It was possible to consider that the extinction ratio degration was mainly caused by the angular misalignment, however the average misalignment was negligibly small, of about 1 degree.

I. Introduction

Some applications of polarization maintaining fibers, such as fiber optic gyroscope(1), have been developed. In the applications the characteristics of cross-talk of the fiber are important. When the polarization maintaining fibers are spliced to each other, fusion splicing technique has been used. As the extinction ratio at the spliced point has much effect on them even if the fiber's characteristics are good, the precision alignment functions are indispensable to the fusion splicer. Several experimental results have been reported about splicing.(2)(3) This paper reports the results about splicing of identical and non-identical fibers and analysis of the misalignment of the principle axis between the fibers spliced to each other.

II. Experiment

1. Spliced fibers

The polarization maintaining fiber (PANDA fiber) shown in Fig. 1 was investigated in the experiment. Geometrical parameters of the fibers are shown in Table. 1. Sample A was spliced with both sample A (identical) and sample B (non-identical) respectively.

2. Experimental setup

Fig. 2 shows the experimental setup. PMF1 and PMF2 denote polarization maintaining fiber. A LD light source with the wavelength of 1.3 μm was used. Light emitted from the light source was linearly polarized through the polarizer and coupled

Springer Proceedings in Physics, Vol. 44 Optical Fiber Sensors
Editors: H.J. Arditty · J.P. Dakin · R.Th. Kersten

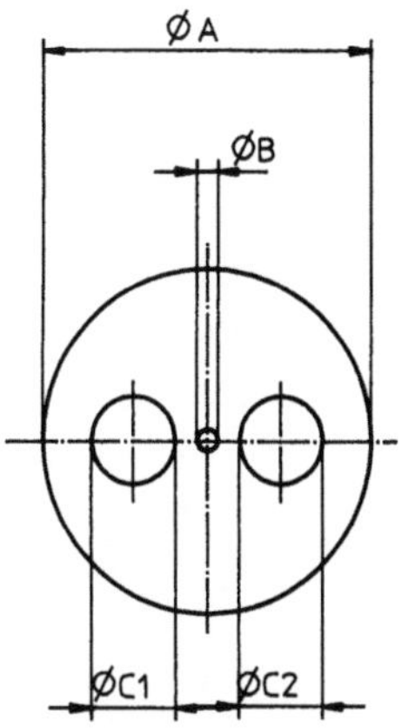

Fig.1 PANDA fiber

Table 1 Geometrical parameters of the fiber

Parameter	Sample A	Sample B
Core diameter :ØA (μm)	7.6	7.7
Clad. diameter :ØB (μm)	124.6	124.9
S.A.P(1) diameter:ØC1 (μm)	33.4	32.6
S.A.P(2) diameter:ØC2 (μm)	34.0	33.2

Clad. :Cladding , S.A.P :Stress Applying Part

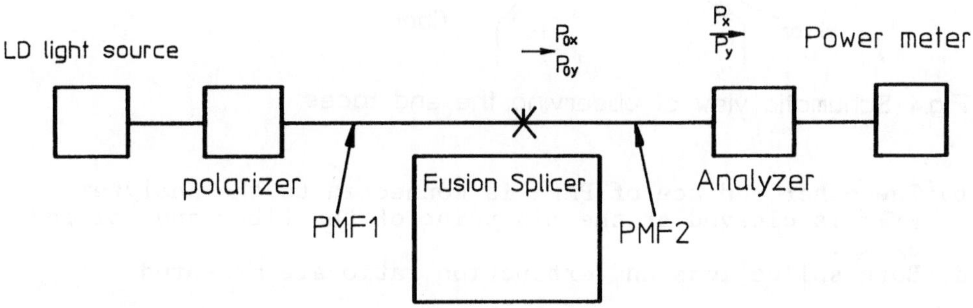

Fig.2 Schematic view of the experimental setup

to PMF1 so that the principal axis of PMF1 agreed with polarized axis. On the other hand the output power from PMF2 was measured by the powermeter through the analyzer. PMF1 was spliced with PMF2 by the fusion splicer.

3.Procedure of experiment

Fig. 3 shows the schematic view of experimental procedure. The procedure was as follows. (identical fiber)

a) PMF1 is connected to the polarizer, so that the principal axis is aligned to the polarized axis by making monitoring power maximum.

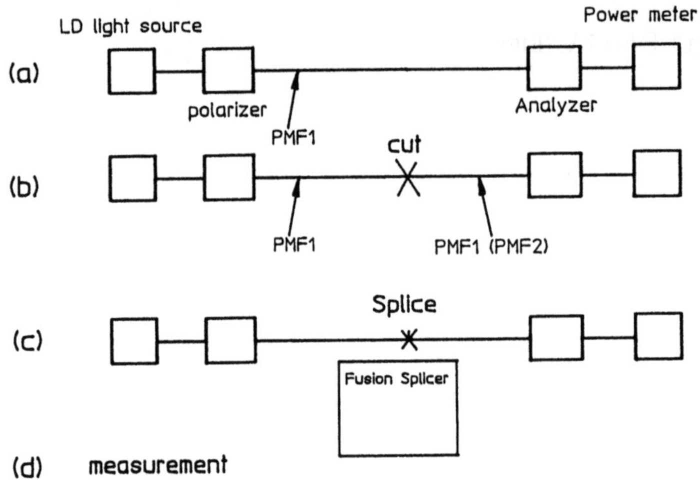

Fig.3 Schematic view of the experimental procedure

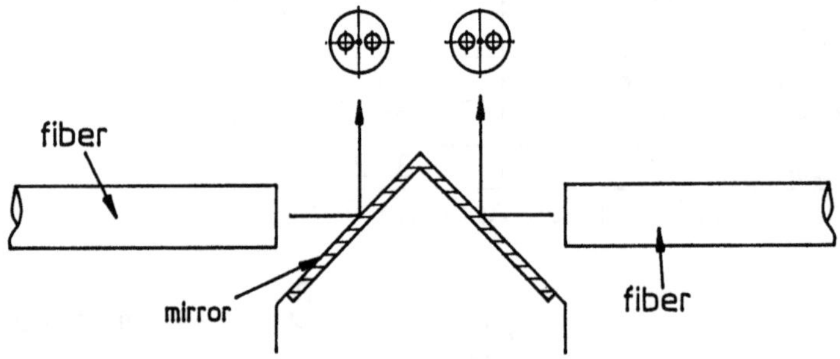

Fig.4 Schematic view of observing the end faces

b) The other endface of PMF1 is connected to the analyzer.
c) PMF1 is cleaved at the mid point of the fiber and spliced again.
d) Both splice loss and extinction ratio are measured.

In case of non-identical fiber's splicing, conventional cutback method was used.

The splicing procedures are expressed simply.

a) The fibers (PMF1, PMF2) are set on the V grooves on the splicer.
b) The end faces are observed by the mirror shown in Fig. 4, the fibers are manipulated not only horizontally and vertically, but also rotatorily to align both the principal axis and the core axis.
c) The end face gap is filled with matching fluid.
d) Both angular and lateral misalignments are corrected by making the monitoring power maximum.
e) The matching fluid is evaporated by discharge.
h) The fibers are spliced.

4.Characteristics of spliced fiber

The polarization maintaining fibers were spliced by the setup described above. Fig. 5 shows the distribution of the splice loss. The average splice loss of 0.04 dB was obtained with the identical fiber (Fig. 5(a)), and the average splice loss of 0.05 dB with the non-identical fiber (Fig. 5(b)). These results mean that the polarization maintaining fibers can be spliced with low splice loss.

Fig. 6 shows the distribution of extinction ratio at the spliced point. The average extinction ratio of 32.7 (dB) was obtained with the identical fiber (Fig.6(a)), and the average extinction ratio of 31.89 (dB) with non-identical fiber (Fig. 6 (b)). Both low splice loss and low extinction ratio were established simultaneously.

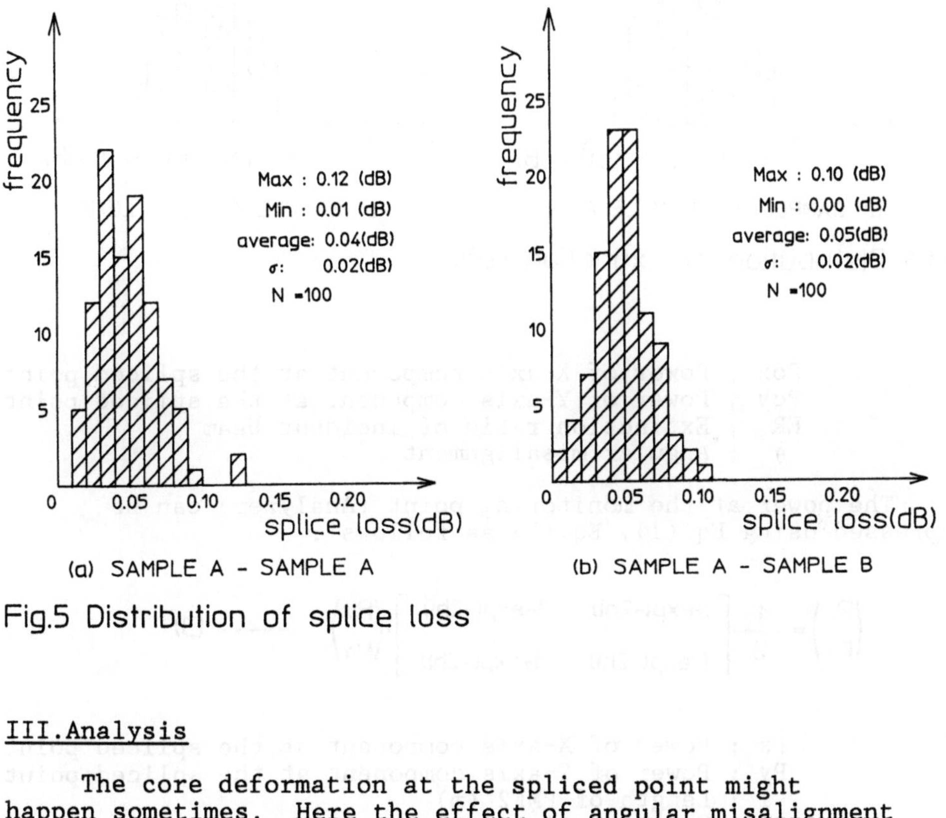

Fig.5 Distribution of splice loss

III.Analysis

The core deformation at the spliced point might happen sometimes. Here the effect of angular misalignment caused by the core deformation on the extinction ratio is analyzed. In case there is the angular misalignment, the power after passing through the spliced point can be expressed as follows :

$$P_{0X} = (1-10^{-\frac{ER}{10}})\cdot\cos^2\theta + 10^{-\frac{ER}{10}}\cdot\sin^2\theta \quad \text{-------- (1)}$$

$$P_{0Y} = 10^{-\frac{ER}{10}}\cdot\cos^2\theta + (1-10^{-\frac{ER}{10}})\cdot\sin^2\theta \quad \text{-------- (2)}$$

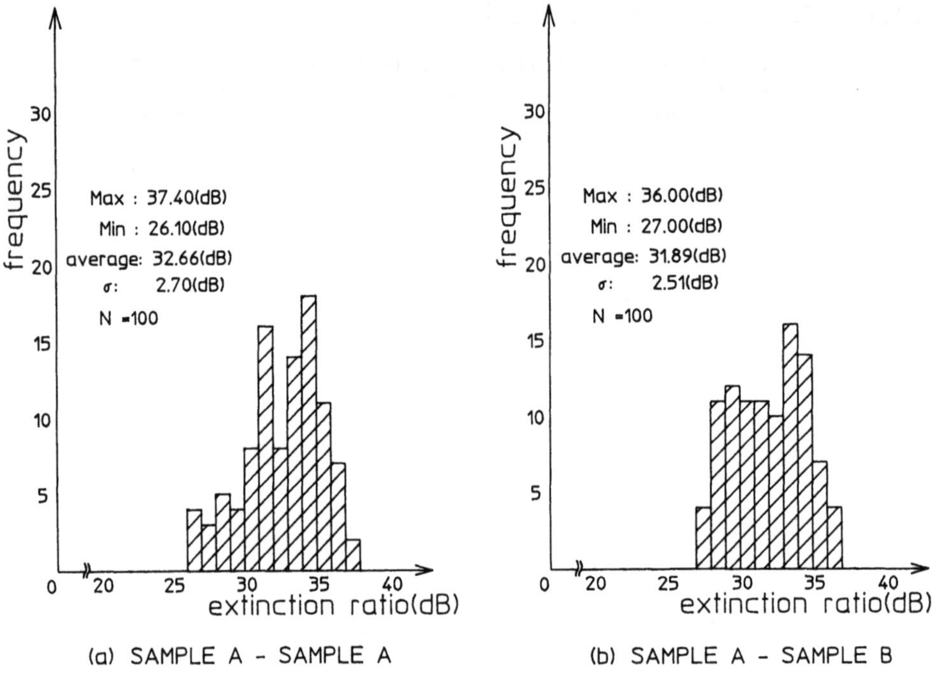

Fig.6 Distribution of extinction ratio

P_{ox} ; Power of X-axis component at the spliced point
P_{oy} ; Power of Y-axis component at the spliced point
ER ; Extinction ratio of incident beam
θ ; Angular misalignment

The power at the monitoring point (Analyzer) can be expressed using Eq.(1), Eq.(2) as follows :

$$\begin{pmatrix} P_X \\ P_Y \end{pmatrix} = \frac{1}{2} \cdot \begin{bmatrix} 1+\exp(-2hl) & 1-\exp(-2hl) \\ 1-\exp(-2hl) & 1+\exp(-2hl) \end{bmatrix} \cdot \begin{pmatrix} P_{0X} \\ P_{0Y} \end{pmatrix} \quad \text{-----} \ (3)$$

Px ; Power of X-axis component at the spliced point
Py ; Power of Y-axis component at the spliced point
l ; Length of PMF2 (m)
h ; Mode coupling parameter

The cross talk of PMF2 (CTo) can be expressed as follows :

$$CT_0 = 10 \cdot \log[\tanh(h \cdot l)] \quad \text{-------------} \ (4)$$

The angular misalignment is defined from these equations as

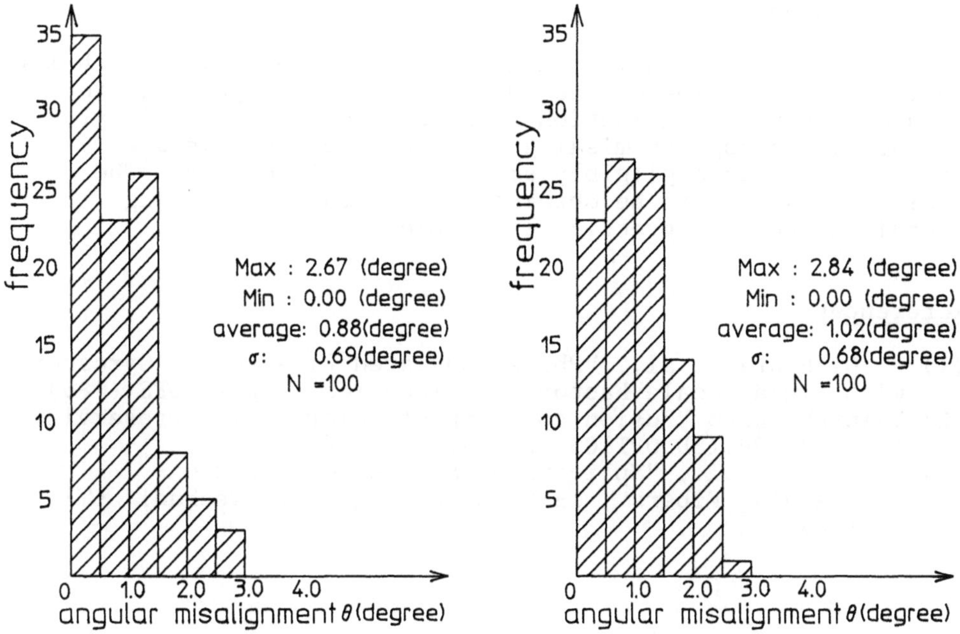

(a) SAMPLE A - SAMPLE A (b) SAMPLE A - SAMPLE B

Fig.7 Distribution of angular misalignment

$$\theta = \arccos \left[\left\{ \frac{(1+a-2ab)-(1-a+2ab)\cdot 10^{\frac{CT}{10}}}{2a(1-2b)\cdot(1+10^{\frac{CT}{10}})} \right\}^{\frac{1}{2}} \right] \text{--- (5)}$$

$$\begin{cases} a=\exp(-2\cdot 10^{-4}) \\ b=10^{-\frac{ER}{10}} \end{cases}$$

with CTo=-40 (dB), l=10(m). ER and CT are experimentally measured coefficients.

Fig. 7 shows the distribution of angular misalignment calculated from Eq.(5). With identical fiber, the average angular misalignment was 0.88 (degree), and with the non-identical fiber, the average angular misalignment was 1.02 (degree). The difference between these two results was caused by the mismatching of the fiber's geometrical parameters. However, that does not affect the extinction ratio, because the difference is negligibly small.

IV. Conclusion

The polarization maintaining fibers were spliced with low splice loss, and low extinction ratio. It was possible to consider that the extinction ratio degradation was mainly caused by the angular misalignment, however, the average misalignment was negligibly small of about 1 degree. The fruitful results can be obtained by the fusion splicing technique, that is practically usable.

References

(1) Y. Nishiura et al., "Phase Modulated Fiber-Optic Gyroscope with Linear Scale Factor" P.9 Technical Digest OFS ('86)

(2) Yasuyuki Kato "Fusion splicing of polarization preserving fibers" APPLIED OPTICS vol.24, No.15 (1985)

(3) J. Noda et al., "Splicing of Single Polarization-Maintaining Fibers", IEEE/OSA J. Lightwave Technol. LT-1, 61 ('83)

Fiber Coupled Analog Rotary Position Sensor

R.W. Huggins, E.C. Goldstick, and B. Van Deventer

Boeing Aerospace and Electronics High Technology Center,
P.O. Box 3999, Seattle, WA98124, USA

A variable optical attenuation analog encoder track has been generated by varying the diameter and density of holes in an opaque film according to a mathematical law. Deviation from linearity is $\pm$ 0.1 deg. over 90 deg., and the rms noise is < 0.1 deg. A two wavelength referenced sensor which uses the encoder is described.

INTRODUCTION

Fiber coupled analog position sensors that have been reported operate in various ways. One method relies upon the interruption of an optical beam by a mask such as a single edge or a wedge-shaped slit[1]. A second method depends upon the variation of the optical coupling between two fibers by a moving reflector[2], and a third method depends upon the position dependent coupling between two axially aligned fibers which move relative to each other[3]. Factors which limit the performance of the above sensors include the necessity for the uniform illumination of the mask or reflector. One method which would reduce the sensitivity to the variation in the illumination across the aperture is to measure the optical density of a variable density track. However, this method does not seem to have enjoyed success presumably due to the inability to produce variable density masks with a linearity of better than 5%. None of the methods reported have sufficient accuracy for aerospace applications.

A new technique has been developed for the fabrication of variable neutral density filters. Neutral density filters are commonly fabricated by evaporating a metallic film of varying thickness on to a glass substrate. The variable thickness film is obtained by controlling the deposition rate as the substrate is moved across an aperture. In the work to be reported, the variable thickness film is replaced by an opaque film with a pattern of circular apertures (holes), whose diameter and spacing are a function of position. The diameter and spacing of the holes can be predicted numerically for any law with the only constraint being that the number of holes in the illumination aperture is sufficiently large so that the noise level on the optical transmission as the filter is moved meets performance specifications. Rotary encoder disks have been fabricated with a deviation from linearity of less than ± 0.1 degrees over 90 degrees, and an RMS noise level of better than 0.1 degrees.

A rotary sensor which uses the encoder disk described above is also reported. Light of two different wavelengths is launched into the outgoing fiber, and is demultiplexed in the sensor. Light of one wavelength passes through the encoder track, and is attenuated by an amount which depends upon the angular position of the disk. Light of the other wavelength is steered around the encoder disk and is unmodulated. The two beams are then recombined and focussed onto the return fiber. With suitable signal processing, an end-to-end attenuation measurement can be corrected for the unknown and variable link losses, such as losses introduced by connectors, to give the attenuation introduced by the variable density encoder track. The principle of two wavelength referencing has previously described by Brogardh[4].

OPTICAL ENCODER DISK

An optical encoder track designed for linear attenuation over 90 degrees is shown in Figure 1. The coordinates of the hole on the i+1th vector and j+1th row are given by

$$\theta_{i+1} = \theta_i + \Delta\theta(\theta_i)$$

and

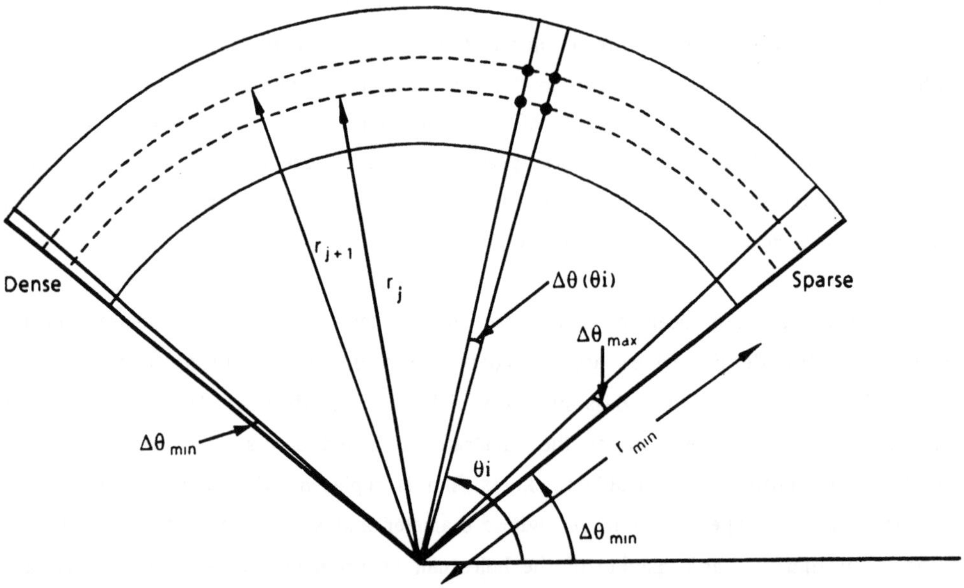

Figure 1. Mathematically Generated Variable Neutral Density Rotary Track

Figure 2. Example of Analog Code Plate

$$r_{j+1} = r_{min}(1 + j\Delta\theta(\theta_i))$$

where

$$\Delta\theta(\theta_i) = \frac{\Delta\theta_{max}\,\Delta\theta_{min}}{\frac{\theta_i - \theta_{min}}{\theta_{max} - \theta_{min}}(\Delta\theta^2_{max} - \Delta\theta^2_{min}) + \Delta\theta^2_{min}}.$$

An example of an early encoder disk is shown in Figure 2. A later design had a track width of 5 mm, a mean track radius of 18 mm, and an attenuation range of 3 to 10 dB. Deviation from linearity in a test fixture was less than ± 0.1 degrees and is shown in Figure 3(a). The mask for the disk was computer generated and fabricated by electron beam lithography.

There exists a tradeoff between the attenuation range and rms noise according to illuminating beam diameter, the hole size, and the hole density. This is apparent in the design phase and can be optimized according to the placement law and desired performance.

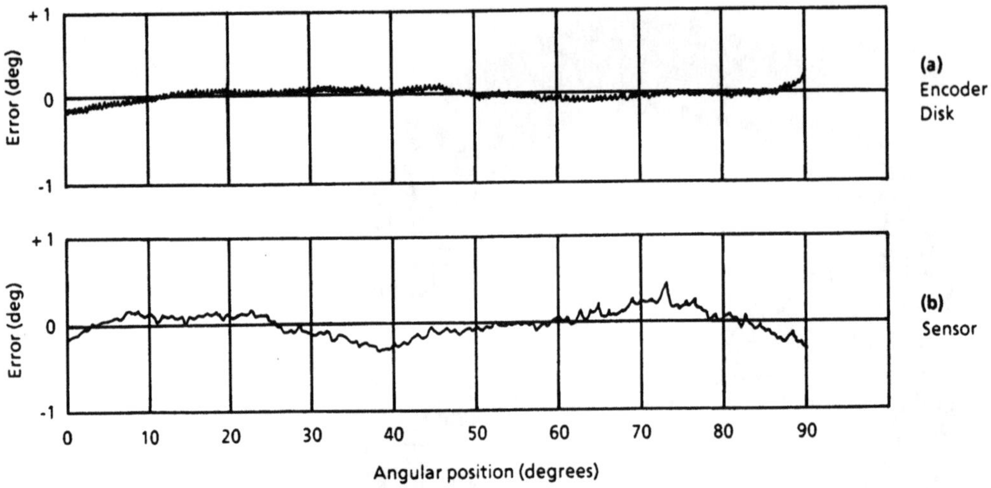

Figure 3. Deviation from Linearity of a) The Encoder Disk, and b) The Sensor

THE TWO WAVELENGTH REFERENCED SENSOR

The sensor is shown schematically in Figure 4. Light at wavelengths λ_1 (735 nm) and λ_2 (865 nm) emerging from the outgoing fiber is focused into a collimated beam by lens L_1, and is incident on a Short Wavelength Pass (SWP) interference filter F_1. The cut off wavelength of the filter is such that light at the shorter wavelength λ_1 (the signal wavelength) is transmitted, and light at the longer wavelength λ_2 (the reference wavelength) is reflected. The transmitted beam passes through the encoder track, is reversed by the Porro prism, and passes through a clear portion of the disk. The beam then passes through SWP filter F_2 and is focused onto the end of the return fiber by lens L_2. As the disk is rotated, light at wavelength λ_1 will be attenuated by an amount depending upon the angle of rotation. The optical path through the encoder track will be referred to as the signal path.

Light at the longer wavelength λ_2 is reflected by filter F_1 and is transmitted by the Long Wavelength Pass filter (LWP) filter F_3. The upper surface of F_3 is silvered, and the transmitted light is reflected back through F_3 and is incident on SWP filter F_2. Light at wavelength λ_2 is reflected by F_2 and recombined with the light which has passed through the encoder disk. Filter F_3 was made from intrinsically short wavelength absorbing glass and blocks any residual light at wavelength λ_1 which is reflected by Filter F_1. This optical path will be referred to as the reference path. The cross talk between the signal and reference paths was measured and found to be < 0.001 for light at wavelength λ_1 along the reference path, and $\cong 0.004$ for light at wavelength λ_2 along the signal path. The latter could be reduced by coating the hypotenuse of the Porro prism with an additional short wavelength pass interference filter.

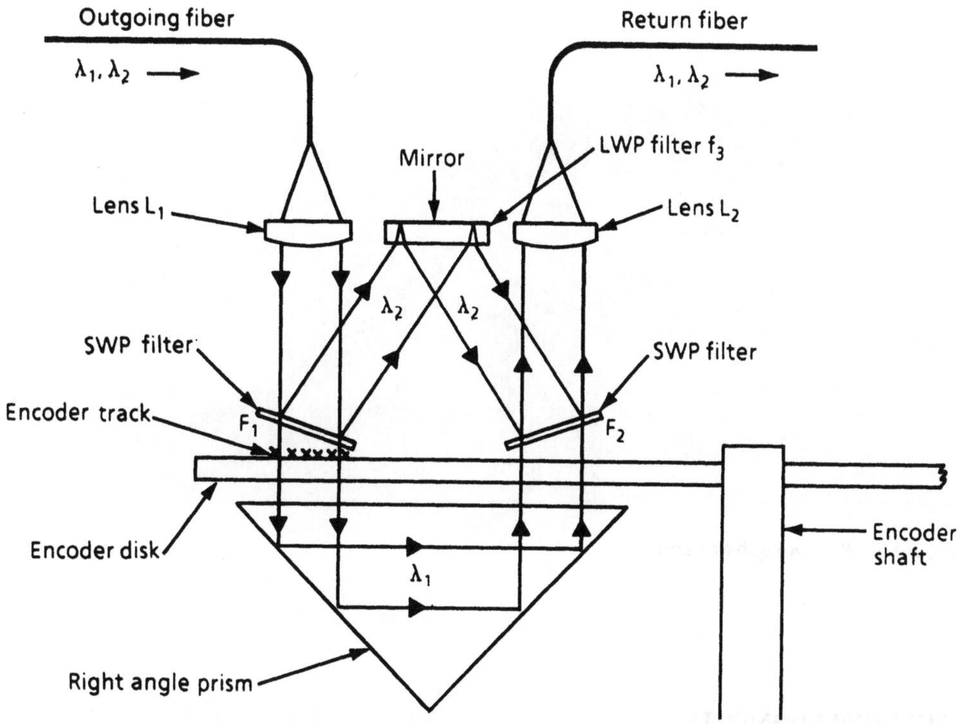

Figure 4. Optical Schematic Diagram of the Sensor

Making the assumption that the link losses are the same for light at the signal and reference wavelengths, it can easily be shown that the angle of rotation is given by

$$\theta = \frac{1}{k}\left[\frac{P_{\lambda_1}}{P_{\lambda_2}} \cdot \frac{P_{0\lambda_2}}{P_{0\lambda_1}} - \delta\right]$$

where $P_{0\lambda_1}$ and $P_{0\lambda_2}$ are the optical powers at the signal and referenced wavelengths launched into the outgoing fiber, P_{λ_1} and P_{λ_2} are the optical powers at the signal and reference wavelengths emerging from the return fiber, k is the law of the track, and δ is the attenuation of the encoder track when $\theta = 0$ degrees.

A photograph of the sensor is shown in Figure 5. The package was designed with a standard resolver mount and had an overall length of 2 inches. The maximum diameter was 1.75 inches. The deviation from linearity of the sensor output after signal processing was less than ± 0.3 degrees and is shown in Figure 3(b).

Figure 5. Photograph of Sensor.

ACKNOWLEDGMENTS

The authors would like to thank Michael Kyne for fabricating the glass components and in particular the analog encoder disk.

REFERENCES

1) R. O. Stanton, "Digital Optical Transducers for Helicopter Flight Control Systems," Procedings of SPIE Vol. 412, pp 122-129, (1983).

2) R. O. Cook and C. W. Hamm, "Fiber Optic Lever Displacement Transducer," Appl. Optics 18, 3230 - 3241 (1979).

3) B. Culshaw, "Optical fibre sensing and signal processing," Peter Perigrinus Ltd., London (1984).

4) T. Brogardh, L. Stensland, U.S. Patent #4,281,245 "Fiber Optic Measurement Apparatus" (1981).

Fiber Optic Current Density Measurement Based on the Faraday Effect

L. Bager[1], *J.E. Schrøder*[1], *and C.J. Nielsen*[2]

[1]Electromagnetics Institute, Building 348,
Technical University of Denmark, DK-2800 Lyngby, Denmark
[2]DANTEC Electronics, Tonsbakken 16-18, DK-2740 Skovlunde, Denmark

A Faraday Ring Magnetometer (FRM) intended for measuring plasma currents in the magnetosphere is described. The FRM is based on the Faraday effect in optical fibers showing negligible intrinsic birefringence.It is demonstrated that the FRM has a very high sensitivity, $52 mA/Hz^{1/2}/m^2$ over a large frequency range, exceeding 20kHz.

I. Introduction

Electrical currents in space are presently detected by their magnetic field or by counting the net flow of charge of either sign in any given direction. Both methods do however suffer from serious drawbacks. In the first case the interpretation of magnetic measurements is geometry-dependent and becomes less sensitive for small scale currents and secondly, detectors counting charged particles always miss the populations of particles with energies above and below the instruments′ measurement range.

The Faraday Ring Magnetometer, however, offers constant sensitivity independent of the geometry of current. The FRM technique has been used for some years for measuring high currents in power transmission lines [1]-[3] but in these applications the sensitivity is not as important as for this space application. We present the results obtained using the FRM as a very sensitive current density detection system. The achieved sensitivity of 52 $mA/Hz^{1/2}/m^2$ is limited by the intensity noise from the laser source.

II. Theory

The Faraday rotation is induced when polarized light propagates in a transparent medium which is located in a magnetic field H. The polarization direction of a linearly polarized optical field will be rotated by the applied magnetic field. The rotation angle θ is given by

$$\theta = V \int_L H_l \cdot dl \qquad (1)$$

where H_l is the magnetic field component in the direction of the optical field and L is the lenght of the transparent material. V is the Verdet constant given by $V = 2.0 \cdot 10^{-35} \cdot \upsilon^2$ rad/A where υ is the frequency of the optical field.

As the optical medium we use a fibercoil of N_f turns, and supply it with a toroidal electrical coil of N_e turns to generate a homogeneous field. The Faraday rotation θ is a function of the electrical current I_t in the coil, or alternatively of the current density J, flowing perpendicular to the plane of the fiber coil

$$\theta = V N_f N_e I_t = 1/2\, V R L J \qquad (2)$$

Here R is the radius of the fiber coil and L the length of the fiber. From (2) we see the importance of having a large coil radius and a long fiber length for this application.

III. Experimental

The experimental set up is shown in Fig. 1. The laser diode (HL7801) emitting at 790 nm wavelength is powered by a battery-based supply, stabilizing the output using the monitor diode in a feedback loop . The beam is collimated with a microscope objective and we use a halfwave plate to control the linearly polarized light before coupling into the fiber coil through a second microscope objective. A special low birefringent fiber (LO-BI 800 from York VSOP) was used in our experiment to reduce the intrinsic linear birefringence [2] as the linear birefringence will decrease the measured Faraday-rotation [6] . The 150 meters of this fiber are coiled in a 0.97 m coil diameter with 50 turns. To simulate the plasma current a toroidal coil N_f = 842 turns was formed on the fiber coil.

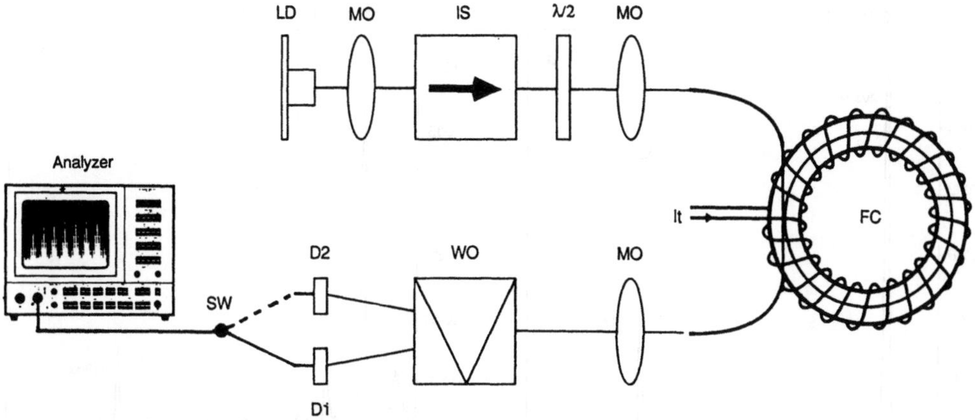

Fig. 1 The experimental setup. LD is the laser source, MO is the microscope objectives, IS the isolator, FC the Faraday coil, WO the analyzing Wollaston prism, D1 and D2 the detectors, SW a switch and the analyzer is a Brüel & Kjaer type 2033.

The output of the fiber is analyzed through a Wollaston prism, adjusted in a 45^0 angle to the output polarisation without test signal. This leads to the best sensitivity and linearity. One of the detector signals is fed to a spectrum analyzer (B&K type 2033) to analyze the response of the Faraday coil.

In Fig. 2a the frequency spectrum of a 205 Hz test-signal of 1.0 V measured across a 50 Ω load in series with the toroidal coil is depicted, giving the test current I_t of 20.0 mA see EQ.2. A response of 1.3 V at 205 Hz is obtained from the corresponding optical baseband spectrum in Fig. 2b. From the achieved signal-to-noise ratio in the optical baseband spectrum, the actual sensitivity I_a of the system can be calculated, setting s/n equal unity. The equivalent noise bandwidth using the Hanning-window of the spectrum analyzer, is for these measurements 1.875 Hz (3-dB value). The signal-to-noise ratio derived from Fig. 2b is 50.3 dB, leading to 53 dB in a 1 Hz bandwidth. The sensitivity I_a of this setup is therefore 45 μA in a 1 Hz bandwidth.

Using (2) one can calculate a corresponding current density sensitivity to J = $52 mA/Hz^{1/2}/m^2$. To test the linearity of the system, a current of I_t 635.6 μA is applied to the coil, resulting in a optical response signal of 41.3 mV, or 30 dB below the signals in Fig. 2 as expected for a linear system.

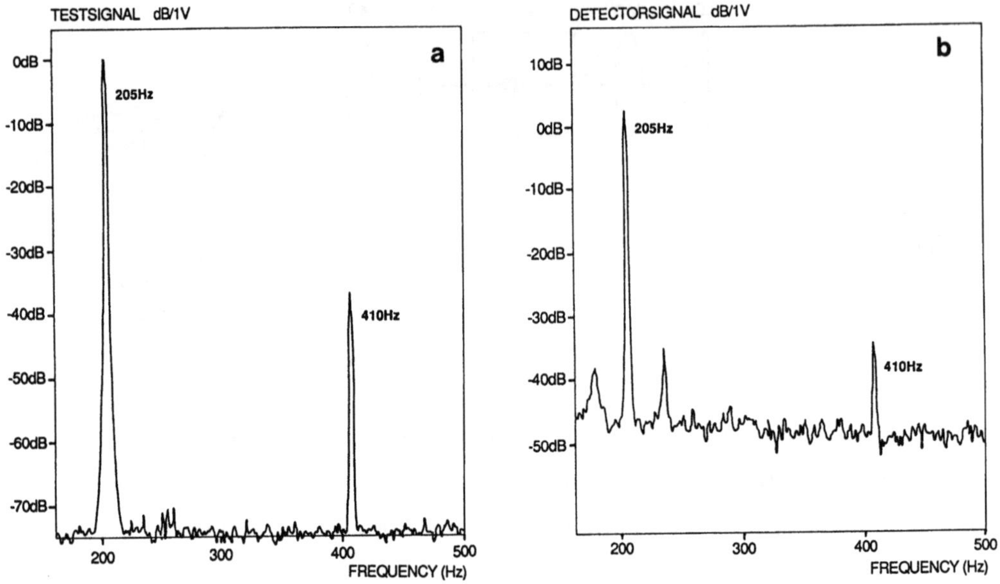

Fig. 2 The measured spectrum from the spectrum analyzer showing the spectrum from 150 Hz to 500 Hz. Fig. 2a shows the applied test signals from the toroidal coil of 20 mA . Fig. 2b illustrates the corresponding baseband signal from detector 1. The components at 150, 180, and 240 Hz are identified as acoustical induced intensity noise.

We have estimated the ideal sensitivity in the absence of linear birefringence: with a received optical power of 3.6 μW, the noise floor of a single channel is measured to 2.2pA/Hz$^{1/2}$ at 200 Hz, which is higher than the detector shotnoise. There may be several sources of this noise, but a measurement of the spectrum of the laser intensity noise density before coupling into the fiber, indicates that a large part of the total noise is originating from the light source see Fig. 3. This is assumed because one can measure the same spectral density without the Wollaston prism analyzer.The ideal sensitivity expressed by the coil current I_i is estimated by using the measured noisefloor and the well known Faraday response together with a detector radiant sensitivity of 0.55 A/W to obtain the minimum detectable Faraday rotation angle $\Delta\theta_{min}$ with unity signal-to-noise ratio. The calculated value $\Delta\theta_{min}$ = 1.1μrad/Hz$^{1/2}$ corresponds to an ideal sensitivity of I_i = 9.1 μA/Hz$^{1/2}$.

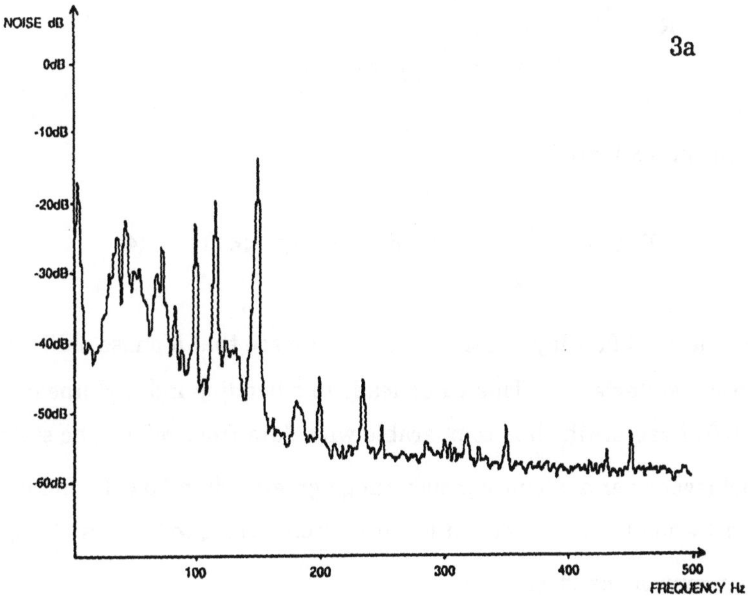

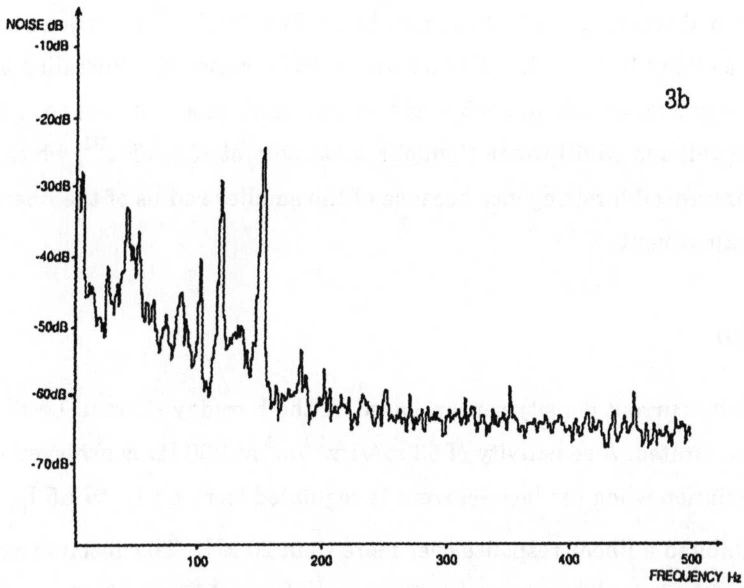

Fig. 3 The measured optical baseband spectrum with no test signal with and without analysing Wollaston prism in a and b respectively.

The difference between I_a and I_i (14 dB) is thought to be due to the linear birefringence introduced by coiling the fiber. From ref. 1 it is known that a linear birefringence δ in the fiber will decreace the sensitivity with a factor R given by [1]

$$R = \frac{I_a}{I_i} = \frac{\mathrm{SIN}\left(2\varphi\sqrt{1 + \left(\frac{\delta}{2\varphi}\right)^2}\right)}{\sqrt{1 + \left(\frac{\delta}{2\varphi}\right)^2}\ \mathrm{SIN}\ 2\varphi} \qquad (3)$$

whitch can be approximated to

$$R \cong \frac{\mathrm{SIN}\ \delta}{\delta}\ ,\ 2\phi << \delta \ \text{and} \ 2\varphi << 1 \qquad (4)$$

here 2φ is the generated faraday rotation. From (4) we see that a phase shift of 150^0 will give the observed deviation . This value leads to a bending induced linear berefringence β_b of 0.017 rad/m which is comparable with data from ref. 6. The stated sensitivity is achieved over a wide frequency range greater than 20 kHz and is constant within 2 dB when the laser current is varied from 1.1 I_{th} to 1.5_{th}, with I_{th} being the threshold current of the laser.

Recently published results for a high-current heterodyne FRM [2] reports a sensitivety of 3 mA for a 40 m fiber length and 800 turns in the toroidal coil. Rescaling to our geometry of 150 m fiber length and 842 toroidal turns leads to a sensitivity of 0.76 $mA/Hz^{1/2}$. This value is 25 dB worse than our sensitivity of 45 $\mu A/Hz^{1/2}$, which might be due to the increased birefringence because of the smaller radius of the fibercoil in the reported experiment.

IV. Conclusion

A high sensitivity current density sensor, based on the Faraday effect in Lo-Bi fibers has been demonstrated. A sensitivity of 52 $mA/Hz^{1/2}/m^2$ at 200 Hz is achieved with only a 2 dB variation when the laser current is regulated from 1.1 I_{th} to 1.5 I_{th} and the system exhibited a linear response over more than 20 kHz. The absolute sensitivity is limited by the laser intensity noise. Future work should concentrate on eliminating the main sources of the disturbing linear birefringence as well as improving the noise characteristics of the laser source.

V. REFERENCES

1. A. Papp and H. Harms "Magnetooptical current Transformer. 1: Principles". Appl. Optics, Vol. 19, pp. 3729-3734,1980.

2. A.D. Kersey and D.A. Jackson: "Current sensing utilizing heterodyne detection of the Faraday effect in a single-mode optical fiber". J. Lightwave Technology, LT-4, pp. 640-644, 1986.

3. H.S. Lassing, W.J. Mastop, A.F.G Vand Der Meer, and A.A.M. Oomens: "Plasma current measurement by Faraday rotation in a single-mode fiber", Appl. Optics, Vol. 26, pp. 2456-2460, 1987.

4. A.M. Smith : "Polarization and magnetooptic properties of single-mode optical fiber". Appl. Optics, Vol. 17, pp. 52-56, 1978.

5. J. Noda, T. Hosaka, Y. Sasaki, and R. Ulrich: "Dispersion of Verdet constant in stressbirefringent silica fibre", Electron Lett., Vol. 20, pp. 906-908, 1984.

6. S. Rasleigh: "Origins and control of polarization effect in single-mode fibers J. Lightwave Technology, Vol. LT-1, pp. 312-331, 1983.

Amorphous Metal Wire Transducers for Fiber Optic Magnetometers

K.P. Koo, F. Bucholtz, D.M. Dagenais*, and A. Dandridge*

Naval Research Laboratory, Optical Science Division, Code 6574, Washington, DC20375, USA

*Permanent address: SFA, Inc., 1401 McCormick Drive, Landover, MD20785, USA

We have investigated a novel amorphous wire transducer for fiber optic magnetic sensor applications. Compared to previous devices, the transducer has significantly smaller volume yet exhibits large magnetostriction (5×10^{-5} Oe^{-2}) and is easily incorporated into fiber interferometers.

I. Introduction

Fiber optic magnetometers based on magnetostriction have employed transducers consisting of either amorphous metallic glass elements in the form of strips or cylinders [1], or nickel-iron alloy jackets deposited directly on the fiber [2]. Recently, amorphous metal wires with diameters of 100 μm or less have become available commercially from Unitika (Japan) and we have investigated this novel material for use as transducers in fiber magnetic sensors. In particular, we anticipated the potential for small, efficient transducers. This paper reports the performance of a short length of amorphous magnetostrictive wire (5 cm x 100 μm dia.) in a fiber- optic magnetometer. Compared to previously reported transducers, wire transducers have significantly smaller volume and exhibit magnetostrictive coupling constant values which are larger by a factor of 10 to 50 . This transducer incorporates the most desirable features of both the jacketed-fiber and bulk metallic glass transducers, namely, low demagnetization and high magnetostriction.

II. Background

In a sensor employing a magnetostrictive transducer, the optical phase shift ϕ measured by the fiber interferometer results from the strain in the transducer produced by a magnetic field. Generally, the magnetostrictive response follows the ideal Coherent Rotation Model (CRM) [3]. The CRM can be used for calculating the magnetostriction parameter (C) from an experimental measurement of the change in phase shift due to a change in dc magnetic field provided the field is large compared to the coercive field (H_c) and small compared to the anisotropy field (H_a). However, predictions of the minimum detectable magnetic field

Springer Proceedings in Physics, Vol. 44 Optical Fiber Sensors
Editors: H.J. Arditty · J.P. Dakin · R.Th. Kersten

(H_{min}) based on the CRM and measured C values can lead to estimates much lower than what are obtained experimentally because H_{min} depends on the total noise of the sensor. Two types of noise are significant in magnetostrictive fiber magnetometers:1) upconverted nonmagnetic noise arising from the presence of large signals on the interferometer and the signal processing electronics (N_{opt}) [4], and 2) magnetic noise from both the magnetic transducer itself and ambient magnetic fluctuations (N_{mag}) [5]. We demonstrate a signal normalization technique to suppress noise of the first type.

According to the CRM, the phase shift ϕ depends quadratically on the total applied magnetic field H_{total}. Typically, H_{total} consists of three components: a dc magnetic field H, a high frequency carrier $h\cos\omega t$, and a low frequency signal field $h_s\cos\Omega t$. ϕ is then given by

$$\phi = \xi\, LC\, (H-H_c + h\cos\omega t + h_s\cos\Omega t)^2 \quad (1)$$

for $|(H-H_c+h+h_s)| < H_a$, $|H-H_c| > 0$. C is the magnetostrictive constant and ξ is the factor relating the change in length between the two arms of the interferometer to the phase shift. The phase shift component at frequency ω is given by

$$\phi_\omega = 2\,\xi\, LC\, h\, (H- H_c) \quad (2)$$

for $|(H- H_c+h+h_s)| < H_a$, $|H-H_c| >0$. Therefore, C can be calculated according to

$$C = \Delta\phi_\omega / (2\,\xi\, L\, h\, \Delta H) \quad (3)$$

where $\Delta\phi_\omega = (\phi_\omega' - \phi_\omega)$ is the change in interferometer phase shift induced by a change in the applied dc magnetic field $\Delta H=(H'-H)$.

III. Experiment and discussion

Fig. 1 shows a schematic of the fiber optic magnetometer. The basic structure is a standard Mach-Zehnder interferometer, path balanced to less than 5mm, and actively stabilized using a piezoelectric (PZT) phase modulator. A short piece of amorphous metal wire (5 cm long x 100 µm dia.) which had been annealed at 400°C in a 1200 Oe transverse magnetic field was bonded to one arm of the interferometer using UV-curable epoxy. The composition of the wire (Unitika type DF-1, $Fe_{77.5}B_{15}Si_{7.5}$) is similar to that of the S2 series of Metglas ribbons (Allied Signal, Inc). During testing the sensor element was excited by two magnetic fields. For measurements of the magnetostriction parameter, an ac field of fixed amplitude at 6.5 kHz provided the magnetic carrier (dither) and a slowly-varying triangular field at 0.01 Hz provided the dc sweep. For signal-to-noise measurements, the triangular magnetic field was replaced by a sinusoidal magnetic calibrating signal at 1 Hz.

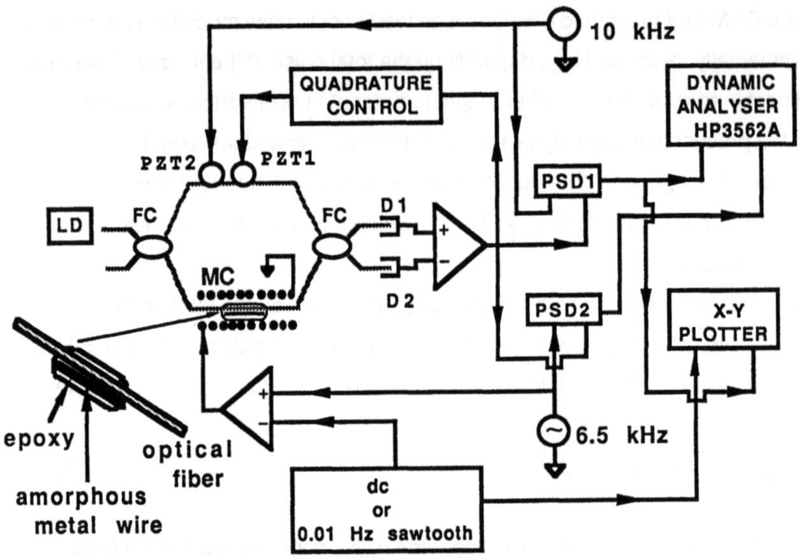

Fig.1. A schematic of the experimental set-up. LD -- laser diode, FC -- fiber coupler, MC -- magnetic coil, PSD1, PSD2 -- phase sensitive detectors, D1,D2 -- detectors, PZT1, PZT2 -- piezoelectric transducers.

To facilitate testing a variety of samples, the sample arm was simply butt coupled, rather than fusion spliced, into the interferometer -- an arrangement which introduced low-frequency intensity noise into the system. However, this noise could be suppressed by putting a pilot tone on the interferometer and processing the output using two phase sensitive detectors (PSD1 and PSD2). PSD1 was locked to the magnetic dither frequency ($\omega/2\pi$). PSD2 was locked to the frequency of the tone applied to a second PZT element which produced a phase signal I_p at 10 kHz. Since both the magnetic dither and the nonmagnetic carrier (I_p) upconvert nonmagnetic noise (in this case intensity noise) in the same fashion [4], dividing the output of PSD1 (magnetic signal) by the output of PSD2 (pilot tone) effectively suppresses the sideband noise due to low-frequency intensity fluctuations. The butt-coupled sample arm together with the normalization technique formed the basis of a system capable of quick and accurate testing of a variety of transducer configurations.

The output of PSD1 as a function of (decreasing) dc magnetic field for h = 50 mOe rms is plotted in Fig.2. The linear portion of this curve is the region where the Coherent Rotation Model is valid and the insert in Fig.2 shows an expanded view of the response near $H = H_C$. Using Eqn.3, the C value for this amorphous metal wire was found to be approximately 5×10^{-5} Oe^{-2} which is approximately 50 times larger than that reported for non-resonant transducers in strip form and approximately 10 times larger than that observed for resonant cylindrical transducers, both made from Metglas ribbons. The value is in good agreement with 6.3×10^{-5} Oe^{-2} which we calculated based on recent independent susceptibility measurements for similar amorphous wires [6]. It should be emphasized that this C value

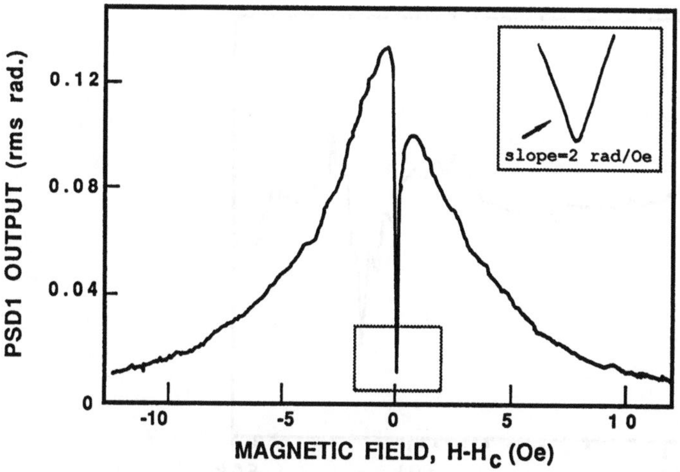

Fig.2. A plot of the interferometer phase shift Φ_ω (PSD1) vs. dc magnetic field H-H $_c$.

was obtained in an operational transducer with a reduction in the volume of magnetostrictive material of over a factor of 450 compared to cylindrical metallic glass transducers. Although it does not appear feasible to incorporate long lengths (>1 m) of fiber in a sensor employing this material, amorphous wire appears to be an ideal transducer configuration for sensors requiring small size and modest performance ($\geq 10^{-6}$ Oe / $\sqrt{Hz}$).

Compared to amorphous ribbon transducers, the larger C value observed for amorphous metal wire is likely due to the substantial decrease in the demagnetization factor (D) resulting from the long, narrow geometry of the wire. The measured C value, which includes the effects of loading and demagnetization, is related to the intrinsic magnetostriction constant $C_{int} = 3\lambda_s/2H_a^2$ by

$$C / C_{int} = (1 + D\chi)^{-2} \, [1 + (E_f A_f)/(E_m A_m)]^{-1} \qquad (4)$$

where λ_s is the saturation magnetostriction, $(1 + D\chi)^{-2}$ is the reduction factor due to demagnetization, $[1+(E_f A_f)/(E_m A_m)]^{-1}$ is the mechanical loading factor, χ is the magnetic susceptibility and $D \approx 8A_m/L^2$ for $L^2 >> A_m$. L is the sample length parallel to the applied magnetic field, (E_f, E_m) and (A_f, A_m) are the Young's moduli and cross-sectional areas associated with the (bare) optical fiber and the metallic glass transducer, respectively. We have neglected the frequency dependence of the susceptibility. Using Eq.(4) and the parameters

E_f	=	7×10^{10} N/m^2	A_f	=	5.0×10^{-9} m^2
E_m	=	1.7×10^{11} N/m^2	$(A_m)_{wire}$	=	7.9×10^{-9} m^2
χ(mks)	=	2×10^4	$(A_m)_{strip}$	=	2.5×10^{-7} m^2

for a metallic glass strip of dimension 25 µm x 1 cm x 5 cm we calculate

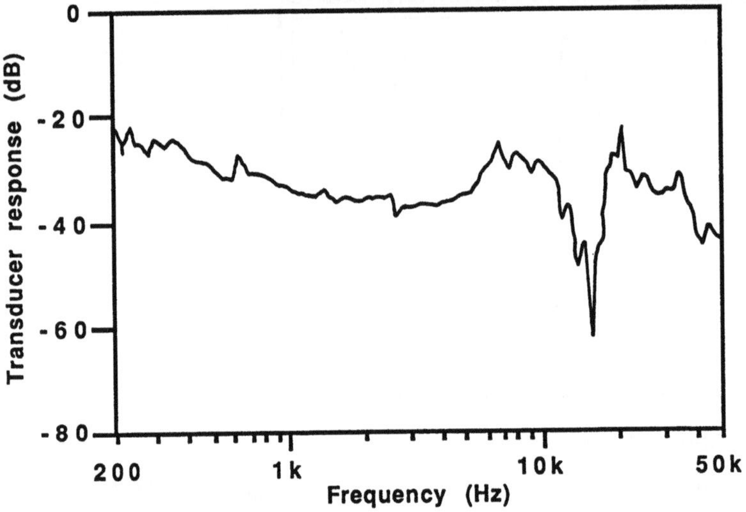

Fig. 3. A normalized spectrum of the transducer response over the frequency range 200 Hz to 50 kHz.

$(C_{wire}/C_{strip}) \approx 100$. This rough estimate is in good agreement with the experimentally observed value $(C_{wire}/C_{strip}) = 50$ for non-resonant strip transducers.

The response of the transducer in the frequency range 200 Hz to 50 kHz, normalized to the dither field amplitude, is shown in Fig.3. The response exhibits resonance peaks near 6.5 kHz and 20 kHz with a relatively flat region between 800 Hz and 5 kHz. A remarkable feature is the deep notch near 17 kHz. By choosing the dither frequency properly, the notch can be used to suppress the 2ω response if necessary.

The minimum detectable field was measured by replacing the sawtooth magnetic field with a 9×10^{-4} Oe rms calibrating sinusoidal magnetic signal at 1 Hz and increasing h to 180 mOe rms. In the absence of carrier signals the interferometer phase noise was approximately 5 μrad rms $/\sqrt{Hz}$ at frequencies above 1 kHz. However, in the presence of the 300 mrad rms residual magnetic signal at 6.5 kHz , the sideband noise 1 Hz from the carrier was equivalent to approximately 800 μrad $/\sqrt{Hz}$ due to upconverted noise [5]. Using the normalization scheme described above, the magnitude of the sideband signal of the magnetic carrier (PSD1) at 6.5 kHz was divided by the the magnitude of the sideband signal of the 300 mrad rms nonmagnetic phase carrier at 10 kHz (PSD2). A dynamic signal analyzer (HP3562A) simultaneously monitored the outputs of PSD1 and PSD2 and performed the normalization. Figure 4 shows the spectrum of the output of PSD1 before (top trace) and after (bottom trace) normalization. The 24 dB improvement in the SNR corresponds to a reduction in the sideband noise to a level equivalent to 50 μrad $/\sqrt{Hz}$ and confirmed that the low-frequency sideband noise in the system was largely nonmagnetic in origin [4]. After normalization, the minimum detectable magnetic field was $H_{min} = 1.35 \times 10^{-5}$ Oe$/\sqrt{Hz}$ at 1 Hz. The remaining noise had a 1/f frequency dependence and was limited not by the

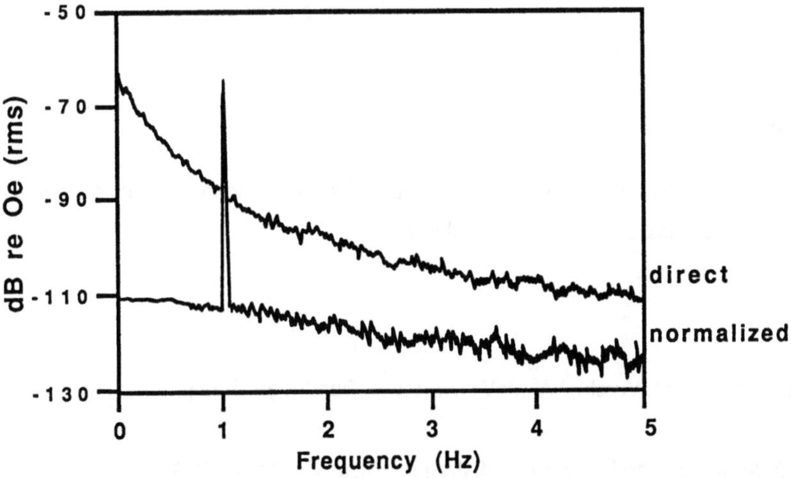

Fig.4. A spectrum showing the 9×10^{-4} Oe(rms) low frequency magnetic signal at 1 Hz and the background noise. Detection bandwidth is 19 mHz implying $H_{min} \approx 10^{-5}$ Oe/√Hz at 1 Hz.

interferometer but rather by the dynamic range of the phase sensitive detectors used to process the output signals.

Significantly better low-frequency minimum detectable field values have been reported but amorphous wire sensors provide an excellent trade-off between resolution and transducer volume. This can be seen by defining a figure-of-merit $\mathbf{M} = (V \cdot H_{min})^{-1}$ where V is the volume of the magnetostrictive element. Although cylindrical transducers having $V = 1.8 \times 10^{-7}$ m^3 have demonstrated H_{min} as low as 1.1×10^{-7} Oe /√Hz [5], **M** for the wire transducer is actually larger than **M** for the cylindrical transducer:

$$\mathbf{M}(\text{cylinder}) = 5.0 \times 10^{13} \ \text{Hz}^{1/2} / \text{Oe} \cdot \text{m}^3,$$
$$\mathbf{M}(\text{wire}) = 1.9 \times 10^{14} \ \text{Hz}^{1/2} / \text{Oe} \cdot \text{m}^3.$$

Independent of the size of the total electro-optical system, the size of the transducer itself is important in configurations such as magnetic heading sensors, remote magnetic sensors and in applications where distortion of the field to be measured must be minimized.

IV. Conclusion

In summary, we have demonstrated the use of magnetostrictive amorphous wire in a fiber optic magnetometer. The magnetostrictive constant was measured to be 5×10^{-5} Oe^{-2}, which is 10 to 50 times larger than previously reported for transducers made from Metglas ribbons having significantly greater volume. A normalization technique was developed to suppress nonmagnetic sideband noise and resulted in a minimum detectable magnetic field of 1.35×10^{-5} Oe /√Hz at 1 Hz for a 5 cm by 100 μm diameter wire transducer. Due to a geometry which is nearly identical to the optical fiber itself, transducers of this type will allow good resolution to be achieved in an extremely compact sensor.

Acknowledgements

The authors would like to acknowledge J.A. McVicker for annealing the magnetic sample and the Office of Naval Technology for support for this work .

References:

[1] F. Bucholtz, K.P. Koo, A.M. Yurek, J.A. McVicker and A. Dandridge, J. Appl. Phys. **61**, 3790 (1987).

[2] J.E. Lenz, C.D. Anderson and L.K. Stranjord, J. Appl. Phys., **36**, 1597, 1984.

[3] J.D. Livingston, Phys. Stat. Sol., **70**, 591, 1982.

[4] K.P. Koo, F. Bucholtz and A. Dandridge, Electron. Lett. **23**, 1062 (1987).

[5] D.M. Dagenais, F. Bucholtz, K.P. Koo and A. Dandridge, J. Lightwave Technol. **7**, 881 (1989)

[6] A.E. Clark and M. Wun-Fogle, paper ED-2, Digest of the 1989 Intermag Conference.

A Digital Tracking Temperature Sensor

A.K. Prewett

Dept. of Electronic Engineering, University College London, Torrington Place, London WC1E 7JE, UK

Application of fringe-counting to an all-fibre polarimetric temperature sensor is reported using two-wavelength interrogation. Such a technique, when combined with an appropriate initialising scheme, is shown to allow substantial increases in multiplexing gain in large TDM sensor systems.

1.Introduction.

The design of digital optical sensors has been addressed as a means of increasing the achievable multiplexing gain in large quasi-distributed sensor systems. By using binary detection, a digital system enables a larger proportion of its optical power budget to be allocated to the Network Loss (the loss associated with power division and losses in the network) hence allowing more sensors to be multiplexed (figs.1&2). By using digital telemetry, the system is rendered insensitive to optical crosstalk levels of up to -12 dB, and for most applications has sufficiently low lead sensitivity to not require self-referencing, hence making a further power saving.

In this report, the information from a simple polarimetric sensor[1] is accessed digitally using fringe-counting (tracking), and is implemented using two-wavelength interrogation. Such a scheme is shown to be capable of providing measurement up to a dynamic range of 40 dB, with a phase sensitivity of $\pi/2$ rads.

The optical power budgeting in a TDM sensor system based on such a technique is compared to systems based on digital (PCM), analog intensity-modulation and white-light Path-Matched Differential Interferometry (PMDI)[1]. The latter would form the initialisation phase of a tracking sensor system in absolute measurement applications.

2.Theory and Experiment:

Digital incremental sensors commonly provide two outputs, each generated from a simple scale (typically of sequence 01010101..etc), though mutually phase shifted by 90 degrees with respect to the measurand (fig.3). By using combinational logic (fig.4)

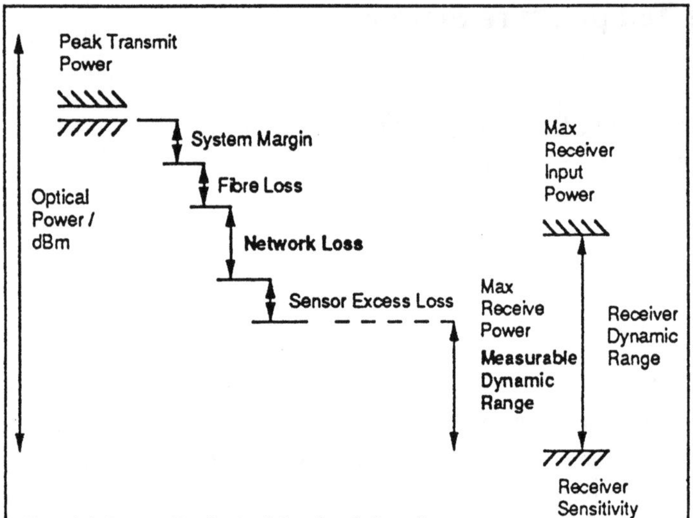

Fig.1(a) Power Budget of Analog Intensity -Modulated Sensor System.

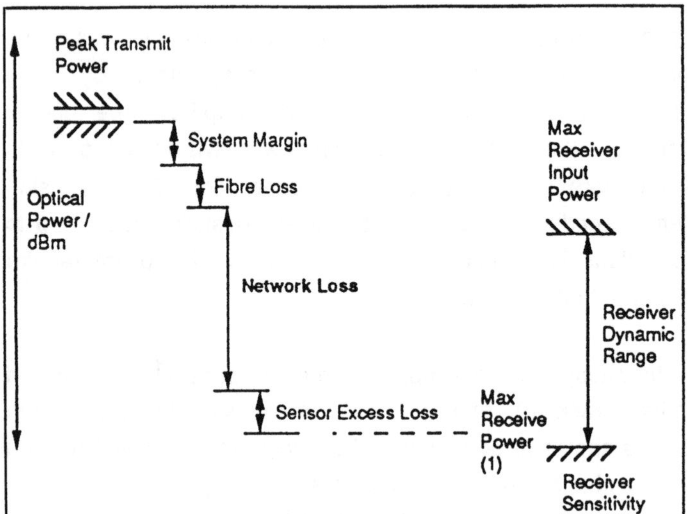

Fig.1(b) Power Budget of Digital Intensity -Modulated Sensor System.

operating on the two outputs, a dedicated counter can be controlled to keep track of scale movement. Use of this quadrature technique enables a simple scale sequence to be used which would otherwise have direction ambiguity.

Extension of this technique to interferometric sensors has been reported [2] using a double-polarisation method [3] to generate the quadrature outputs. A much simpler two-wavelength method was chosen for the temperature sensor to enable the use of a standard polarimetric sensing head. This technique utilises the dependence of the optical phase

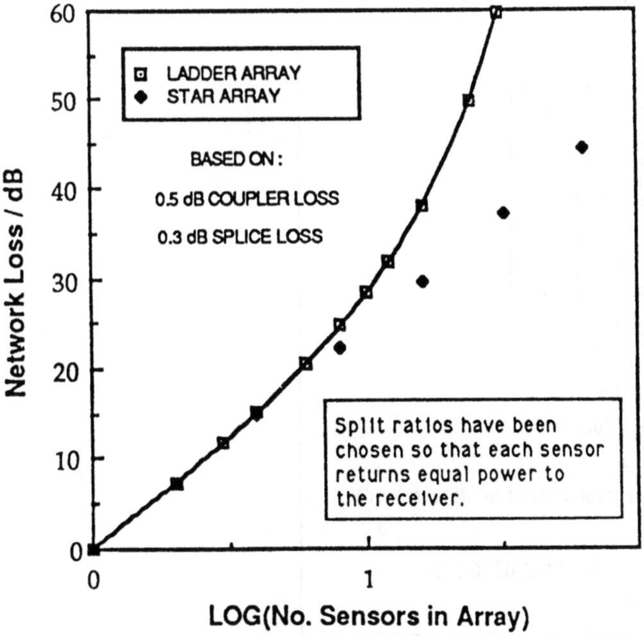

Fig.2 Network Losses of single Star and Ladder Arrays (using 2*2 Couplers)

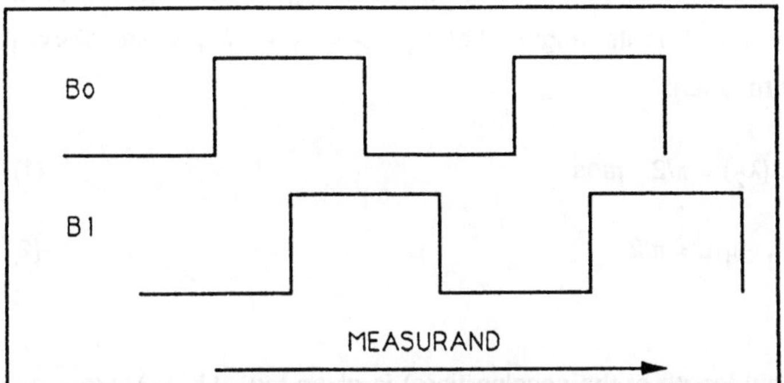

Bo	0	1	1	0	0	1	1	0
B1	0	0	1	1	0	0	1	1

Fig3.Outputs from a typical incremental encoder.

shift in a polarimeter on source wavelength. When a polarimeter is illuminated at wavelengths λ_1 and λ_2 successively, the optical phase shifts are given by:

$$\phi(\lambda_1) = \frac{2\pi}{\lambda_1}(n_s - n_f)L \quad \text{and} \quad \phi(\lambda_2) = \frac{2\pi}{\lambda_2}(n_s - n_f)L$$

Bo B1 \ Bo' B1' (Previous state)	00	01	11	10
Current state 00	NC	U	E	D
01	D	NC	U	E
11	E	D	NC	U
10	U	E	D	NC

U Increment Counter

D Decrement Counter

NC No Change

E Error (direction cannot be determined)

Fig.4 Karnaugh Diagram of Combinational Logic to control tracking.

where n_f and n_s are the refractive indices of the birefringent sensing fibre's fast and slow axes respectively, and L is its length. Letting $\lambda_2 = \lambda_1 + \Delta\lambda$, where $\Delta\lambda \ll \lambda_1$, the outputs are in quadrature when:

$$\Delta\phi = \phi(\lambda_1) - \phi(\lambda_2) = \pi/2 \text{ rads.} \qquad -(1)$$

ie. $$\Delta\phi = \frac{2\pi\Delta\lambda}{(\lambda_1)^2}(n_s - n_f)L = \pi/2 \qquad -(2)$$

Noting that Lb (the beat length of the sensing fibre) is given by: $Lb = \lambda_1/(n_s - n_f)$, this occurs when:

$$\frac{\Delta\lambda}{\lambda_1} = \frac{Lb}{4L} \qquad -(3)$$

For York HB800 fibre with a beat length of 1mm and a temperature phase sensitivity of 6 rads/m.K, a sensor temperature resolution of 0.25 °C (corresponding to a phase change of $\pi/2$ rads) is obtained using a sensing length of 2m. A wavelength difference of $\Delta\lambda = 0.1$ nm is required, which can be produced by a current step of 15 mA in a chirped Hitachi HLP1400 laser diode.

Note that $\Delta\phi$ varies proportionally to the optical path difference,$(n_s - n_f)L$, in the polarimeter (eqn.-(2)). Since the path difference is modulated by the measurand, care should be taken to ensure tracking can be maintained over the measurement range. From fig.3, it should be clear that tracking direction is interpreted correctly provided $0<\Delta\phi<\pi$ ie. so that Bo will always lag (lead) B1 for an increase (decrease) in the measurand. Since the wavelength change has been chosen to make $\Delta\phi = \pi/2$ at the initial optical path difference, OPD_i, tracking can be maintained from $0 < OPD < 2\,OPD_i$ ie. over a path difference range of $2\,OPD_i$. As the digital tracking system has a resolution of $\lambda/4$, the tracking dynamic range is:

$$D_T = 10\,\mathrm{Log}_{10}\left(\frac{2\,OPD_i}{\lambda/4}\right) = 10\,\mathrm{Log}_{10}\,(2\,\lambda/\Delta\lambda) \qquad \text{-(4)}$$

from eqn.(2)

For the prototype, this gives a predicted dynamic range of 42 dB.

Fig.5 shows the experimental set-up. Optically, the system comprised a temperature-controlled HLP1400 single longitudinal-mode laser diode, launching polarised 830 nm radiation into one axis of a 30 m Hi-Bi delivery fibre via an isolator

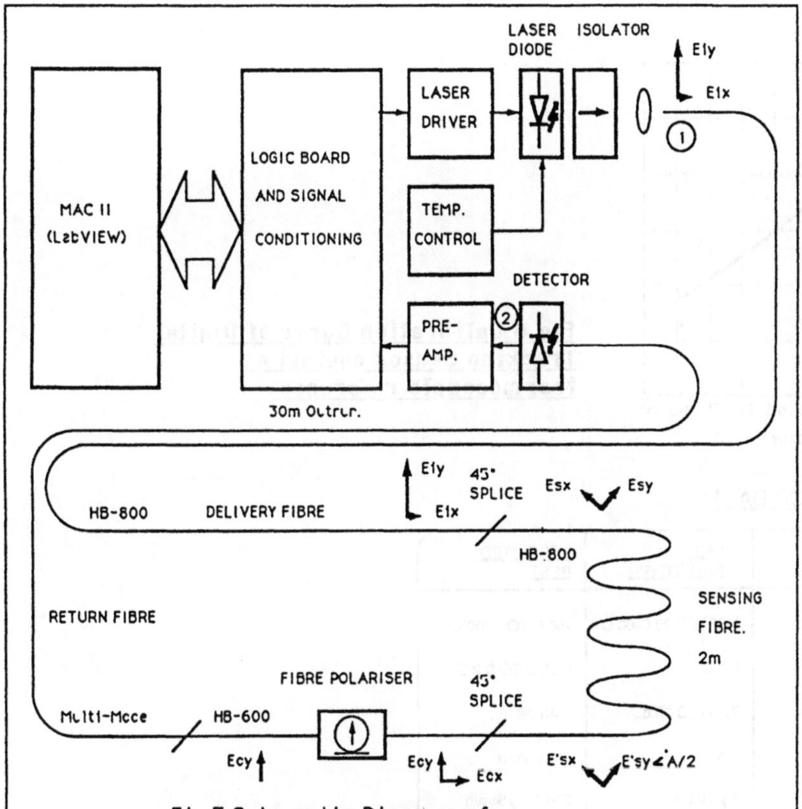

Fig.5.Schematic Diagram of Tracking Temperature Sensor

and objective. The polarimeter, formed using 45 degree fusion splices, was followed by a wound fibre polariser which was connected via a multi-mode return fibre to a silicon p.i.n. diode. Electronically, the system was controlled by a logic board, one function of which was to alternately switch the laser drive current, at 1 KHz, between 70 mA and 85 mA (adjustable levels), to provide succesive interrogation at the two wavelengths. A receiver pre-amplifier was followed by a switched attenuator (to eliminate the power modulation caused by laser current chirping) and a comparator, which performed logic level decision. The logic board performed fringe-counting and error detection (fig.4) using a Texas Instruments THCT2000 (Incremental Encoder Interface) and combinational logic. The board supplied an 8-bit fringe-count to an Apple MAC II, which data logged using LabVIEW.

The system was tested against a Type K thermocouple using a heated water bath. A temperature cycle was performed between 20 °C and 60 °C, and results indicated sensitivity of 0.2 °C (fig.6). Linearity was good but hard to quantify amidst errors introduced by the calibration set-up (chiefly temperature gradients between the point reference sensor and the line-averaged polarimetric sensor). Error free operation was achieved over an 8 hour period. Table 1 summarises the sensor performance.

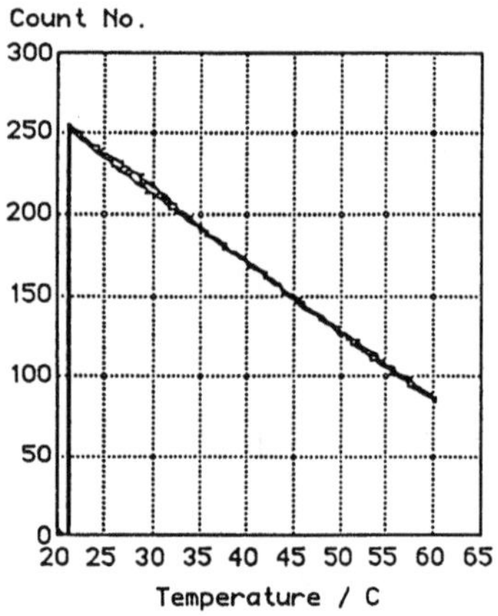

Fig.6 Calibration Curve of Digital Tracking Sensor against a thermocouple reference

TABLE 1. SENSOR DATA:

PARAMETER	LAB. PROTOTYPE	PREDICTED BEST
MEASUREMENT RANGE	TESTED 20 C-60 C	-50 C TO +400 C
TEMP RESOLUTION	0.2 C	0.05 C TO 0.2 C
DYNAMIC RANGE	TESTED 25 dB	40 dB
SENSING LENGTH	2m	1m to 20m
STRAIN SENSITIVITY	1 μStrain	2 to 0.1 μStrain
EXCESS LOSS	6 dB	1 dB

3. Implications for Multiplexing:

The performance of the tracking technique in large TDM sensor arrays was investigated by considering the limitations imposed by optical power budgeting. The tracking/initialing approach was compared to systems based on digital (PCM), analog intensity-modulation and PMDI.

To do this, the relevant terms in the optical budgets of each system were identified (Table 2); these were based on a requirement for N sensors, a measurement dynamic range of D dB and a measurement bandwidth of B Hz. Consideration of the minimum update/sampling rates and frame widths in each system enabled the receiver processing gain , achieved by signal averaging , to be identified for each case, allowing accurate determination of receiver sensitivities.

For each system, the maximum number of sensors which could be supported in a star-configured network were calculated, based on a pulse width of 20 nS (inter-sensor time-delay 40 nS), dynamic ranges of 10 to 40 dB, and bandwidths of 0.1,10, and 1000 Hz (the latter is of significance in polarimetric strain sensing) . The results are detailed in Table 3.

Results from this relatively simple analysis indicate that in all cases considered, digital tracking will allow the highest sensor capacities; typically 60-80% higher than digital (PCM), and over 100% higher than analog (whose capacities will be further restricted by optical crosstalk and self-referencing). In relation to white-light PMDI (which would form the initialisation phase in an integrated system), digital tracking allows a 10,000 fold increase in sampling rate for the same system requirement.

The improvement over digital(PCM) is achieved in the tracking system by the use of a single modulator at the sensor head; a digital (PCM) sensor is required to divide delivered optical power between ' D/3' bit modulators (and then recombine modulated beams afterwards), resulting in a high sensor excess loss. The improvement over analog intensity-modulated sensors is achieved by the use of digital detection ; the maximum received power in an analog system must be D dB greater than the receiver sensitivity, thus reducing the maximum network loss which the system can tolerate. The improvement in sampling rate over PMDI is due to the high level of signal-averaging which the PMDI approach needs to recover the budget deficit caused by low launch power.

4. Conclusions.

Quadrature-phase tracking has been successfully applied to a simple fibre polarimeter, allowing differential measurement over a dynamic range of up to 40 dB. The sensor system's use of a single modulator at the sensing head, coupled with digital telemetry has been shown, by relatively simple analysis, to alieviate some of the constraints imposed by optical power budgeting on the sensor capacities of large TDM sensor systems.

TABLE 2 : OPTICAL POWER BUDGETS OF TDM SENSOR SYSTEMS BASED ON VARIOUS SENSOR TYPES.

BUDGET TERMS:	ANALOG (INTENSITY-MOD)	DIGITAL (PCM/ENCODED)	DIGITAL (TRACKED)	WHITE-LIGHT PMDI.
LAUNCH POWER (dBm) (peak pulse)	0	0	0	-15 Δ1
FIBRE LOSS (dB)	1	1	1	1
NETWORK LOSS (dB) (see Δ 0)	25 LOG(N)	25 LOG(N)	25 LOG(N)	25 LOG(N)
SENSOR EXCESS LOSS (dB) (on-state/min. loss)	2	2 + 20 LOG(Nbits) Δ2	2	2
SENSOR DYNAMIC RANGE(dB)	D	Not Applicable(DIGITAL DET)	Not Applicable(DIGITAL DET)	10 Δ3
SYSTEM MARGIN (dB)	3	3	3	3
RECEIVER SENSITIVITY(dBm) (at 50 MBits/s)	-40	-40	-40	-40
LESS: PROCESSING GAIN (dB) (see Δ5)	$10\ LOG \sqrt{\frac{1}{4BWN}}$	$10\ LOG \sqrt{\frac{1}{4BWNNbits}}$	$10\ LOG \sqrt{\frac{1}{4\pi BWNS}}$ Δ4	$10\ LOG \sqrt{\frac{1}{4BWNS}}$ Δ4
MIN UPDATE TIME:	$\frac{1}{2B}$ Δ6	$\frac{1}{2B}$ Δ6	$\frac{1}{\pi BS}$ Δ7	$\frac{1}{2B}$ Δ6
FRAME WIDTH:	2WN	2WNNbits	4WN	2WNS Δ8

NOTES:

Δ0 For a Star Configured Network using 0.5 dB insertion loss 2*2 couplers and 0.3 dB splice losses. (see fig.2)

Δ1 For 60 nm line-width LED in Single-Mode Fibre.

Δ2 Sensor head divides power between 'Nbits' bit modulators where Nbits = D/3.01

Δ3 Allows discrimination of centre of visibility function

Δ4 S is the number of quantising levels given by: S = 10^(D/10)

Δ5 Due to the averaging of M frame returns where: $M = \frac{\text{Min Update Time}}{\text{Frame Width.}}$ (for M int. >1)

Δ6 Nyquist Rate

Δ7 To intercept each state change and hence maintain tracking. This assumes full range sweep at B Hz.

Δ8 A frame comprises an interrogation at each of S optical path difference settings of the matching polarimeter.

TABLE 3. ESTIMATED SENSOR CAPACITIES OF TDM SENSOR SYSTEMS BASED ON VARIOUS SENSOR TYPES.

(A) ANALOG (INTENSITY-MOD.)

Bandwidth \ Dynamic Range	10 dB	20 dB	30 dB	40 dB
0.1 Hz	141	65	30	14
10 Hz	65	30	14	6
1 kHz	30	14	6	3

(B) DIGITAL (PCM/ENCODED)

Bandwidth \ Dynamic Range	10 dB	20 dB	30 dB	40 dB
0.1 Hz	96	60	44	33
10 Hz	44	27	20	15
1 kHz	20	12	9	7

(C) DIGITAL (TRACKED)

Bandwidth \ Dynamic Range	10 dB	20 dB	30 dB	40 dB
0.1 Hz	171	116	79	54
10 Hz	79	54	36	25
1 kHz	36	25	3*	-

(D) WHITE-LIGHT PMDI

Bandwidth \ Dynamic Range	10 dB	20 dB	30 dB	40 dB
0.1 Hz	30	20	14	9
10 Hz	14	9	6	4
1 kHz	6	4	3	1*

* Denotes Frame-Width limited rather than power limited

5. Acknowledgements.

The author would like to thank Dr. G Parry for his assistance, and SERC & BICC(SDC) for funding this project.

6. References.

1. Jackson D.A., Monomode Optical Fibre Interferometers for precision measurement. J.Phys E,Sci.Instruments,18,1985,P.981

2. Furstenau N., Double-polarisation interferometer for digital force sensing by fibre tension bending. Optics Letters,13,no.3,1988,P.242

3. Sheem S.K. et al, Optical techniques to solve the signal fading problem in fibre interferometers.Applied Optics,21,no.4,1982,P.689

High Performance Polarization Maintaining Fiber Coils for Fiber Optic Gyroscopes

E. Sasaoka, Y. Kubo, O. Kawado, H. Suganuma, M. Takagi, and S. Tanaka

Sumitomo Electric Industries, Ltd., 1, Taya-cho, Sakae-ku,
Yokohama 244, Japan

Compact polarization maintaining fiber coils (PMF coils) were fabricated with conventional diameter (125μm) PMF and small diameter (75μm) PMF. These coils showed low crosstalk (less than -25dB at 500m) over a wide temperature range (-50°C to 90°C), and are therefore suitable for high performance fiber optic gyroscopes (FOGs).

I. Introduction

Polarization maintaining fiber coils (PMF coils) are essential components of fiber optic gyroscopes (FOGs). For interferometric FOGs (I-FOGs), PMF coils with a long length of PMFs are needed to achieve high sensitivity [1], PMF coils should have low crosstalk to get high stability [2] and compact PMF coils are desirable. To satisfy these requirements on PMF coils, high performance PMF and coiling technique are indispensable. Already the first trial to this aim was reported [3]. However, in this study, crosstalk degradation was observed at low temperature.

In this paper, we describe the PMF coils which satisfy the above requirements, including the temperature performance, and demonstrate the high stability of optical properties.

II. Fiber for PMF coil

One of the serious problems of PMF coils is crosstalk degradation induced by winding PMF onto small diameter bobbins. It is known that crosstalk stability against winding is improved by increasing the refractive index difference between core and cladding (Δn) [4]. Hence, Δn was increased considering the mode field diameter reduction and core dopant (GeO_2) increment, which will cause connection difficulties and increased loss respectively.

Coating of PMFs has a significant effect on optical properties of PMFs, especially on crosstalk at low temperature. Dual coating of UV resin was employed and coating materials are selected according to experimental results to get low crosstalk even at low temperature.

Fiber diameter is an important parameter when compact PMF coils are required. If the coating diameter of PMF is 125μm, the volume which is shared with PMF is one fourth of that of 250 μm diameter PMF. The fiber breakage probability, which depends on the amount of static fatigue on fiber surface, becomes about 10^{-5} times smaller than the 125 μm cladding diameter PMF when the cladding diameter is reduced to 75 μm. Hence, small diameter PMFs have advantages both in size and failure probability.

Considering the points described above, PMFs with the conventional cladding diameter of 125μm (Fiber A) and a small cladding diameter of 75 μm (Fiber B) were fabricated. Structural parameters of the PMFs are shown in Table 1. The PMFs are PANDA type PMFs and designed to operate at wavelengths around 0.85μm, where low cost laser diodes are easily available. Optical properties of the PMFs are summarized in Table 2. Crosstalk of both PMFs 1km long is extrapolated to be less than -25dB. Hence, the PMFs are expected to be suitable for high sensitivity and high stability FOGs.

Table.1. Structural parameters of PMFs

	Fiber A	Fiber B
Core diameter	4.4 μm	2.9 μm
Cladding diameter	125 μm	73 μm
Distance between SAP*s	18.6 μm	4.8 μm
SAP* diameter	34.1 μm	21.0 μm
Coating diameter	250 μm	130 μm

* SAP : Stress applying part

Table.2. Characteristics of PMFs

	Fiber A	Fiber B
Cutoff wavelength	0.83 μm	0.66 μm
Birefringence*	4.7x10-4	7.8x10-4
Crosstalk*	-39.0 dB (100 m)	-35.6 dB (95 m)
Mode coupling coefficient	1.26x10-6	2.90x10-6

* Measured at 0.85 μm

III. Coiling techniques

A photograph of the PMF coils is shown in Fig.1. Even when compact FOGs are required, the bobbin diameter of PMF coils in not necessarily small, because the important parameter is the total volume of the FOGs and we assumed that optical and electronics parts are mounted in the inner cavity of the bobbin. The bobbins are made from bakelite and ceramic.

Low tension and precise fiber winding are important to get high performance PMF coils [3], because high tension and rough fiber winding cause crosstalk degradation and bending loss. In winding the fiber into the bobbin, fiber tension was carefully controlled at a low level.

Fig.1 Photograph of the PMF coil (Coil A)

IV. Coil performance

PMF coils were fabricated with Fiber A and Fiber B. Optical properties were measured at a wavelength of 0.85μm and the results are shown in Table 3. There is little degradation of crosstalk after coiling. Low crosstalk PMF coils were obtained as expected. The length of PMFs is sufficiently long to achieve high sensitivity FOGs. Fig.2 shows the wavelength dependence of transmission loss of Fiber A, before and after coiling. No coiling induced loss is observed. FOGs are required to have high stability against large temperature changes because they are expected to be used in airplanes and automobiles. Hence, crosstalk stability of PMF coils against temperature change was examined. Fig.3 shows the crosstalk dependence on temperature of PMF coils. Crosstalk degradation is only around 3dB even at -50°C. Stable and low crosstalk

Table.3. Charateristics of PMF coils

	Coil A	Coil B
Fiber length	740 m	819 m
Bobbin material	Bakelite	Ceramic
Outer diameter of coil	66.9 mm	51.9 mm
Crosstalk	-26.3 dB	-23.9 dB
Mode coupling coefficient	3.17x10-6	4.97x10-6

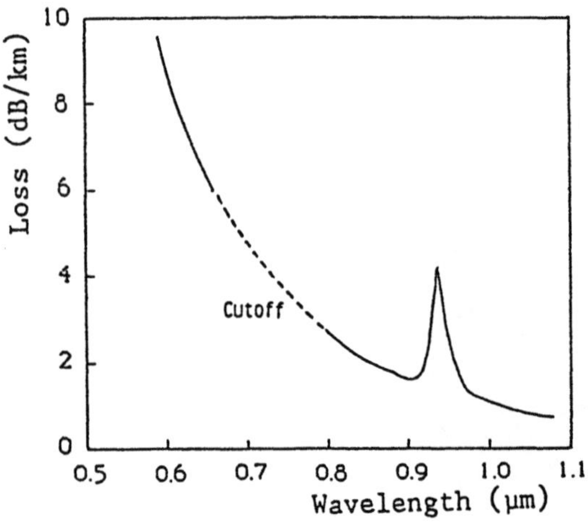

Fig.2 Wavelength dependence of transmission loss (Fiber A and Coil A)

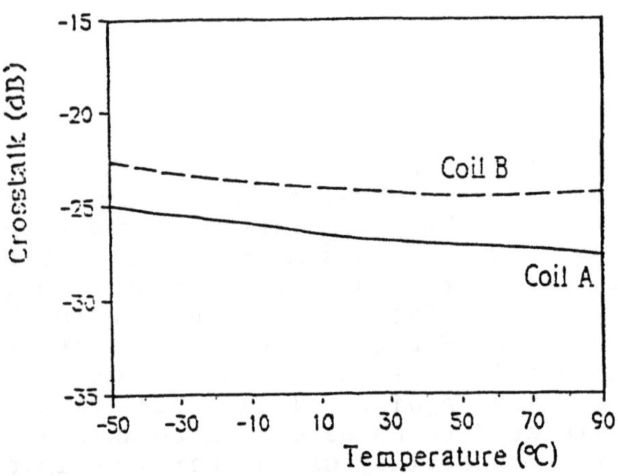

Fig.3 Crosstalk dependence on temperature of PMF coils

and confirmed over the wide temperature range from -50 ℃ to +90 °C.

V. Conclusion

High performance PMF coils have been fabricated. Fiber lengths were 740m and 819m, which are long enough for the high sensitivity FOGs. There coils showed low crosstalk over the wide range of temperature of -50 C to +90 C. coupled with the dimensional compactness, these PMF coils are suitable for high performance FOGs.

Acknowledgement

The present research effort is part of a national research and development project (large scale project) on advanced robot technology, conducted under a program set up by the Agency of Industrial Science and Technology of the Ministry of International Trade and Industry.

References

[1] T. G. Giallorenzi et al., IEEE J.Q.E, QE-18-4, pp.626-665, Apr. 1982.

[2] K. Okamoto et al., Electron. Lett., 20-10, pp.428-430, May 1984.

[3] D. Courtney et al., OFS'88, WBB4, pp.25-18, Jan. 1988.

[4] E. Sasaoka et al., OFS'88, WBB3, pp.18-24, Jan. 1988.

International Standardization Activities for Optical Fiber Sensors

L.B. Jeunhomme

Photonetics SA, 52, Avenue de l'Europe, BP 39, F-78160 Marly-le-Roi, France

Providing recognized international standards is a key element for any technology-driven industry, in order to give users confidence in the proposed products.
A working group in charge of preparing standards for optical fiber sensors has been recently created by the International Electrotechnical Commission. It will hold its first meeting in September 1989, in Paris.

INTRODUCTION

Highly innovative technologies and products developed by small companies, like the fiber optic sensors industry, often experience difficulties in finding access to the marketplace. Among others, one reason is the combination of "Fear, Uncertainty, and Doubt" of the potential users concerning the operational reliability of both the technology and the manufacturer. Moreover it turns out that in many industries, especially the process industry, products are acceptable only on a multivendor basis.
The ability of the fiber optic sensors community to create recognized and widely accepted standards concerning its technology and its products is thus a key element for transforming this technology-driven activity into a real industrial business.

STANDARDS

An official international standard defines and specifies as accurately as possible what a given class of products shall perform, which level of performance shall be met, how it shall be tested against its specifications, and how it shall be interfaced with its environment.

Products said to conform to a particular standard are thus normally interchangeable and they meet the requirements and specifications of their reference standard.

Springer Proceedings in Physics, Vol. 44 **Optical Fiber Sensors**
Editors: H.J. Arditty · J.P. Dakin · R.Th. Kersten

The first document established to form the basis of a standard is the Generic Specification : it describes and defines precisely the type of products to which the standard shall apply, establishes the particular terminology, and indicates the general characteristics, methods for quality control, product classes, and general test procedures. For example, concerning the fiber optic sensors, the generic specification will have to define what a fiber optic sensor is : does this domain or type of product include only sensors in which the fiber itself is the transducer (intrinsic sensors), or also sensors in which the fiber is limited to carrying the light to and from an all optical sensor (extrinsic sensors), etc... ? The generic specification will also have to describe the general characteristics of a fiber optic sensor, how they shall be tested, and define classes of products.

The next step in the standardization activity is to establish several sectional specifications, normally one for each class defined in the generic specification. The sectional specification of a given class describes the normal conditions under which the products shall be tested, establishes the qualification and agreement procedures that a product shall follow in order to be recognized as meeting the requirements of the standard.

Finally a product specification is established for each particular product, following a common basis for an entire class. The product specification describes the model and the options, and gives all the characteristics particular to this product, with the appropriate values.

STATUS OF WORK

The international standardization organization to which fiber optics is related is the International Electrotechnical Commission, based in Geneva, Switzerland. Inside the IEC, fiber optics is covered by Technical Committee TC 86, which is organized in dedicated subcommittees and working groups (Fibres and Cables, Fiber Optics Interconnecting and Passive Components, Terminology, Safety, Fiber Optic Sensors).

At its meeting held in Paris in February 1989, TC 86 of the IEC decided to set up Working Group 5 on Fiber Optic sensors. The task of WG5 is to develop specifications and test procedures for fiber optic sensors. France, Italy, Japan, Sweden, United Kingdom, and U.S.A. expressed interest in appointing one or two experts to serve on this Group. The secretary of this Group has been appointed by the French National Committee. Presently, Belgium, France, Italy and Japan have effectively appointed experts to this Group. The first meeting will be held in Paris on September 21 and 22, 1989 just after OFS'6, with the purpose of starting effectively the work toward the establishment of a generic specification for fiber optic sensors.

The goal is to attempt to have the generic specification adopted at the next TC 86 meeting in October 1990.

HOW TO CONTRIBUTE

Most of the work is based on the submission to the Group of documents proposed by national experts. The normal way of contributing is thus to be appointed as an expert to this Group by the National Committee of its own country, or to contribute to a similar standardization work inside its own National Committee. The names and addresses of the respective chairmen of the various National Committees active in TC 86 can be obtained from the IEC [1] or from the chairman of TC 86 [2].

REFERENCES

[1] 3 rue de Varenbé, case postale 131, 1211 GENEVE 20, SWITZERLAND

[2] Monsieur R. BOUILLIE, Société FILERGIE, 6 Bd du Général Leclerc, 92115 CLICHY CEDEX, FRANCE

Part X

Smart Structures and Strain Monitoring

Fiber Sensors for Smart Structures

E. Udd

McDonnell Douglas Electronic Systems Company,
Huntington Beach, CA92647, USA

Invited Paper: Advanced vehicles and structures are making increased use of new materials that have the ability to withstand increasingly severe environments over prolonged periods of time. Fiber sensors offer a means to enhance this process by (1) monitoring the manufacturing process, (2) augmenting nondestructive evaluation techniques, (3) enabling vehicle health monitoring and damage assessment systems and (4) supporting control systems.

Introduction

Many advanced applications require materials to perform to stringent specifications under extremely hostile environmental conditions. Successful performance under these conditions often entails a detailed knowledge of the materials used. Fiber sensors are uniquely suited to the task of extending the knowledge base on materials and enabling in situ monitoring capability of the state of the material. The functions supported by fiber sensors fall into four functional categories. The first use of fibers for smart structures is to monitor the manufacturing process itself. This can be done by embedding or attaching fibers to the material to be processed and using fiber sensors to monitor key parameters such as temperature, pressure, strain, viscosity and water vapor content. Once the fibers are in place and the parts manufactured the same fiber sensors may be used to augment nondestructive evaluation techniques. These techniques can be used to assess the integrity of parts prior to installation and to assess damage due to such factors as abusive handling. After the individual parts are checked the entire vehicle may be assembled and the fiber sensors integrated into a vehicle health management or damage assessment system. Finally the integrated fiber sensors may be used as part of control systems for active damping of structures, engine performance monitoring, adaptive flight control and other applications. This paper provides an overview of the usage of fiber sensors in the rapidly emerging area of smart structures[1, 2].

Fiber Sensors for Manufacturing Composite Materials

Fiber sensors have a series of unique advantages that make them extremely amenable to monitoring composite materials during the manufacturing process. These advantages include small size, capacity to withstand high operating temperatures and pressures, immunity to electromagnetic interference, and intrinsically passive all dielectric configurations. Small size is necessary to allow consolidation of the fibers into the material without degrading structural integrity. High operating temperatures and pressures are characteristic of the manufacturing process. The feature of immunity from electromagnetic interference reduces the need for shielding and greatly lowers cost. The requirement for passive all dielectric elements results from the need to eliminate conductive paths through the material which could result in catastrophic failure due to lightning strikes and other high discharge phenomena.

Springer Proceedings in Physics, Vol. 44 Optical Fiber Sensors
Editors: H.J. Arditty · J.P. Dakin · R.Th. Kersten

While some fiber sensors are used only during the manufacturing process most are potentially useful throughout the life of the material. Particularly interesting examples of simple sensors are a sensor that is made of resin that disappears during the manufacturing process demonstrated by Afromowitz[3] and the optrode which has been used by Levi[4] to monitor water vapor and viscosity of resins in composite materials during the manufacture. Virtually any type of intrinsic fiber sensor may be used to support manufacturing.

Nondestructive Evaluation

Once the fibers have been installed into parts they may be used to augment nondestructive evaluation techniques prior to installation and integration. There are several methods that have been employed to accomplish this. Measures[5] has developed techniques to weaken fibers that are embedded in Kevlar. Upon impact the fiber breaks resulting in greatly increased attenuation. Measures[6] has also embedded fiber sensors in an array to measure strain field changes induced by impact. Smith[7] has used fiber acoustic sensors embedded in coupons to assess impact damage.

Vehicle Health Monitoring

The parts can then be integrated and the fiber sensors multiplexed as the vehicle is assembled into a health management system. Examples of these systems could include systems that routinely schedule structural maintenance for aircraft, systems that warn of excess stress on aircraft due to factors such as icing, launch control systems and leak monitoring. The types of sensors that are required for these applications are strain sensors, temperature sensors and pressure sensors. There is also an acute need to multiplex these fiber sensors effectively. Promising options are distributed fiber sensors and methods that can be used to multiplex in line discrete sensors.

Control Systems

Future advanced aircraft and spacecraft may be designed to fly close to the edge of the vehicle performance envelope. Fiber optic smart structures can be used to continually redefine the flight envelope which may be degraded due to structural aging, damage or changing external environmental parameters. Examples of situations where these systems could be used involve space structures and advanced launch vehicles.

Fiber Optic Smart Structure Technology

The technology needed to support the implementation of fiber optic smart structures is shown in Figure 1. The first elements of the process involve embedding or attaching the fibers to the material to be monitored. The materials of interest can range from relatively low temperature carbon epoxy to high temperature metals, ceramics and carbon carbon. Most work on embedding fiber sensors into materials has been performed using carbon epoxy or thermoplastics. It has been shown that key considerations for successfully integrating fibers into these materials are (1) the orientation of the fibers relative to the strength members of the material (they should be parallel) and (2) compatibility of the fiber coating material with the structure into which it is to be embedded. For organic materials the epoxy resin of the composite can be used as the fiber coating[8]. The issues for embedding fibers into metals are similar in that the metallurgy of the coatings and materials must be compatible, in analogy to chemical compatibility for the case of organic materials.

Once the fiber coating and material compatibility issues have been resolved the next major technology area involves establishing fiber optic leads. There are two distinct approaches. In one case the fibers are embedded directly into the material and means must be devised to make connections to other structures and or the outside world. This can be handled by

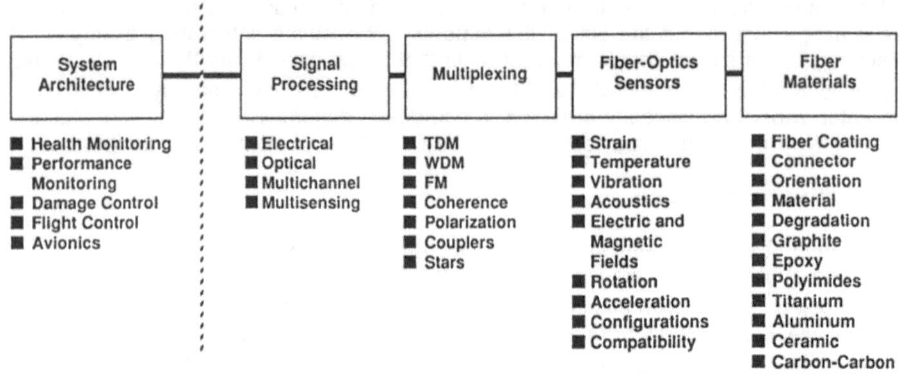

Figure 1. Fiber Optic Smart Structure Technology

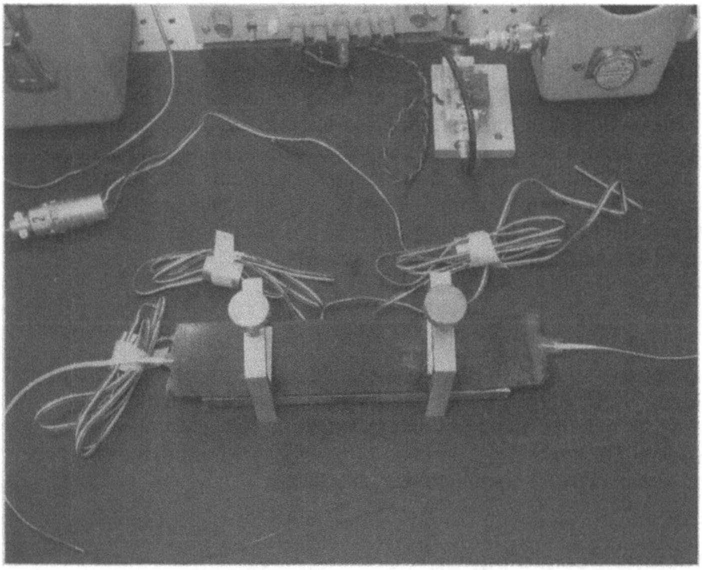

Figure 2. Carbon Epoxy Coupon with Embedded Fiber Optic Sensors

forming notches in the material in combination with strain relief or by embedding connectors directly into the structure[9] . shows a carbon epoxy coupon with twelve fibers embedded in it to support strain measurement tests. RTV has been used at the edges of the coupon to provide strain relief. An alternative approach is to forgo embedding fibers into the material and simply bond the fiber sensors to the surface of the material. This approach greatly simplifies the interconnection problem. Maintenance and repair procedures are also improved since patches can be made by dealing solely with the surface of the structure. Figure 3 illustrates an aluminum coupon that has fiber optic strain sensors epoxied to its surface to monitor strain around a simulated rivet hole.

The next step in implementing a fiber optic smart structure system is to make the appropriate choice of fiber sensors. If the sensors are to be embedded directly into the material it is highly desirable that sensors not exceed the diameter of the fiber itself. This situation arises because of the possibilities of structural degradation and sensor failure during processing and service due to excess stress induced by discontinuities. Similar

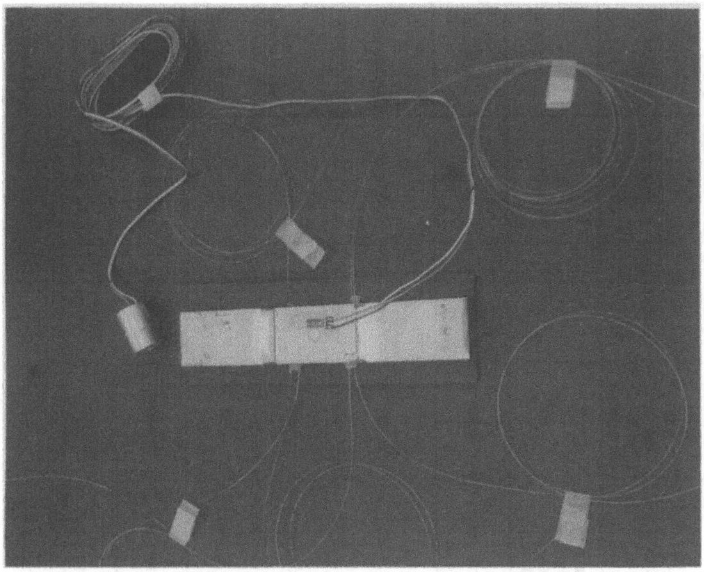

Figure 3. Aluminum Coupon with Simulated Rivet Hole, Fiber Sensors are Bonded to Surface to Measure Strain

considerations apply, although perhaps not as strictly to the case of fiber sensors bonded onto the surface of the structure. Fortunately there are a wide variety of appropriate sensor choices available as is illustrated by Figure 4. The choice of sensor is highly dependent on the multiplexing approach. Figure 5 is a table of tradeoffs for multiplexing interferometric fiber optic sensors using the most common multiplexing techniques. The important point is that the overall performance of the system is highly dependent upon the interaction between fiber sensor performance and the multiplexing technique to be implemented.

The last step to complete the system after the fibers have been embedded or attached and suitable multiplexing methods established is to develop the optical and electronic signal processing necessary to transfer the resulting information to the control system. Dependent on the complexity of the system this step may range from a trivial exercise to making breakthroughs in the state of the art. The use of fiber optic elements developed for sensors as signal processing elements may play a key role here.

Fiber Optic Smart Structure Applications

There are many examples of the applications of fiber optic smart structure technology to aerospace, bridges and buildings. Figures 5, 7 and 8 show three application areas. In the case of Figure 6 fiber sensors are embedded or attached to struts that are used to form a structure. These sensors may be used to monitor the history of stress and strain on a structure as an aid to maintenance scheduling. Another application would be to use the fiber sensors in combination with embedded actuators that would form active strut elements capable of damping out vibrations due to maneuvering or docking[10, 11]. Figure 7 shows a space based habitat that could be subject to a series of damage and or partial failure scenarios. Arrays of fiber sensors could be used as monitors for many potential problem areas. Figure 8 illustrates how fiber sensors could be wound into a liquid hydrogen tank formed with composite material. In this application the tank is wound using composite filament and fiber could be wound in during the manufacturing process. The fiber could be used in combination with sensing techniques to act as a cure monitor for the tank. The same

Characteristic / FO Sensor	Scale Factor Linearity	Multisensing Potential	Sensor Types	Temp Range	Gauge Length	Strain and Temperature Sensitivity	Dynamic Range	Multiplex Potential	Ability To Be Embedded	Risk
SAGNAC Interferometer	High	High	Strain Temperature	High >1000°C	>10m to <1 cm	<0.5 μm Demo. <10A Possible	$>10^8$	High	High	Low
Mach-Zehnder Interferometer	Moderate	Moderate	Strain Temperature	High >1000°C	>10m to <1 cm	<0.1A	$>10^6$	Very High	High	Low
Blackbody Sensor	Low-Moderate	Low	Temperature	Very High >2000°C	>10m to ~1 cm	~3°C Demo <1°C Possible	50°C-1200°C Silicon 2400°C Sapphire	Low	High	Low
Evanescent Sensor	Low	High	Strain Temperature	High >1000°C	~5 cm to <1 cm	4×10^{-4} Gauge Length Demo	$\sim 10^4$	Low	Moderate	Moderate
Polarization Sensor	Low	High	Strain Temperature	Med >700°C	>10m to <1 cm	<5 μm	$\sim 10^6$	Low	High	Moderate
Optrode	Low	Moderate	Cure Monitoring For Soluble Water Strain Temperature	Low-Med	<1 mm	<1°C	10^4	Low	Med	Moderate High
Michelson Interferometer	Moderate	Moderate	Strain Temperature	High >1000°C	>10m <1 cm Possible	<10A	$>10^6$	High	High	Low-Moderate
Dual Mode Sensor	Low	Moderate	Strain Temperature	High >1000°C	>10 cm to >1 cm	<5 μm	10^3	Low	High	Moderate
OTDR	Low	Moderate	Strain Temperature	High >1000°C	>10 m to <1 cm	<10 μm	10^3-10^4	Moderate	Moderate	Moderate
Etalon	Low	Moderate	Strain Temperature	High >1000°C	<1 cm	<10 μm	$\sim 10^4$	Moderate High	Moderate	Moderate

Figure 4. Fiber Optic Sensor Candidates for Strain and Temperature Monitoring

	Ease of Extracting Muxed Signals	Component Complexity	Phase Noise	Reduction In No. of Leads	Power Efficiency	Channel/ Channel Separation
WDM	Excellent	Poor	Excellent	Poor	Excellent	Fair/Good
TDM	Fair	Good	Fair	Excellent	Fair/Good	Good
Coherence	Good	Good	Poor	Good	Poor	Good/ Excellent
Polarization	Excellent	Poor	Excellent	Poorest	Good	Poor/Fair
Frequency	Good	Good	Good	Good	Fair	Good

Figure 5. Fiber Optic Sensor Multiplexing Tradeoffs

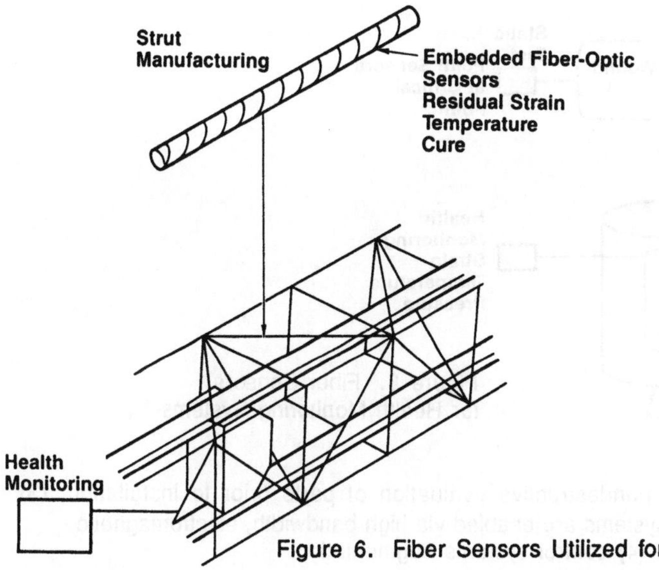

Figure 6. Fiber Sensors Utilized for Structural Damping Systems

fiber sensors could then be used to augment nondestructive evaluation of the tank prior to installation and finally integrated into a health monitoring system.

Conclusions

Fiber sensors in smart structures offer the prospect of making these advanced structures a reality. Significant improvements enabled by this technology are (1) reductions in manufacturing cost by improving yield through cure monitoring via embedded

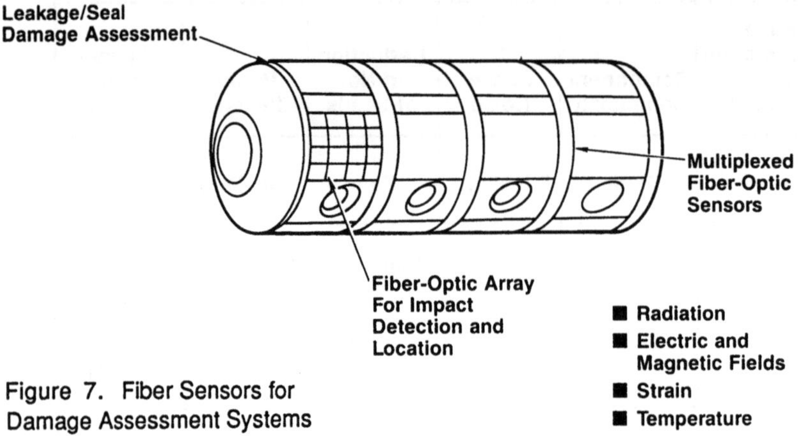

Figure 7. Fiber Sensors for Damage Assessment Systems

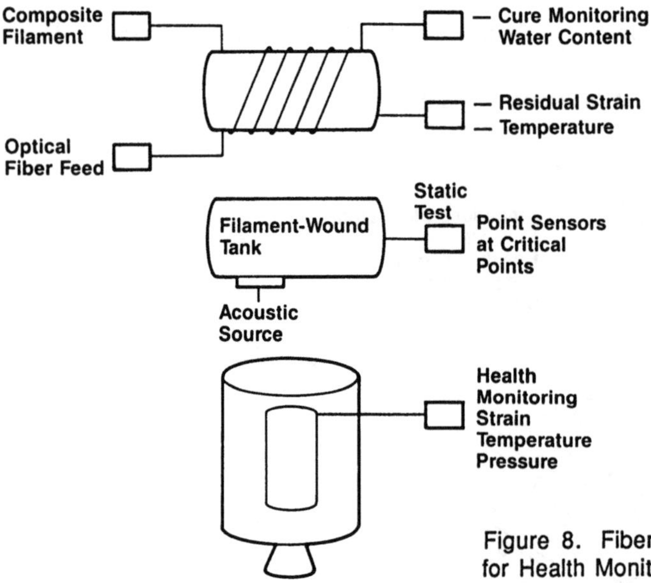

Figure 8. Fiber Sensors for Health Monitoring Systems

fiber sensors, (2) improved nondestructive evaluation of parts prior to installation, (3) vehicle health management systems are enabled via high bandwidth, electromagnetic immunity and small size and (4) control systems augmentation.

References

1. E. Udd, "Embedded Sensors make Structures "Smart", Laser Focus, p. 135, 1988.
2. E. Udd, Editor, "Fiber Optic Smart Structures and Skins", Proceeding of SPIE, Volume 986, 1988.
3. M. A. Afromowitz and K. Y. Lam, "Fiber Optic Cure Sensor for Thermoset Composites", Proceedings of SPIE, 986, p. 135, 1988.
4. R. L. Levy and S. D. Schwab, "Advances in the Development of the Fiber Optrode Cure Sensor", Polymer and Material Science and Engineering, 59, p. 591, 1988.

5. R. M. Measures, et. al., "Structurally Integrated Fiber Optic Damage Assessment System for Composite Materials", Proceedings of SPIE, 986, p. 120, 1988.
6. R. M. Measures, D. Hogg, R. d. Turner, T. Vallis and M. J. Gilberto, Proceedings of SPIE, 986, p. 32, 1988.
7. H. Smith, "Smart Structures Concept Study", Proceedings of SPIE, 1170, to be published.
8. E. Urruti, R. M. Hawk, and P. E. Blaszyk, 'Optical Fibers for Structural Sensing Applications", Proceedings of SPIE, 986, p. 158, 1988.
9. W. B. Spillman, Jr., "Fiber Optic Sensors for Composite Monitoring", Proceedings of SPIE, 986, p. 6, 1988.
10. R. S. Rogowski et. al., "Fiber Optic Strain Measurements in Filament-Wound Graphite-Epoxy Tubes Containing Embedded Fibers", Proceedings of SPIE, 986, p. 194, 1988.
11. C. A. Rodgers et. al., "Demonstration of Smart Material with Embedded Actuators and Sensors for Active Control", Proceedings of SPIE, 986, p. 90, 1988.

Static Strain Monitoring OFS Using a FM Laser Diode

R. Ohba, S. Kakuma, H. Yamane, and I. Uehira

Department of Applied Physics, Faculty of Engineering,
Hokkaido University, Sapporo 060, Japan

Abstract Absolute value of an optical path length is determined by a new interferomertic method using a frequency modulated laser diode. The principle of a static strain monitoring OFS applying the method is proposed. Results of the experimental study of the static strain OFS based on the principle are also presented.

1. Introduction

It is well known that the frequency or wavelength modulation, as well as the duration modulation, is superior to the intensity modulation for the modulation scheme of optical fibre sensors (OFSs) and is promising for realising a robust OFS for the practical use(1). There have been a number of reports on the practical fibre-optic sensors using the frequency modulation; most of them utilize an analogue spectrum analyzing means, a few uses a computer controlled digital one(2). The former is simple and practical but is limited in accuracy, whereas the latter has enough accuracy and is robust for the practical environment at the expense of the computer controlled spectrometer. The latter, however, seems impractical because of its cost. It may be extremely useful for practical OFSs that an economic digital spectrum analyzing means is developed to realise a frequency modulated OFS.

An interferometric method inherently provides digital nature and it can be applied to realize a digital spectrum analyzing means. We have devised an interferometric spectrum analysing means for OFSs(3). Conventional interferometric measuring method uses a constant wavelength light source. It detects changes of a measurand by incremental fringe-counting, hence a continuous observation is needed in order to detect the change of the path difference. It is not always very easy to maintain the performance stable or repeatable. Furthermore, it is impossible for the method to measure a static quantity without a mechanical scanning scheme. An interferometric OFS for a static or very slowly changing measurand is difficult for the conventional interferometric technique because of problems which are due to the incremental nature(4).

Recently we have proposed an interferometric digital measuring method to determine an optical path difference using a frequency modulated laser diode(3). The results of a basic experiment were positive for the method to be applied to OFSs.

Springer Proceedings in Physics, Vol. 44 **Optical Fiber Sensors**
Editors: H.J. Arditty · J.P. Dakin · R.Th. Kersten

The present paper describes both the principle and some results of advanced experiments of the method.

2. Principle

A mono mode laser diode is a monochromatic coherent source if both the driving current and the temperature of the diode chip are appropriately controlled. Suppose that a light beam of frequency ν from a laser diode is divided by amplitude into two beams of amplitude E_i, $i = 1, 2$ as shown in Fig. 1. Interference fringes of (1) can be observed when the two beams are recombined after travels of respective path of length δ_i, $i = 1, 2$.

$$I = |E_1 \exp(j2\pi\nu\delta_1/c) + E_2 \exp(j2\pi\nu\delta_2/c)|^2$$
$$= (E_1{}^2+E_2{}^2)\{1+[2E_1E_2/(E_1{}^2+E_2{}^2)]\cos[2\pi\nu(\delta_1-\delta_2)/c]\}, \quad (1)$$

where c is the light speed in the vacuum. This equation shows that fringes vary depending not only on the path difference but also on the frequency ν, and that they can be changed even for a constant path difference by sweeping the frequency.

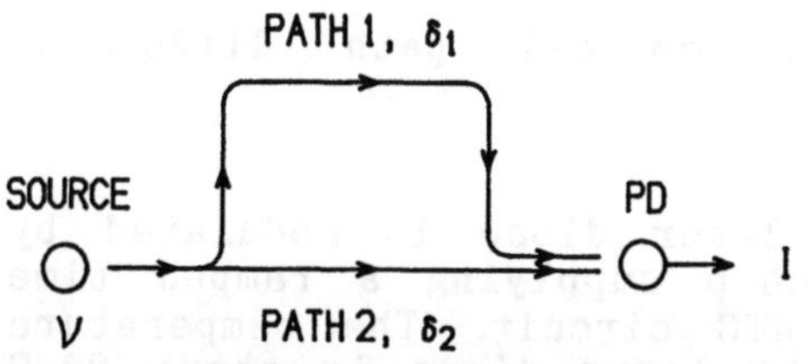

Fig. 1 Recombination of two beams from the identical source after travels of different optical paths.

Now let a light beam be introduced into an interferometer with path difference l_i, then $N_i = l_i(\nu_2-\nu_1)/c$ times of intensity change may be observed in its transmitted light when the frequency is swept from ν_1 to ν_2. If a pair of light beams from the identical source are simultaneously launched into two interferometers with path difference l_1 and l_2, respectively, either of those path differences can be determined as follows by both the other path difference and the ratio N_1/N_2,

$$l_1 = (N_1/N_2)l_2. \quad (2)$$

If l_2 is known, the other optical path difference l_1 can be determined by (2) independently on the frequency sweep range of the source, $(\nu_2-\nu_1)$. Therefore, even if the optical path difference l_1 is constant (that is, a static path difference), it is possible to determine l_1 in terms of a known path difference l_2 via a frequency modulated laser diode irrespective to the absolute value of its wavelength. The resulting measurement is given as a rational-number multiple of the known optical path difference.

3. Experiments

Two experiments were performed; one was to confirm validity of the present principle and the other to apply it to a static strain monitoring OFS.

3.1 Optical Path Difference Determination

Figure 2 shows a schematic diagram of the experimental setup. A laser diode (HITACHI, HL7801E, λ = 780 nm) stabilized both in the optical output power and the chip temperature by an auto power control- (APC) and an auto temperature control- (ATC) circuit, respectively, was used for the light source.

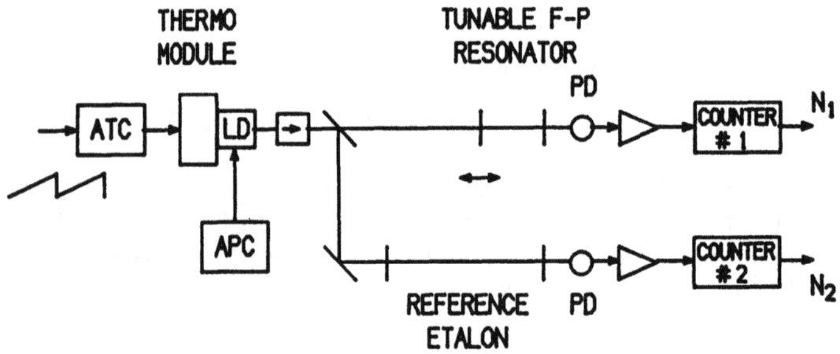

Fig. 2 Experimental setup for optical path difference determination.

The light frequency of the laser diode is modulated by varying the temperature of the chip supplying a ramped time signal as the reference to the ATC circuit. The temperature dependence of the frequency of the laser diode is about 28.9 GHz/℃ and is almost constant for the temperature range of 10-30 ℃ and for optical output power of 2.5-5 mW. The output light beam from the laser is then introduced into an optical isolator consisted of a combination of a polarizing beam splitter and a quarter wave plate and is further divided by a beam splitter. One of the divided beam is launched into a tunable Fabry-Perot resonator, transmitted light of the resonator is detected by a silicon pin photo-diode (PD). The intensity changes of the output light from the tunable Fabry-Perot resonator are counted while the temperature change range of the laser diode is fixed to 7 ℃.

The number of the intensity change of the transmitted light due to the frequency modulation of the light source is counted by the counter #1 and gives a count N_1. While the other half beam is introduced to a reference etalon, and transmitted light of which is also detected by another Si pin photo-diode. The number of its intensity change is also counted by the counter #2 and gives a count N_2. The numbers N_1 and N_2 of intensity changes were counted simultaneously for both the tunable F-P resonator and the reference etalon. Figure 3 shows a plot of the resultant ratios N_1/N_2 for various values of the path difference of the tunable Fabry-Perot resonator.

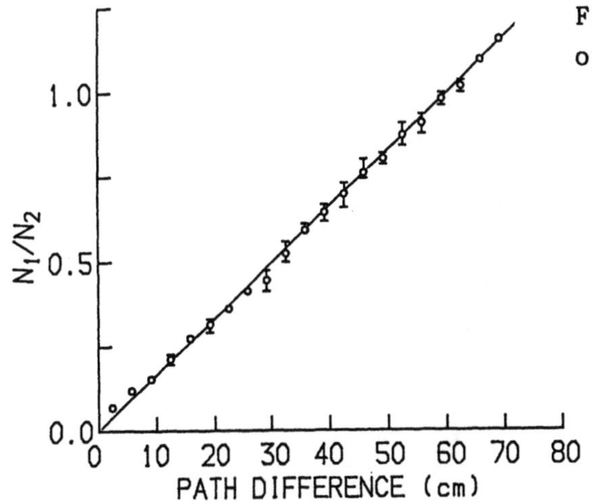

Fig. 3 Measured N_1/N_2 vs. optical path difference.

3.2 Static Strain Monitoring

The present method was applied to monitor quasi-static path length changes of the OFS strain gauge. A schematic diagram of the experimental setup for the fibre optic strain sensor is shown in Fig. 4; the same LD as of the previous experiment was used and its temperature was changed over 9.8 ℃. Frequency modulated output light beam from the laser was launched into the cascade of two Fabry-Perot resonators; one was a fused scilica etalon of 5.5 cm thick and the other a monomode fibre of 91 cm long. A part of the fibre (62.5 cm) was wound on a polyethylene cylinder and set in a heat bath so as to apply a longitudinal strain on the fibre via the thermal expansion of the cylinder by varying the temperature of the bath, T_B.

Figure 5 shows example traces of the observed intensity changes. The lower and the higher frequency component of each trace correspond to the intensity change caused by the silica etalon and the fibre Fabry-Perot resonator, respectively. Since the refractive index of the fibre core is 1.5 and the cavity length of the fibre resonator is 91 cm, the optical path difference of the resonator is equivalent to 273 cm, so that its free spectral range is 110 MHz. The intensity changes more

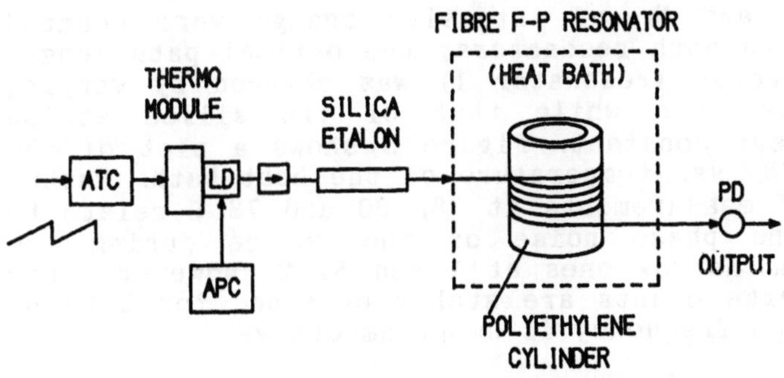

Fig. 4 Experimental setup for quasi-static strain monitoring.

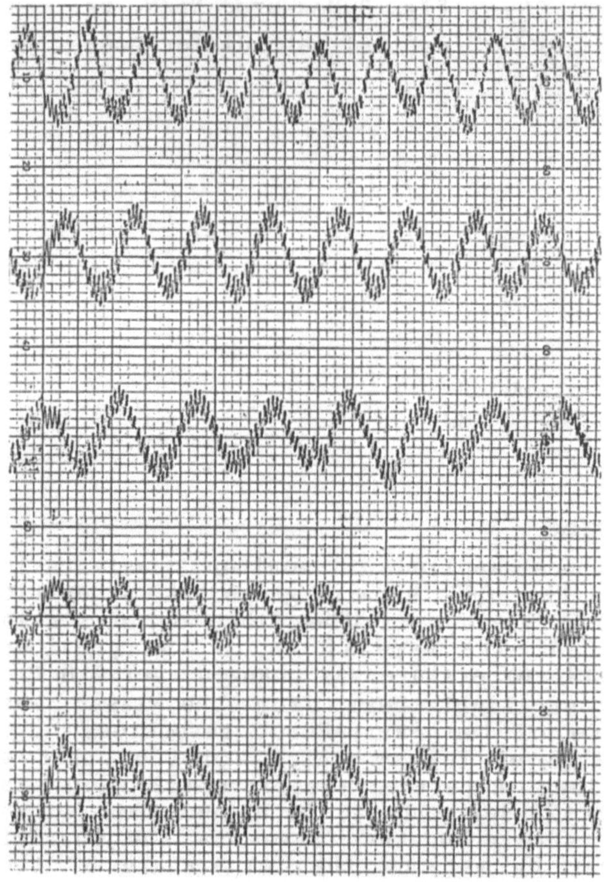

Fig. 5 Example traces of the output signal; the lower and the higher frequency components of each trace correspond to the reference silica etalon and the fibre-optic Fabry-Perot resonator. Optical path length of the latter is varied by the longitudinal strain effected by the thermal expansion of the polyethylene cylinder.

than 2,300 times during the temperature change, suggesting continuous frequency modulation over 250 GHz.

The numbers N_1 and N_2 of intensity change were counted simultaneously for the both resonators; the optical path length of the fibre Fabry-Perot (resonator 1) was changed by varying the heat bath temperature while that of the silica etalon (resonator 2) was kept constant. Figure 6 shows a plot of the resultant ratio, N_1/N_2 vs. temperature of the heat bath. Rather wide distribution of measurements at 38, 60 and 73 ℃ relate to the mode hop or the phase noise of the source during the frequency modulation. Narrow ones at 1 and 51 ℃, however, show that sufficiently large counts are stably obtained for both N_1 and N_2 when the source frequency is swept smoothly.

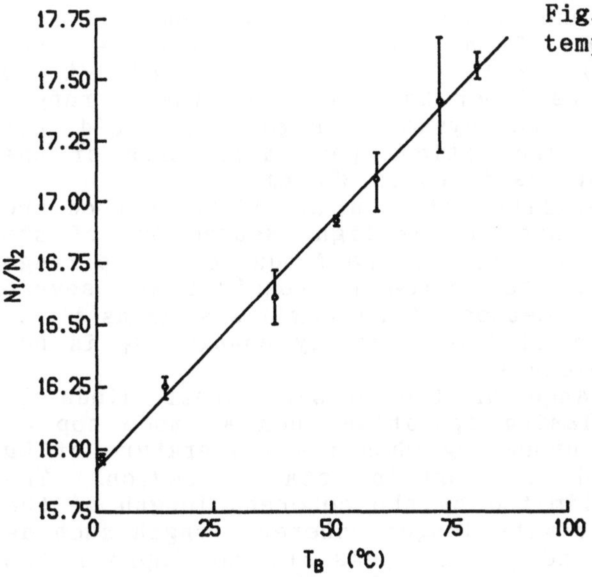

Fig. 6 Measured N_1/N_2 vs. temperature of the heat bath.

4. Discussions and Conclusions

Basically the sensitivity/resolution of the present method is determined by the inverse of the maximum count, $N_{max} = \max(N_1, N_2)$, and which is related to both the frequency sweep range of the source and the free spectral range of the maximum path difference, $l_{max} = \max(l_1, l_2)$. The dynamic range of the method is also determined by l_{max} and which is then determined by the coherent length of the source. That is, the coherent length determines both the upper bound of the path difference and the free spectral range of the resonator.

Practically, however, the upper bound of the dynamic range is related to the maximum optical path difference that gives a good SNR visible interference fringes. So the upper bound is governed not only by the coherent length of the source but also by the noise performance of the photo-detector. The lower bound, also, is determined by the condition to detect at least one count for N_1. The condition depends on the width of the spectral range on which the frequency of the source is swept stably. It is apparent that for the higher resolution and the wider dynamic range, both the longer coherent length or narrower spectral width and wider frequency sweep range are required for the source.

In the static strain detection experiment, the frequency sweep range of the source can be estimated as about 283 GHz since the temperature coefficient of its frequency shift is 28.9 GHz/℃ and the temperature change of the LD chip is about 9.8 ℃ during the frequency sweep. The lower bound 0.106 mm of the dynamic range is determined as the path length of the resonator whose free spectral range is 283 GHz. On the other hand, the optical path difference of the fibre Fabry-Perot resonator is equivalent to 273 cm because the length and refractive index of the fibre are 91 cm and 1.5, respectively, and it corresponds to 110 MHz of the free spectral range.

Therefore N_{1max}, the maximum value of N_1, is estimated nearly 2,570 for the experiment. Then this value determines the resolution as 1/2,570 = 3.9 x 10^{-4}. The resolution of 1/2,570 corresponds to 41 nm at the lower bound of the dynamic range. For the upper bound of the dynamic range, we could not determine the value since the optical path difference of the fibre Fabry-Perot resonator was fixed to 273 cm.

The repeatability/stability and accuracy of the method are solely dependent on those not of the light source but of the reference path difference l_2. Neither the frequency sweep range nor the spectral width of the source is required any severe conditions for the present method if an auxiliary means(5) to exterpolate fractional part of the intensity changes N_2 is not adopted to increase the resolution.

The maximum sweep range of the present laser diode is limited by the unstable lasing operation such as mode hop or phase noise of the laser appearing when the temperature of LD chip is changed beyond a certain stable region. The sensitivity, however, is limited by the coherent length of the laser light and if a laser with longer coherent length such as a DFB laser diode is used it is possible to improve the sensitivity by increasing the cavity length of the reference etalon applying such as a polarization maintaining optical fibre.

In practice, it is easier for the reference etalon than an optical path difference to be measured to elongate the path length. Therefore the sensitivity or resolution of the method is determined by the free spectral range of the reference etalon; the narrower the free spectral range is the higher the sensitivity is realised. So an etalon with a longer cavity length is needed in order to realize a higher resolution since the width of the free spectral range is inversely proportional to the cavity length of the etalon. The maximum value of the length, however, is constrained by the stable detection of intensity changes and relates to the visibility of the interference fringes. Then the maximum value of the cavity length can be determined by the coherent length and spectral width of the light source.

The present method is very useful and applicable to various kinds of the interferometric measurements and sensors. It can be well expected that the method is applied to multiplex several OFSs, each of which is composed of a fibre Fabry-Perot resonator, by a simple cascade connection of them. We are developing a fibre-optic multiplexed sensors system for static strain monitoring, in which a group of fibre-optic interferometric strain gauges buried in a coal tunnel are monitored intermittently from a remote station.

5. Acknowledgements

The authors are grateful to both the Suhara memorial foundation and the Ichimura foundation for their support to this work.

References

(1) **Jones, B. E.**: Optical fibre sensors and systems for industry, J. Phys. E: Sci. Instrum., 18, pp. 770-782 (1985).

(2) **Jones, B. E. and Collier S. J.**: Optical fibre sensors using wavelength modulation J. Phys. E: Sci. Instrum., 17, pp. 1240-1241 (1984).

(3) **Ohba, R., Hamano, Y. and Uehira, I.**: Digital measurement of an optical path difference by using frequency modulated laser diode, Proc. of the 5th Workshop on Optical Fibre Sensors, Soc. of Appl. Phys. of Jpn., pp. 185-190 (1987).

(4) **Dakin, J. P.**: Temperature measurement using intrinsic optical fibre sensors, J. of Optical Sensors, 1, 2, pp. 101-114 (1986).

(5) **Kobayashi, T., Tanaka, K., Ito, M. and Ohtaka, M.**: High resolution FM heterodyne measurement of distance using a semiconductor laser, Proc. of the 4th Workshop on Optical Fibre Sensors, Soc. of Appl. Phys. of Jpn., pp. 53-59 (1987) *in Japanese.*

Interferometric Displacement Sensor with Lateral Resolution Employing Fiber Tension Bending

N. Fürstenau

German Aerospace Research Establishment (DLR),
Institute for Flight Guidance, D-3300 Braunschweig, Fed. Rep. of Germany

A double - polarization interferometer with digital readout by fringe counting allows for discrimination between strain applied to the 'reference' and 'signal' arms respectively under transverse force induced tension bending. This is utilized for laterally resolved force or displacement sensing by employing an array of interferometer arms placed under an elastic skin. Experimental results of displacement measurements employing the four parallel arms of two Michelson interferometers are compared with the theoretically expected strain - displacement characteristic.

Tactile sensors for laterally resolved displacement and force sensing are required for advanced man - machine interfaces and robotic systems with sensory feedback. A number of different approaches employing fiber optics have been suggested in recent years e.g. [1]. To our knowledge up to now only setups employing multimode optical fibers have been investigated for this application. In this paper we propose a concept based on an array of single mode fibers as arms of fiber optic Michelson interferometers, arranged under an elastic skin. Displacement is transferred to the sensing fibers via transverse - force - induced fiber - tension - bending and the resulting fiber strain is measured via fringe counting [2]. For eliminating the ambiguity of the counting procedure in remote readout of the interferometers two phase - shifted orthogonally polarized interference signals are created employing the double - polarization method [3] [4]. With two - arm interferometers usually the sensing-(S-) arm is exposed to the measurand while the reference-(R-) arm is isolated for the sake of creating a measurand induced phase difference. The aim of the investigations reported here is to demonstrate the possibility of laterally resolved displacement sensing employing both interferometer arms as sensitive fibers. The experimentally obtained field of strain - displacement characteristics for a setup with four parallel fibers of two Michelson interferometers arranged under an elastic skin is compared with the characteristic of a single fiber without a skin and with the theoretically expected, to first order quadratic dependence of strain on displacement.

The displacement resolution for a bare section of fiber is determined by the strain induced change in the optical path length [5]. For wavelength $\lambda_0 = 632.8\,nm$ the theoretical sensitivity in the fringe counting mode of operation is given by $S_e = \Delta\Phi/\Delta L = 3.64\ \pi rad/\mu m$[2] . If displacement H is transferred into fiber strain $\varepsilon_z = \Delta L/L$ by means of a transverse force $F_t \approx 2\pi r^2 E\,(H/L)\,\varepsilon_z$[4] (with fiber length L, radius r and modulus of elasticity E) the strain is to first order a quadratic function of displacement [6]. Higher order terms have to correct for the finite bend radius R of the fiber (see figure 1):

$$\varepsilon_z \approx 2\left(\frac{H}{L}\right)^2 \left(1+\frac{8\,H\,R}{L^2}\right) . \tag{1}$$

With L = 40 mm a strain of ≈ 0.1 % (corresponding ΔL = 50 μm) is obtained by a transverse displacement of H = 1 mm. From (1) we obtain the theoretical scale factor model $\Delta\Phi = (2S_e/L)\,H^2 + (16S_eR/L^3)\,H^3 = S_t H^2 + (8R/L^2)\,S_t H^3$ with $S_t = 364\,(\pi/2) rad/mm^2$. With $2R = R_1 + R_2 = 17.5$ mm (see fig.1) we get a coefficient of $32\,(\pi/2) rad/mm^3$ for the correction term. It has been shown earlier [4] that the strain - force characteristic $\varepsilon_z(F_t)$ exhibits a $F_t^{2/3}$- dependence if $\varepsilon_z(H)$ is approximated by the quadratic characteristic.

The output signal of a fiber optic two - arm interferometer is given by (e.g.[4][8])

$$P_{\pm} = \frac{P_0 T}{4(1+\mu)} \left[1 \pm \mu \cos(\Delta\Phi + \varphi)\right] , \tag{2}$$

Springer Proceedings in Physics, Vol. 44 **Optical Fiber Sensors**
Editors: H.J. Arditty · J.P. Dakin · R.Th. Kersten

where the sign indicates the respective output port of the interferometer. P_0 is the input power and $\Delta\Phi = \Phi_R - \Phi_S$ is the measurand induced phase term, where Φ_R and Φ_S are the mean phase delays in the LP_{01} -fundamental modes of the orthogonal fiber - eigenaxes in the R- and S- arm respectively. Transmissivity T, fringe visibility μ and phase bias φ are functions of the length of R- and S- arm, fiber birefringence, input and output polarization. The double - polarization method employed for eliminating the ambiguity of the counting procedure is described in previous publications [4] [7]. It utilizes the residual birefringence β of the non polarization - preserving single mode fibers for creating two orthogonally polarized (e.g. horizontal H, vertical V) phase shifted $\Delta\varphi = \varphi_H - \varphi_V \approx \pi/2$), though otherwise identical interference signals by means of a polarization beam splitter at the output port. The sign of $\Delta\varphi$ changes with the sign of the change in measurand and allows for identification of the latter by means of conventional electronics. It has been demonstrated earlier [4] that the double polarization method is applicable under transverse force sensing via fiber tension bending despite the large changes in birefringence.

Lateral resolution of force sensing can be achieved in one dimension by a single interferometer, because the sign of the change in measurand, as defined by the sign of $\Delta\varphi$, also changes when exchanging the R- and S - arm. Placing both fibers under an elastic skin, this arrangement e.g. allows for determining the center position between both fibers where the interferometer exhibits zero sensitivity (zero relative strain between both fibers), while forces acting to either side of the center yield + and - -signals respectively. The elastic skin gives rise to coupling between both arms depending on the elastic parameters and the geometry of the body exerting the transverse force. Transverse force (displacement) measurement with higher lateral resolution can be achieved by placing an appropriate number (2N) of interferometer arms beneath the skin at positions x_i, (i = 1,...,2N) with constant intervals Δx_i , and with distance $N\Delta x_i$ between the two arms of a single interferometer. Two - dimensional lateral resolution is achieved by providing two groups of fibers arranged as a matrix with crossing points at coordinates (x_i, y_i) .

Schematic drawings of the tactile sensor and of a single double - polarization Michelson interferometer are shown in figure 1. Figure 1a) shows an arrangement of two groups of Michelson - interferometers crossing each other at right angles, with four sensing fibers at positions x_i, $i = 1,...,4$ and four fibers at positions y_i, $i = 1,...,4$. In our experimental setup only the x_i- fibers were present with distances $\Delta x_i = 5$ to 6 *mm*. While interferometer 1 (IF1, detector D_{x13}) measures the relative strain of fibers 1 and 3, IF2 (detector D_{x23}) measures that one of fibers 2 and 4. The dashed line indicates the elastic rubber - skin, which can be put on top of the fibers. S and D denote light source and detector arrangements, including additional optics and readout electronics.

Figure 1b) shows details of a double - polarization Michelson interferometer with an arrangement for transverse force (F_t) or displacement (H) sensing, corresponding to the setup in the tactile sensor. Linearly polarized light from a He-Ne - Laser is focused into a polarization maintaining fiber which is spliced to the input arm of a 3-dB directional coupler. Input - and output arms are mounted to polarization controllers (fiber sqeezers) for the sake of adjusting the required phase shift $\Delta\varphi = \pi/2$ between H - and V - interference signals. A length L of the interferometer arms is fixed to a linear translation stage (in the experimental setup L = 39.5 to 41.6 mm) with an adhesive as used for electrical strain gauges. The plastic jacket of the fibers is removed at the fixing points. The arms may be exposed to a lateral force either directly by means of a cylindric body (radius $R_1 = 5$ mm, thickness 2 mm) affixed to a translation stage arranged orthogonally with respect to the fiber or via deformation of the elastic skin. The stage is displaced by means of a μm - screw (1 μm resolution) and forces the cylinder against the taut skin on top of the fibers. The vertical stage can be moved to different x - positions by means of a third translation stage. The output wave is split into its H and V components which are detected by PIN-photodiodes after passing several meters of plastic fiber. TTL - level counting pulses are created electronically at the zero - crossings of the symmetrized interference signal. After determination of the sign of the phase shift between H-and V-signals in the logic unit of the readout electronics counting pulses are fed into an up-down counter which are read via a PC - controlled interface system.

Measurements are performed by stepwise straining the fibers via displacement of the (vertical) translation stage. The stage is moved from zero to a maximum displacement of 1 mm with 0.1 mm intervals and than returned back to the initial value for determining possible hysteresis effects. Maximum displacement is chosen corresponding to a maximum strain of 0.1 %. Each run of 21 displacement measurements is repeated five times with standard deviations of measured values in general < 1 count. Looking at a single run, no significant hysteresis is observed with regard to increasing and decreasing displacement H. The difference between the two strain measurements at the same H does not exceed 1 count (= 0.14 μm). This is valid for the case with as well as without the skin. Experimental results are evaluated by means of least - squares analysis of the

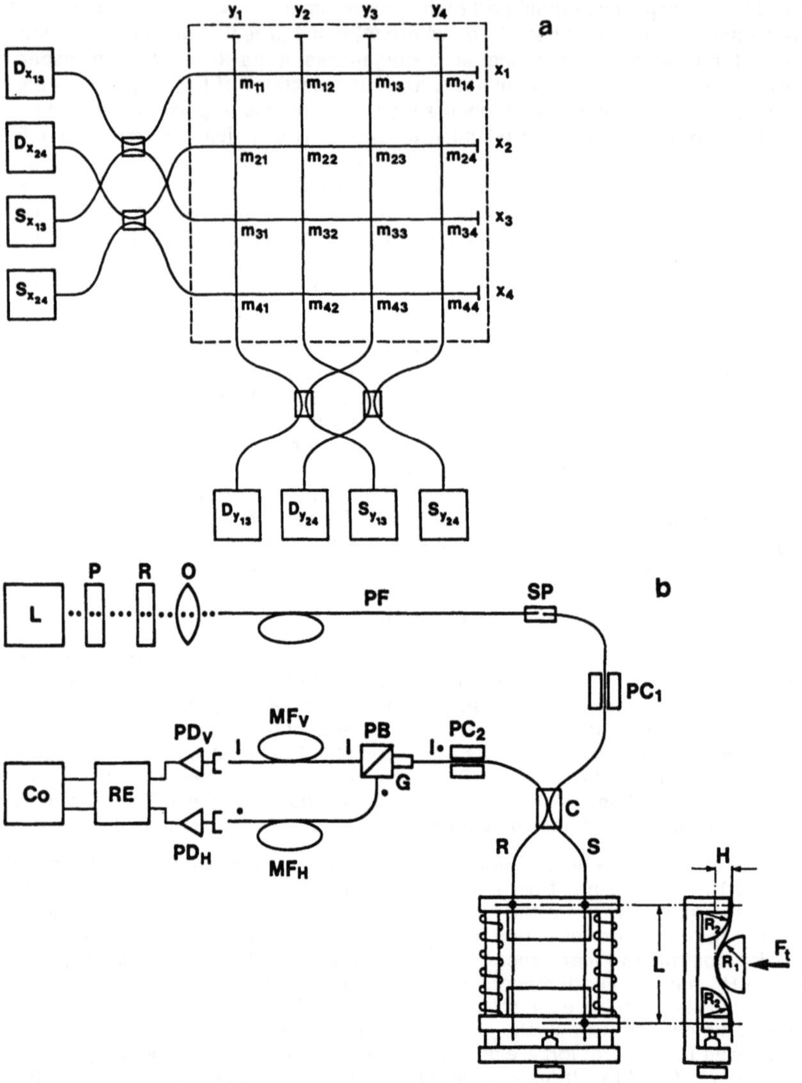

Figure 1 a) Schematic drawing of the tactile sensing array: two groups of Michelson interferometers with four sensing fibers (positions x_i, $i = 1,...,4$) crossing the four other fibers (positions y_j, $j = 1,...,4$) at right angles at points m_{ij}. Dashed line indicates elastic skin. S_{xij}, S_{yij}, D_{xij}, D_{yij} denote light source and detector units as described in fig.1 b). b) Double - polarization Michelson interferometer. L = He-Ne Laser, P = polarizer, R = retarder, O = 20x microscope objective, PF = polarization maintaining fiber, SP = mechanical splice, PC = polarization controller , C = 3 dB single mode coupler, R, S = reference and sensing arm, $L \approx 40\ mm$ = strained section of S - arm, H = fiber displacement by transverse force F_t , G = GRIN-lens, PB = polarization beam splitter, MF = multimode output fibers for H and V components of interference signal, PD = photodiodes, RE = readout electronics, Co = up-down counter.

mean values of five runs demonstrating the repeatability of the measurements. The model for fitting the experimental data has to take into account a periodic nonlinearity of the μm - screw with frequency $\omega = 2\pi/H_p$ corresponding to one turn of the screw: $H_p = 1.0\ mm$ For a nominal transverse displacement H the characteristic is approximated by

$$N = a_0 + a_1 H + a_2 H^2 + a_3 \sin(\omega H) + a_4 \cos(\omega H) \quad , \qquad (3)$$

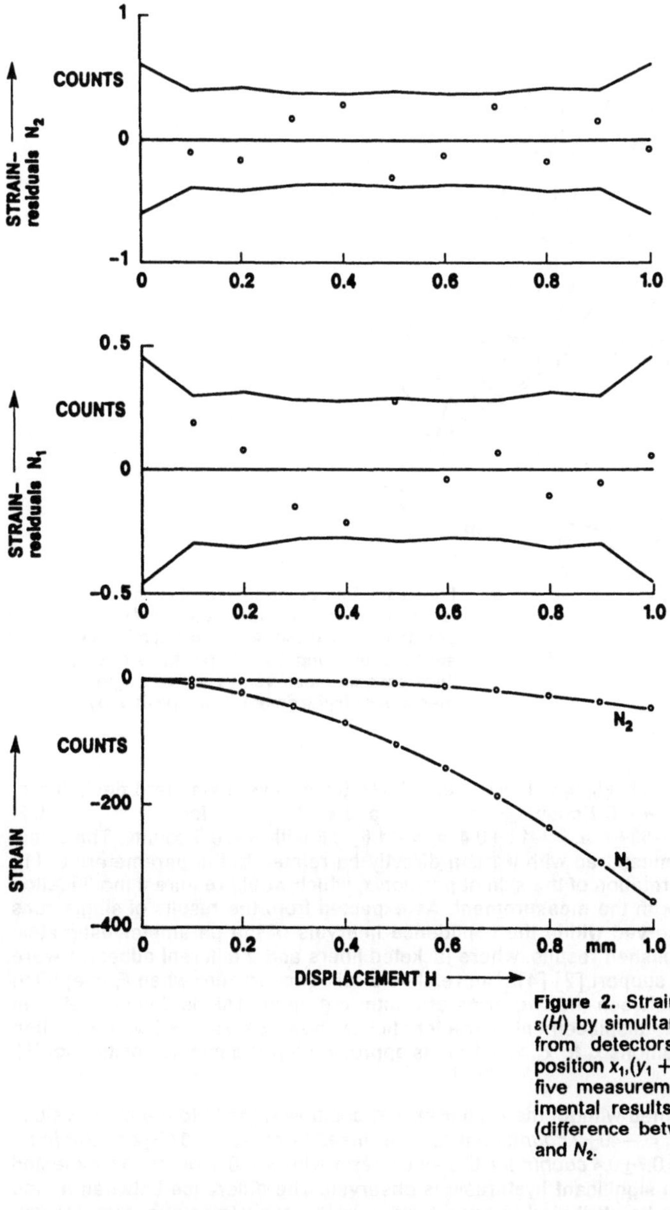

Figure 2. Strain - displacement characteristics $\varepsilon(H)$ by simultaneously recording signals (counts) from detectors D_{x13} : N_1 and D_{x24} : N_2 with F_t at position $x_1,(y_1+y_4)/2$. Circles represent mean of five measurements. From bottom to top: Experimental results with least squares fit, residuals (difference between measurement and fit) of N_1 and N_2.

with number of counts $N = \Delta\Phi/(\pi/2)$ being proportional to axial strain ε_z. In contrast to (1) constant and linear terms (coefficients a_0 and a_1) have to be included in the model. They are nonzero if the nominal displacement H differs from the effective one by a constant offset H_0. Because of these additional model parameters the small third order term of (1), as expected theoretically has not been considered in the fits to the experimental results as shown below. Estimates of the coefficients are given with 95% - confidence.

Results of strain measurements with displacement of the vertical translation stage at location $x_1,(y_1+y_4)/2$ and the rubber skin on top of the fibers are presented in figure 2. Shown is the measured strain via the nominal transverse displacement H as simultaneously recorded with both interferometers. While IF1 indicates the effective displacement of the stage, IF2 measures the deformation of the elastic skin at position $x_2 = (x_1+x_3)/2$. The parameter estimates of the fit for increasing displacement yield for IF1:

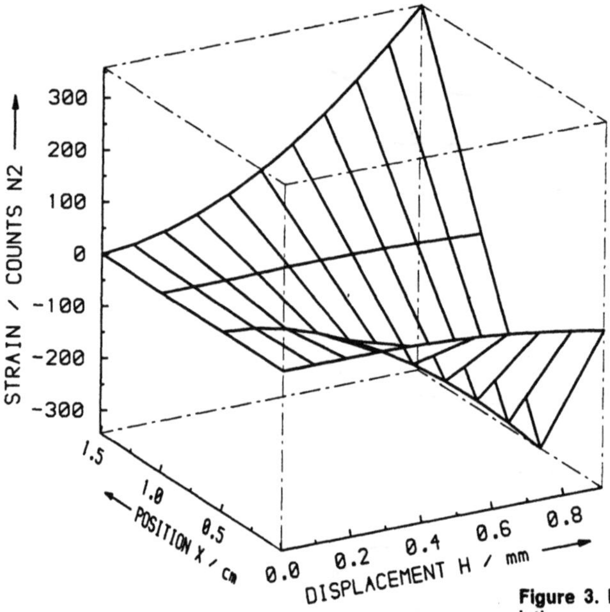

Figure 3. Field of strain - displacement characteristics as recorded with detector D_{x24} (fibers at positions $x_2 = 0.5cm$ and $x_4 = 1.5cm$) when F_t is successively applied at positions x_1 to x_4. Solid lines connect mean values of five single measurements per displacement H and position x_i.

$a_0 = -1.2 \pm 0.9$, $a_1 = -40 \pm 5$, $a_2 = -315 \pm 5$, $a_3 = 1.2 \pm 0.3$, $a_4 = 1.3 \pm 0.6$ with mean standard deviation of the fit $\sigma < 0.2$ counts and for IF2: $a_0 = +1.3 \pm 1.2$, $a_1 = +16 \pm 7$, $a_2 = -59 \pm 7$, $a_3 = -1.6 \pm 0.4$, $a_4 = -1.6 \pm 0.8$ with $\sigma < 0.5$ counts. The strain - displacement relationship as measured with IF1 can directly be related to the parameters of (1), whereas IF2 measures the deformation of the skin at position x_2 which would require a modification of (1) for a direct comparison with the measurement. As expected from the results of single runs no significant hysteresis is observed within the confidence intervals of the parameter estimates. This contrasts to previously published results, where jacketed fibers and a different adhesive were used for fixing the fibers to the support [2] [4]. The relative strain goes to zero when F_t is applied to the skin just in the middle between the two arms of a interferometer. This is demonstrated in figure 3 where the field of strain - displacement characteristics is shown as obtained with IF2, when F_t is applied successively at positions x_1 to x_4. $x_3 = 1$ cm is approximately the middle position for IF2 where sensitivity goes to zero.

If the measurements are performed without the rubber skin at position x_1 the fit to the mean values yields significant coefficients $a_1 = -40 \pm 2\ counts/mm$ for the linear term, $a_2 = -375 \pm 2\ counts/mm^2$ for the quadratic term and $a_3 = 0.7 \pm 0.4\ counts$ for the sine - term with $\sigma < 0.4$counts. As expected also in these measurements no significant hysteresis is observed. The difference between a_2 and the theoretical sensitivity S_t can be attributed to uncertainties in the experimental parameter values (e.g. length L) and the neglection of the H^3 - term in the fit which contributes to a_1. The lower value of a_2 in the case of measurements with the skin on top of the fibers is attributed to the skin deformation which lowers the relative strain between fibers 1 and 3 of IF1.

We have demonstrated that conventional single mode fibers as arms of two - arm interferometers, arranged in parallel under an elastic skin can be utilized for simultaneously measuring the value and the location of a pointlike force displacing the skin. The center position between the two arms of a single double - polarization interferometer separates the positive and negative signal regions. For determining both measurands at positions between the adjacent fibers of two interferometers the ratio of the corresponding two interferometer signals as well as the sum have to be evaluated. For two - dimensional localization the signals from the second group of interferometers have to be measured simultaneously. If in addition to displacement and position also the shape of objects has to be determined, either the number of interferometers has to be increased and/or the skin has to be replaced by a more sophisticated displacement - to - strain transducer. A different transducer also has to be chosen if a linear strain - displacement characteristic is required. Preliminary experiments with fibers atttached to flexible solid plates (diaphragms) for measuring the surface strain under elastic deformation yields the expected linear strain - displacement relationship [9].

[1]J.S. Schoenwald, A.W. Thiele, D.E. Gjellum *A Novel Fiber Optic Tactile Array Sensor* Proc. IEEE Conf.. on Robotics and Automation, IEEE Computer Soc. Press 1987, pp. 1792 - 1797

[2] N. Fürstenau *Double - Polarization Interferometer for Digital Displacement and Force Sensing by Fiber Tension - Bending* , Proc. Int. Conf. Optical Fiber Sensors OFS'88, Opt. Soc. Am. New Orleans 1988, pp. 191 - 194

[3] S.K. Sheem, T.G. Giallorenzi, K. Koo *Optical Techniques to Solve the Signal Fading Problem* Appl. Opt. 21 (1982) 689

[4]N.Fürstenau, *Double Polarization Interferometer for Digital Force Sensing by Fiber Tension Bending* , Opt. Lett., vol. 5, (1988), pp 242-244

[5]C.D. Butter, G.B. Hocker, *Fiber Optics Strain Gage* , Appl. Phys. B 35, (1984), pp 59-64

[6]A. Kersey, F. Bucholtz, A. Dandridge *New Nonlinear Phase Transduction Method for DC Measurand Interferometric Fibre Sensors* , Electron. Lett. 22, (1986), 75-76

[7]N. Fürstenau, *Performance of the Double-Polarization Method in Interferometeric Digital Force Sensing by Fiber Tension Bending* , Conference O-E LASE '88: 'Polarization Considerations in Optical Systems', SPIE Proc. vol.891, (1988), pp 209-214

[8]J. Dakin, B. Culshaw, ed. *Optical Fiber Sensors, vol. 1* , pp. 76 -78, Artech House, Boston and London 1988

[9]N. Fürstenau, *to be published*

Part XI

Chemical and Biochemical Sensors

Novel Techniques and Materials for Fiber Optic Chemical Sensing

O.S. Wolfbeis

Analytical Division, Institute of Organic Chemistry,
Karl Franzens University, A-8010 Graz, Austria

Abstract: Continuous methods for optically sensing chemical species rely on monitoring the optical properties of the analyte, preferentially via an optical waveguide, or by immobilizing some sort of indicator chemistry on a solid support or at the end of an optical fiber. The resulting devices are called *optodes* or *optrodes*. Because the reagent is immobilized, its optical properties can be monitored continuously and can give an indication of the concentration of chemical parameters such as pH, oxygen partial pressure, and various electrolytes. By coupling enzymatic reactions to oxygen-consuming or producing reactions, transducer-based biosensors are obtained. Aside from established methods that essentially are based in intensity measurements, new techniques have introduced in the past few years which include lifetime measurements, ET studies, and enzyme-coupled sensing methods. In addition, the use of new polymer materials can provide perm-selectivity, improved biocompatibility, and ease of dye and protein immobilization.

1. Introduction

Chemical sensors are devices which allow the continuous and reversible measurement of chemical parameters. Ideally, such sensors do not require sample treatment but are employed in direct contact with it. Hence, all errors associated with sampling and sample preparation play no role in true sensors. Moreover, sensing (by definition) is a real time method and, ideally, lends itself to unattended operation.

The two largest groups of chemical sensors are the electrodes and the so-called optodes. The word *optode* is derived from the Greek and means "the optical way". Optrode (from optical electrode) is also popular. While optical methods have been used in chemical analysis for many years (e.g., in the well-known pH paper strips), it was only during the past couple of years that optical methods have been applied to on-line sensing. Optical sensor technology is a merger between various specific technologies including optics, spectroscopy, electronics, analytical chemistry and polymer chemistry. A number of representative reviews is available.[1-5]

2. Optode Features

Optical fiber sensing can offer some of the following advantages over other chemical sensor types:

(1) Optical sensors do not require a reference signal as in all electrochemical methods.

(2) The ease of miniaturization allows the development of very small, light, and flexible fiber sensors that lend themselves to invasive methods.

(3) Low-loss optical fibers allow transmittance of optical signals over wide distances, typically 0.5 - 1000 meters, which allows inaccessible sites in production plants to be sensed.

(4) Because the primary signal is optical, it is not subject to electrical interferences.

Springer Proceedings in Physics, Vol. 44 Optical Fiber Sensors
Editors: H.J. Arditty · J.P. Dakin · R.Th. Kersten

The disadvantages of fiber sensing mainly result from the sensitivity toward ambient light, the limited stability of the materials and indicators that frequently have to be used, the limited dynamic ranges, and the lack of optimized components for chemical sensing purposes.

Typical fields of applications are in biomedical sciences, process control (including biofermentation), and in environmental sciences. Because most optodes are made from inexpensive polymer materials and traces of reagents immobilized thereon, sensing layers can be fabricated at very low costs. As a result, they are ideally suited for single-shot assays using disposible devices.

3. Chemical Sensor Types

The majority of existing optodes for chemical sensing are based on two fundamental sensing schemes. In the first, a plain (bare-ended) fiber is used to measure the optical properties of the analyte. In the second, an indicator and/or enzyme is immobilized on a fiber or a planar waveguide.

3.1. Plain Fiber Sensors

These can be used when the analyte of interest has a measurable optical property. A typical representative of the plain fiber sensor type is the one for methane which is based on the measurement of the NIR absorption of methane.[6] Other plain fiber applications include on-line measurement of water in organic solvents, remote detection of the fluorescence of uranium species, and monitoring the NADH fluorescence in bioreactors. A fluorosensor for monitoring NADH in bioreactors has become available commercially. Fig. 1 shows various sensing configurations that have been applied for remote fiber sensing via absorption measurements. Similar arrangements have been designed for use in combination with fluorometry, reflectometry, and Raman spectroscopy. A popular fiber arrangement is the one shown in Fig. 2, where the inner fiber transports the incoming light, and the outer fiber ring collects scattered or re-emitted light.

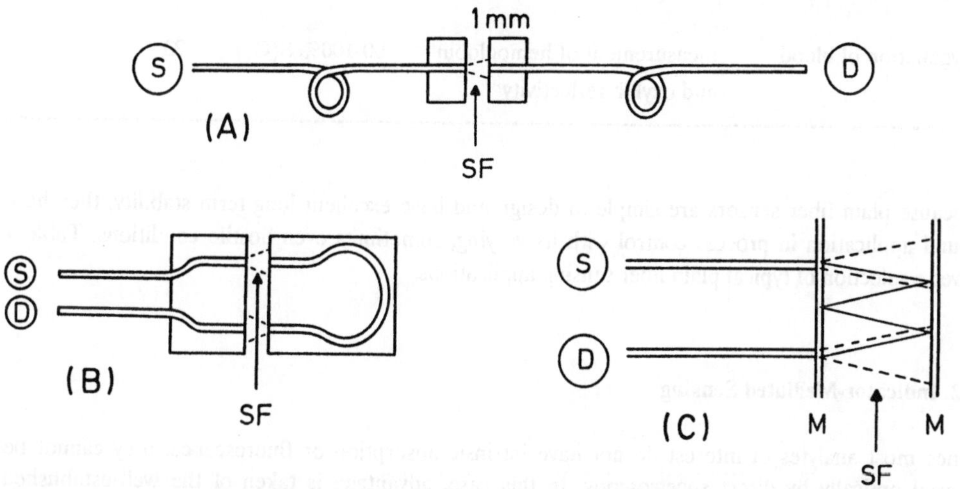

Fig. 1. Plain fiber sensing configurations for fiber absorptiometry. S, light source; SF, sample flow; D, detector; M, mirror. The pathlength increases in going from configurations (A) to (C). Method (C) is frequently applied to sense gases with their low absorbance in the near infra-red.

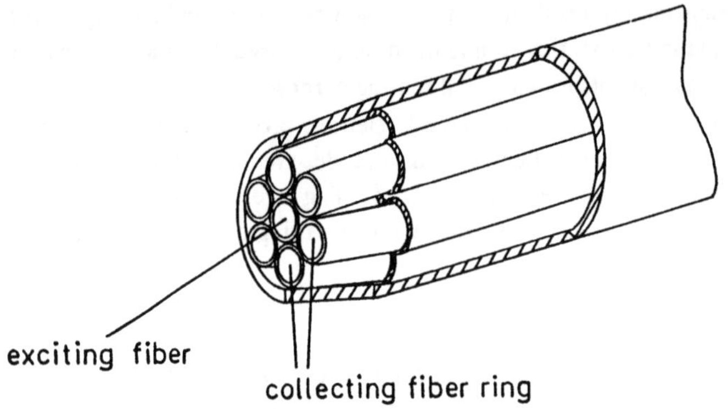

Fig. 2. Typical fiber bundle arrangement for performing remote fluorescence or Raman spectrometry using plain fibers.

Table 1. Typical examples of Plain Fiber Chemical Sensors

Species	*Detection Principle*	*Typical Range*	*Ref.*
uranium ion	remote absorptiometry	50 - 500 mM	19
methane gas	remote detection of the NIR absorption at 1666 nm	10 - 1000 ppm	20
phenolic groundwater contaminants	detection of the UV fluorescence of phenols	10 - 500 ppm	21
water in organic solvent	measurement of the NIR absorption of water at 1430 nm	0.01 - 3.0 %	22
oxygenation of blood	measurement of hemoglobin and oxyHb reflectivity	60-100% $S(O_2)$	23

Because plain fiber sensors are simple in design and have excellent long term stability, they have found application in process control with its varying, sometimes even hostile conditions. Table 1 gives a selection of typical plain fiber sensing applications.

3.2. Indicator-Mediated Sensing

Since most analytes of interest do not have intrinsic absorption or fluorescence, they cannot be sensed optically by direct spectroscopy. In this case, advantage is taken of the well-established indicator chemistry and existing methods of dye immobilization on optically transparent supports. A typical example is the oxygen optode which is based on the quenching of the fluorescence of certain indicators by molecular oxygen. The indicator is dissolved in a polymer such as silicone

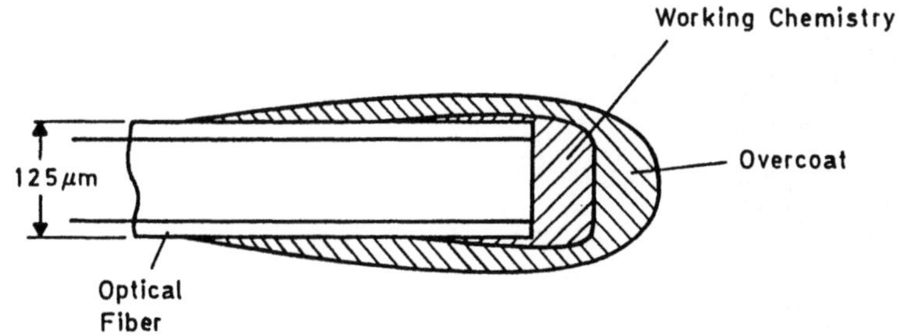

Fig. 3. Schematic of a fiber optic oxygen sensor with an indicator placed in a silicone drop at the fiber end (the so-called working chemistry) and covered with a protective coat.

rubber and placed at the end of the fiber optic (Fig. 3). The relation between pO_2 and fluorescence intensity is given by the Stern-Volmer equation.

We have developed, in the past 10 years, numerous kinds of oxygen sensors based on fluorescence quenching, the more recent ones being compatible with modern solid-state electronics. Also, it was demonstrated that both intensity and, preferentially, fluorescence lifetime[7] can be used as the analytical information that can be related to oxygen partial pressure.

Similarly, by immobilizing a pH-sensitive dye on a hydrophilic polymer at the end of an optical fiber or on a planar surface, a pH sensor is obtained. In practice, the dyes are chosen so to make them compatible with existing solid state components like LEDs and photodiodes. Major problems associated with pH optodes result from the ionic-strength sensitivity of all pH indicators which make them sensitive to salt effects. On the other side, pH optodes have distinct advantages over electrodes in case of high pH or when measuring within a small pH range only.

Carbon dioxide and related acidic or basic gases can be determined indirectly by entrapping a buffer solution plus a pH-sensitive dye in a silicone polymer as shown in Fig. 4. CO_2 from the sample passes the polymer membrane and changes the pH of the internal buffer. This is indicated by the pH indicator. Table 2 gives a selection of existing optodes for gases while Table 3 summarizes typical sensors for electrolyte species.

Table 2. Selected Gas Optodes

Species	*Detection Principle*	*Typical Range*	*Ref.*
oxygen	fluorescence quenching	0.5 - 300 Torr	7,14,15,24
	phosphorescence quenching	0.0005 - 0.1 Torr	24
CO_2	via pH measurement in an internal buffer	0.5 - 100 Torr	14,25
SO_2	fluorescence quenching	70 ppm - 5 %	26
toxic gases	absorption in the visible	0.01 - 5 %	27
hydrogen	interferometrically	5 - 2000 ppm	28

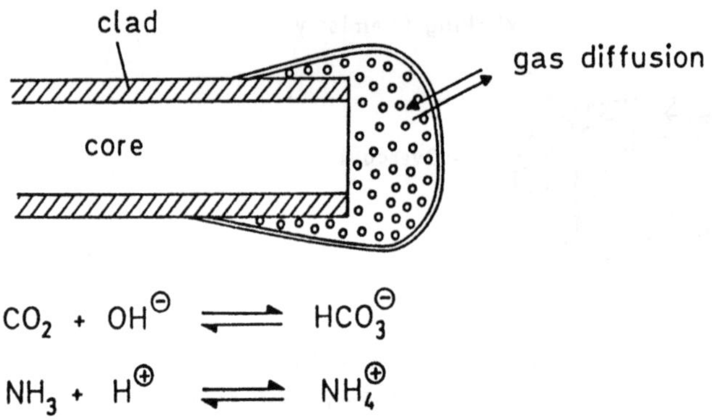

Fig. 4. Optode for sensing acidic or basic gases. The gas penetrates the polymer and changes the pH of the aqueous emulsion in the polymer. This is indicated by a pH-sensitive indicator.

Table 3. Selected Sensors for Electrolytes

Species	*Detection principle*	*Typical Range*	*Ref.*
H^+	reflectometry (via indicator)	2.5 - 3 pH units	3,6,29,30
	fluorescence (2 indicators)	4 - 5 pH units	31
Cu^{++}	self-absorption at 820 nm	50 - 500 mM	32
Al^{+++}	fluorescence of the Al(III)/morin complex	1 - 100 uM	33
K^+, Ca^{++}	optical measurement of membrane potential;	0.001 - 100 mM	16,34
	absorbance or reflectometry (indicator)	5 - 50 mM	17,35
Halides	fluorescence quenching	15 - 200 mM	36

A major field of application of optical sensors is in clinical and biochemical sciences. Essentially based on the sensor chemistry we had developed in the early 1980's, the CDI Corp. (Irvine, CA) has developed two types of fiber optic sensors for continuous measurement of pO_2, pCO_2 and pH. The first type (called CDI System 400 in its most recent version) is suitable for on-line analysis of the three parameters in an extracorporeal loop. The second one is an invasive catheter, 1 mm thick, with three single fibers forming a bundle, each equipped with a suitable working chemistry at its end[9]. The bundle can be inserted into an artery. Figure 5 shows a schematic of this sensor type.

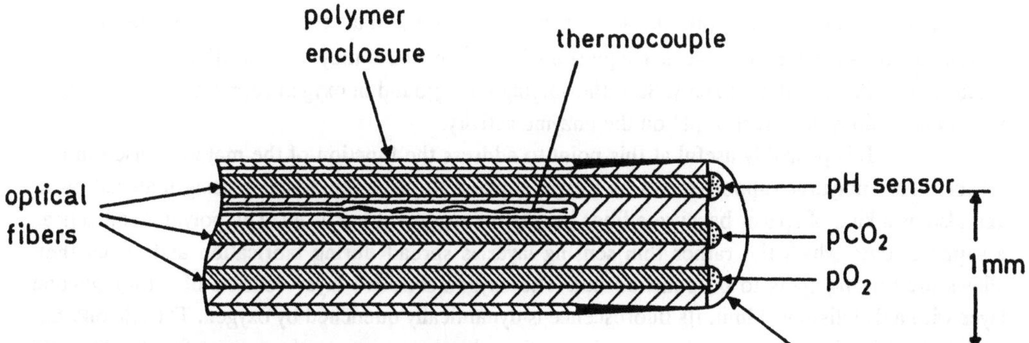

Fig. 5. Cross section through an invasive catheter for measurement of pH, pCO_2 and pO_2 in an artery.

4. Enzyme-Based Biosensors and Immunosensors

For many species of significant analytical interest including glucose, there are no indicators known which would allow their direct optical determination in a complex matrix. These species can be determined, however, by employing a catalytic transducer such as an enzyme which converts the analyte with high selectivity into a species that can be determined. Alternatively, a species may be consumed during enzymatic action and its decrease in concentration may be detected as well.

When, for instance, glucose is oxidized to give gluconic acid (under consumption of oxygen and production of hydrogen peroxide), three species can be detected: The first is the increased proton concentration, the second the decreased oxygen partial pressure, and the third is the hydrogen peroxide thus produced. The third approach has not been realized in optical sensors yet because of the lack of a sensitive and reversible optode for hydrogen peroxide, but the former two have been studied by us. Fig. 6 shows a cross section through a glucose-sensing membrane with

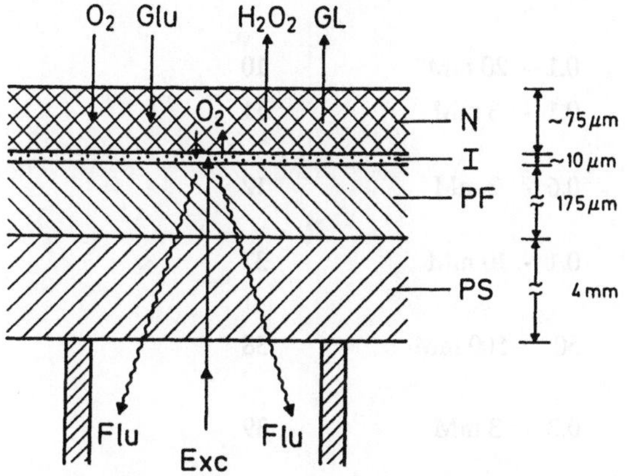

Fig. 6. Cross section through a planar glucose-sensitive membrane with an oxygen optode as a transducer. N, nylon net with immobilized glucose oxidase; I, oxygen-sensitive indicator layer; PF, polymer foil; PS, polymer support. Also shown are the diffusional processes involved, with glucose (Glu) and oxygen diffusing into the membrane, and hydrogen peroxide and gluconolactone (GL) diffusing out. Exc and Flu symbolize the directions of the exciting and fluorescent light within the fiber bundle.

an oxygen optode as the transducer[10]. It was concluded that the oxygen transducer-based biosensors are superior to pH sensor type transducers[11] because they are not affected so much by changes in pH and buffer capacity. Still, the varying background in oxygen supply can impose some problems, as does the effect of pH on the enzyme activity.

It is probably useful at this point to address the function of the materials used in this sensor. The polymer support serves as a rigid support to place the actual sensing material on it, and also as a kind of spacer because a fiber bundle arrangement was used. The polymer foil acts as a support onto which the rather thin sensing layer is spread during fabrication and which then allows the sensing spots to be punched out. The oxygen-sensitive layer consists of a thin silicone layer with a dye dissolved in it. Its fluorescence is dynamically quenched by oxygen. The silicone has a triple function in that it provides a mechanically stable layer, acts as the solvent for the dye, and provides some selectivity by preventing interfering gases to enter the membrane. A nylon net of well-defined thickness is loaded with enzyme and acts as the reactive layer where glucose is oxidized. Sometimes, the top layer is covered with a black analyte-permeable membrane that serves as an optical isolation, so to prevent ambient light to enter the optical system of the sensor.

Only recently, we have discovered a quite new possibility for sensing enzyme substrates such as lactate[12] or glucose by simply measuring the intrinsic fluorescence of an oxidase-type enzyme which, during oxidation, fully reversibly changes fluorescence intensity. The major advantage of this device relies on the fact that in this case the biochemical recognition element is identical with the transduction element. Therefore, no classical transducer such as an oxygen or pH optode is needed any more and the design can be extremely simple (Fig. 7). Still, however, the response depends on oxygen supply. Table 4 gives a selection of existing optical biosensor types.

Table 4. Selected Optical Biosensors

Species	*Detection Principle*	*Typical Range*	*Ref.*
glucose	enzymatically via O_2 optode	0.1 - 20 mM	10
	enzymatically via pH optode	0.1 - 3 mM	11
lactate	intrinsic fluorescence of the enzyme	0.6 - 2 mM	12
	enzymatically via O_2 optode	0.1 - 10 mM	37
ethanol	enzymatically via O_2 optode	50 - 500 mM	38
cholesterol	enzymatically via O_2 optode	0.2 - 3 mM	39
pesticides	inhibition of enzyme activity	>20 nMol/l	40

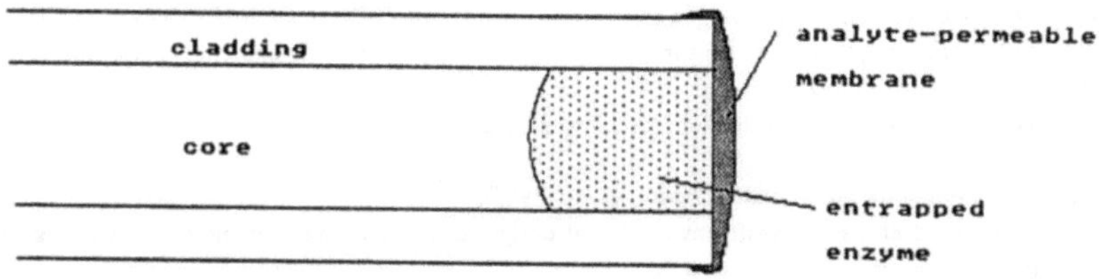

Fig. 7. Schematic of a fiber optic biosensor where the enzyme is both the recognition and transduction element.

5. Current Trends in Sensing Techniques

It is generally accepted now that the combination of established spectroscopic techniques with the fascinating possibilities provided by planar or fiber waveguides and the process made in opto-electronics, along with recent developments in enzymology, have led to a technology that is considered to be one of the fastest growing in analytical sciences. Current activities are directed towards a better biocompatibility of invasive sensors, the improvement of the long-term stability of sensors, for instance by making use of lifetime (phase) measurements instead of intensity (amplitude) measurements[7]. The construction of multi-sensor bundles or multi-analyte fiber sensors, the increase in information intensity by multiplexing and demultiplexing light signals[4], and the use of new materials and techniques such as the Langmuir-Blodgett[16] and other thin film technologies are among the current activities, some of which are performed in our group. The outstanding binding selectivity of ionophors is now being used in quite new sensing schemes for alkali metal ions, for instance in elegant methods of charge cycling[17] or in new fluoro-ionophors[18].

Research is carried out with great enthusiasm at both academia and industry and it is hoped that this article can mediate some of the enthusiasm that is shared by the researchers active in this field.

6. References

1. Hirschfeld T., Callis J. B. und Kowalski B. R., Chemical Sensing in Process Analysis, *Science* **226**, 312 (1984).

2. Wolfbeis O. S., Analytical Chemistry with Optical Sensors, *Fresenius Z. Anal. Chem.* **325**, 387 (1986).

3. Boisdé G. and Perez, J. J., Une nouvelle generation de capteurs: les optodes, *La Vie de Sciences, Compt. Rend.* **5**, 303 (1988).

4. Narayanaswamy R. und Sevilla F., Optical Fibre Sensors for Chemical Species, *J. Phys. E.: Sci. Instrum.* **21**, 10 (1988).

5. Wolfbeis O. S., Fiber Optic Fluorosensors in Analytical and Clinical Chemistry, in: *Molecular Luminescence Spectroscopy: Methods and Applications*, Schulman S. G. (ed.), Wiley, New York, 1988, chapter 3, pp. 129-281.

6. Seitz W. R., Chemical Sensors Based on Immobilized Indicators and Fiber Optics, *CRC Crit. Rev. Anal. Chem.* **19**, 135 (1988).

7. Lippitsch M. et al., Fiber optic oxygen sensor with the fluorescence decay time as the information carrier, *Anal. Chim. Acta*, **205**, 1 (1988).

8. Janata J. Do optrodes really measure pH?, *Anal. Chem.* **59**, 1351, 1987.

9. Miller W. W. et al., Performance of an in-vivo, continuous blood gas monitor with disposable probe, *Clin. Chem.* **33**, 1528 (1987).

10. Trettnak W., Fiber optic glucose biosensor with an oxygen optrode as the transducer, *Analyst*, **113**, 1519 (1988).

11. Trettnak W., Leiner M. J. P. and Wolfbeis O. S., Fibre optic glucose sensor with a pH optrode as the transducer, *Biosensors*, **4**, 15 (1988).

12. Trettnak W. and Wolfbeis O. S., A fully reversible fiber optic lactate biosensor based on the intrinsic fluorescence of lactate mono-oxygenase, *Fresenius Z. Anal. Chem.*, in press (1989).

13. Walters B. S., et al., Fiber optic biosensor for ethanol based on an internal enzyme concept, *Talanta* **35**, 151 (1988).

14. Wolfbeis O. S. et al., Fiber optic fluorosensor for oxygen and carbon dioxide, *Anal. Chem.* **60**, 2018 (1988).
15. Sharma A. and Wolfbeis O. S., Fiber optic oxygen sensor based on fluorescence quenching and energy transfer, *Appl. Spectrosc.* **42**, 1009 (1988).
16. Schaffar B. P. H. and Wolfbeis O. S., New optical chemical sensors based on the LB technique, *Proc. SPIE* **990**, 122 (1989).
17. Perisset P. N. J. et al., An ion-selective photodiode, *Chimia*, **43**, 10 (1989).
18. Bryan A. J., et al., Photo-induced electron transfer as a general design logic for fluorescent molecular sensors for cations, *Biosensors* **4**, 169 (1989).
19. Groll P. et al., An optical fibre laser photometer for on-line measurements, *Anal. Chim. Acta* **190**, 265 (1986).
20. Alarcon M. C. et al., All optical remote sensing of city gas through CH_4 absorption, *Appl. Phys. B.*, **43**, 79 (1987).
21. Chudyk W. A. et al., Remote detection of groundwater contaminants using UV laser induced fluorescence, *Anal. Chem.* **57**, 1237 (1987).
22. N. N., The measurement of low levels of water in solvents, *Appl. Note A1-387*, Guided Wave Inc., El Dorado Hills, CA, 1988.
23. Baele P. L. et al., Continuous monitoring of venous oxygen saturation in critically ill patients, *Anesth. Analg.* **61**, 513 (1982).
24. Wolfbeis O. S. and Leiner M. J. P., Recent progress in optical oxygen sensing, *Proc. SPIE* **906**, 42 (1988).
25. Hirschfeld T. et al., Laser fiber optic optrode for reall-time in vivo blood CO_2 level monitoring, *J. Lightwave Technol.* **5**, 1027 (1987).
26. Wolfbeis O. S. and Sharma A., Fiber optic fluorosensor for sulfur dioxide, *Anal. Chim. Acta* **208**, 53 (1988).
27. Inaba H., Optical remote sensing of environmental pollution and danger, *Springer Ser. Opt. Sci.* **39**, 288 (1983).
28. Butler M. A. and Ginsley D. S., Hydrogen sensing with Pd-coated optical fibers, *J. Appl. Phys.* **64**, 3706 (1988).
29. Grattan K. T. V. et al., Dual wavelength optical fiber sensor for pH measurement, *Biosensors* **2**, 17 (1987/1988).
30. Scheggi A. M. et al., pH sensing via fiber optics, *Optica Acta* **33**, 1587 (1986).
31. Posch H. E. et al., Towards a gastric pH sensor: an optrode for the pH 0 - 7 range, *Fresensius Z. Anal. Chem.* **334**, 162 (1989).
32. Freeman J. E. et al., A fiber optic absorption cell for remote determination of copper in industrial electroplating baths, *Anal. Chim. Acta* **177**, 121 (1985); and refs. cited therein.
33. Saari L. A. and Seitz W. R., Immobilized morin as fluorescence sensor for determination of aluminium, *Anal. Chem.* **55**, 667 (1983).
34. Wolfbeis O. S. and Schaffar B. P. H., An calcium selective optrode based on fluorimetric measurement of membrane potential, *Anal. Chim. Acta* **217**, 1 (1989).
35. Ashworth D. C. et al., An optical calcium ion sensor, *Anal. Chim. Acta* **213**. 251 (1988).
36. Urbano E. et al., Optical sensor for continuous determination of halides, *Anal. Chem.* **56**, 427 (1984).
37. Trettnak W. and Wolfbeis O. S., Fiber optic lactate biosensor with an oxygen optrode as the transducer, *Anal. Lett.*, in press (Sep. 89).
38. Wolfbeis O. S. and Posch H. E., A fiber optic ethanol biosensor, *Fresenius Z. Anal. Chem.* **332**, 255 (1988).
39. Trettnak W. and Wolfbeis O. S., A fiber optic cholesterol sensor, *Anal. Biochem.*, submitted (1989).
40. Wolfbeis O. S. and Koller E., Fiber optic detection of pesticides in drinking water, in: *Proc. Intl. Workshop on Biosensors: Applications in Medicine, Environmental Protection, and Process Control*, Schmid R. E. (Ed.), Braunschweig, FRG, May 21-23, 1989.

Probe Performance Optimization for pH Continuous Monitoring*

M. Bacci, F. Baldini, F. Cosi, G. Conforti, and A.M. Scheggi

Istituto di Ricerca sulle Onde Elettromagnetiche del C.N.R.,
Via Panciatichi, 64, I-50125 Firenze, Italy

The structure and performance of different pH probes are described, finalized to in vivo measurements. Particular care was devoted to the requirements necessary for the insertion in the human body. Sensitivity of 0.02 pH units and response time of about one minute for steps of 0.25 pH units are reached.

1. Introduction

The detection of pH by optical fibers finds a noticeable application either in chemical industry or in biomedical field [1-6]. It is important to outline that a pH sensor, based on phenol red absorption, was the first developed optical fiber chemical sensor [7]. Since then on, several pH sensor models were realized, making use of fluorescent or absorbing substances [8-10].

For what concerns the application in the clinical field, the use of optical fibers allows, due to their dimensions, an in vivo insertion and a continuous monitoring, which in some illnesses result very important due to the necessity of very accurate and reliable values of the parameter under test.

The knowledge of pH is very important in many cases: for instance monitoring blood pH can reveal respiratory, renal or metabolic problems, while pH measurement in the stomach can help diagnosis in some entero-gastric diseases. It is apparent that in these applications the sensor probe , inserted in the human body, must respond to very strict requirements: reliability, mechanical resistance, non-toxicity, biocompatibility, no changes or inhibition of the optrode colorimetric properties due to sterilization. Many times, moreover, a parallel measurement of other parameters can result very useful if not necessary, and so a high miniaturization is necessary. It must also be taken into account that pH measurements are usually made in "complex" biological liquids and then a particular care must be devoted to avoiding interferences by other substances; for this reason it is often more suitable the use of indicators working on absorption because, also if less sensitive in comparison with fluorescent substances, generally they do not undergo interferences by other substances.

Subject of the present paper is the description of the structure and performance of pH probes finalized to direct in vivo measurements.

*This work has been developed and supported by CNR Finalized Program on Material and Devices for Solid State Electronics

2. pH probes design and set up

Since the interest was devoted to the development of a sensor for pH measurements in the blood, the attention was directed to dye indicators, working on absorption and sensitive in the range 7.0÷7.7. In previous works a spectrophotometric investigation on different dye indicators was carried out; a careful study with spectrophotometer allowed to point out phenol red like one of the best indicators for pH measurement in the physiological range, in terms of sensitivity and absence of hysteresis, after the immobilization on XAD-2 microspheres [11]. This immobilization procedure is very simple and is widely described in the literature [12]. Then a first optical fiber pH sensing model with a solid optrode was proposed [13]. Here the microspheres, with phenol red adsorbed on them, were fixed at the end of two PCS fibers (200 μm) by means of a gel (agarose) cylinder, inserted into a stainless steel capillary (external diameter: 1.8 mm, length: 20 mm) thus obtaining a highly compact and practically "frozen" structure. As a consequence, problems related to signal fluctuations due to micromovements of the microspheres were avoided and a more stable response was achieved. Moreover as the microspheres resulted to be placed not abutted on the fibers, better signal levels could be obtained.

The optoelectronic system was already described [13-14]. An improvement in the level signals was carried out by a proper design of the coupling between LED and optical fiber. The LED was polished in the upper part, up to emission filament: in order to obtain the best coupling, a commercial type connector (Radiall) was adapted so to house the polished LED and was connected to the fiber inserted into an aluminum case by placing the LED and the fiber practically in contact.

2.1 Probe with gel

Since the first measurements with agarose had given good results [12], different gels, transparent to the visible light and highly hydrophilic, were systematically investigated so to verify the possibility of optimization of this type of probes: acrylamide, gelatin and the already used agarose were tested. For all these substances the gelification takes place during the cooling of the gel solution after heating between 60 and 100 °C [15]. Acrylamide was not suitable for our purposes because the construction of the gel cylinder embedding the microspheres resulted impossible, due to the very high velocity of the gelification process. On the contrary the probe construction with gelatin and agarose resulted simple and easily reproducible: accordingly probes making use of these two gels were tested and their performance for different pH steps was characterized in terms of sensitivity and response time.

As for gelatin probe, although the construction is very simple, gelatin has a negative drawback due to its chemical structure: its isoelectric point is about 4.85 pH units. Hence it is clear that, in physiological range tests, gelatin molecules are characterized by negative charges due to the presence of the carboxylic group COO^-, as it can be noted in Figure 1 where the gelatin structure is schematically shown in acid ambient (a), at the isoelectric point (b) and in alkaline ambient (c): R represents the remaining part of the proteinic structure. This fact gives rise to a completely different response time of the probe when the acidity of the sample is increased or decreased as it can be observed in Figure 2, where the response curve of the probe is shown for different pH steps: very fast in the first case (about one minute) and very slow in the second (several minutes). It is reasonable to ascribe this behaviour to the gelatin structure: the presence of a local electric field effect, caused by the gelatin molecules, makes

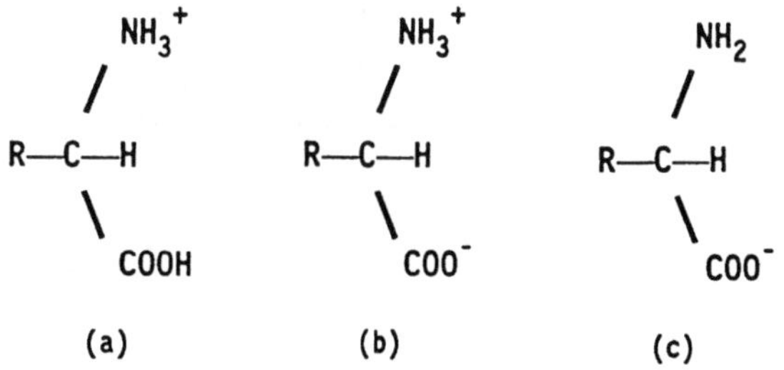

Figure 1 Schematic sketch of gelatin structure (a) in acid ambient, (b) at the isoelectric point, (c) in alkaline ambient.

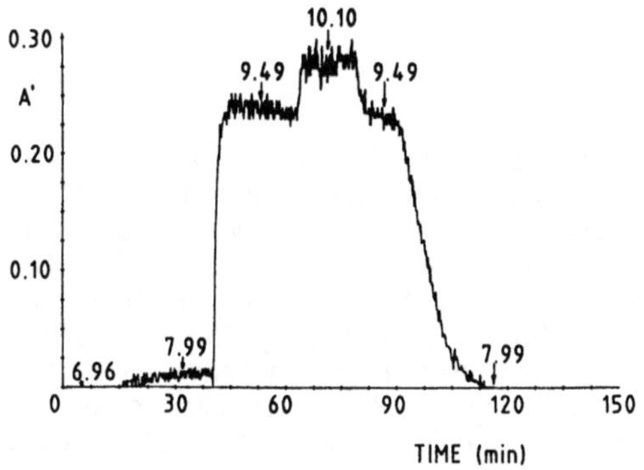

Figure 2 Response curve of the gelatin probe for different pH steps.

easier or hinders the hydrogen ions diffusion according to the pH step. Another fact to be pointed out is that this gel, due to the presence of several NH_3^+ and COOH groups, slightly alters the pH of the external solution and acts itself as a buffer: phenol red, immobilized on XAD-2, does not "see" the hydrogen concentration of the sample but that of the gel which is slightly shifted towards smaller values. However it is clear that this effect can be practically neglected because the dimension of the gel cylinder in the optrode is much more smaller than the sample solution volume. The worst drawback of the gelatin is due to its solubility at a temperature of about 40 °C, which is unacceptable in view of the insertion of the probe inside the human body. For this reason the gelatin probes were subjected to a insolubilization procedure by dipping them in a 1% formaldehyde solution for ten minutes. The so treated probes present the same optical behaviour, but are characterized by much more longer response time (several tens of minutes), and then they result not suitable for an application in the human body.

Better results were obtained with the agarose probe. First of all there is no solubility problem at the temperature ranges of the human body. Also in this case the influence of the agarose on the external buffer is quite negligible. Figure 3 shows the response curve obtained with this type of probe for different pH steps. Here the response time is not characterized by a different behaviour according to a pH increase or decrease, and, also if it is different for different pH steps, it is always of about few minutes in the physiological range. The sensitivity was calculated starting from the noise band in the steady state for all pH steps and remains always less than 0.04 pH units in the physiological range.

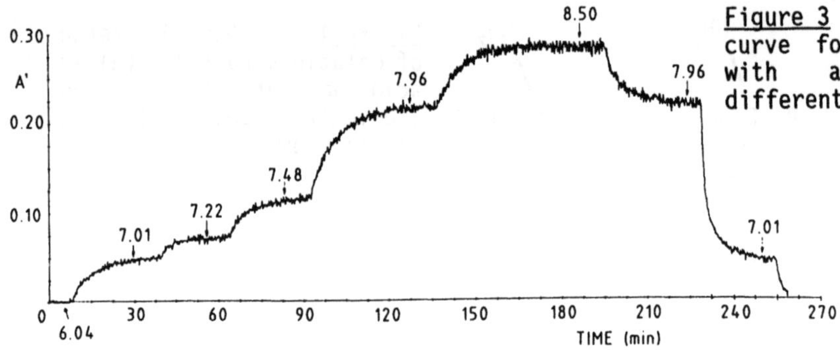

Figure 3 Response curve for the probe with agarose for different pH steps.

2.2 Probe without gel

The immobilization of the XAD-2 microspheres inside a gel cylinder allowed to construct a compact and very stable probe. However this solution was not perfectly satisfactory from a medical point of view because the response time was too long, due to the diffusion of the H^+ ions inside the gel. It is important, indeed, to outline that in a typical in-vivo blood pH monitoring, response time of less than one minute is necessary for the safety of the patient. A second probe was designed and constructed in order to obtain a better performance especially in terms of response time. In this case the probe (Fig.4) is constituted by a stainless steel capillary, smaller than the previous one (external diameter: 1.4 mm, length: 8 mm), inside which the XAD-2 microspheres, with phenol red adsorbed on them, are inserted. The capillary is terminated by a threaded mirrored cap which allows to adjust the volume containing the microspheres. Very small slits, made on the lateral surface of the capillary, warrant the exchange with the external environment. The microspheres, before the adsorption of the dye, were selected by a squared mesh sieve so to have a diameter between 430 and 780 μm, larger than the slit dimension. The two optical fibers, connected with the sources and with the detector, were stripped of their jacket and placed inside a stainless steel cylinder (external diameter: 1 mm), inserted into the capillary above described. So doing the cores of the fibers are closer and the overlapping zone of the fiber acceptance cones is at a shorter distance from the fibers: this warrants that a higher level of the light scattered by the microspheres is collected by the fiber connected with the detector. Figure 5 shows a photo of the probe coupled with two optical fibers. It is important to point out that the chosen microspheres are properly sized so to be inserted inside the stainless steel capillary where they cannot move, due to the pressure produced by the threaded cap; hence the response of the probe is very stable.

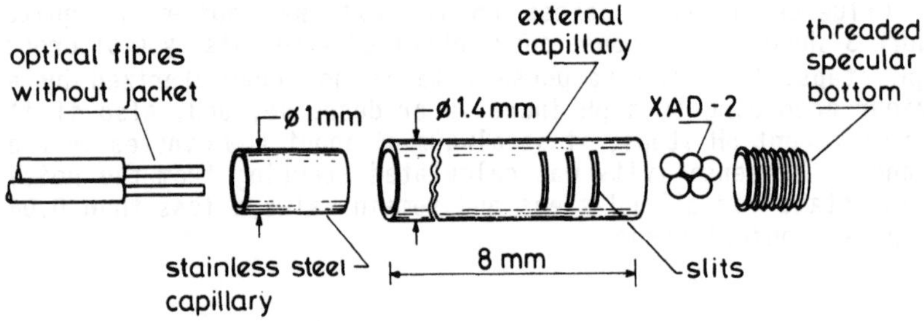

Figure 4 Sketch of the probe without gel

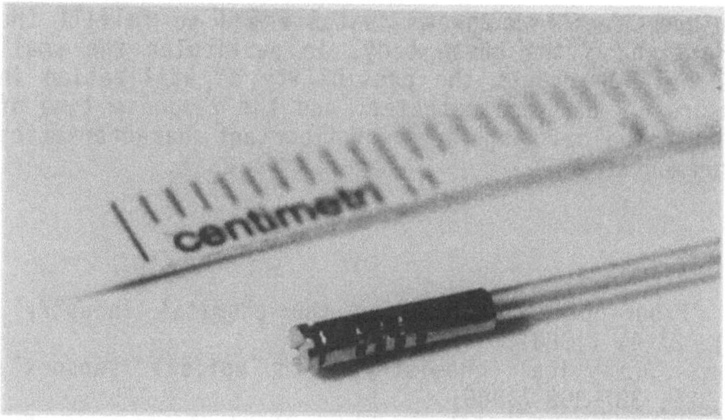

Figure 5   Photo of the pH probe without gel.

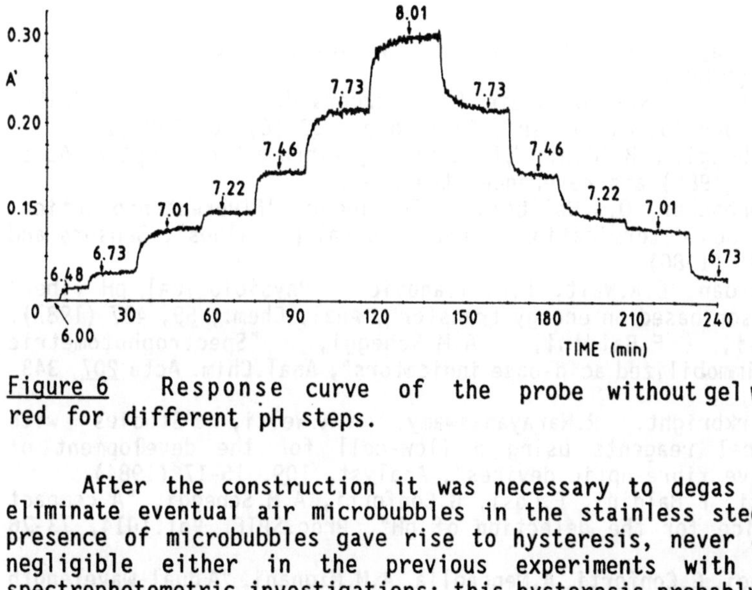

Figure 6   Response curve of the probe without gel with phenol red for different pH steps.

After the construction it was necessary to degas the probe, so to eliminate eventual air microbubbles in the stainless steel capillary. The presence of microbubbles gave rise to hysteresis, never appeared or quite negligible either in the previous experiments with gel or in the spectrophotometric investigations; this hysteresis probably was due to light movements of the microbubbles or to a partial degradation of the adsorbed phenol red in contact with air.

Figure 6 shows the response curve of the probe after the outgasing is showed. The sensitivity is less than 0.02 pH units and the response time is about one minute for 0.25 pH unit steps in the physiological range.

Another important result is that the colorimetric properties of the phenol red adsorbed in XAD-2 did not change after a process of sterilization by formaldehyde or by glutaraldehyde, necessary for in vivo tests.

3. Conclusion

An in vivo application of optical fiber sensors is of great interest and finds more and more growing reply in the medical ambient. The new probe for pH sensing, above described, seems to appear suitable and to fulfill the requirements for an insertion in the human body. In particular the small dimension of the probe, which warrants the possibility of utilization in parallel with other sensors on the same catheter, and the response time of about one minute in the physiological range are very important characteristics in view of in vivo measurements.

4. References

1. R.T.Murray, "Fiber-optic sensors for the chemical industry", J.Optical Sensors, 1 (1), 27-41 (1986).
2. O.S.Wolfbeis, "Analytical chemistry with optical sensors", Fresenius Z.Anal.Chem., 325, 387-392 (1986).
3. H.H.Miller, T.B.Hirschfeld, "Fiber optic chemical sensors for industrial and process control", Proc.SPIE, Vol.718, 39-45 (1986).
4. A.M.Scheggi "Optical fiber sensors in medicine", Proc. OFS'84 (Berlin, Offenbach:VDE-Verlag), 93-104 (1984).
5. G.G.Vurek, "In vivo optical chemical sensors", Proc.SPIE, 494, 2-6 (1984).
6. A.M.Scheggi, F.Baldini, "Sensing by optical fibers in medicine", DOPS-NYT, 1, 9-13 (1988).
7. J.I.Peterson, S.R.Goldstein, R.V.Fitzgerald, D.K. Buckhold, "Fiber optic pH probe for physiological use", Anal.Chem., 52 (6) 864-869 (1980).
8. A.M.Scheggi, F.Baldini, "pH sensing by fibre optics", Optica Acta, 33 (12), 1587-1597 (1986) and references therein..
9. H.Offenbacher, O.S.Wolfbeis, E.Furlinger, "Fluorescence optical sensors for continuous determination of near-neutral pH values", Sensors and Actuators, 9, 73-84 (1986).
10. D.M.Jordan, D.R.Walt, F.P.Milanovich, "Physiological pH fiber-optic chemical sensor based on energy transfer", Anal. Chem., 59, 437 (1987).
11. M.Bacci, F.Baldini, A.M.Scheggi, "Spectrophotometric investigations on immobilized acid-base indicators", Anal.Chim. Acta 207, 343-348 (1988).
12. G.F.Kirkbright, R.Narayanaswamy, N.A.Welti, "Studies with immobilised chemical reagents using a flow-cell for the development of chemically sensitive fibre-optic devices", Analyst, 109, 15-17 (1984).
13. M.Bacci, F.Baldini, F.Cosi, G.Conforti, A.M.Scheggi, "A compact optical fibre device for the detection of pH", Proc.SPIE, Vol.1014, 73-76 (1989).
14. M.Brenci, G.Conforti, A.Mencaglia, A.M.Mignani, "A dual-wavelength optoelectronic system for intensity-modulated fibre-optic sensors", Int.J.Optoelectronics, 3, 257 (1988).
15. The Merck Index (IXth Ed.), Ed.Merck & Co., Rahway, N.J., 1976.

High Sensitivity Fiber Optic Evanescent Wave Sensing for Fluoroimmunoassay

W.F. Love[1], *I.M. Walczak*[2], *and R.E. Slovacek*[2]

[1]Corning Incorporated, Corning, NY 14831, USA
[2]Ciba Corning Diagnostic Corporation, 63 North Street, Medfield, MA 02052, USA

A high sensitivity fiber optic evanescent wave sensing system has been developed with application to fluoroimmunoassay. This B-phycoerythrin dye, laser system demonstrates sensitivity of 10^{-12} molar free solution. Launch angle experiments agree qualitatively with the model and verify that evanescent wave interaction is responsible for the measured fluorescence.

The application of optical fiber evanescent sensing to fluorescence based immunoassay has been demonstrated in previous studies.[1-4] Two prototype assays, viz., digoxin and ferritin were demonstrated over the clinical range of interest.[4] The limiting sensitivity corresponded to approximately 10^{-9} molar free solution of fluorescein isothiocyanate (FITC).

In this paper, we present results which demonstrate an increase in sensitivity that is approximately four orders of magnitude, viz., 10^4, greater than previous studies.[4,5] A systematic variation for the input launch angle was completed and verifies that evanescent interaction is the underlying mechanism for the received signal. Preliminary experiments with the enzyme, CKMB, have been made and these results will be presented in more detail at the conference.

Figure 1 shows a block diagram of the characterization apparatus. A single-mode, TEM_{00}, He-Ne laser operating at 543.5 nm is focused on an unclad optical fiber immersed in a dilute buffered dye solution contained within a flow-thru cell. A bandpass filter in the input optics reduces the plasma tube fluorescence from this 0.4 milliwatt laser. B-phycoerythrin (BPE) dye is used since its optical absorption is large, some 30 times greater than FITC on a molar basis, and it closely matches the laser wavelength. This particular phycobiliprotein has been used as a fluorescent label in immunoassay.[6] Light emanating from the fiber is collected in the backward direction, collimated, and passed through two high-contrast interference filters. Light at 580 nm corresponding to the fluorescence from BPE is transmitted, whereas laser light at 543.5 nm is rejected by approximately 10^{12}. An image of the fiber input end is

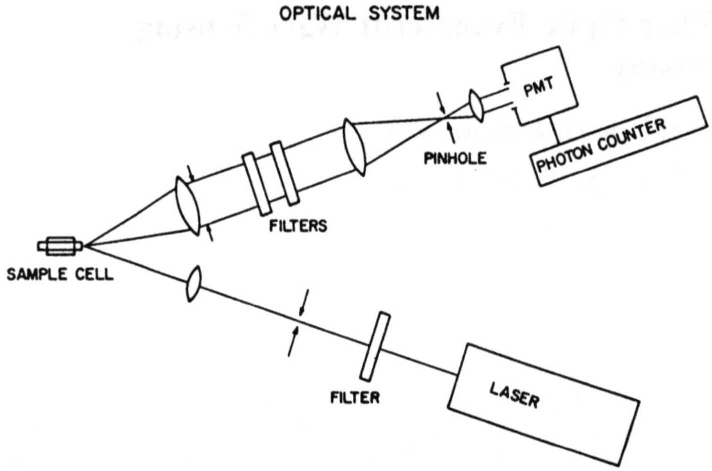

Figure 1. Optical research apparatus.

formed at the pinhole which serves to reduce stray light. A sensitive low-noise photomultiplier is used for detection along with photon counting.

The launch angle for the incident laser light is varied over the range, $\Theta = 0°$ to $40°$, whereas the collection angle remains fixed at 25° relative to the fiber axis. The fluorescent signal at the detector due to evanescent sensing is expected to be proportional to N, the number of reflections in the optical fiber and $|t|^2$, the Fresnel transmission coefficient. Thus for meridional rays, the fluorescent signal is given by,

$$\mathrm{Sig} \propto N\ |t|^2 \propto \sin^3\Theta_i/\cos\Theta_i$$

where Θ_i is the internal angle with respect to the fiber axis.

The practical limits to achieving high sensitivity depend critically upon reducing background fluorescence not directly associated with evanescent interaction with the BPE dye. The apparatus has been characterized by substituting a silica flat in place of the fiber. The resulting photon count rate is approximately 20 counts per second (cps) from $\Theta = 0°$ to $40°$ with the dark count equal to 2 cps. Figure 2 shows the background fluorescence from a 1 mm diameter silica fiber in trizma hydrochloride (tris) buffer solution. The variation with launch angle is close to $\tan\Theta_i$, which is proportional to N. This indicates a surface-related fluorescence since a bulk-related fluorescence should scale as the optical path length, viz., $L/\cos\Theta_i$. The decrease in background fluorescence at 40° is due to the launching of unbound light rays.

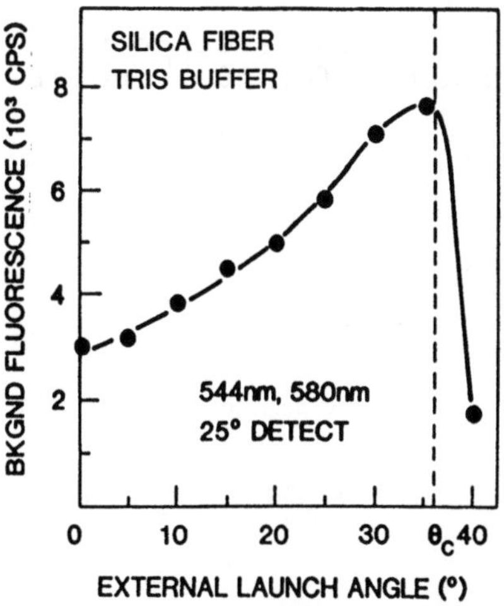

Figure 2.
Background fluorescence from silica fiber.

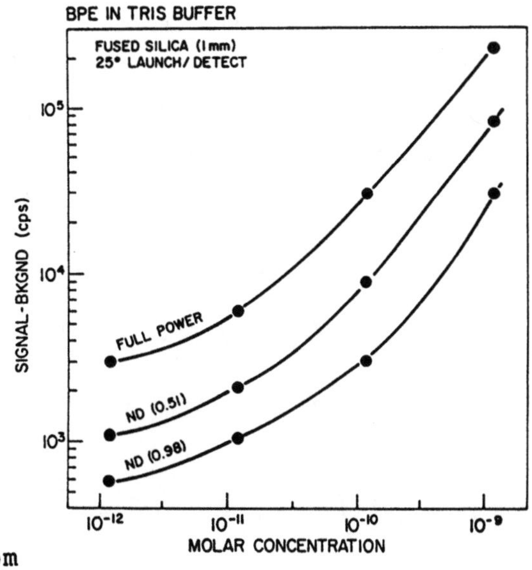

Figure 3.
Variation of fluorescent signal minus background versus concentration of BPE for three power levels.

Sensitivity curves are shown in Figure 3 for launch and detection optics at 25°. Neutral density filters were inserted into the incident light optics to check for linearity and possible photobleaching effects. The response was approximately linear with slight compression at the lower concentrations. No photobleaching was observed during the experiments. Sensitivity down to 10^{-12} molar BPE was obtained. The calculated signal-to-noise ratio at 10^{-12}M, based on photon counting statistics equals 100 for an integration time of 10 seconds. Thus, at 10^{-14}M solution which is just barely detectable, it is informative to calculate the number of fluorescent molecules within the evanescent wave zone. Using an estimated value of λ/10 for the effective evanescent penetration depth, approximately 100 molecules are detectable. Although single chemical species, containing eight rhodamine-6G fluorophores have been detected previously by D. C. Nguyen of Los Alamos Labs[7], the sample volume of 50 femtoliters gives a minimum detectable concentration of ~2 x 10^{-10}M.

The angular dependence for the fluorescent signal minus background normalized to the distal end thruput optical power is shown in Figure 4. Because the signal varied with time during the course of these experiments, perhaps due to adsorption, these preliminary data should be viewed qualitatively and therefore not in disagreement with the model. The normalized fluorescence does, however, change significantly with launch angle, viz., ~100x and this does indicate evanescent wave interaction.

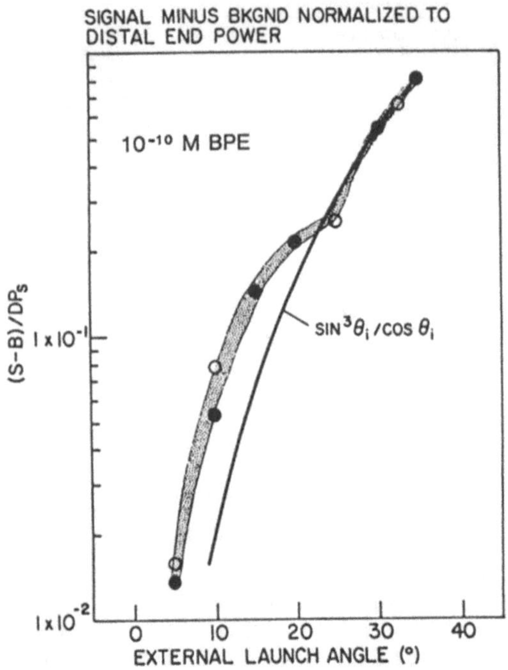

Figure 4.

Normalized fluorescence versus launch angle for 10^{-10} molar BPE.

A comparison of the sensitivity achieved in this paper for dilute BPE solution with previous studies using FITC solution[4,5] has been made. Approximately four orders of magnitude, viz., 10^4x, increase in sensitivity at equal S/N ratios has been obtained. Fluoroimmunoassay for the enzyme, CKMB coupled with BPE, has been measured using a tungsten-halogen lamp and apparatus similar to reference 4. The data appears to scale correctly relative to CKMB-FITC based on extinction coefficients and optical filtration, viz., ~10x. Results for CKMB-BPE fluoroimmunoassay using the optical system of Figure 1 will be given at the conference.

In summary, a high sensitivity fiber optic evanescent wave sensing system has been developed with application to fluorescence immunoassay. This BPE dye, laser system demonstrates sensitivity of 10^{-12} molar solution. Finally, the launch angle experiments agree qualitatively with the model and indicate that evanescent wave interaction is responsible for the measured fluorescence.

References

1. T. B. Hirschfeld and M. J. Block, paper presented at FACSS Eleventh Annual Meeting, September 1984.

2. J. D. Andrade, R. A. VanWagenan, D. E. Gregonis, K. Newby and J. N. Lin, IEEE Trans. Electron Devices ED-32, 1175 (1985).

3. R. M. Sutherland, C. Dahne, J. F. Place, and A. S. Ringrose, Clinical Chemistry 30, 1533 (1984).

4. W. F. Love and R. E. Slovacek, paper 6.6 presented at the Fourth International Conference on Optical Fiber Sensors OFS '86, Tokyo, Japan (1986).

5. W. F. Love and L. J. Button, paper presented at SPIE O/E Fiber Lase, Boston, MA, September 1988, (Volume 990).

6. M. N. Kronick, J. Immunological Methods 92, 1 (1986).

7. Sensor Technology, P. 3, June 1987.

Development of pH-Sensitive Substrates for Optical Sensor Applications

G.E. Badini[1], *K.T.V. Grattan*[1], *A.W. Palmer*[1], *and A.C.C. Tseung*[2]

[1]Measurement and Instrumentation Centre, Dept. of Electrical, Electronic and Information Engineering, Northampton Square, London EC1V 0HB, UK
[2]Chemical Energy Research Centre, Dept. of Chemistry and Biological Chemistry, University of Essex, Colchester, Essex CO4 3SQ, UK

The sol-gel process is discussed as applied to preparing a suitable support for optical fibre pH sensing technology. Two preliminary approaches are described; incorporating FITC (a pH-sensitive fluorophore) into a sol which is coated on a glass substrate and also a method where FITC is impregnated into the structure of porous gel monoliths. Fluorescence properties thus produced are reported.

INTRODUCTION

The use of optical techniques for the monitoring of pH by colourimetry has a long history, but in recent times fibre optic techniques have been applied to this particular sensor problem. Fluorimetric techniques have long been recognized as superior to those relying upon absorption but so far have been applied little to fibre optic sensor schemes for pH determination. Although the colourimetric techniques may be applied to the monitoring of pH changes and end points in titration analysis with relative ease[1], the development of a transducer with the pH sensitive material incorporated has presented more difficulties. Several patents have been issued and many schemes, mostly absorptiometric, have been discussed in the literature. An excellent review of these has been carried out by Wolfbeis[2] where many of the applications of optical pH sensors to medical analysis, water and pollution, process control and biosensing,amongst others are discussed. However, in spite of the considerable research effort in the field, as yet few effective fibre optic pH sensors have been discussed. Problems of high cost, sensitivity, stability and response time need to be addressed.

Porous glass has often been used as a support for chemically sensitive indicators. Essentially, it provides many of the properties which are desired in a sensor substrate, including ruggedness, high specific area and a surface chemistry which lends itself to simple immobilization routes[3]. Unfortunately, it is fairly expensive (about £100 per gram) and is only readily available in granular form; the shape of the glass would be tailored to a particular requirement. Alternatives to glass include many types of polymers which are often capable of providing a useful sensing layer.

Springer Proceedings in Physics, Vol. 44 Optical Fiber Sensors
Editors: H.J. Arditty · J.P. Dakin · R.Th. Kersten

However, it is quite common for the polymers themselves to exhibit optical properties which can interfere with the detection procedure, as for example both polystyrenes and polyethylenes fluoresce in the blue or violet under UV excitation. In the novel approach in this work, an attempt has been made to apply sol-gel technology to develop a support which can offer the combined advantages of porous glass and polymers without some of their limitations.

THE SOL-GEL PROCESS

Although sol-gel synthesis was reported as early as last century, it was not until the mid-seventies that interest in this technology really started to take root. There is no formal widespread definition of the sol-gel process, although it can be summarized it as a chemical phenomenon in which a single or multicomponent metal oxide solution undergoes gelation to form a coherent rigid network of the oxides present. The reason it is of interest is that it has the potential to be used to produce "tailor-made" glasses and ceramics with wide-ranging applications and also it allows for the lower temperature treatments than is normally possible in the production of conventional glasses. One of the reasons of the hitherto limited number of uses of the sol-gel process is the lack of understanding and hence controllability of the reaction parameters.

There are three distinct stages to the sol-gel process:-

(i) Mixing of the precursors in solution to achieve intimacy on an atomic scale.

(ii) Geling of the solution in a manner which will retain the chemical homogeneity achieved in step (i).

(iii) Thermal treatment of the gel to convert it to a glass.

In a typical sol-gel process, there are four main reaction components : metal oxide source, water, solvent and catalyst. The most common source of metal oxides are the corresponding alkoxides although other metal organics as well as inorganic compounds have also been used. The preparation of shaped glasses is illustrated[4] in figure 1. Crystalline an non-crystalline ceramics can also be produced by altering one or more of the reaction conditions.

The properties of an oxide gel that make it useful in this instance are not only its potential to be advantageously shaped, but also the feature of its porosity and hydroxylated surface. Hence, the objective of forming porous glass-like coatings or monoliths seems attainable in principle and herein work has been carried-out to try and achieve this goal.

In most cases alkoxy compounds of the type $R_nSi(OR')_{4-n}$ are used as raw materials and the process can normally be described by two types of reactions:-

FIGURE 1. - THE SOL-GEL PROCESS.

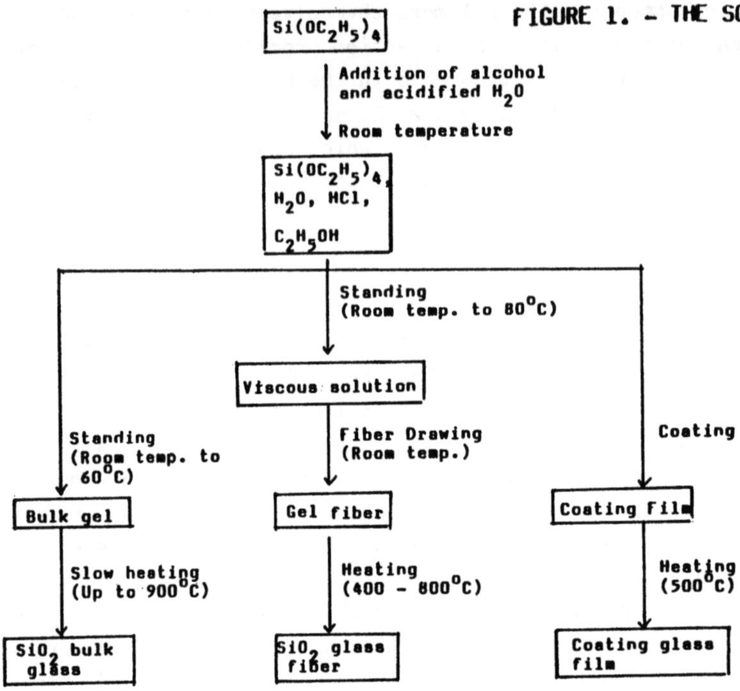

$$\text{Si-OR} + H_2O \underset{\text{esterfication}}{\overset{\text{hydrolysis}}{\rightleftharpoons}} \text{Si-OH} + \text{ROH} \qquad (1)$$

$$\text{Si-OR} + \text{HO-Si} \underset{\text{alcoholysis}}{\overset{\text{alcohol condensation}}{\rightleftharpoons}} \text{Si-O-Si} + \text{ROH} \qquad (2a)$$

$$\text{Si-OH} + \text{HO-Si} \underset{\text{hydrolysis}}{\overset{\text{water condensation}}{\rightleftharpoons}} \text{Si-O-Si} + H_2O \qquad (2b)$$

The hydrolysis and condensation polymerisation reactions occur simultaneously. Figure 2 summarizes some of the most important features of the process. In fact, there are many other factors which influence the growth and subsequent physical properties of a gel, resulting in a process which is quite difficult to understand and is still effectively in its infancy of usefulness.

EXPERIMENTAL ASPECTS

Two investigation have been undertaken to date with a view of preparing a porous glass-like support on which can be immobilized a pH sensitive fluorophore and which can be utilized in an optical fibre sensor

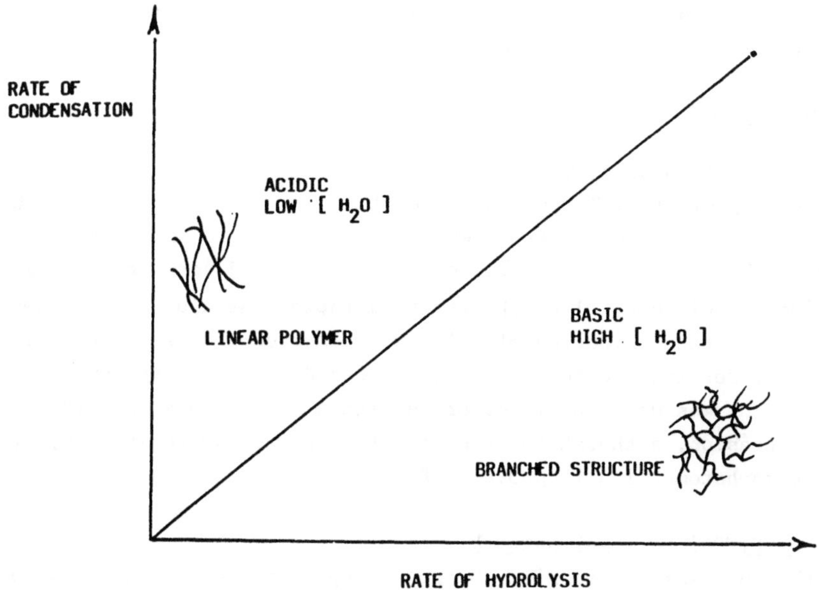

FIGURE 2. AFFECT OF VARIOUS PARAMETERS ON SOL-GEL PROCESS.

configuration. The reagent used is fluorescein isothiocyanate [3326-32-7] which has a pK_a value of around 6.5.

Preparation of sol-gel coatings incorporating FITC

Sol-gel mixtures incorporating FITC were prepared and untreated microscope slips were used as the substrates. The coatings were prepared by mechanically withdrawing the glass slips form the mixtures at a constant rate ($0.29 cms^{-1}$) and allowing them to dry at room temperature. UV-Visible spectral analysis was carried out on some of the slides in order to see if it was possible to detect the presence of FITC in the coatings.

Preparation of FITC containing monoliths

A coating at the fibre tip or surface is probably preferable in most instances, but there are situations in which it can be envisaged that a monolithic substrate could also be useful. For example, this could be to provide a larger signal intensity or to increase the longevity of the sensor by immobilising a larger quantity of the fluorophore. Ideally, it is desirable to be able to cast the derivatized gel in a configuration which would be ready to incorporate in the probe, but as an alternative, it must at least be possible to cut the glass-like gel to size.

Two approaches were used to prepare the FITC-containing monoliths; in the first, the gel was prepared and then the FITC was impregnated into its network, and in the second the FITC was incorporated at the sol stage.

Fluorescence analysis of these gels were carried out using an optical fibre link to a commercial spectrophotometer.

RESULTS AND DISCUSSION

Sol-gel coatings containing FITC

The wetting ability of the FITC-containing mixture was very dependent on the concentration of water used; better wetting was achieved with the lower concentrations. Most of the coatings prepared exhibited very few surface flaws at ambient conditions, although cracking rapidly developed if heated. The characteristic green colouring of FITC was observed in the coated films and some evidence for its presence could be detected by UV-VIS spectroscopy. A similar approach was used to incorporate methyl red into a coating and even on that occasion, although the red colouration was quite visible, its detection spectrophotometrically proved difficult.

Impregnation of FITC into a gel network

Gelation of the sols was accompanied by phase separation and occurred after 24 to 48 hours, depending on the compositions of the mixtures. However, ageing in the sealed containers at 80°C was continued for a total of 11 days prior to the commencement of slow drying. During the ageing period no noticeable shrinkage of the gel occurred. Subsequently, the volumes decreased significantly with drying and although no quantitative measurements were made, it was apparent that as the gels became dryer, they became more and more glass-like. For example, it became progressively difficult to cut the gels as they contracted in size, especially in the case of the sample having the longest drying time. However, it should be pointed out that the gel-glass transition would not be expected to occur at the low temperatures used in these preparations.

As the FITC impregnated the pores of the polymeric network, the gels developed a strong green fluorescent appearance. At this stage the gels which had not been exposed to 3-aminopropyltriethoxysilane (3APTS) [919-30-2] seemed to fluoresce more strongly than their counterparts. However, the FITC from these gels had a tendency to wash-out much more quickly and to a much greater degree than from the gels which had been treated with the silylating agent. This seems to indicate that FITC can be irreversibly bound to a gel by a method similar to that used for its immobilization onto a conventional silica surface[5]. Fluorescence spectra are shown of the some of the two types of gels (Fig. 3) indicating that the FITC is in fact less likely to wash out from gels in which 3APTS was used. However, it was noticed that a significant amount of FITC was nevertheless removed after prolonged washings. Also, a fluorescence response of one such gel under conditions of varying pH is also shown in Fig. 4. The curve is very similar to that expected for FITC in solution. It may be possible to use the impregnated gels in their present formats for some limited

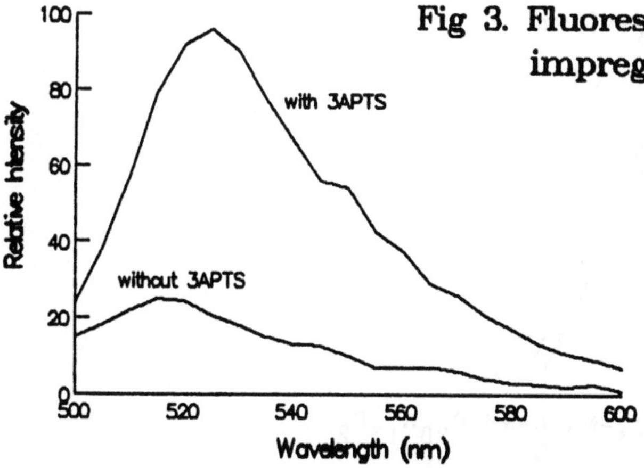

Fig 3. Fluorescence of FITC-impregnated gels

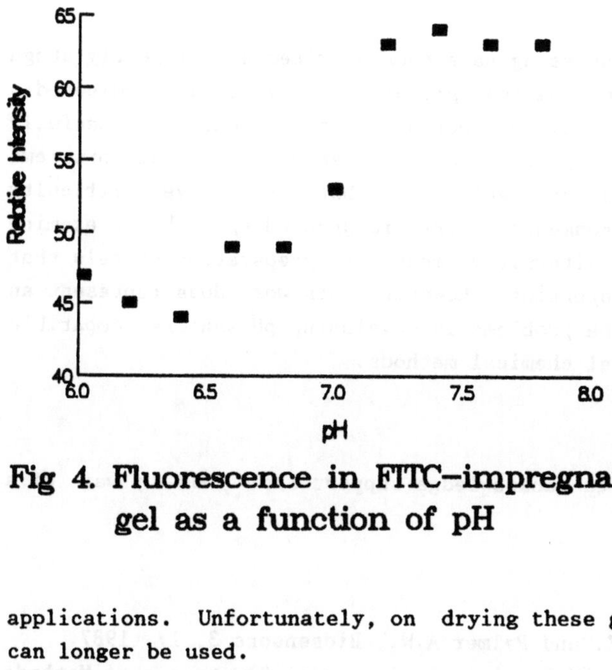

Fig 4. Fluorescence in FITC-impregnated gel as a function of pH

applications. Unfortunately, on drying these gels become very friable and can longer be used.

The above observations can be explained by considering the structure and surface chemistry of a gel which has aged and partially dried. Basically, it consists of a polysiloxane [(Si-O-Si)$_n$] network with hydroxyl groups (-OH) and ethoxy groups (-OEt) along the chain. The fact that the FITC penetrates throughout the gel indicates that the porosity is continuous, and the reduced tendency of the dye to be washed out when 3APTS is used suggests that the 3APTS can act as a linkage group by covalently bonding one end of its molecule to the gel surface and the other end to the FITC. This is illustrated in the scheme below.

$$-OH \quad + \quad H_2N(CH_2)_3Si(OEt)_3$$

[3APTS]

$$\rightleftharpoons$$

$$-O-Si-(CH_2)_3NH_2 \quad + \quad FITC$$

$$\rightleftharpoons$$

$$-O-Si-(CH_2)_3-NH-\underset{S}{C}-NH-C_{20}H_{13}O_2$$

SUMMARY AND COMMENTS

Several aspects of sol-gel processing have been examined at this early stage in the work. To date, gels and coatings prepared with FITC have indicated a novel approach to preparing a suitable porous substrate which can usefully be incorporated into a fibre optical chemical sensor. The main problems encountered with both types of gels, is that they have difficulty withstanding changes of environments. Work is proceeding both to examine these effects and to look at alternative means of preparation of gels that may have stronger physical properties. However, this work does represent an attempt to overcome some of the problems in developing pH sensors compatible with optical fibres using novel chemical methods.

ACKNOWLEDGEMENT

One of us (GEB) would like to acknowledge support from SERC by way of a studentship.

RERFERENCES

1. Grattan K.T.V., Mouaziz Z. and Palmer A.W., Biosensors 3, 17, 1987.
2. Wolbeis O.S., Chapter in "Molecular Luminescence Spectroscopy: Methods and Applications - Part II", John Wiley & Sons, 1988.
3. Badini G.E., Grattan K.T.V., Palmer A.W. and Tseung A.C.C., Proc. CLEO'89, Pub: IEEE/OSA, New York, NY USA, 1989.
4. Sakka S., Ceramic Bulletin 64(11), 163(1985).
5. Matlin S.A. et al., J. HRC&CC, 81(1981).

A Refractometer with a Fully Packaged Integrated Optical Sensor Head

W. Konz, A. Brandenburg, R. Edelhäuser, W. Ott, and H. Wölfelschneider

Fraunhofer-Institut für Physikalische Meßtechnik, Heidenhofstr. 8
D-7800 Freiburg, Fed. Rep. of Germany

A fully packaged refractometer in integrated-optics technology on glass has been developed. The sensor element is formed by two integrated-optical Fabry-Perot (IOFP) resonators to measure n and T. The complete sensor consists of the sensor head with the connecting fibers, the signal processing unit and a data recorder. The fibers and the housing are mounted at one end of the sensor chip with the other end being exposed to the medium to be measured.

Introduction:

One of the most important fields in optical measuring techniques is refractometry which is also needed for industrial process control applications.

Fiber optic refractometers with U-shaped or conically polished fiber tips have been proposed and realized earlier (Harmer, /1/; Spenner, et al /2/). These approaches did not allow the compensation of temperature fluctuations and errors due to stray light produced by turbidity. The refractive index is measured by evaluating the phase shift of the guided light which is influenced via the evanescent field by the measured liquid. We use a Fabry-Perot interferometer formed by a straight waveguide with mirrored end faces.

The refractometer for measuring refractive index changes in liquids consists of the sensor head with the connecting fibers, the signal processing unit and a data recorder as shown in Fig. 1. The integrated optical Fabry-Perot (IOFP) resonator is shown in Fig. 2. When operated with a coherent light source such as a laser diode, the transmitted intensity is given by Airy's formula:

$$I_T(\phi) = I_0 \cdot (1-R)^2/((1-R)^2 + 4R\sin^2(\phi/2)) \qquad (1)$$

The complementary intensity of the reflected light is correspondingly

$$I_R(\phi) = I_0 \cdot (1-(1-R)^2)/((1-R)^2 + 4R\sin^2(\phi/2)) \qquad (2)$$

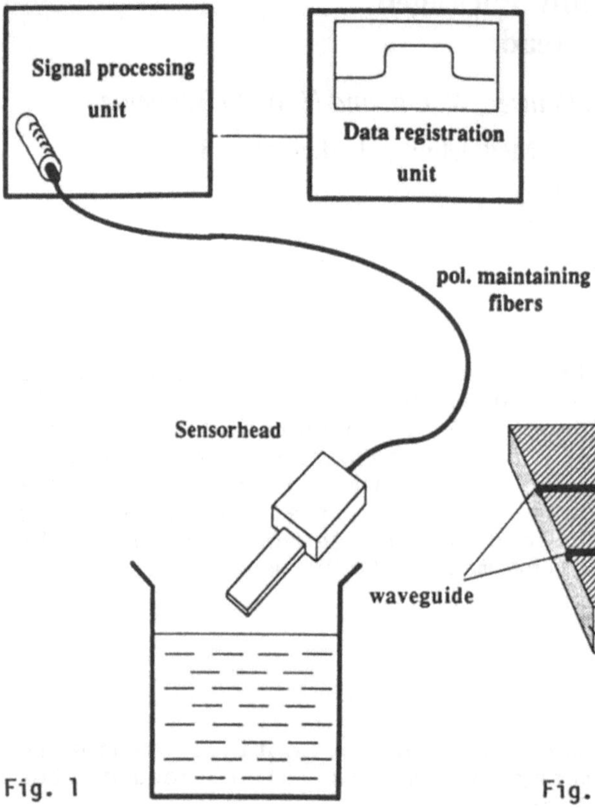

Fig. 1

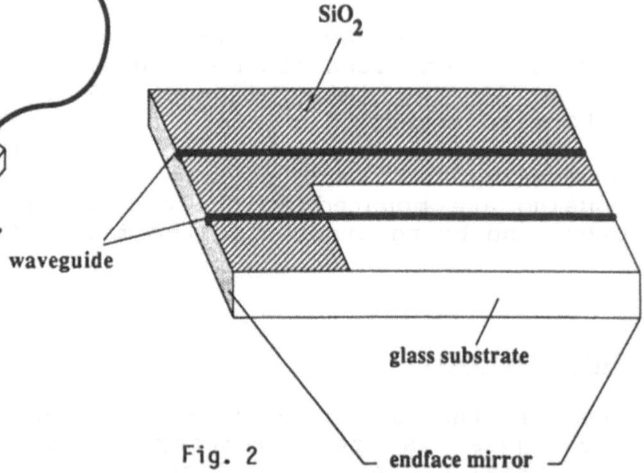

Fig. 2

R is the reflectivity of the mirrors. The internal round trip phase shift is

$$\phi = 4\pi \cdot n_{eff} L / \qquad (3)$$

where L is the resonator length and n_{eff} the effective refractive index of the guided mode.

According to (3) we get for the phase change $d\phi$:

$$d\phi = 4\pi \; {}^{-1} (n_{eff} \cdot dL + l \cdot dn_{eff}) - 4\pi \cdot n_{eff} \cdot L \cdot \; {}^{-2} d \qquad (4)$$

l is the interaction length with the superstrate.

The phase shift due to d can be neglegted since the laser diode source is sufficiently stabilized with respect to the injection current and temperature.

The measured phase change $d\phi$ originates from a change of the superstrate refractive index n_{ss}:

$$d\phi / dn_{ss} = 4\pi \; {}^{-1} \cdot (dn_{eff} / dn_{ss}) \cdot l \qquad (5)$$

i.e. via the evanescent wave and from any temperature change dT:

$$d\phi/dT = 4\pi \quad {}^{-1} (n_{eff} (dL/dT) + L(dn_{eff}/dT)) \qquad (6)$$

i.e. via the thermal expansion and the thermooptic effects.

In order to discriminate between dn_{ss}- and dT-induced influences, a second IOFP resonator spaced by 200 µm with respect to the first one and disconnected from the outer medium by a protecting layer, is implemented into the chip.

Sensor head

The essential part of the sensor head is the BK7 glass chip with the two integrated optical Fabry-Perot resonators as shown in Fig. 2. The waveguides were fabricated by K^+ ion exchange. One of the resonators is entirely covered by a SiO_2 film of 700 nm thickness. It is used for measuring changes of the chip temperature. The second resonator is uncovered over the length l and detects both parameters, n and T.

Both end faces of the chip are polished and mirrored with TiO_2-layers deposited by evaporation. The reflectance of these mirrors has been measured to be about 23 percent, and the resonator has a corresponding finesse of about 2. The length of the resonators is 36 mm, the length of the window for refractive index measurement is 22 mm.

At one side both IOFP-waveguides are coupled to polarization maintaining fibers using a Si-V-groove technique. The chip as well as the fiber coupling element are mounted and packaged in a small and handy housing.

Signal processing unit

An important feature of this refractometer in Fabry-Perot technique is the fact that it is downlead insensitive, i.e. independent of intensity variations (bending of the fibers, connector losses etc.). The measurement of the phase shift is realized with an evaluation method described by Kist et al. /3/. The cosine-shaped signal of the low finesse Fabry-Perot sensor is two times electronically differentiated. Fig. 3a shows the intensity I(t) of the Fabry-Perot as function of the time dependent parameter P(t) to be measured. The two derivatives are symmetric with respect to the zero line and hence independent on the initial DC intensity offset. According to Fig. 3b the zero crossings are used to generate pulses of rectangular shape that increment or decrement the counters V and R (Fig. 3c), respectively, depending on the sign of the change of P(t). As shown in Fig. 3d the counting pulses produce via a D/A-converter a step voltage U_A. Each step corresponds to an increment P(t) that produces a quarter

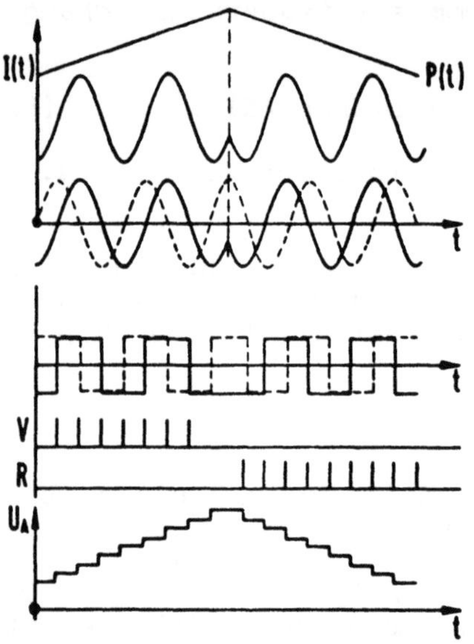

Fig. 3

a)

b)

c)

d)

fringe variation of the Fabry-Perot signal. By applying an additional low frequency triangular phase modulation and following low-pass filtering technique the output signal U_A provides a phase resolution better than 10 mrad.

Measurements

Measurements with the IOFP chip were made using various test oils of known refractive indices. The results are shown in Fig. 4. The refractive index values of the oils are indicated

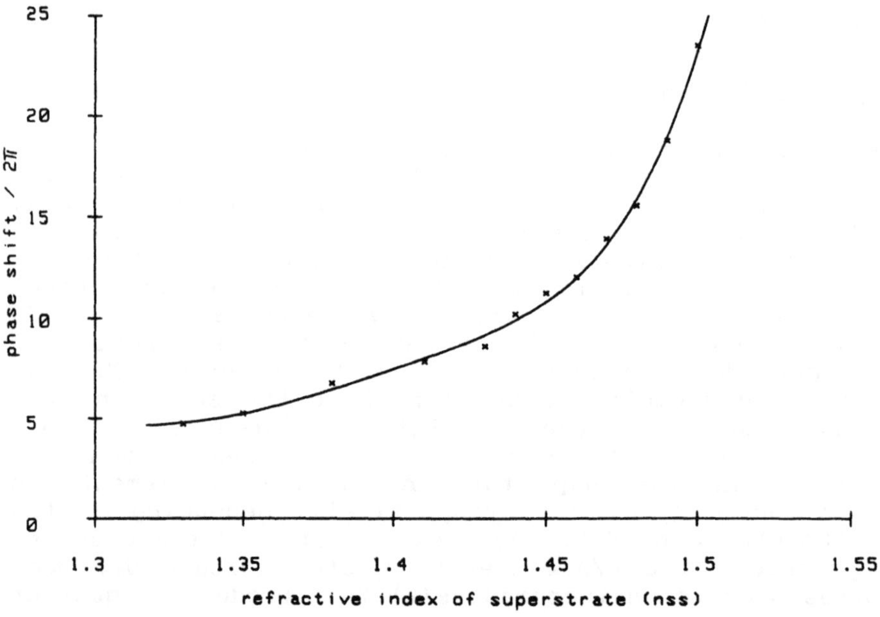

Fig. 4

by the supplier for the wavelength = 656.3 nm without any information concerning the dispersion curve, while the refractometer operates at = 840 nm.

The resolution of the refractometer chip obtained so far is $d\phi/dn_{ss}$ = 167 rad at n = 1.33 and 1818 rad at n = 1.5. Further improvements are aimed to reduce the noise floor of the sensor.

Acknowledgement

The authors wish to acknowledge contributions of various coworkers of the fiber and integrated-optics division, especially A. Fischer, M. Kublin, M. Stiefvater, Ch. Vogel.
This work was supported by the Bundesministerium für Forschung und Technologie (BMFT) of the Federal Republic of Germany.

References

/1/ A.L. Harmer, "Optical Fiber Refractometer Using Attenuation of Cladding Modes", in: First internat. Conf. opt. Fiber Sensors, April 1983. IEE London, pp. 104-108.

/2/ K. Spenner, et al., "Experimental Investigations on Fiber Optic Liquid Level Sensors and Refractometers", in: First internat. Conf. opt. Fiber Sensors, April 1983, IEE London, pp. 96-99.

/3/ R. Kist, S. Ramakrishnan, H. Wölfelschneider, "The Fiber-Fabry-Perot and its applications as a fiber-optic sensor element ", 2nd. Internat. Technical Sympos. on Optical and Electro-Optical Applied Science and Engineering, Cannes (Nov. 1985), pp. 126-133.

Part XII

Pressure Sensors

A Review of Fabry-Perot Interferometric Sensors

G.L. Mitchell

Metricor, Inc., 18800-142 Avenue Northeast, Woodinville, WA 98072, USA

Spectral modulation is a popular sensing mechanism because it provides some independence from "real world" problems such as fiber bending and connector effects. Fabry-Perot interferometers provide a popular form of spectral modulation because they can be coupled to many physical parameters (pressure, temperature, ...). Both single and multimode Fabry-Perot fiber sensors have been developed in the last decade.

Introduction

During the years since the first OFS was held substantial changes have become evident in the technology called fiber optic sensors. In the first OFS meeting, held more than six years ago, a large fraction of the papers presented concepts that could be called intensity modulation. The current OFS meeting has almost no mention of microbend, loss, or reflection sensors that only change signal intensity without implementing a reference of some sort to stabilize sensor drift.

One common method of obtaining a reference as well as a signal is to couple the parameter being measured, e.g., pressure, to a change in sensor spectral reflectance. A number of methods can be used to accomplish this, the one described here is a Fabry-Perot interferometer that has its optical "thickness" changed.

The history of Fabry-Perot sensors began at the turn of the century with derivatives of a parallel plate interferometer. Sensors for voltage and pressure were described by Perot and Fabry [1]; Meggers and Peters [2] measured refractive index. Of course, all of these sensors were constructed without the benefit of fibers or solid state light sources.

Multimode Fabry-Perot Sensors

Background

Multimode sensors offer a number of general advantages.

- light coupling efficiently with inexpensive, long-lived LED sources

Springer Proceedings in Physics, Vol. 44 **Optical Fiber Sensors**
Editors: H.J. Arditty · J.P. Dakin · R.Th. Kersten

- use of commonly available connectors, couplers, and other fiber components

- easy fiber termination.

For Fabry-Perot sensors a number of additional benefits from the use of multimode fiber can be realized:

- high volume production of sensing elements with integrated circuit fabrication technology.

- wide variety of sensor types use similar electronic and optical instrumentation.

Fabry-Perot multimode sensors have less sensitivity than single mode equivalents, however, this is frequently not a problem for industrial measurements. Multimode temperature sensors, for example, manufactured by MetriCor[3] have resolution on the order of 0.1 °C. A single mode interferometric temperature sensor can have more sensitivity. With fibers 10 cm long it can have a resolution of 10^{-5} °C using Hocker's [4] value for fiber temperature sensitivity, 100 radian $°C^{-1}$ m^{-1} and assuming an interferometer with 10^{-4} radian sensitivity.

This multimode sensitivity disadvantage is not of much concern because the dynamic ranges of all open loop interferometric sensors are of the same order of magnitude. Having more resolution than needed in this situation only decreases the effective operating range of the sensor. It is possible to overcome the dynamic range limitation if fringe counting or an equivalent technique can be employed in the instrument. Fringe counting methods generally have readout ambiguity problems when electrical power is interrupted and cannot be used for many applications.

Progress in Multimode Fabry-Perot sensors

Fabry-Perot interferometers are constructed of two reflectors on either side of an optically transparent medium. The theory of these devices is reviewed in detail by Born and Wolf[13] and Hernandez[14]. With proper spacing of the reflectors, the transmittance of the interferometer is high. Changing the spacing causes transmittance to drop. With high reflectivity reflectors, the transmittance is very sensitive to changes in wavelength or reflector spacing.

Overall interferometer performance is frequently characterized by finesse $F = 4R / (1 - R)^2$. Where R is the reflectance of lossless reflectors. High finesse interferometers are useful because they have precisely located spectral features; low finesse interferometers allow linear operation over a larger range of the sensed parameter without complex feedback schemes.

Early work with fiber optic Fabry-Perot sensors at the University of Washington by Christensen[5] demonstrated 0.5% accuracy for a temperature sensor with 150°C range. Others have developed a variety of sensors using both low and high finesse etalons. Cox and Jones[6] comprehensively describe the key Fabry-Perot fiber sensor concepts that are currently popular.

Commercialization of fiber Fabry-Perot sensors has progressed from individually assembled prototypes to devices that can operate over an extended temperature range (James, et al.[7], Quick, et al.[8], Berthold and Ghering[9]. Other refinements in sensor designs have produced a family of sensors that are fabricated using integrated circuit technology to achieve low cost and sensor-to-sensor uniformity (Saaski, et al.[10] , Hartl, et al.[11]. Some temperature sensors have been designed that can be deposited on the end of the fiber simply by evaporation of a 0.7 micron thick layer of silicon (Schultheis, et al.[12]).

A variety of light sources have been used for multimode fiber Fabry-Perot sensors. Generally the source width is the critical parameter of interest and this is determined by the specific readout scheme being used. A closed loop readout allows the use of a high finesse interferometer and laser source. In this example, the laser wavelength tracks interferometer resonance shifts. White light sources can be used if a spectrophotometer or equivalent is available. LED sources are the most common and can be applied to a variety of interferometer and readout designs. For some sensors, two sources can be used to reduce errors from system losses or interferometer ambiguities (Dakin, et al.[15, 16]).

Typical sources are GaAlAs LEDs with center wavelengths in the 850 nm region. Longer wavelength LEDs have greater spectral widths. Several different interferometers can be used with this source. One has a large (say 30 micron) micron reflector separation; another may have only a 1.0 micron separation. The 30 micron separation case has many resonance cycles across the LED width. In this case the LED is merely a broadband source that allows tracking of the resonance peaks and valleys. James, et al.[7], Quick, et al.[8], Belsley, et al.[17], and Beheim, et al.[18] have demonstrated that this arrangement is useful for a number of physical measurements.

The 1.0 micron reflector separation case is typical of sensors which have small separation yielding spectral width comparable to or larger than a LED source width (Cox and Jones[6]). These systems can probe the interferometer resonance using the LED and readout analyzers to linearize the sensor performance (Saaski, et al.[10]).

Multimode sensor applications

Fabry-Perot interferometers are attractive for sensor applications because they can be easily coupled to the physical or chemical variable being sensed. This is illustrated in the case of a temperature sensor by the following list. Each item represents a different optical parameter that can be the link between a temperature change and optical resonance of the interferometer.

- Linear expansion of a spacer ring
- Refractive index of the medium betweeen reflectors.
- Expansion of medium between reflectors
- Bowing of a reflector
- Reflector absorption of reflectivity
- Spectral absorption or scattering in the medium

Fortunately, it is possible to design the interferometer to only sense one or two of the above, eliminating response to interfering parameters. This allows construction of, for example, temperature sensors that are not pressure sensitive.

Fabrication of sensors

Fabrication methods for multimode sensors vary from large assemblies of machined quartz to thin films deposited on the end of a fiber. The variety of designs is illustrated in figures 1 and 2. Two temperature sensors are shown in figure 1. Here a spacer is used to vary the interferometer characteristics. In (a) the spacer is in the light path affecting optical spacing by both its thickness and index change with temperature. Figure 1 (b) shows a spacer that only affects the spacing by its expansion characteristics.

A family of Fabry-Perot sensors is shown in figure 2. All of these sensors are fabricated with the tools used by integrated circuit manufacturers. Typically the overall sensor size is 0.5 x 0.5 x 0.2 mm. The pressure, temperature, and refractive index sensor all operate with the same readout optics and electronics.

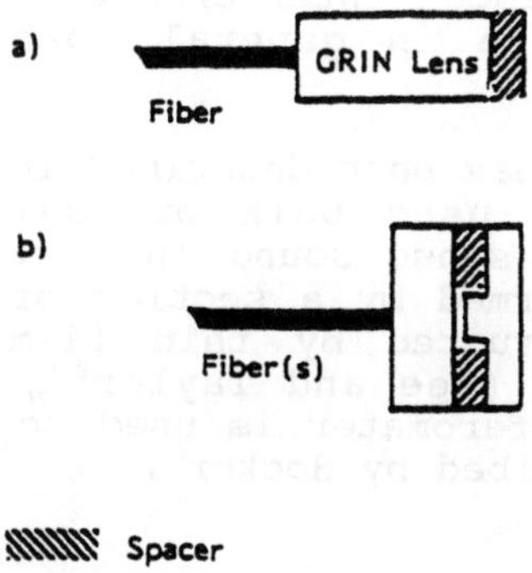

Figure 1

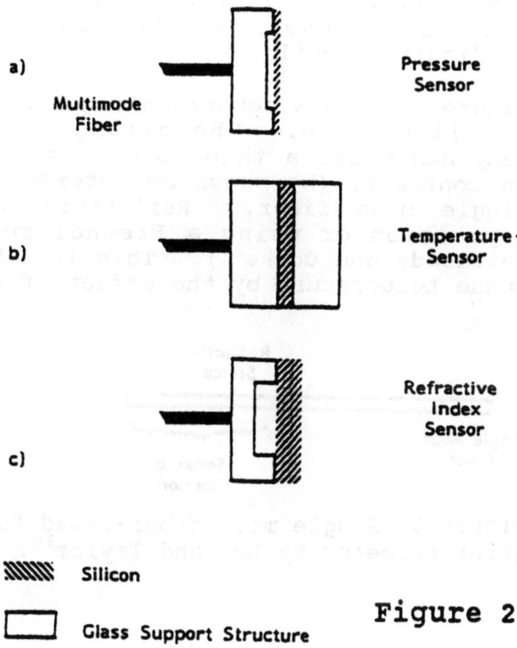

Figure 2

Figure 1 Temperature sensors using expansion of a spacer to vary interferometer spacing. These sensors were described in OFS 1 by Cox and Jones[6], Quick, et al.[8].

Figure 2 Fabry-Perot sensors for (a) pressure (b) temperature (c) liquid refractive index from Saaski, et al.[10] presented in OFS 4.

Readout techniques

Interferometric sensors can use a single photodiode as a simple intensity readout, however, performance is limited by the nonlinear nature of sensor reflectance response. This is a particularly difficult situation with high finesse Fabry-Perot devices. The most flexible readout using hundreds of photodiodes and dispersive optics can be designed to provide noise cancellation and linear operation over many cycles of interferometer displacement (James, et al.[7], Cox, et al.[6], Belsley, et al.[19a]).

Other multimode Fabry-Perot readout arrangements include servo driven interferometers (Beheim, et al.[18], Fritsch and Beheim[19b]) and two-wavelength source decoding readouts (Dakin, et al.[15]).

Single mode Fabry-Perot sensors

Background

Single mode Fabry-Perot sensors generally use laser sources and can form the sensing interferometer within the fiber itself. The long coherence length of laser sources allows some design flexibility that is not available with multimode interferometric sensors, however, single mode systems are in general more difficult to build.

Figure 3 shows a sensor configurations that has been described in the literature. The arrangement in (a) uses bulk optical components and a thin polymer diaphragm to sense sound in air. In contrast, (b) shows an interferometer formed by a section of single mode fiber. Reflectors are constructed by thin film deposition or using a Fresnel reflection (Lee and Taylor[20], Leilabady and Corke[21]). This in-fiber interferometer is used to sense temperature by the effects first described by Hocker[4].

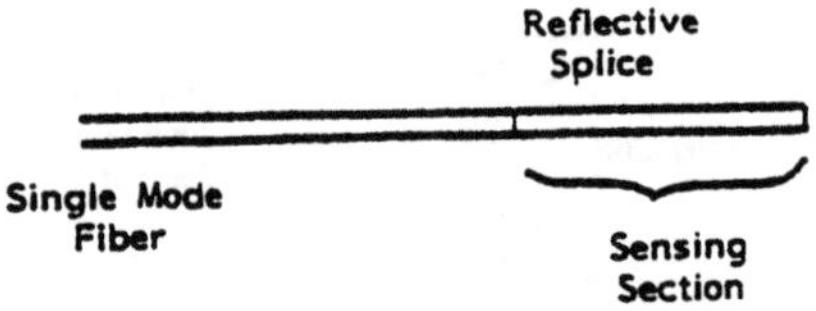

Figure 3 Single mode fiber-based temperature sensing interferometer by Lee and Taylor[20].

With few exceptions (Kersey, et al.[22]), in-fiber Fabry-Perot interferometers are fundamentally different from Mach-Zehnder interferometers. Fabry-Perot interferometers do not require servo control of a fiber stretcher or management of a reference fiber but they do not allow much more than a coherence length of sensor length. The nonlinearities of high finesse Fabry-Perot sensors require special readout techniques when they are used for practical applications.

Single mode readout options

Simple intensity readouts have been used (Matsumoto, et al.[23], Lee, et al.[24a], Lee and Taylor[20]) as well as more complex arrangements using Michaelson interferometers (Farahi, et al.[24b]). Various servo control systems have been employed (Leilabady, et al. [25], Leilabady and Corke[21]). In this readout, laser wavelength is swept through the range of sensor resonance observing reflected light at the signal photodiode. This has the flexibility of a spectrophotometer readout with much less complexity.

With a bulk interferometer readout system it is possible to use sensors that have a length somewhat greater than the laser coherence length. This allows a number of different-length sensors to be formed on the same fiber and multiplexed by the readout interferometer.

Specialty fibers can be used for simultaneous detection of two polarizations (Leilabady, et al.[26], Blake, et al.[27]) in applications where polarization maintaining fibers are used. These fibers can allow simultaneous measurement of two related parameters such as strain and temperature.

References

1. Perot, A., Fabry, C. H., Sur un voltmètre électrostatique interférentiel pour étalonnage, **J Physics**, 7, 650-659, (1898)

2. Meggers, W. F., Peters, C. G., Measurements on the index of refraction of air for wavelengths from 2218 Å to 9000 Å**Bull Bur Stds**, 14, 697-740, (1918)

3. Metricor, Inc. Data sheet for model 1400 multisensor system, Woodinville, WA, USA, (1988)

4. Hocker, G. B., Fiber-optic sensing of pressure and temperature, **Applied optics**, 18, 1445-1448, (May 1, 1979)

5. Christensen, D. A., Temperature measurement using optical etalons, Annual Meeting Optical Society of America, (October 1974)

6. Cox, E. R., and Jones, B. E., Fibre optic colour sensors based on Fabry-Perot interferometry, First International Conference on Optical Fibre Sensors, London, England, (April 1983)

7. James, K. A., Quick, W. H., Strahan, V. H., Optical temperature sensors for propulsion control systems, NASA Technical Report #C-81-341/201, (1981)

8. Quick, W. H., James, K. A., and Coker, J. E., Fiber optics sensing techniques, First International Conference on Optical Fibre Sensors, London, England, (April 1983)

9. Berthold, J. W., Ghering, W. L., Varshneya, D., **IEEE J. of Lightwave Technology** LT-5, 1-6, (July 1987)

10. Saaski, E. W., Hartl, J. C., Mitchell, G. L., Wolthuis, R. A., Afromowitz, M. A., A family of fiber optic sensors using cavity resonator microshifts, Fourth International Conference on Optical Fiber Sensors, Tokyo, Japan, (1986)

11. Hartl, J. C., Saaski, E. W., Mitchell, G. L., Fiber optic temperature sensor using spectral modulation, **Proceedings of SPIE**, 838, 257-261, (1988)

12. Valente, L. C. G., Camargo, S. S., Jr., Soares, E. A., Temperature measurement utilizing hydrogenated amorphous silicon films, SPIE Fiber Optic & Laser Sensors V, **838**, 247, (1987)

13. Born, M., Wolf, E. **Principles of Optics**, Pergamon Press, Oxford, (1964)

14. Hernandez, G. **Fabry-Perot Interferometers**, Cambridge University Press, Cambridge, (1986)

15. Dakin, J. P., Wade, C. A., Withers, P. B., An optical fibre pressure sensor, **International J. of Optical Sensors**, 2, 191-199, (June 1987)

16. Dakin, J. P., Wade, C. A., Withers, P. B., An optical fibre pressure sensor, **Proceedings of SPIE**, 734, 194-201, (1987)

17. Belsley, K. L., Huber, D. R., and Goodman, J., All-passive interferometric fiber-optic pressure sensor, Instrument Society of America Annual Meeting, Paper #86-2801, (1986a)

18. Beheim, G., Fritsch, K., and Anthan, D. J., Fiber-optic temperature sensor using a spectrum-modulating semiconductor etalon, SPIE O-E/Fibers '87, also NASA Technical Memorandum 100153 SPIE, San Diego, California, (August 1987)

19a. Belsley, K. L., Huber, D. R., and Goodman, J., All-passive fiber-optic pressure sensor, **InTech**, 33, 39-42. (Dec 1986)

19b. Fritsch, K, Beheim, G., Wavelength-division multiplexed digital optical position transducer, **Optics Letters**, 11, 1 - 3, (1986)

20. Lee, C. E., Taylor, H. F., Interferometric opticsl fibre sensors using internal mirrors, **Electronics Letters**, 24, 193-194, (Feb 1988)

21. Leilabady, P. A., Corke, M., All-fiber-optic remote sensing of temperature employing interferometric techniques, **Optics Letters**, 12, 772-774, (Oct 1987)

22. Kersey, A. D., Jackson, D. A., Corke, M., **Optics Communications**, 45, 71-74, (March 1983)

23. Matsumoto, M., Fujimura, K., Hattori, K., Naono, H., Fiber-optic acoustic sensor based on Fabry-Perot interferometer, Fourth International Conference on Fiber Optic Sensors, Tokyo, Japan, (1986)

24a. Lee, C. E., Atkins, R. A., and Taylor, H. F., Performance of a fiber-optic temperature sensor from -200 to 1050°C, **Optics Letters**, 13, 1038-1040, (Nov 1988)

24b. Farahi, F., Newson, T. P., Jones, J. D. C., Jackson, D. A., Coherence multiplexing of remote fibre optic Fabry-Perot sensing system, **Optics Communications**, 65, 319-321, (March 1988)

25. Leilabady, P. A., Jones, J. D. C., Jackson, D. A., **Optica Acta**, 32, 233-240, (Feb 1985)

26. Leilabady, P. A., Jones, J. D. C., Jackson, D. A., Monomode fiber-optic strain gauge with simultaneous phase- and polarization-state detection, **Optics Letters**, 10, 576-578, (Nov 1985)

27. Blake, J. N., Li, Q., Kim, B. Y. Fifth International Conference on Fiber Optic Sensors, New Orleans, (January 1988)

A Photo-elastic Pressure Sensor with Loss-Compensated Fiber Link

G. Martens, J. Kordts, and G. Weidinger

Philips GmbH Forschungslaboratorium Hamburg, Vogt-Koelln Str. 30, D-2000 Hamburg 54, Fed. Rep. of Germany

A photoelastic fiber optic 6 bar pressure sensor has been set up in a four port polariscope configuration. The down lead insensitivity is achieved by the application of the balanced bridge referencing technique. The sensor has an excellent insensitivity to source intensity fluctuations as well as to mode dependent and independent changes of the down leads damping. The sensor has a linearity of +-0.1 % and a temperature stability of +-1 % within a range of 0 to 60 °C.

I. Introduction

The balanced bridge referencing technique is a promising technique to yield down lead insensitive signals from fiber optic intensity modulated sensors [1-5]. This technique makes use of a fiber optic four port sensor configuration. According to Fig. 1 two ports act as fiber optic inputs which are connected to two light sources S1 and S2. The other two ports act as outputs which are connected to two optical receivers E1 and E2. In the general case this set up is a variable four port coupler, whose splitting ratio K(P) is determined by the magnitude of a measurand P. The primary light intensity I_{01} is split into the two partial light intensities I_{11} and I_{21}. I_{02} is split into I_{12} and I_{22}. A distinction between I_{11} and I_{12} in the receiver E1 and between I_{21} and I_{22} in the receiver E2 can be made by time division or frequency multiplexing of the light sources. The loss compensated signal Q is defined by the following ratio:

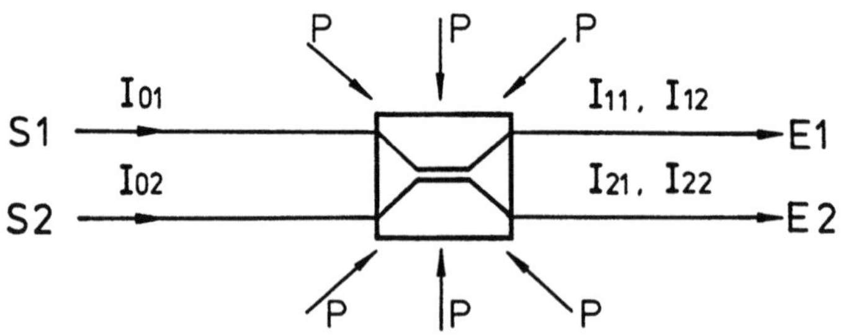

Fig. 1: Schematic set up of a four port variable coupler sensor adapted to the balanced bridge referencing technique

Springer Proceedings in Physics, Vol. 44 Optical Fiber Sensors
Editors: H.J. Arditty · J.P. Dakin · R.Th. Kersten

$$Q = I_{11}*I_{22}/(I_{21}*I_{12}) = K^2(P) \quad (1)$$

One can easily calculate that Q is independent of all damping values of the four fiber links, independent of variations of light source intensities and of the sensitivities of the receivers E. Q only depends on K and thus on the measurand P. In the following we will restrict to this variable coupler sensor set up and will discuss a photo elastic sensor configuration for pressure or differential pressure.

II. Sensor Set Up

The general set up of a four port polarimetric sensor is displayed in Fig. 2. It mainly consists of two polarizing beam splitters (beam splitting cubes) with the polarimetric sensing element in between them. The light is provided by two fibers, which after collimation to a parallel light beam by sphere or SELFOC lenses enter the polarizer at its two input ports. The light components leaving the polarizer at the common output port are linearily polarized with their polarization directions normal with respect to each other. Both light components pass the $\lambda/4$-retardation plate and the polarimetric sensing element, which changes their state of polarization due to the interaction with the measurand. The state of polarization of each light component then is analyzed by the second polarizing beam splitter giving rise to a total of four light intensity components, which are picked up by two fibers (via some focussing optics) and are guided to the two receivers.

The primary measurand of the sensor element in a photo elastic set up is a stress σ, which depends on pressures, forces or the temperature. In two and three dimensional photo-elastic applications the difference $\Delta\sigma$ of the stresses of the principal axes normal to the vector of light propagation is the primary measurand. The characteristic $Q(\Delta\sigma)$ of the sensor is given by:

$$Q = Ko*\tan^4[(1+\varepsilon)*\pi/4+\pi*C*l*\Delta\sigma/\lambda] \quad (2)$$

with C the photo-elastic constant of the sensor material, l the length of the light path under stress $\Delta\sigma$, λ the vacuum wavelength of the light and ε the mismatch of the quater wave retardation plate with respect to the wavelength. Ko is a figure describing the optical coupling of the two transmitting to the two receiving fibers when no stress is applied to the sensing element. In a well adjusted configuration Ko is equal to 1.

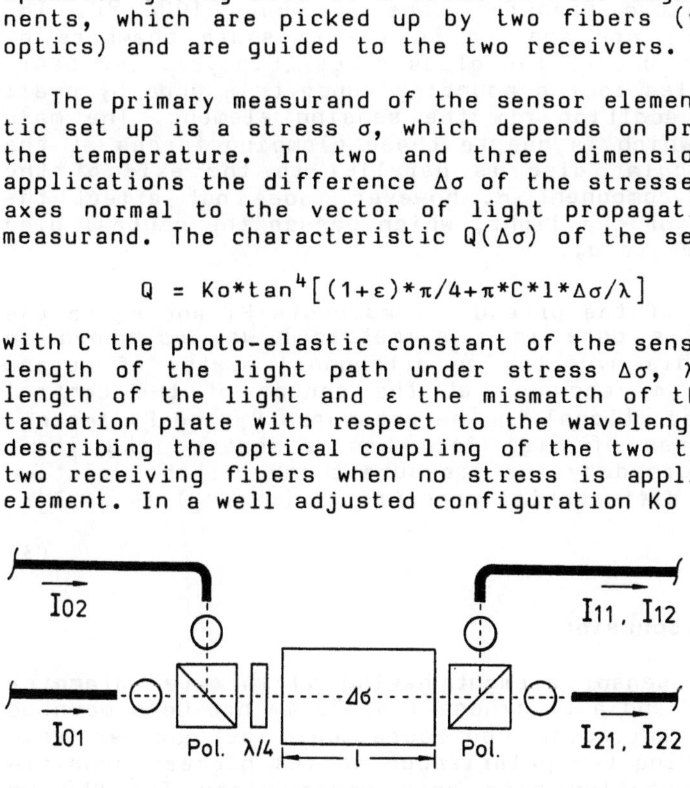

Fig. 2: Set up of the fiber optic four port polariscope

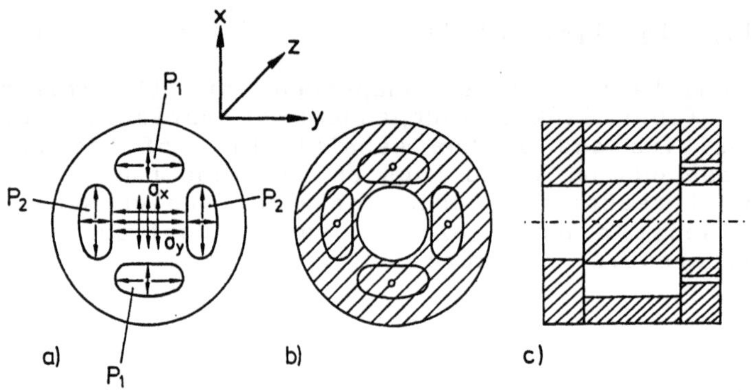

Fig. 3: Set up of the photo-elastic sensing element for pressure and differential pressure. a) cross-section of the central glass disc, b) sealing of pressure chambers by glass rings, c) axial cut of the sensor element

The photo-elastic sensor element for pressure and differential pressure is displayed in Fig. 3. The sensor element consists of a circular disc of silica glass. Four pressure chambers have been milled symmetrically into the disc (Fig. 3a). The back and front faces of the chambers have been sealed with glass rings (Fig. 3b), which have been sintered onto the central disc by a glass transfer tape at about 1000 °C. The pressures P_1 and P_2 are applied to the pressure chambers by small bore holes in one of the glass rings (Fig.3c). The sealing of the bore holes to the mounting support is done by small O-rings, which in addition fix the sensing element. The main stress component, which is due to these clamping forces at the center of the circular disc is parallel to the axis of the disc. This stress component σ_z however, does not affect the polarization state of the light, which passes the central disc at its axis parallel to σ_z.

The application of the pressure components P_1 and P_2 to the pressure chambers is done in a symmetrical way as shown in Fig. 3a. Due to this symmetry an orthogonal system of stress components σ_x and σ_y show up at the center of the central disk, which are proportional to the pressures P_1 and P_2 respectively. For the case of Fig. 3a, when a square like inner cross-section is formed by the pressure chambers, the relation between $\Delta\sigma$ and the differential pressure ΔP ($=P_1-P_2$) is [6,7]:

$$\Delta\sigma = \Delta P \tag{3}$$

III. Results and Discussion

A photo elastic sensor element having 60 mm outer diameter of the central disc and a thickness l of 30 mm has been mounted in a four port polariscope configuration. The quarter wave plate used for setting the polariscope at its highest sensitivity is a self-supporting mica zero order plate for 850 nm

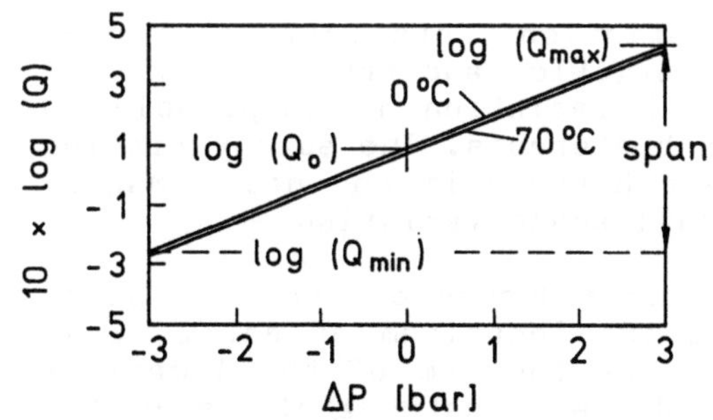

Fig. 4: Measured characteristic Q(ΔP) of the sensor in the logarithmic presentation at the temperatures of 0 and 70 °C

wavelength. The fibers (50/125 µm GI) have been adapted to the set up by commercial available ball lens connectors (series 6000, PKI/F&G-Cologne). The 850 nm light is provided by an LED via a fiber optic switch (for time division multiplexing of I01 and I02) and a mandrel wrap filter followed by a cladding mode stripper in each transmitting line. The detection of the light intensities has been performed by a dual channel power meter (Anritsu) with a .001 dB reading of the input ratio.

According to eq. 2 the characteristic Q(ΔP) of the sensor is nonlinear. For linearization the logarithm is applied to the measured characteristic yielding a straight line within a measuring range of 6 bar (see Fig. 4). The linearity with respect to the span of 6 bar is within +-0.1 %. A temperature dependence mainly shows up in a small offset.

The temperature dependence of log(Q) is analyzed in more detail in Fig. 5. The variation of the full-scale deflection log Qmax (at +3 bar) normalized to the span of 6 bar (Fig. 5a) exhibits a small hysteresis but in the temperature range of 0 to 60 °C still is within the +-1 % limits. The variation of the full-scale deflection is the superposition of the zero point

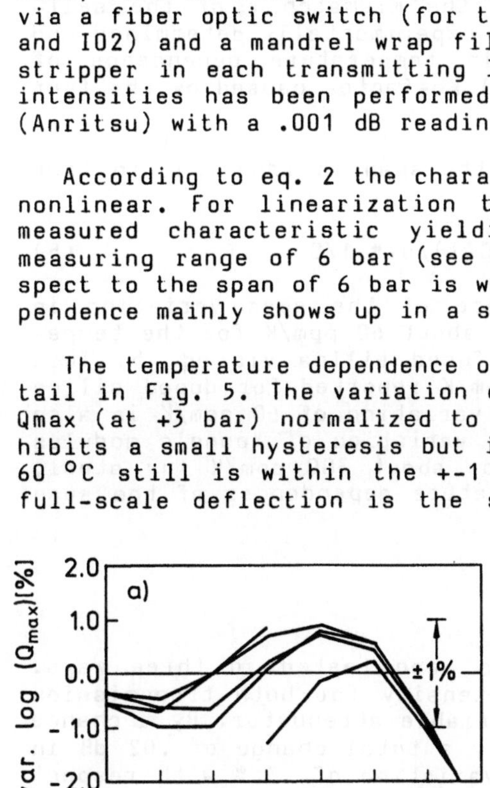

Fig. 5: Measured temperature dependence of a) the full-scale deflection log(Qmax) and b) the span with respect to the span of 6 bar

log(Qo) variation and the span variation. Both terms have been derived from the measured characteristics and are normalized to the span of 6 bar. The zero point variation is comparable to that of Fig. 5a in magnitude and hysteresis. The span variation (Fig. 5b) exhibits no significant hysteresis and has a magnitude 10 times smaller than the full scale variation.

The origin of the temperature dependences of the zero point and the span is found by taking the logarithm of eq. 2. Easy calculation yields the zero point as the sum of the logarithm of the proportionaly factor Ko and the mismatch ε of the quarter wave plate.

$$\log(Qo) = \log(Ko)+2*\pi*\varepsilon \tag{4}$$

The temperature dependence of the mismatch ε of the self-supporting glimmer plate has been experimentally determined and is smaller than .6 %. The main temperature dependence of log(Qo) is due to the hysteretical Ko-factor caused by the four port fiber to fiber coupling.

The temperature dependence of the span is identical to that of the photo-elastic constant C.

$$d(span)/dT*1/span = dC(T)/dT* 1/C \tag{5}$$

From the temperature dependence of the span variation in Fig. 5b one can derive a value of about 60 ppm/K for the temperature variation of C for pure fused silica via eq. 5. This value is small compared to 130 ppm/K reported for doped silica [8] in fibers. The observed span variation of 60 ppm/K is also small compared to the temperature variation of Young's modulus of about 100 ppm/K for silicon or about 300 ppm/K for steel, which mainly determine the temperature dependences of the span in classical pressure sensors.

IV. Down lead sensitivity

The down lead sensitivity has been tested in three ways. First the common source light intensity for both transmission channels has been changed by a variable attenuator. By a change of 6 dB in primary light intensity a total change of .02 dB in Q has been observed, which is a variation of .3 % with respect to the span of 6 bar. Second the attenuation of one transmission line between the fiber optic switch and the mandrel wrap filter has been altered up to 6dB by the variable attenuator. No significant change within the power meter reading of +-.001 dB in Q has been observed. In the third experiment a fiber coil of 15 windings at a diameter of 40 mm was inserted between the fiber optic switch/mandrel wrap filter unit and the sensor in one of the transmission lines. This bending diameter is smaller than the minimum bending diameter of 60 mm specified by the fiber cable manufacturer. As a result a total change of 0.003 dB in Q has been observed, which is identical to an error of smaller than .05 % with respect to the full span.

Acknowledgement

The support by Mr. M. LIEHR concerning the mechanical design is gratefully acknowledged. The work has been financially supported by the Bundesministerium für Forschung und Technologie BMFT.

References:

[1] B. Culshaw, F. Fowley, and I.P. Giles, "A balancing technique for optical fibre intensity modulated transducers", Proc. of the 2nd Int. Conf. on Optical Fiber Sensors (OFS'84), Stuttgart 1984, p. 117-120 and Proc. SPIE Vol. 514 (1984)

[2] I.P. Giles, S. McNeill, and B. Culshaw, "A stable remote intensity based optical fibre sensor", J. Phys. E: Sci. Instrum., Vol. 18, 502-504 (1985)

[3] I.P. Giles, D. Uttam, S. McNeill, and B. Culshaw, "Self-compensating technique for remote fibre optic intensity modulated transducers", Proc. SPIE Vol. 522, 233-239 (1985)

[4] G. Beheim and D. J. Anthan, "Loss-compensation of intensity modulating fiber optic sensors", Proc. SPIE Vol. 718 (Fiber Optic and Laser Sensors IV), 259-265 (1986)

[5] G. Beheim, "Loss-compensation technique for fiber-optic sensors and its application to displacement measurements", Applied Optics 26, 452-455 (1987)

[6] G. Martens, "A new photoelastic pressure sensor", Proceedings of EFOC/LAN85 (Inform. Gatekeepers), 222-224 (1985)

[7] G. Martens, "Faseroptische Sensoren mit spannungoptischem Sensorelement für die Kraft- und Druckmessung", Technisches Messen tm 53, 331-338 (1986)

[8] A. J. Barlow, and D. N. Payne, "The stress-optic effect in optical fibers", IEEE Journal of Quantum Electronics QE-19, 834-839 (1983)

High-Pressure Fiber-Optic Sensor Based on Polarization-Rotated Reflection

*W.J.Bock and T.R. Wolinski**

Laboratory of Optoelectronics, University of Quebec at Hull,
P.O. Box 1250, Station B, Hull, Quebec, Canada J8X3X7
*On leave from the Institute of Physics, Warsaw University of Technology,
Koszykowa 75, PL-00-662 Warsaw, Poland

A novel structure for a high-pressure fiber-optic sensor incorporating polarization-rotated reflection has been demonstrated. The sensor consists of two identical parts of the highly birefringent polarization-maintaining fiber, spliced at 90 degrees in relation to their polarization axes, and of a reflective mirror deposited on the pressure-sensing part of the sensor. In this way insensitivity to any ambient distortions equally acting on both parts of the sensor has been achieved.

1. Introduction

High hydrostatic pressure is commonly recognized as an important environmental parameter in many industrial processes and in laboratory investigations. Some attempts have been made to apply fiber-optic sensors in various configurations to the task of measuring low acoustic pressures, differential pressures and strain [1]. However, very little attention has been paid to the possibility of using fiber-optic sensors for high pressure measurements. So far, only one such sensor has been reported with measurement characteristics provided for pressures up to about 70 MPa [2].

Recently, we developed a new fiber-optic sensor for pressures up to 200 MPa based on the direct effect of hydrostatic stress on polarization-mode coupling occuring in highly birefringent (HB) polarization-maintaining fibers (PMF) [3]. The new sensor can detect high pressure in two basic configurations: transmission or reflection. In the transmission configuration, the sensing element consists of a HB PMF of pre-determined length. The fiber is fusion-spliced at both ends with two single-mode (SM) fibers of the same diameter and cut-off wavelength as the HB fiber designed to input and output the appropriate light signals to and from the HB fiber. In the reflection configuration only one input/output SM fiber is required and it is spliced with a HB sensing element. Linearly polarized light was launched into the input SM fiber then aligned at 45 degrees to the principal polarization axes of a HB PMF allowing the incoming light to excite both polarization modes in the sensing element.

In both configurations, two significant problems had to be overcome: temperature desensitization of the sensor output signal and elimination of polarization fluctuations caused by lead SM fibers. Up to now the fiber-optic high pressure sensors suffered from temperature artifact and required time-consuming termostatization. This paper describes how we applied the method of polarization-rotated reflection to construct a temperature-insensitive high-pressure fiber-optic sensor in the reflection configuration, insensitive as well to polarization fluctuations in the lead fibers.

Springer Proceedings in Physics, Vol. 44 **Optical Fiber Sensors**
Editors: H.J. Arditty · J.P. Dakin · R.Th. Kersten

2. Polarization-rotated reflection

The method of polarization-rotated reflection (PRR) was initially proposed for stabilizing temperature performance of light intensity modulators, through bidirectional use of an electrooptic crystal for the effective cancellation of its birefringence. The working mechanism of a fiber-optic sensor based on PRR was introduced in [4] and some experimental results of magnetic field and temperature sensors were presented at that time. The idea of the proposed sensor with the PRR method is based on the 90 degrees rotation of the input polarization of light when the beam is reflected back by the combination of a quarter-wave plate and a reflective mirror. In this way, two cross-polarization eigenmodes of a PMF exchange their polarization states with each other for the forward and the backward transmission through the fiber.

The PRR method was also applied to the sensor head with a birefringent crystal for the measurement by reciprocal effects [4]. By using a suitable birefringent crystal this sensor head was able to measure other physical values as well as temperature. Application of two birefringent crystals of equal lenght with principal axes of one crystal rotated by 90 degrees relative to those of the other one enabled the cancellation of the natural birefringence in light modulators of the transverse type.

3. Sensor construction

Based on PRR method and on the idea of compensated polarimetric sensor [5], we have constructed a novel polarimetric high-pressure fiber-optic sensor in compensation-reflective configuration, as presented schematically in Fig.1.

A linearly polarized light beam is launched into one of two principal axes of a HB input/output lead fiber, spliced at 45 degrees with a compensation section (L_1) of a HB fiber and spliced again at 90 degrees with a sensing section (L_2) of the same HB fiber. The sensor tip is covered by a gold compound obtained from Engelhardt Chemicals, New Jersey, which after some heating and cooling processes made a stable layer acting as a mirror with a reasonable reflective coefficient.

Let us consider the action of this type of high-pressure sensor. The phase retardation for each mode propagating in an optical fiber is given by

$$\Phi = \frac{2\pi}{\lambda} n L \qquad (1)$$

where λ is the wavelength used and n is the effective refractive index of a given mode. For a reflective configuration the effective length of a HB sensor is twice its actual length so

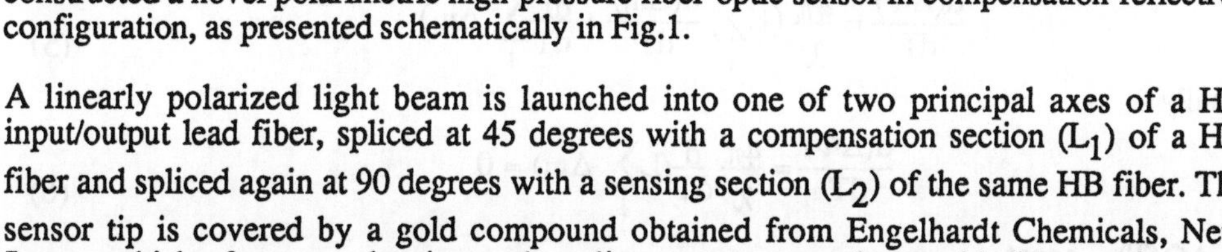

Fig. 1 Fiber-optic sensor head in compensation-reflective configuration for high-pressure measurements

that $L = 2L_i$ $(i = 1,2)$. The total relative phase retardation between the two perpendicularly polarized eigenmodes propagating in a HB fiber will be given by

$$\Delta\Phi = \Delta\Phi(L_1) + \Delta\Phi(L_2) \qquad (2)$$

where

$$\Delta\Phi(L_i) = \frac{2\pi}{\lambda}\Delta n_i\, 2L_i = \frac{4\pi}{\lambda}\Delta n_i\, L_i$$

Temperature-induced phase retardation can be expressed by

$$\frac{d(\Delta\Phi)}{dT} = \frac{4\pi}{\lambda}\sum_i \frac{d}{dT}(\Delta n_i L_i) = \frac{4\pi}{\lambda}\sum_i \left(L_i\frac{d(\Delta n_i)}{dT} + \Delta n_i\frac{dL_i}{dT}\right) \qquad (3)$$

where $\Delta n_i = n_{xi} - n_{yi}$ and n_{xi}, n_{yi} are the refractive indices for two orthogonal linear polarization modes of an i-section.

Since $L_1 = L_2 := L$ and both parts are rotated one fiber relative to the other by 90 degrees about its axis we obtain

$$\Delta n_2 = -\Delta n_1 \qquad \text{or} \qquad \Sigma\Delta n_i = 0 \qquad (4)$$

Hence

$$\frac{d(\Delta\Phi)}{dT} = \frac{4\pi}{\lambda}\left(L\sum_i \frac{d(\Delta n_i)}{dT} + \frac{dL}{dT}\sum_i \Delta n_i\right) \qquad (5)$$

Finally

$$\frac{d(\Delta\Phi)}{dT} = \frac{4\pi}{\lambda}\frac{d}{dT}\left(L\sum_i \Delta n_i\right) = 0 \qquad (6)$$

or $\Delta\Phi(T) = \text{const.}$

From (6) it follows that the new sensor will be immune to ambient temperature or any other physical parameter acting on both its parts. However, if only one part of the sensor is exposed to the influence of a physical parameter the sensor will measure this parameter. In this way we were able to construct the high-pressure fiber-optic sensor with temperature compensation.

4. Experimental set-up

The measurement set-up shown in Fig.2 represents a modification of a set-up described elsewhere [3], [6], [7]. Here, the pressure sensing part of the fiber-optic head was placed inside a standard high-pressure chamber designed to sustain pressure up to 500 MPa. The chamber and the sensor assembly were thermally stabilized with 0.01% accuracy by using an automatic computer-controlled regulator. A Harwood DWT-35 deadweight tester was used to generate and calibrate high pressure up to 105 MPa inside the chamber with a reading accuracy of 0.01% traceable to NBS.

Linearly polarized monochromatic light at wavelength of 632.8 nm was launched from a 3 mW He-Ne laser into one input of a single-mode fiber-optic bidirectional coupler, one of two outputs of which was directly spliced with a HB fiber leading down to the reflection sensor. Adjustment for the input optical power polarization along a principal axis of the HB input/output fiber was made by half-wave plate and appropriate polarization-controlling

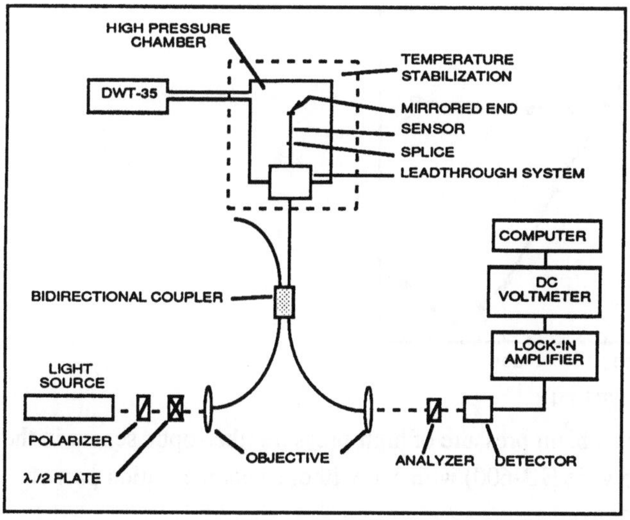

Fig.2 Experimental set-up to measure high-pressure effects in HB fibers using the reflection configuration

equipment. The set-up included a synchronous detection system composed of a silicon photodetector, polarization controllers, a lock-in amplifier and a computer-controlled digital DC voltmeter. This has been described in detail elsewhere [6].

5. Results

Fig.3 shows the initial results obtained for a high hydrostatic pressure fiber-optic sensor in the reflection configuration without thermal compensation, for two different values of sensing length: 2 mm and 8 mm. The sensor was made of York bow-tie HB-600 highly

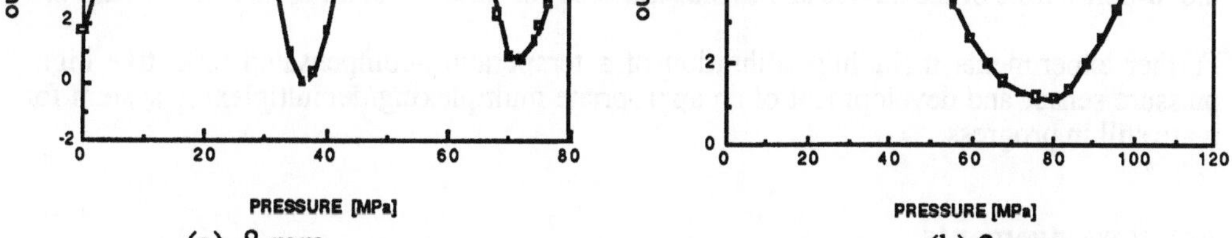

(a) 8 mm (b) 2 mm

Fig.3 Detected output signal for two different lengths of York bow-tie HB-600 fiber in the reflection configuration without thermal compensation, λ = 632.8 nm

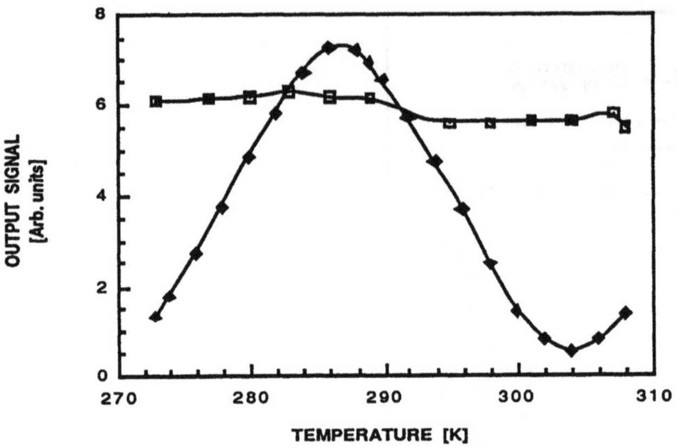

Fig.4 Thermal characteristics at ambient pressure of high-pressure fiber-optic sensor in the reflection configuration (York bow-tie HB-600) with and without a compensation part of the same HB fiber, $\lambda = 632.8$ nm

birefringent fiber, having an external diameter of 125 μm and a beat length of $L_b = 1.2$ mm at 632.8 nm and 20°C. The output signal of the sensor was found to be a periodic function of high hydrostatic pressure and was dependent on the length of the sensing fiber, similar to the other high-pressure sensors in transmission configurations described earlier [3]. Nearly the same behavior was found for the reflection sensor when a compensation part was built in and the same length of fiber was exposed to the influence of high pressure. However, a significant difference was found in the thermal characteristics of these two types of high-pressure fiber-optic sensors. The compensated sensor reveals no thermal dependence - within error limits - while the non-compensated sensor is strongly temperature-dependent (see Fig.4). Such a behavior results from the direct application of the PRR method to the construction of the high-pressure fiber-optic sensor.

6. Conclusions

We have demonstrated a novel structure for a high hydrostatic pressure fiber-optic sensor incorporating polarization-rotated reflection. The sensor consists of two identical parts of PMF of HB type spliced at 90 degrees in relation to their polarization axes and of a reflective mirror deposited on the pressure-sensing part of the sensor. The input polarization is launched at 45 deg exciting both eigenmodes of PMF. In this way we obtain a stable high-pressure fiber-optic sensor immune to any ambient distortions generally caused by temperature drift. The compensated high-pressure sensor considerably improves the response time of the device and avoids the need for time-consuming thermostatization.

Further experiments including calibration of a temperature-compensated reflective high-pressure sensor and development of an appropriate multiplexing/demultiplexing system for it are still in progress.

Acknowledgements

The authors gratefully acknowledge support for this work by the Natural Sciences and Engineering Research Council of Canada.

References

1. R.P.De Paula, E.L.Moore, "Fiber-optic sensor overview", Proc. SPIE, vol. 556, pp 2-15, 1985.

2. K.Jansen, P.Dabkiewicz, "High-pressure fiber-optic sensor with side-hole fiber", Proc.SPIE, vol. 798, pp 56-60, 1987.

3. W.J.Bock, A.W.Domanski, "High Hydrostatic Pressure Effects in Highly Birefringent Optical Fibers", J.Lightwave Technol., vol. 7, 1989 (in press).

4. A.Enokihara, M.Izutsu, T.Sueta, "Optical Fiber Sensors Using the Method of Polarization-Rotated Reflection", J.Lightwave Technol., vol. 5, pp 1584-90, 1987.

5. J.P.Dakin, C.A.Wade, "Compensated polarimetric sensor using polarization-maintaining fiber in a differential configuration", Electr.Lett., vol. 20, pp 51-3, 1984.

6. W.J.Bock, "High Pressure Polarimetric Sensor Using Birefringent Optical Fibers", Proceedings of IEEE Instr.Meas.Tech. Conf, pp 414-17, Washington DC, April 25-27, 1989.

7. W.J.Bock, J.Chrostowski, "High-pressure Fibre-optic Leadthrough System", J.Phys.E Sci.Instrum., vol. 21, 9, pp 839-41, 1988.

Optically Powered Silicon Microresonator Pressure Sensor

L.M. Zhang, D. Uttamchandani, and B. Culshaw

Department of Electronic and Electrical Engineering, University of Strathclyde, Royal College Building, 204 George Street, Glasgow G1 1XW, UK

A pressure sensor based on the optically excited self-oscillation of a silicon microresonator is reported. The sensor is set in transverse vibration by absorption of incident light. To maintain the vibration, internal optical positive feed back is provided by means of a Fabry-Perot interferometer. When the pressure applied to the resonator changes from 0.0 to 0.6 bar(gauge) a shift in the resonant frequency of the sensor from 56.3kHz to 150kHz is observed.

INTRODUCTION

Recently, great interest has been shown in the use of optically excited and interrogated silicon micromechanical resonators as sensors [1,2]. The resonant structures, which are fabricated by standard silicon processing including photolithography, boron diffusion and anisotropic chemical etching [3], can be activated into vibration at their natural frequencies by using an optical source intensity modulated at the same frequency. To optically excite the sensor, various schemes have been proposed. For example, light whose intensity is modulated by a sinusoidal waveform, a single pulse and band-limited analogue noise has been demonstrated for excitation of the resonator. In each case, the frequency component of the

Springer Proceedings in Physics, Vol. 44 Optical Fiber Sensors
Editors: H.J. Arditty · J.P. Dakin · R.Th. Kersten

excitation light corresponding to the natural resonant frequency will excite the silicon resonator into transverse vibration, and this vibration can be easily detected by an optical interferometer. Physical parameters such as pressure and temperature and any others which can be made to induce stress in the resonator can therefore be measured by observing the corresponding changes in resonant frequency which occur due to stress. The main advantages of this family of sensors are that they can be directly connected to modern digital processing equipment without A-D converters since the signals are inherently 'digital'; that they have the capability of making high precision measurements and that the sensor readings can be transmitted over long distances when combined with optical fibres whilst avoiding transmission errors which are very common in intensity based optical sensors.

In order to simplify the requirements of an optical micromechanical resonator sensor, it would be very attractive to demonstrate such a device operating with a single, unmodulated optical source. An experiment showing this so-called "self-oscillation" has been demonstrated to drive and detect the vibration of a resonator by using only one unmodulated laser [4]. In that experiment, the oscillation was sustained by optical feedback provided by a Fabry-Perot interferometer formed between the surfaces of a polyester cantilever resonator and a partially reflecting mirror. Later, Stokes et al [5] reported a self excited resonator by using a Fabry-Perot cavity formed between the surface of a metal coated silica microresonator and the end of a single mode fiber.

In this paper, a pressure sensor based on this self oscillation process is reported. A thin silicon bridge is fabricated on a pressure sensitive silicon diaphragm. Changes in pressure will generate a mechanical force in the silicon bridge, and this force will change its natural resonant frequency which can subsequently be detected by the optically powered self-oscillation mechanism. This kind of sensor has the advantage of easy optical excitation and detection by using only one unmodulated laser. It can also be expanded to a multisensor system by using an optical fibre network.

CONDITION OF SELF-OSCILLATION

The minimum optical power which is needed to maintain the self-oscillation depends on the optical response of the bridge and the optical characteristics of the Fabry-Perot interferometer as explained below. The transmission characteristics, shown in Figure 1, is also the energy storage in the Fabry-Perot cavity, peaking at a value of the order of FP_{in} where F is the finesse and P_{in} the input power. Suppose we put a vibrating bridge at A as shown in Figure 1, in which A is a positive feed back position. If the optical response of the bridge (here meaning the deflection of the bridge for a given input power) is X(um/mW), then the device should remain in self-oscillation when:

$$XY > 1$$

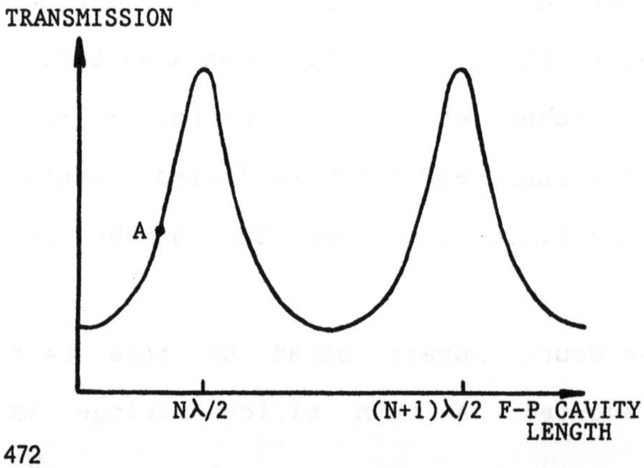

Figure 1.
Transmission characteristic of Fabry-Perot interferometer

where Y is the slope of the Fabry-Perot characteristic expressed in (mW/μm). If the most sensitive point is chosen, the slope will be approximately (FP_{in})/λ. In this case we have

$$FP_{in}X/\lambda > 1 \qquad (1)$$

Based on experiment data, X is approximatly 1μm/mW, F=3, and λ=0.63um, we can get the minimum optical power needed for maintaining the vibration

$$P = \lambda/(XF) = 210\ \mu W$$

The oscillation can only occur when the working point is chosen to satisfy equation (1). Thus, in order to obtain a stable self oscillation, the Fabry-Perot cavity must be adjusted to an accuracy of less than one optical wavelength.

EXPERIMENT ARRANGEMENT

A silicon bridge resonator described in [6] is first used to observe self-oscillation. Figure 2 shows the experimental arrangement. A Fabry-Perot interferometer is configured between the silicon bridge and a 60% semi-transparent film. The bridge is coated with a 20nm thick aluminium film by vacuum deposition. The He-Ne laser(average power ~ 0.8mW) passes though a 10/90 beam splitter to the Fabry-Perot interferometer which has a finesse of approximately 5. A 20X lens inside the cavity focuses the laser

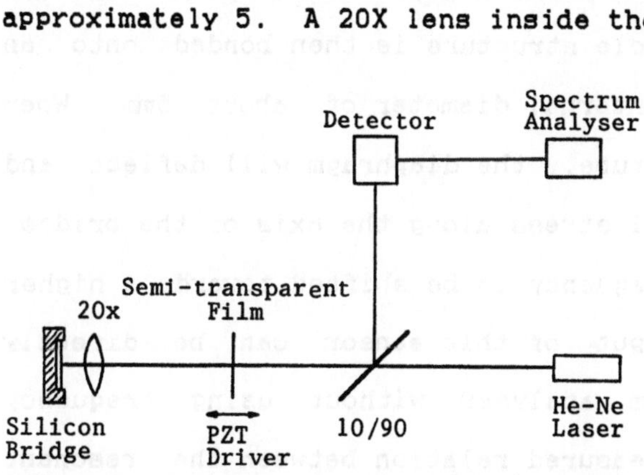

Figure 2.
Experimental arrangement

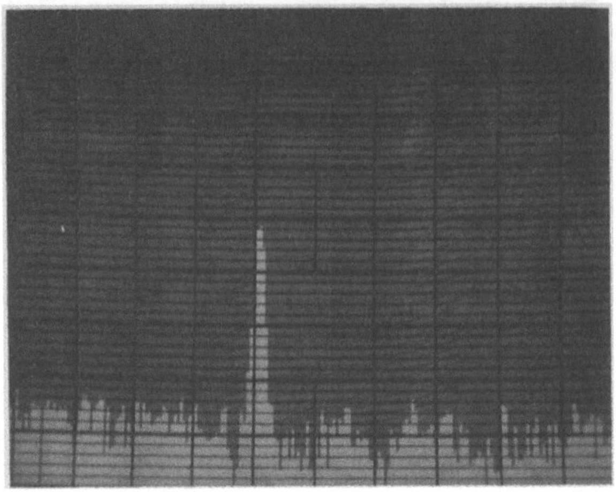

Figure 3.
The self-oscillation signal displayed on a spectrum analyser

Horizontal: 5.0kHz/div.
Vertical : 10.0dB/div.
Bandwidth : 500Hz

beam onto the center of the bridge, and a PZT micro positioner is used for fine tuning of the Fabry-Perot cavity length. Light returning from the interferometer is detected by a PIN photodetector, whose output signal is analysed by a spectrum analyser. The experimental result is shown in figure 3. The peak frequency of the displayed spectrum, which represents the natural resonant frequency of the bridge, is 85.6KHz. The Q factor of the resonator is around 170.

Figure 4 shows the resonator sensor used for pressure measurement. The 0.7mm long 20μm wide 2μm thick silicon bridge is fabricated on a square silicon diaphragm with side of 2mm and thickness of 20μm. The whole structure is then bonded onto an aluminium tube with an internal diameter of about 5mm. When pressure is applied to the tube, the diaphragm will deflect and this will induce an internal stress along the axis of the bridge. This causes the resonant frequency to be shifted towards a higher value. The frequency output of this sensor can be directly observed on the spectrum analyser without using frequency tracking techniques. The measured relation between the resonant

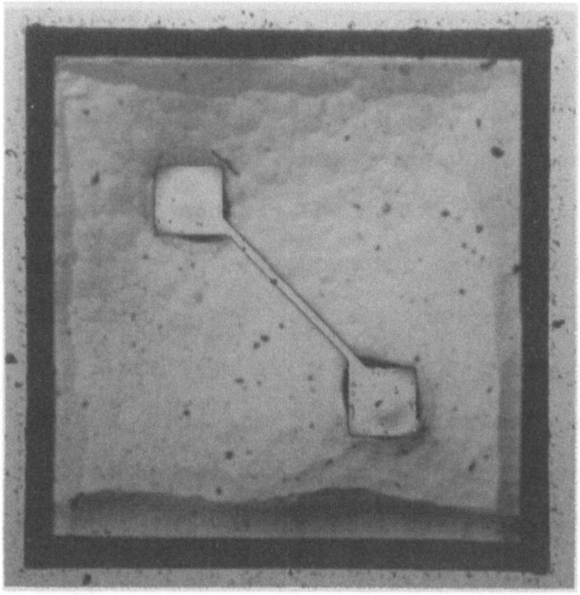

Figure 4.
The micrograph of the silicon resonator

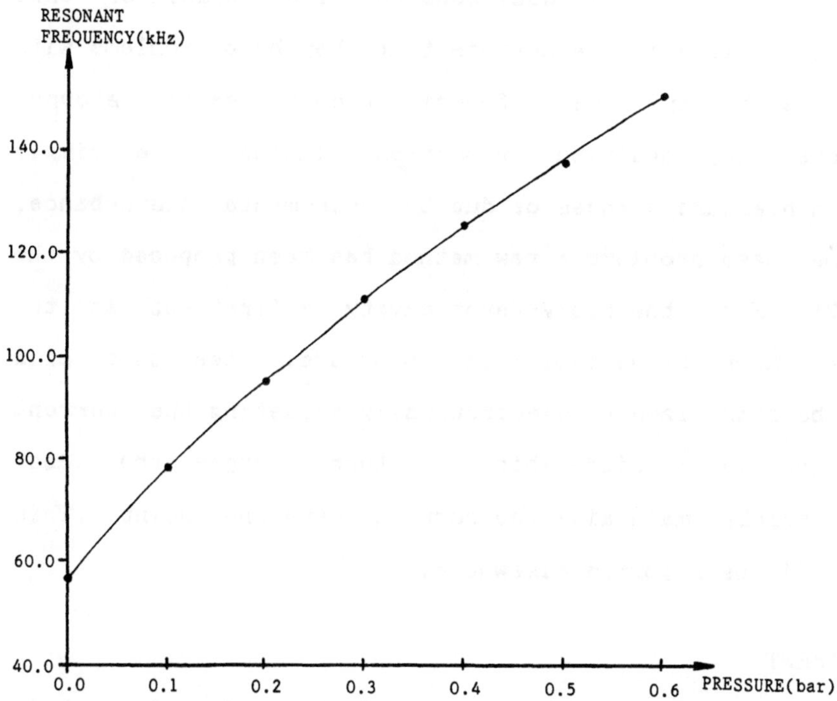

Figure 5. The resonant frequency of the silicon resonator vs. applied pressure

frequency and applied pressure is shown in Figure 5. The sensitivity of this device is 140Hz/mbar over the range shown.

DISCUSSION AND CONCLUSION

An optically excited silicon pressure sensor with a frequency output is reported. Since the sensor operates on a self-oscillation process, laser modulation and frequency tracking are no longer needed. A large increase in the resonant frequency (from 56.3kHz to 150.0kHz) has been observed when the applied pressure increased from 0.0 to 0.6 bar (gauge). The sensitivity is 140Hz/mbar. Such sensors can be fabricated in very small size for use with optical fibres. And, if needed, several sensors working at different resonant frequencies can be linked into a fibre optic bus with different measurands being detected by observing the frequency bands corresponding to the different resonant frequency of individual sensors. The drawback of this kind of self-oscillating sensors is that they have problems with bias point stability. The self-oscillation is easily stopped since the operating condition for self-oscillation can be simply changed with pressure changes or due to enviromental disturbance. To overcome these problems a new method has been proposed by the authors [7]. Here the Fabry-Perot cavity is first set at the right bias where the self-oscillation occurs. Then this bias point can be stabilized by electronically adjusting the current fed into the laser diode which in turn changes the laser frequency thereby maintains the correct operating point. This experiment will be reported elsewhere.

ACKNOWLEDGEMENT

L. M. Zhang would like to thank Foxboro Co. for their financial support and the authors would like to thank Dr. P. Dobson of Birmingham University for fabrication of the silicon bridge.

REFERENCES

1. VENKATESH, S., and CULSHAW, B.'Optically excited vibration in a micromachined silicon structure', Electron. Lett., 1985, 21, pp315-317

2 ANDRES, M. V., FOULDS, K. W. H., and TUDOR, M. J.,:'Optical activation of silicon vibrating sensors', Electron. Lett. 1986, 22, pp 1097-1099

3 PETERSON, K. E.,:'Silicon as a mechanical material', Proc. IEEE, 1982, 70, pp 420-457

4. LANGDON, R. M., and LYNCH, B. J.'Photoacoustics in optical sensors', Proc. of SPIE, 1987, 798, pp86-93

5. STOKES, N. A. D., FATAH, R. M. A., and VENKATESH, S.,'Self-excited vibrations of optical microresonator', Electron. Lett., 1988, 24, pp 777-778

6. ZHANG, L. M., UTTAMCHANDANI, D., and CULSHAW, B.'Transient excitation of silicon microresonator', Electron. Lett., 1989, 25, pp 149-150

7. ZHANG, L. M., UTTAMCHANDANI, D., and CULSHAW, B.'The stabilization of optically excited self oscillation', submitted to Electron. Lett.

A Hybrid Resonant Differential Pressure Transmitter with Wavelength Multiplexed Power and Data Channels

E. Bois[1], *R.C. Spooncer*[2], *and B.E. Jones*[2]

[1]Slick Optique Electronique, BP 42,
F-77312 Marne la Vallée Cedex 2, France
[2]Brunel Centre for Manufacturing Metrology, Brunel University,
Uxbridge, Middlesex, UB8 3PH, UK

The force beam of a differential pressure transmitter tensions a double-ended tuning fork to provide a pulsed optical output,the frequency of which represents the measurand . A single optical fibre link carries two wavelengths ; one excites the vibrating element through photovoltaic conversion , the second determines the resonant frequency .

1_INTRODUCTION

The advantages of optical fibre sensors compared with conventional sensors have been well documented : - freedom from E.M.I. - intrinsically safe operation in hazardous environments and a capability of using classical optical measuring methods such as interference to give a resolution in some applications not possible with conventional sensors .
The process industries , however , whilst recognizing the advantages of intrinsically safe and E.M.I. free optical telemetry are reluctant to exchange established measuring methods for untried optical sensing and as a result all-optical OFS have as yet made few inroads into the sensing market .
A compromise has emerged in which conventional sensing elements are retained but electrical telemetry is replaced by optical fibre links . Such sensors are known as hybrids and consist essentially of an established transduction element which may be excited in its normal manner either by batteries or by photovoltaic conversion of optical power but which produces an optical output signal . Hybrid sensors thus combine the established reliability and performances of established sensors with the benefit of optical fibre technology without recourse to immature technology .
Hybrid sensors may use an analogue or digital output but the digital format which is intensity indifferent and corruption free compared with analogue modulation is preferred. The benefit is enhanced if the primary element is intrinsically digital as , for instance , in vibrating elements whose resonant frequency is a function of the measurand .
Jones and Philp (1983) described a hybrid vibrating wire force sensor which has been commercialized ; an improved version (Halliwell and Jones , 1987) uses a double ended tuning fork (DETF) as the resonant element in which the tines vibrate in antiphase so that the momentum is cancelled at the roots and there is no loss of energy at the supports . A quality factor (Q) as high as 10,000 has been measured .
Both the applications use a three fibre configuration ; one fibre excites the vibrating element through a photovoltaic conversion , the other pair forms a reflective sensor to determine the resonant frequency .
This paper describes the use of a DETF to provide an optical output from a differential pressure cell in a dual wavelength configuration which requires only a single fibre . The reflective sensor in which the tines of the DETF acts as a partial shutter is based on an autocollimation principle (Bois, Huard and Boisdé , 1989) .

Springer Proceedings in Physics, Vol. 44 **Optical Fiber Sensors**
Editors: H.J. Arditty · J.P. Dakin · R.Th. Kersten

2_ RESONANT PROPERTIES OF A DOUBLE ENDED TUNING FORK (DETF)

Figure 1 shows the outlines of a typical double-ended tuning fork . It may be considered as two beams in tension , vibrating transversaly . The resonant frequency under zero tension **Fn(T=0)** is given by (Chesmedoost and al , 1989) :

$$Fn = \frac{Kn^2}{2 . \pi . L^2} . \sqrt{\frac{E . I}{\rho . S}} \quad \text{in which} \quad I = \frac{W . H^3}{12} \text{ : second moment of area} \qquad (1)$$

n : order of the vibrational mode (for **n**=1, **Kn**=4.73 ; for **n**=2, **Kn**=7.85....)
E : Young's modulus of the material
ρ : density of the material
S = W . H : cross-section area of the beam

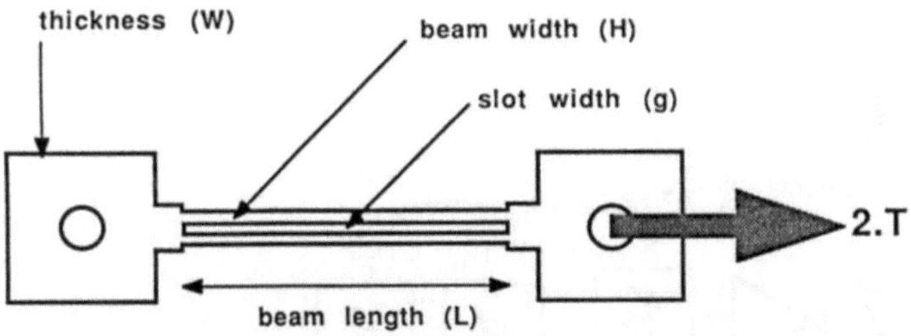

Figure 1 - Geometrical description of a DETF

For a given tension **2.T** , the equation of motion may be solved by numerical analysis and the resonant frequency F1 for the fundamental mode is given by :

$$\left.\begin{aligned} &F1 = 1.028 . C_0 . \frac{H}{L^2} . [1 + 0.710 . B^{1.75}] \\ &\text{where B (the tension function) is defined by :} \\ &B = \frac{L}{H} . \left[\frac{3 .T}{2 . \pi^2 . E . S}\right]^{\frac{1}{2}} \end{aligned}\right\} \qquad (2)$$

This expression may be used to design a DETF (dimensions , material) to meet the requirements of a given application . The DETF used in this investigation was chemically etched from stainless steel and has dimensions : L = 11.6 mm , H = 0.4 mm , W = 0.3 mm ; the measured resonant frequency are :

F1 (T=0) = 12,940 Hz
F1 (T=3Kgf) = 14,200 Hz

3_ DESCRIPTION OF THE TRANSMITTER

A Taylor type 392TD differential pressure sensor from which the force balance system has been removed leaving the force beam exposed provides the conventional measuring element of

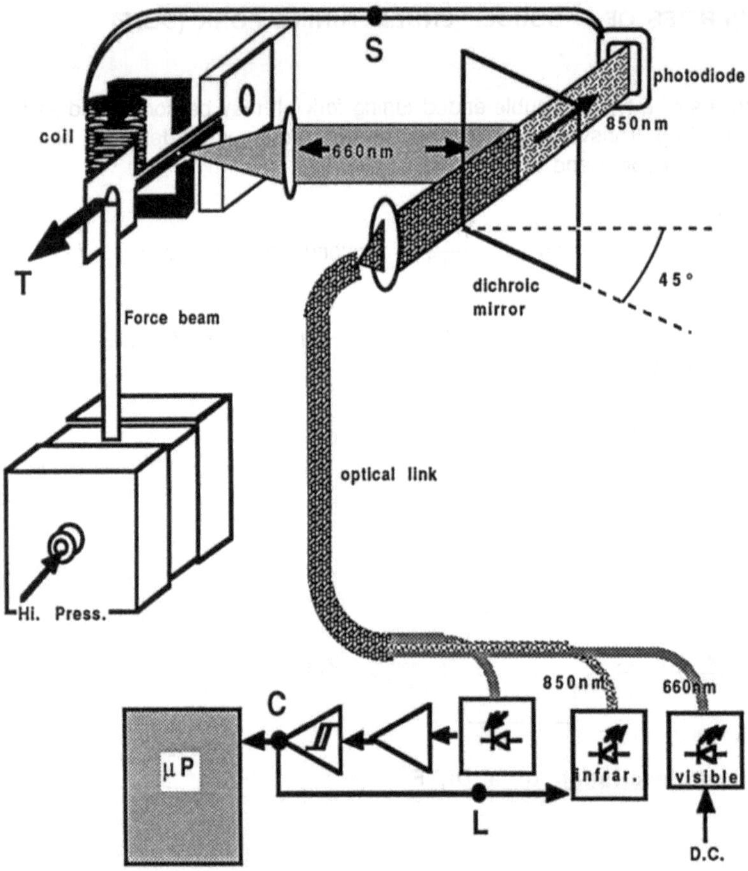

Figure 2 - Transmitter schematic

the transmitter . The force balance system is replaced by a mini optical table which carries the different optical components required to excite the resonant element and to monitor its resonant frequency (Fig. 2) .

The double ended tuning fork is mounted between the poles of a permanent magnet , suspended between a fixed anchorage point and the force beam which tensions the element . The permanent magnet generates a transverse field across the air gap which attracts the tines outwards .

A coil connected across the output of a photodiode in the sensor head is wrapped around the yoke of the magnet ; variations of the photo current modulates the attractive force on the tines. The DETF may be maintained in resonance by pulsing the photocurrent at the natural frequency of the oscillatory element .

The vibration of the tines is monitored by an optical fibre reflective sensor illuminated by a 660nm source . The pulsed return signal from the sensor is detected at the measuring head and used to drive a 850nm LED , the optical output of which is fed back to the photodiode in the sensor completing a feedback loop oscillating at the natural frequency of the DETF .

Light from each source is launched into a single 400µm fibre via one leg of a 3:1 coupler and the two wavelengths are separated at the transmitter by a dichroic mirror or a grating .The pulsed return signal from the reflective sensor backtracks the illuminating light and is tapped off at the measuring head by the third leg of the coupler .

4_ EXPERIMENTAL RESULTS

The transfer function of the loop can be recorded by connecting a spectrum analyser between points S and C (see Fig.2 , S : output of the S.A. , C : input of the S.A.) . A typical scan is shown in Figure 3 from which two important loop parameters , the quality factor **Q** , and the electrical power **Pe** required to maintain the resonator in oscillation , may be determined . The quality factor **Q** of the complete system is 3705 compared with the higher figure of 10,000 for the resonator itself but still high enough to ensure the feasibility of optical powering . The lower **Q** may be caused by unperfect clamping of the DETF to the force beam . A voltage across the coil of 0.2 Vrms , corresponding to an electrical power **Pe** of 1.5 μW is supplied to maintain the resonator in oscillation .
The frequency characteristic of the transmitter was measured initially by direct loading of the force beam with weights . The output characteristic is shown in Figure 4 . (sensitivity : 242 Hz/ kg.f ; linearity : 0.2% of 3 kg.f scale ; Hysteresis : < 1% of 3 kg.f scale) . The pressure was then supplied with a load bias of 2 kg . The range obtained using pressure is limited to about 10% of that achieved with direct loading with weights because of pressure limitations in the experimental rig . To get an appropriate sensitivity over the extended range , a redesigned DETF with a higher resonant range would be necessary. The following table summarizes the characteristics of the transmitter :

Table 1 - Characteristics of the transmitter

	Direct loading	Pressure loading
Span	726 Hz	69.3 Hz
Sensitivity	242 Hz/kg	12.6 Hz/bar
Hysteresis	<1% full scale	< 5Hz
Linearity	0.2 % full scale	3.4% of the 69.3Hz scale

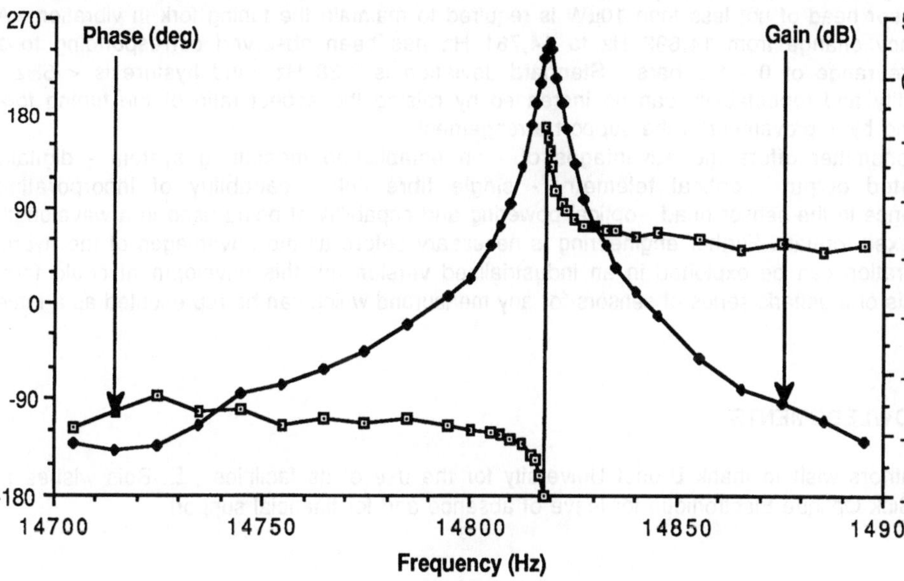

Figure 3 - Transfer function of resonant loop (2.T=3.5Kg.f)

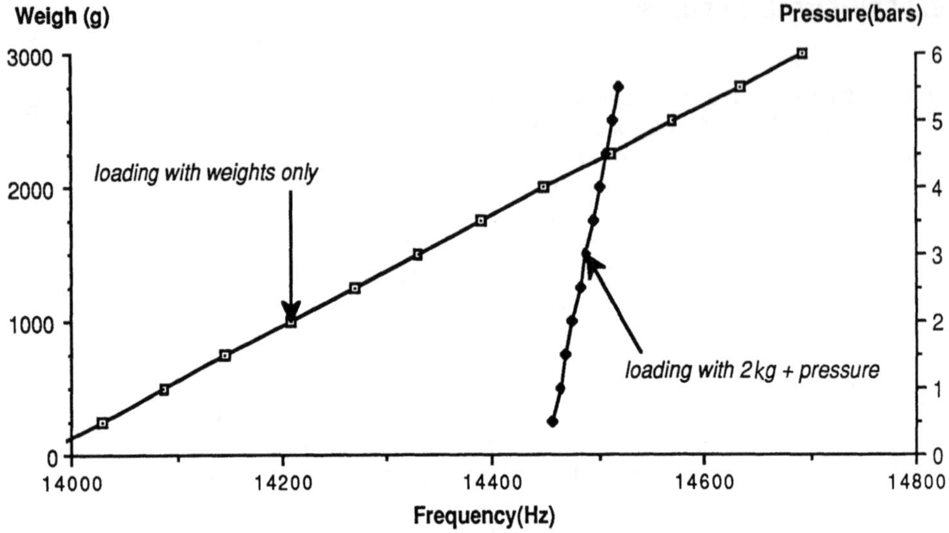

Figure 4 - Frequency characteristic of the transmitter

5_ CONCLUSIONS

A hybrid resonant differential pressure transmitter has been demonstrated in which a double ended tuning fork optically excited through photovoltaic conversion provides the resonant element . The resonant frequency which is a function of the measurand is determined by a reflective sensor . A two-wavelength single fibre link provides exciting power at one wavelength (850nm) ; the frequency is readout at the second (633nm) . Received power at the sensor head of not less than 10µW is required to maintain the tuning fork in vibration . A frequency change from 14,692 Hz to 14,761 Hz has been observed corresponding to a pressure range of 0 - 5.5 bars . Standard deviation is 2.38 Hz and hysteresis < 5Hz . Sensibility and repeatability can be increased by raising the aspect ratio of the tuning fork tines and by improvement in the support arrangement .

The transmitter offers the advantages of - an established measuring system - digitally modulated output - optical telemetry - single fibre link - capability of incorporating intelligence in the sensor head - optical powering and capability of being used in a wavelength multiplexed format . Further engineering is necessary before all the advantages of the hybrid configuration can be exploited in an industrialized version but this development could form the basis of a generic series of sensors for any measurand which can be represented as a force input .

ACKNOWLEDGMENTS

The authors wish to thank Brunel University for the use of its facilities ; E. Bois wishes to thank Sick Optique Electronique for leave of absence and for financial support .

REFERENCES

Bois,E. ; Huard ,S.J. and Boisdé G. (1989) : " Loss compensated fiber optic displacement sensor including a lens " - Applied Optics **28** , 419-420 ,1989.

Cheshmehdoost,A. ; Halliwell,M.J. ; Jones,B.E. (1989) : "Design of a miniature high Q double-ended tuning fork resonator employing non contact drive and pick-up arrangements for force measurements" - BSSM , Int. Weighing Conference , July 1989 , paper 16 .

Halliwell,M.J. ; Jones,B.E. (1987): "Force transducer using a double-ended tuning fork with optical fibre links " - Proc.Conf. on Sensors and their Applications , Cambridge 22-24th September 1987 , pp145-147 ,Institute of Physics ,Bristol, Eurosensor 1987 .

Jones,B.E. and Philp ,G.S. (1983) : "A vibrating wire sensor with optical links for force measurements" - Proc. Conf. on Sensors and their Applications , Manchester 20-22th September 1983 ,pp86-88 ,Institute of Physics ,Bristol .

Part XIII

Multiplexed Sensors

Ten-Element Time-Division Multiplexed Interferometric Fiber Sensor Array

A.D. Kersey and A. Dandridge

US Naval Research Laboratory, Optical Techniques Branch, Code 6574, Washington, DC20375, USA

Abstract: This paper reports the operation and performance of a ten-element all-fiber time-division multiplexed sensor array based on Mach-Zehnder interferometer sensors arranged in a ladder topology. Minimum detectable phase shift sensitivities of 12 μrad/√Hz at a frequency of 1 kHz and sensor crosstalk levels of ~ -67 dB are reported.

Introduction: Interferometric fiber sensors for a variety of physical fields have been successfully packaged and tested in recent years. Presently, an area of particular interest in interferometric fiber sensor research is that of multiplexing. Many of the applications for fiber sensors require multi-sensor systems; for example, sensor networks designed to monitor a variety of measurands in process control-type applications such as those found in manufacturing, chemical and power plants. A second area of interest for multiplexing is where a large number of single-measurand sensors are required, such as encountered in geophysical surveying applications for example.

A number of addressing formats are conceptually possible for use with interferometric fiber sensor arrays, such as time-, frequency-, wavelength-, coherence-, and polarization-division addressing [1-7]. Of these, only certain time and frequency division based schemes have demonstrated performance levels comparable to those obtained with 'single-sensor' systems. Recent work has concentrated on the development of these schemes, which seem capable of supporting a reasonably large number of sensors without serious performance degradation due to excess noise, poor power budget or significant crosstalk effects.

Array topology: In this paper we report the operation and performance of a ten-element all-fiber time-division multiplexed sensor array based on Mach-Zehnder interferometer sensors arranged in a ladder topology [5]. The system, which is shown schematically in Figure 1, utilizes sensing interferometers which are only slightly unbalanced to allow for passive demodulation via laser-FM based phase generated carrier or synthetic heterodyne techniques [8]. The input pulse width, τ, is set $\leq T$ (delay in fiber delay coils), such that at the output of the N sensor array a series of N output pulses is generated for each input pulse. As the path imbalance in the sensors is small the pulses from the signal and reference arms overlap and interfere at the output of each interferometer, and the output pulse train simply represents samples of the interferometer outputs interleaved in time sequence. In the experimental system, the light from a CW operated 830 nm diode laser (Hitachi HLP1400) was passed through an acousto-optic modulator (AOM) which was used to pulse the input light to the array. The delay coils were 30 m in length, corresponding to a 145 ns group delay, and an input pulse width of ~ 115 ns FWHH was used to avoid overlapping of the non-rectangular optical pulses produced by the input AOM. At the array output, the optical pulse train was monitored using an avalanche photodiode (APD) with a rated responsivity of 77 A/W. The detected pulses were then amplified using a wide-bandwidth preamplifier (DC - 300 MHz), before being fed to a gating circuit which was used to select the pulse corresponding to a particular sensor in the array.

Springer Proceedings in Physics, Vol. 44 **Optical Fiber Sensors**
Editors: H.J. Arditty · J.P. Dakin · R.Th. Kersten

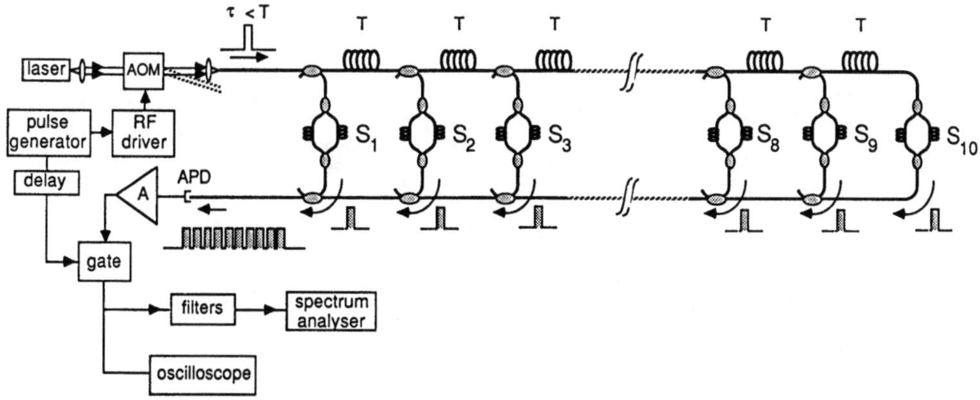

FIGURE 1. Schematic of the ten-sensor TDM Mach-Zehnder ladder array

Clearly, in order to balance the performance of the array it is desirable to obtain an equal amount of power from each sensor. This can be accomplished by tailoring the coupling ratios of the 2 x 2 couplers used in the input feed and output fibers. For an ideal system (i.e., neglecting splice, coupler and fiber loss) the power splitting ratio of the n th coupler (numbered 1 to N-1, where N is the number of sensors) is given by

$$\kappa_n = 1/(N - n - 1). \tag{1}$$

The returned peak power in each pulse from the array relative to the input pulse is then

$$\{P_n/P_o\} = 1/N^2 \tag{2}$$

When system losses are taken into account, the coupling ratios can be tailored to compensate for the loss. By numerical analysis of this array configuration with losses due to intrinsic fiber loss, couplers, and splices, we have derived the following expression for the coupling ratio, κ'_n, required to equalize the power (± 1 dB) returned from each sensor,

$$\kappa'_n = \kappa_n \cdot \alpha^{(N-n)} = \{\alpha^{(N-n)}\}/(N - n - 1), \tag{3}$$

where α is a transmission factor determined by the loss in the couplers, delay coil and splices associated with each sensor-section of the array. Figure 2 shows the results of calculations based on Equations 1 and 3 for various losses, and a schematic of the loss model used to characterize the array.

Results: Figure 3.a shows the output optical pulse train detected at the APD. The peak power in each pulse corresponds to ~ 1.5 μW. The input power to the array was 1.5 mW. With the system operating at an optimum repetition rate of 695 kHz, the average optical power from each sensor was ~ 50 nW. This accounts for the duty cycle of 0.1, a factor of approximately 0.75 for the non-rectangular pulse shape, and a factor 0.5 for the averaging of the interference signal. Figure 3.b shows an example of the operation of the gating circuit used to selectively address each sensor output (sensor # 4 selected). The isolation of this gate directly affects the crosstalk level in the system; an on/off isolation > 70 dB was measured for the gates used in these experiments.

Figure 4 and Table 1 show the phase detection sensitivities obtained with the array. In Figure 4, the noise floor of sensor # 5 is shown. Here, a test phase shift of 10 millirad rms at 1 kHz was applied to the sensor via a piezoelectric phase shifter. The signal to noise of the detected output held set at quadrature was 58.6 dB/√Hz, corresponding to a minimum detectable phase

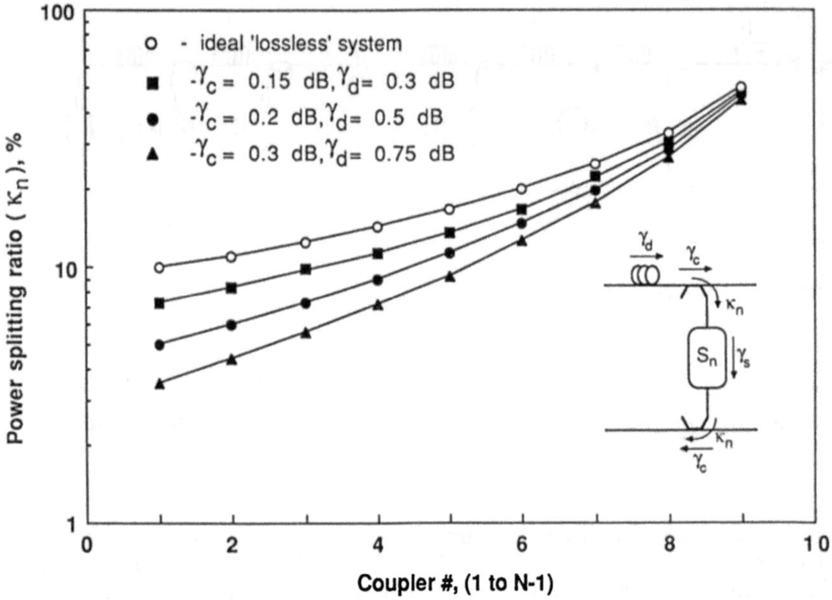

FIGURE 2. Theoretical curves showing power splitting ratio of couplers required for equal pulse amplitudes at array output for various loss levels. Inset shows loss model for each sensor section of array

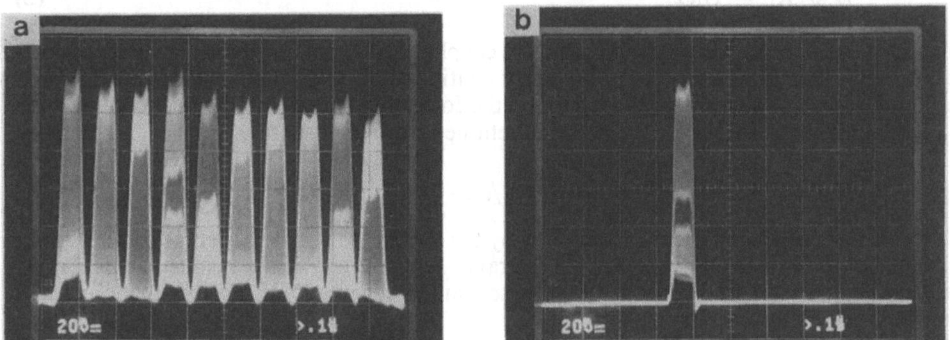

FIGURE 3. Output optical pulse train (a) directly at APD output, and (b) after gating (note sensor # 4 selected)

shift sensitivity of 12 μrad/√Hz. As shown in Table 1, the phase detection sensitivities of the other nine sensors ranged from 13 to 18 μrad/√Hz (at 1 kHz). As the noise floor of the detection electronics was ~ 8 μrad/√Hz, these noise levels are attributed to laser frequency jitter induced phase noise. An example of the crosstalk between sensors in the array is shown in Figure 5. Here, the crosstalk between two non-adjacent sensors is shown (sensor #5 to #7); the effective crosstalk is ~ -67 dB. For adjacent sensors, crosstalk levels of ~ -50 to -55 dB have been observed. This higher crosstalk between adjacent sensors is felt to be due to settling in the photodetector.

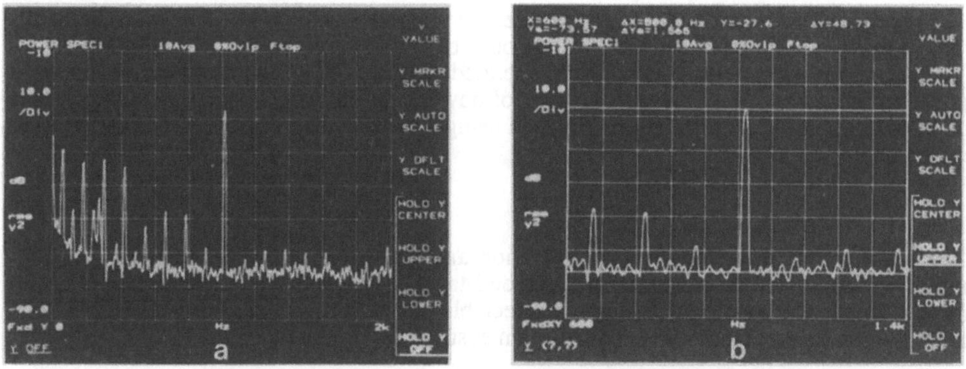

FIGURE 4. ***System noise floor:*** ***(a)*** ***0 - 2 kHz with 10 mrad rms signal at 1 kHz,***
(b) ***S/N of test signal = 48.7 dB in 9.55 Hz BW (= 58.6 dB/√Hz).***
Noise floor at 1kHz corresponds to 12 μrad/√Hz.

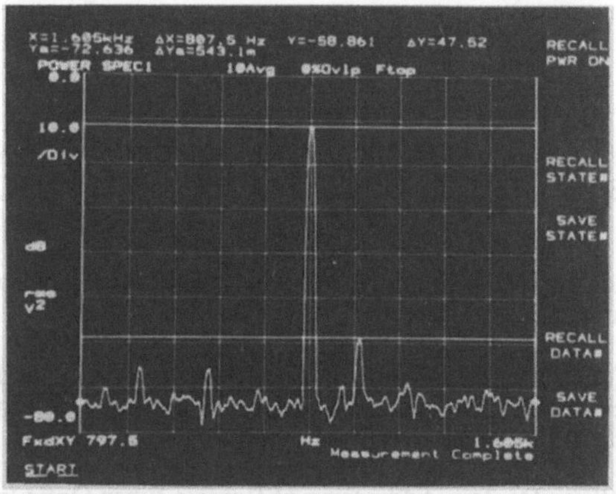

FIGURE 5 Example of crosstalk between sensor elements:
Sensor # 7 addressed, 100 mrad signal applied to sensor # 7 at 1.2 kHz (center),
1 rad applied to sensor # 5 at ~1.3 kHz. Effective crosstalk ~ -67 dB.

TABLE 1. Measured minimum detectable phase shift sensitivities for the ten-sensor TDM Mach-Zehnder ladder array

Sensor	1	2	3	4	5	6	7	8	9	10
Phase detection sensitivity, μrad/√Hz	17	14	13	18	12	17	16	13	17	18

In a TDM system, in order to avoid aliasing effects the sampling rate must satisfy the Nyquist theorem for all significant frequency components in the sensor outputs. In our system, the maximum repetition rate, or sensor sampling rate, was 695 kHz. To demodulate the sensor outputs we have used the laser-FM based phase generated carrier technique. This approach produces significant frequency components in each of the sensor outputs at several harmonics of the laser FM modulation frequency (f_m- $5f_m$). In order to avoid aliasing effects the sampling rate must be an order of magnitude greater than f_m. Assuming a 40 kHz laser FM, this sets a pulse repetition rate requirement of > 400 kHz. For the pulse widths (~ 150 ns) used here, this would limit the number of sensors which could be multiplexed using this approach to 16. However, in recent experiments we have operated a source-AOM input optical system with optical pulse widths of ~ 50 ns and ~ 3 mW of power (pulse peak) coupled into a fiber. This would enable > 40 sensors to be multiplexed using a phase-carrier frequency of 40 kHz and repetition rate of > 400 kHz.

Conclusions: We have reported the operation and performance of a ten-element all-fiber time-division multiplexed sensor array based on Mach-Zehnder interferometer sensor elements arranged in a ladder topology. Minimum detectable phase shift sensitivities between 12 to 18 $\mu rad/\sqrt{Hz}$ at a frequency of 1 kHz have been measured for the array.

Acknowledgements: This work was supported by the Office of Naval Technology program on Electro-Optics.

References:

1. A. D. Kersey *et. al.* ," Overview of Multiplexing Techniques for Interferometric Fiber Sensors," Proc. SPIE Vol. 838, Fiber-Optic and Laser Sensor V, p. 184, San Diego, CA, 1987.

2. J.P. Dakin, C.A. Wade and M.L. Henning, "Novel Optical Fibre Hydrophone Array Using a Single Laser Source and Detector", Electron. Lett., 20, p. 53, 1984.

3. J.L. Brooks, R.H. Wentworth, R.C. Youngquist, M. Tur, B.Y. Kim, and H.J. Shaw, "Coherence Multiplexing of Fiber-Optic Interferometric Sensors", J. Lightwave Technology, LT-3, p. 1062, 1985.

4. J. L. Brooks *et. al.*," Time-Domain Addressing of Remote Fiber-Optic Interferometric Sensor Arrays," IEEE J. Lightwave Technol., LT-5, p.1014, 1987.

5. A.D. Kersey, A. Dandridge and A.B. Tveten,"Multiplexing of Interferometric Fiber Sensors Using Time Division Addressing and Phase Generated Carrier Demodulation", Optics Letters, 12, p. 775, 1987.

6. A. Dandridge, A.B. Tveten, A.D. Kersey and A.M. Yurek, "Multiplexing of Interferometric Sensors using Phase Generated Carrier Techniques", IEEE J. Lightwave Technology, LT-5, p. 947, 1987.

7. A. D. Kersey and A. Dandridge, " Transmissive Serial Interferometric Fiber Sensor Array," J. Lightwave Technol., 7, p. 846, 1989.

8. A. Dandridge, A.B. Tveten and T.G. Giallorenzi, "Homodyne Demodulation Scheme for Fiber-Optic Sensor Using Phase Generated Carrier", IEEE J. Quantum Electron., 18, p. 647, 1982.

A New Approach to Digital Optical Sensor Networking

A. Pervez

Dept. of Electronic and Electrical Engineering, University College London, Torrington Place, London WC1E 7JE, UK

Digital optical sensor multiplexing and networking problem has been considered from a communication systems view point .As a result ,a hierarchical multiplexing solution has been derived .This scheme allows a very large number of digital optical sensors to be implemented in a ring network topology. A network prototype has been developed , constructed and evaluated.

1. Introduction

A considerable number of optical sensor design and multiplexing concepts have been proposed in recent years [1-3]. Most of the work to date has been concentrated on the design and multiplexing problems of high quality analogue sensors . However , there are situations in which only a particular value, or a threshold of a measurand, needs to be monitored , for example: in smoke detection , temperature or pressure threshold detection, and numerous process control applications . In these situations a periodically interrogated on-off type (binary digital) , all-optical sensor is all that is necessary . In addition , a complex industrial process or,the simple monitoring over an extensive area of environmental conditions, may require a large number of sensors spatially distributed within the area concerned . In these situations, hierarchical multiplexing , which combines a large number of sensing signals , becomes attractive . The sensing signals organised in this way represent a sensor network , which may then be controlled and sequentially interrogated by a remote control centre . A different 'digital' approach is required from those approaches adopted for analogue sensor systems , as the system constraints for the two cases are not the same .

Almost all the sensor systems proposed to date in the literature have been based on radar system design principles , consequently , well understood radar concepts and techniques have been extensively applied to analogue sensor systems . In contrast , for a binary digital sensor network it seems appropriate to view the system design analogous to a digital communication network design problem , since the similarities between TDM binary sensor signals and a digital communication data stream are remarkable .

It is the design concept and development of such a network that is the *central theme* of our work.

We are proposing a novel sensor network topology , which explores the similarities between digital sensor networks and data communication networks (such as LANs) . Furthermore , the proposed scheme tackles the sensor networking problem from a digital communication viewpoint , hence digital communication techniques can be applied to multiplexing , signal processing and networking of optical digital sensors . This approach allows a very large number of sensors to be implemented in a network.
A network prototype has been developed and constructed . The network design concept and implementation is discussed . The system performance is examined and theoretical and measured results are presented .

2. Network Design Concept

The network topology is based on a centralised control/interrogation , a regenerative token power ring and a signal response bus , as shown in figure 1 .

A sensor array of N binary digital transmissive sensors may be constructed as shown in figure 2. The network may have any number (M) of such arrays , installed at the required sensing locations . A token pulse of duration , T, is generated,under the control of the system clock , by the network control centre (NCC) . The corresponding optical pulse is transmitted to the first sensor array(SA1) . As the pulse ripples through the array interrogating each sensor in turn , the corresponding sensor output , describing its state , appears on the response bus . Thus , at the end of a complete interrogation period of an array of N sensors , a frame of (N +1) bits is produced as shown in figure 3 . The token is then regenerated by the TR and passed on to the next sensor array . The token regeneration removes the transmit power constraint which usually exists in most sensor systems.Thus, a network having M sensor arrays , each containing N sensors , would produce a TDM super frame (hierarchical scheme) as shown in figure 4 . A super frame is generated for each token cycle . The frame bits may be

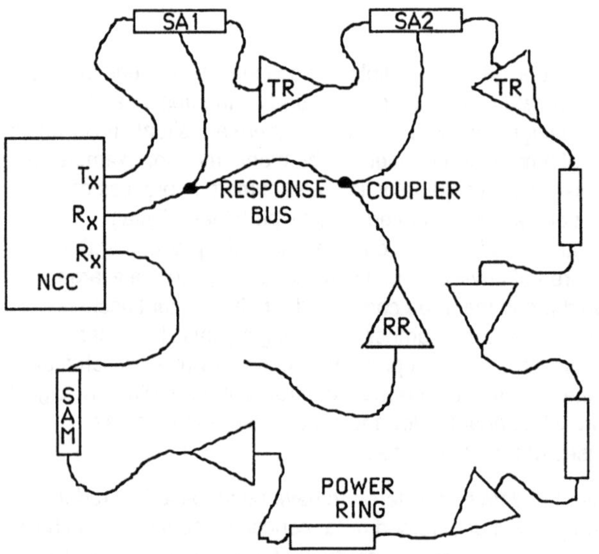

FIGURE 1 NETWORK TOPOLOGY

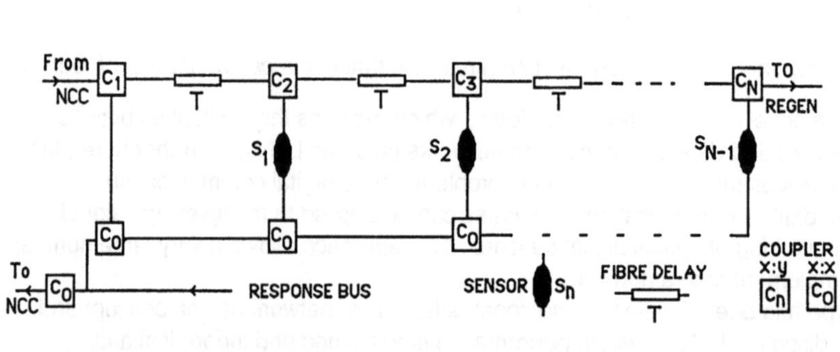

FIGURE 2 SENSOR ARRAY

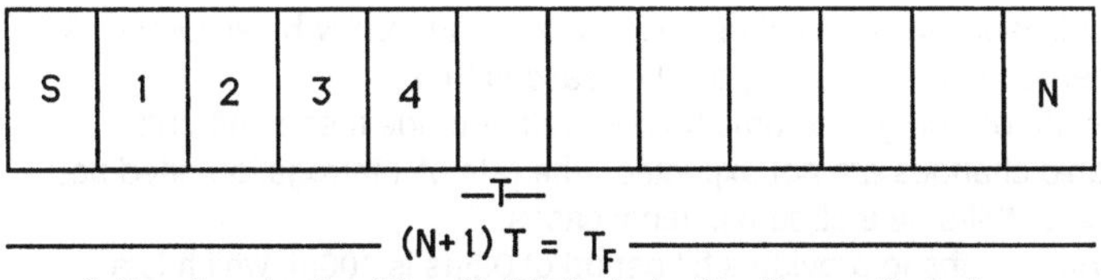

S-Synchronisation Bit; T- Bit Period; T_F- Frame Period

FIGURE 3 DATA FRAME

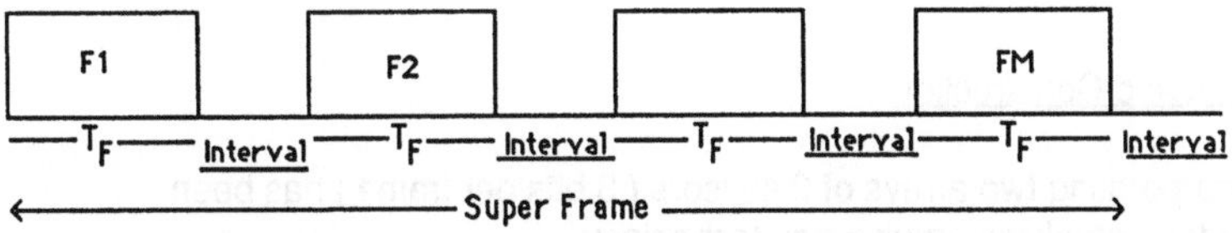

FIGURE 4

regenerated by the successive regenerative repeaters (RR) along the response bus , depending upon the location of the particular array in question .The bit stream on the bus can be regarded as serial data at a rate of 1/T bit/s. The token passes through each sensor array in succession . Each array possesses the token for a well defined and predetermined period of time in which the array can transfer information on to the response bus . The response bus transmits data , received from a particular array , to the NCC for processing .

The first bit(S) in each frame is a dummy bit (always logic 1) , and is used for frame synchronisation at the NCC .The receiver monitors a '0' to '1' transition , (logic '0' being the idling state) . The transition edge is used to synchronise receiver timing to the incoming data frame . Following the frame synchronisation a counter records N bits , arriving from the transmitting array . The bit number identifies sensor location within the transmitting array and the state of the bit conveys sensor information . A separate counter records the arrival of frames and hence identifies the location of the transmitting sensor array .

Thus , the state of each sensor within the network is readily detectable , subject to interrogation cycle period , at the NCC .

The returning token is detected by a separate optical receiver at the NCC. This information is fed to the token generator, which then and only then transmits another token.In this way the token cycle is kept under a regular check and more importantly, the system becomes completely independent of the spatial distribution of the sensor arrays within the network.This feature makes the scheme conceptually different from the previously proposed TDM sensor system schemes in the literature, as they rely entirely on spatial distribution of sensors in order to calculate the time delay of the reflected pulses. This aspect of the scheme offers a practical advantage that is , sensor arrays can be spatially redistributed without the need to modify NCC circuitry or timing .

The network information rate is governed by the choice of the token pulse width which in turn depends on the maximum acceptable length for fibre inter-coupler delay lines . Thus , there exists a trade off between information rate and the length of fibre delay lines . More importantly , the chosen data rate must satisfy sensor interrogation requirements for a given measurand .

Since the network design concept is based on digital communication systems , it is appropriate to adopt well established parameters which are used in conventional digital communication systems . As a result , the achievable values for those parameters can readily be assumed .A data rate of 2Mbit/s is considered to be satisfactory for the reasons that:

a) Binary digital sensors are basically recommended for rather crude measurements (applications in which very rapid changes are not expected), the rate of interrogation need not be higher than that given by a 2 Mbit/s data stream in most cases.

b) The corresponding fibre lengths to provide a bit period of 0.5μs is 100m, which is a fairly manageable length when carefully coiled on a reel.

For a 2 Mbit/s system a receiver sensitivity better than -63dBm for 10^{-9} BER [4] has been achieved in telecommunication systems . In this system a receiver sensitivity of -53 dBm for 10^{-9} BER will be assumed

3. Network Development and Construction

A network prototype incorporating two arrays of 2 sensors (3 bits per frame) has been developed and constructed , employing multimode technology.

The network block diagram is shown in figure 5 .

The timing for the control token and the corresponding data frames generated by the arrays are shown in figure 6 .

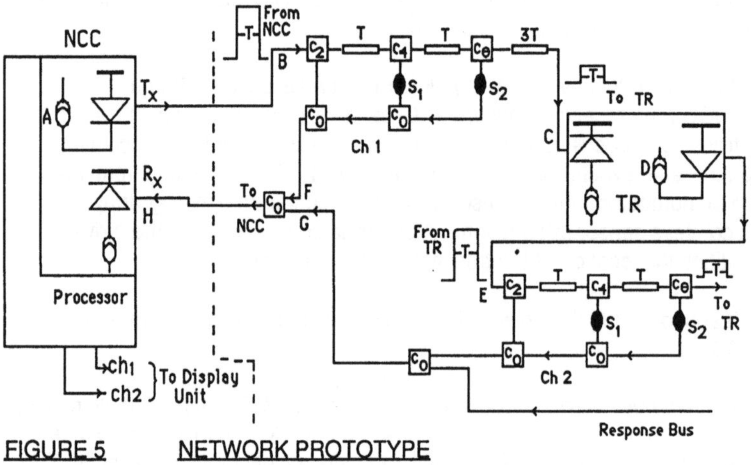

FIGURE 5 NETWORK PROTOTYPE

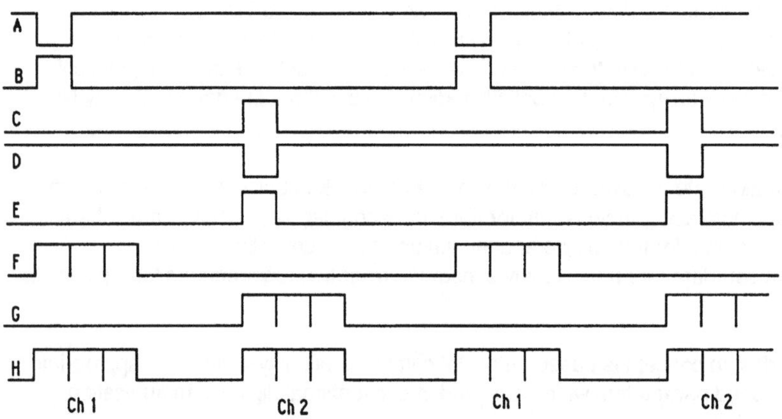

FIGURE 6

Figure 6 represents the following information :

Sensor array no.	Sensor no.	Sensor state
1	1	ON
1	2	ON
2	1	ON
2	2	ON

4. NCC Design

The network control centre is the focal point of the network.It performs the following functions :

a) provides optical power to sensor array 1 .
b) interrogates sensors sequentially by transmitting an optical token .
c) receives , synchronises and processes data frames sent by sensor arrays.
d) identifies array locations within the network .
e) identifies sensor location within an array and the state of that sensor .
f) synchronises the token cycle.
g) converts serial frames into parallel .

4.1. Operation

With reference to figure 5,the NCC transmits a control token pulse(B) of duration 0.5μs to the first sensor array (channel 1) . Channel1 is permitted to place a single frame of information (3 bits in this case) on the response bus(F) during the period for which it possesses the control token . The token is subsequently regenerated(E) by the TR and passed on to the next sensor array (channel2) . Similarly , a 3 bit frame is generated and placed on the response bus(G) by channel 2. The two frames appear on the response bus as two serial channels(H) separated by 3 time slots (this separation is necessary for an unambiguous synchronisation of NCC timing) . Since the separation slots are regarded as "low" bits by the NCC , the transmission is considered to be continuous at a rate of 2 Mbit/s .
Each token cycle results in the arrival of two frames at NCC (or equal to the no. of arrays , in general) , at which point the next token pulse is generated, delayed by three time slots , and then transmitted to initiate the next interrogation cycle . Thus , each sensor is sequentially interrogated every 6μs .

In figures 7 a and b photographs of the recovered data , are shown .From the display,the required sensing information is readily available.
The network can be expanded to incorporate a large number of sensors by increasing the number of sensor arrays.

5. System Evaluation

The system impairment sources which need to be considered at a data rate of 2Mbit/s are :

a) Receiver Noise
b) Dispersion
c) Modal Noise
d) Modal Distribution in couplers

The receiver noise aspect can be fully accounted for by designing a receiver to give the specified required sensitivity .

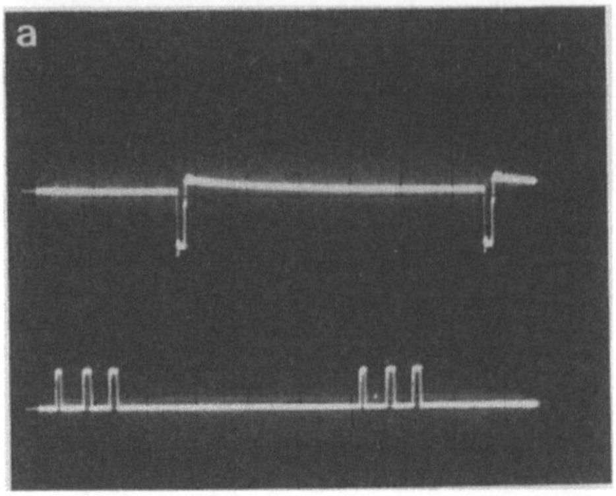

TOKEN (D)

DATA (CH1)

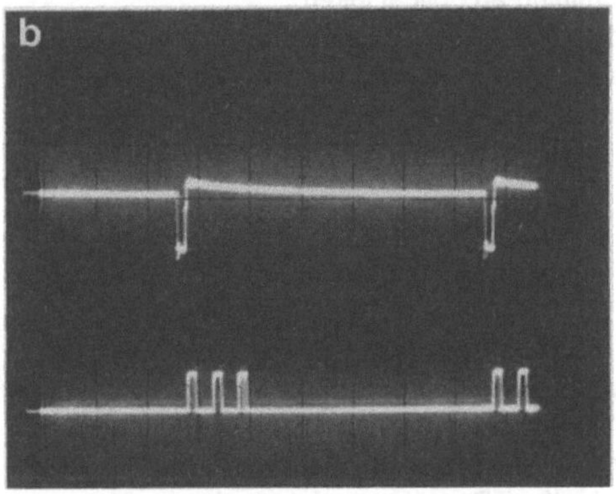

TOKEN (D)

DATA (CH2)

FIGURE 7 a and b

The power penalty due to dispersion was found to be negligible in the prototype .

Modal noise degradation can be eliminated by using a non-coherent (eg. LED) light source.

When light is launched into a multimode optical fibre coupler , the corresponding light at the output ports may be seriously affected by the launch conditions at the input port ie., if the launched light is not in a state of modal equilibrium then any imbalance between the coupler split ratio would give rise to random fluctuations, in the outputs, as a function of the launch number.This is due to the fact that for each launch attempt a new set of modes may be distributed between the output branches . This behaviour introduces a degree of uncertainty between the input/output relationship of a coupler .The problem becomes severe for J x J couplers with a large J,or the simplest (2 x 2) coupler with a highly unequal split ratio, such as 5 : 95, or even 20 : 80 . For 50 : 50 multimode couplers the modal distribution is much more uniform , and hence the resultant output is predictable with an acceptable degree of confidence even when the launched light is not in a state of modal equilibrium . However ,in a ladder array design ,50: 50 couplers would result in an unacceptable level of power loss. On the other hand , an array employing unequal split ratio couplers(limited to 5:95) in conjunction with an LED (-13dBm) drive ,and assuming an equilibrium modal state of the launched light, would allow a maximum of 11 sensors (12 bits/frame). It is interesting to note that any increase in the transmit power would not significantly improve the situation. This is due to the constraint on the coupler split ratios imposed by practical limitations.This implies that a new sensor array design is needed.

6. Symmetrical Array Design

A solution has been found which combines the advantages of 50 : 50 couplers and an LED drive. Unlike the ladder array , the symmetrical sensor array performs parallel power distribution . In this way, the power is optimally distributed throughout the array structure, consequently, the number of sensors per array is increased for the same available power.Furthermore, the total number of couplers are the same as that required for a ladder array having the same number of sensors . A symmetrical array of N sensors is shown in figure 8.

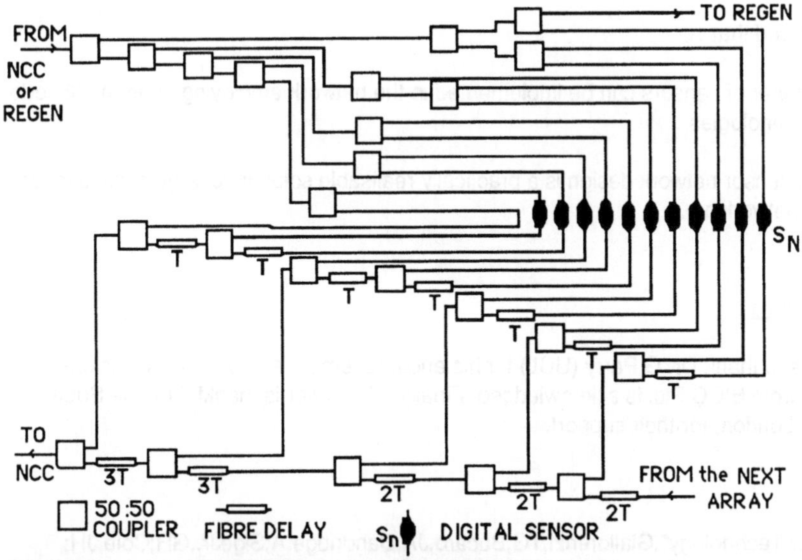

FIGURE 8 SYMMETRICAL SENSOR ARRAY

The symmetrical sensor array minimises the modal distribution problem as well as removes the modal noise (assuming LED drive) constraint, enabling the power budget to be determined by more straightforward means.Under these conditions , it is appropriate to use the excess loss values (provided by the manufacturers) for the couplers to calculate the power requirements.

As the proposed network is technology transparent ie. ,can be realised either in multimode or single mode technologies , it is of interest to consider the symmetrical array design in the single mode for comparison .
The maximum number(N) of sensors per array in each case(considering multimode{MM} and single mode{SM} technologies) is given in the following :

	N	Cex	Tx	Rx
{MM}------	19-20	0.25 dB	-13dBm(LED)-	-53dBm
{SM}------	30-31	0.1 dB	0dBm(LASER)	-53dBm

Where

Cex - Coupler Excess Loss ; Tx - Transmit Power ; Rx - Receiver Sensitivity .

However,there is no limit ,in principle,on the number of sensors in the overall network.

7. Conclusions

A novel sensor network topology based on a hierarchical multiplexing scheme has been derived . The sensor networking problem has been considered from a digital communications view point. Digital communication principles and techniques have been applied to multiplexing , signal processing and networking of binary digital optical sensors . The design concept of the proposed network topology has been confirmed by theoretical and experimental evaluations .
A symmetrical sensor array design , which combines the advantages of balanced (50 : 50) couplers, parallel power distribution , and an LED drive, has been derived .

Finally it is concluded that :

1) Very large number of sensors can be implemented in the network employing either multimode or single mode technologies .

2) The proposed sensor network design is a practically realisable solution to large scale sensor multiplexing and networking .

8. Acknowledgements

The author wishes to thank Dr. G.Parry (UCL) for his encouragement and useful discussions . Financial support from BICC Plc. is acknowledged. Finally ,the author is thankful to The South Bank Polytechnic,London, for their support.

9. References

1) "Optical Sensor Technology",Giallorenzi,TG;Bucaro,JA;Dandridge,A;SigeJr.,GH;Cole,JH; Rashleigh,SC; and Priest,RG; IEEE J.of Quantum Electronic,Vol.QE-18,no.4,April 82 , pp. 626-665.
2) "Review of Modulating Techniques for Optical Fibre Sensors",Medlock,R; Int. J. of Optical Sensors,1(1986),pp.43-68.
3) "Fibre Optic Sensors for Networks", Kist,R; Proc.of the 4th. Int. Conf. on Optical fibre Sensors, Tokyo Oct.1986, pp.209-225.
4) "High Performance Digital Optical Receivers with Pin Photodiodes", Smith,DR;Hooper,RC; Webb,RP; Proc. of 1979 ISCAS, pp. 511-514.

Multi-frequency Modulation for Passive Multiplexing of Fibre Optic Sensors

H.S. Al-Raweshidy, D. Uttamchandani, and B. Culshaw

Department of Electronic and Electrical Engineering,
University of Strathclyde, 204 Geroge Street, Glasgow G1 1XW, UK

The passive multiplexing of intensity modulating fibre optic sensors using a semiconductor laser source modulated at different sub-carrier frequencies is described. The use of a synchronised gating signal at the detector output enables the simultaneous demultiplexing of signals from different sensors. Experiments on a network of three sensors are reported.

In recent years there has been considerable interest shown in the electronically passive multiplexing of fibre optic sensors. Fibre optic sensors offer the advantages of high sensitivity and dynamic range, small size and weight, immunity from electromagnetic interference and compatibility with fibre optic transmission media. In order to fully exploit the features of fibre optic sensors it is necessary to consider techniques for electronically passive multiplexing of fibre optic sensors. Multiplexing of sensors is important because in applications such as process control or aerospace it is envisaged that several tens of sensors will be required either for measurement of different parameters or as Back up devices in case of failure of one or more sensors. To fully exploit their advantages the fibre optic sensors should be connected into a passive network, also of optical fibre, which conveys all the data from the sensors to a central measurement and control station.

Passive multiplexing of fibre optic sensors can be divided into a number of categories depending on whether the sensors are of the interferometrics on non-interferometric type [1]. The work reported in this paper is applicable to non-interferometric, that is intensity modulating sensors. Multiplexing techniques for such sensors can be broadly divided into three categories which are potentially of greatest use. These are time division multiplexing, wavelength division and spatial multiplexing. In time division multiplexing each sensor in the network is characterised by a unique optical propagation time (or strictly speaking, group delay). In wavelength division multiplexing each sensor is characterised by a unique wavelength, whilst in spatial multiplexing each sensor is characterised by a unique optical propagation channel.

Time division multiplexing is the most commonly used technique for passive multiplexing and can be implemented in either the time or frequency domains. In the time domain, the procedure is to inject a short pulse of optical energy into the optical network and wait for separate pulses to return from each sensor in the network [2]. This involves relatively large peak power injection into the optical fibre at a low duty cycle and consequently a long integration time of the return signals is required to achieve a good signal to noise ratio. An alternative approach is to work in the frequency domain using techniques like sub-carrier frequency modulated continuous wave (FMCW) and synthetic pulse generation. The high duty cycle

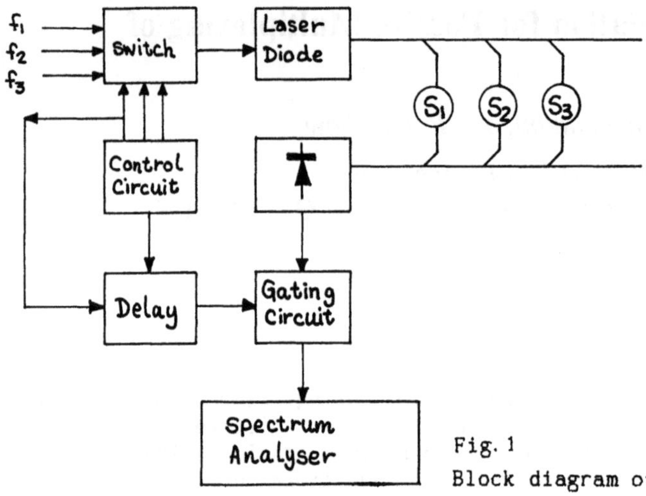

Fig. 1
Block diagram of multiplexing system

(and therefore average power) available using the above techniques make them attractive for use in sensor multiplexing. In this paper we report preliminary results from a novel passive multiplexing system suitable for intensity sensors which uses time division multiplexing and operates in the frequency domain.

The fibre optic sensor system is shown schematically in Fig 1. It comprises the fibre optic network which is passive, and the transmitting and receiving opto-electronic circuits. The network has the feature that the propagation delay from input to output for path 1 is T, the propagation delay for path 2 is 2T and so on for subsequent paths. With this arrangement in place, a multi-frequency sub-carrier signal is generated to modulate the intensity of the optical source. The sub-carrier signal comprises a constant amplitude burst of N different frequencies, each frequency lasting for a duration of T seconds. N is the number of sensors in the network. In our experiments a network of three sensors has been assembled using single mode optical fibre and directional couplers design for 1.3 μm operation. The optical source is a diode laser operating at 1.3 μm. The wave form generated to modulate the laser is shown in Fig 2 (a). The period T is determined by the minimum propagation delay between sensors and in our network this equals 0.25 μs. The reception period NT equals 0.75 μs since N is 3. Note that 1/NT equals 1.33 MHz. The values of the frequencies f_1, f_2 and f_3 were chosen to be 12, 24 and 40 MHz which are all integer multiples of 1.33 MHz.

The frequency spectrum of the modulation waveform thus consist of a discrete line spectrum with components at 1.33 MHz under three (sin x)/x where $x=\omega T/2$ envelopes whose peaks are at 12, 24 and 40 MHz. Figures 2(b), 2(c) and 2(d) show the signals arriving at the detector from the three paths where the 'sensors' are located. The 'sensors' are required to produce intensity modulation of light and work through bending loss of the fibre. The timing diagrams show that the signals arriving at the detector have a relative delay of T seconds between them. A gating pulse of duration T and repetition period NT is now applied to the detector output Fig 2(e). It can be seen that the gated output signal has three frequency components. The gated signal is observed on the spectrum analyser where for the position of the gating pulse illustrated, the signal amplitudes at frequencies f_1, f_2 and f_3 are proportional to the optical power returned from sensors S_1, S_2 and S_3 respectively.

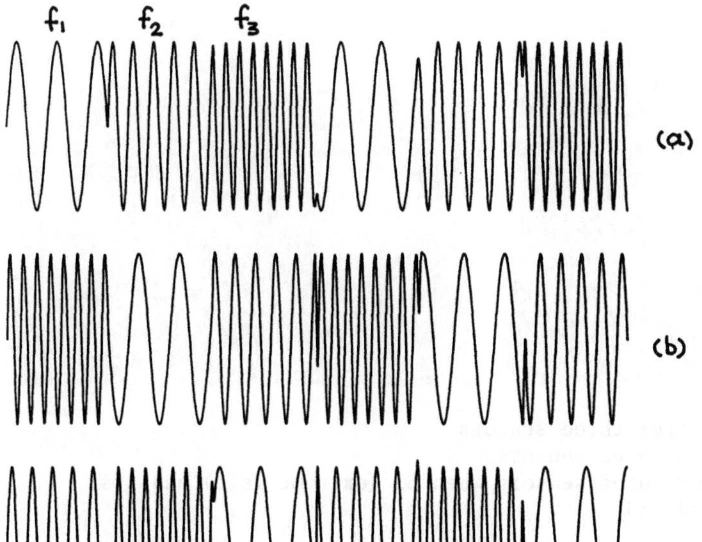

Fig. 2 Input and Output Signals of Multiplexing System
(a)- input signal. (b), (c) and (d) - delayed output signals from sensors S1, S2 and S3 respectively. (e) -gating signal

Thus by using the above technique, the system operates in the frequency domain where each of the frequencies corresponds to an individual sensor in the network. A band pass filter centred at each frequency can be used to separate the signal corresponding to each sensors. The present system therefore produces an output spectrum similar to that reported by Mallalieu et al[2]. Their system required the generation of a continuous frequency chirped sub-carrier signal to modulate the laser source whereas our system just uses a set of discrete oscillators and a switch.

The main advantage of using a continous frequency chirp over our technique is that the former makes use of correlation detection between a reference signal and the signal from the network of sensors. Correlation detection offers the best signal to noise ratio. On the other hand, a large

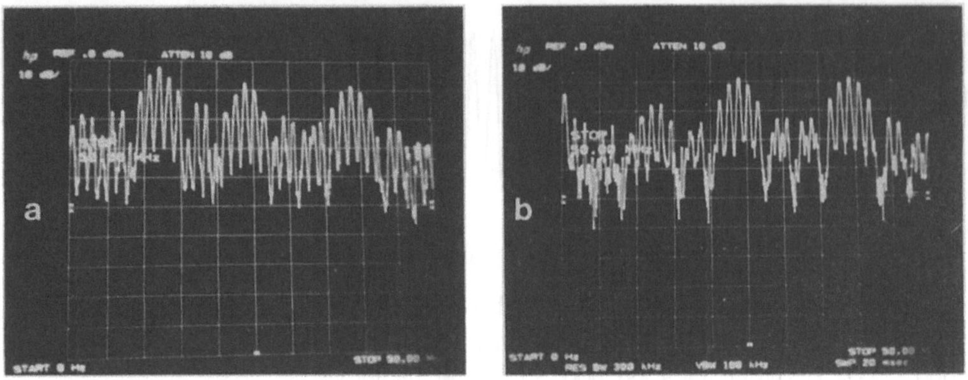

Fig.3 System response from three sensors
(a) response from three sensors
(b) response from three sensors when S3 (extreme left) has its intensity reduced

frequency sweep rate of constant gradient is required in the frequency chirp technique and this demands the use of more complex electronics when compared to our system.

Figure 3(a) shows the spectrum of the gated signal from the system when there is no intensity modulation on the sensors. The peaks at 12, 24 and 40 MHz correspond to sensors S_1, S_2 and S_3 respectively. The power in the three signals are unequal because of the unequal losses in each sensor are caused by different splice losses in the whole system. Figure 3(b) shows the same spectrum when only sensor S_1 has its throughput intensity reduced. The signal from S_1 drops by 13 dB as expected, but note that there is also a drop in amplitude at the frequencies corresponding to the two other sensors. Amplitudes corresponding to sensors S_2 and S_3 drop by 2 and 1 dB respectively. Crosstalk between sensors is reduced if two conditions are satisfied. Firstly, the optical delay difference between successive sensors must be equal to the pulse duration T described earlier. Secondly, exact synchronisation is required between the detector output signal and the gating pulse. Departure from these two conditions gives rise to what we refer to as primary crosstalk which can be understood by referring to figure 4. This

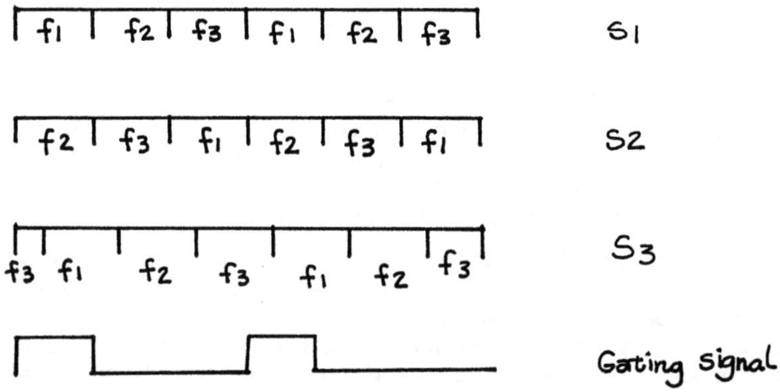

Fig.4 Output signals from multiplexing system when the delay line of sensor three is not an integral of T

diagram shows the situation when the sensors do not have integral values of the delay T associated with them. Clearly, the frequency components of the gated signals whose amplitude should uniquely represent the throughpoint of the sensor have an additional amplitude component arising from an adjacent sensor.

A second order crosstalk also exists due to the fact that the gated signals have a discrete frequency spectrum under a sinx/x envelope. The discrete spectra centred at adjacent frequencies will overlap, also generating crosstalk. However by selecting suitably large frequency spacings or by locating one frequency in one of the nulls of the spectrum of an adjacent sensor, the electrical crosstalk can be reduced to below a percent.

The principal feature of the present system is that the gated output signal provides information simultaneously on all sensors whilst, for example, in the OTDR type multiplexing technique[3] the sequential arrival of pulses implies that in principle only one sensor can be interrogated at a time.

In conclusion, a novel technique for passive multiplexing of fibre-optic sensors has been proposed and demonstrated. The technique is based on time division multiplexing, but operates in the frequency domain. In the practical system the magnitude of the cross-talk has been measured whilst consideration has been given to factors which contribute to the cross-talk. Also, the technique allows simultaneous interrogation of all sensors in the network.

References

1. Dakin J P, " Multiplexed and Distributed Optical Fibre Sensor Systems", J.Phys. E., 1987, 20, pp 954-967.

2. Mallalieu, K I, Youngquist R, Davies D E N, "FMCW of Optical Source Envelope Modulation for Passive Multiplexing of Frequency-Based Fibre-Optic Sensors", Electronics Letters, 1986, 17, pp 809-810.

3. Nelson A R, Mahon D H, Gravel R L, "Passive Multiplexing Systems for Fibre-Optic Sensors", Applied Optics, 1980, pp 2917-2920.

Specific Design of Optical Fiber Sensor Systems for Wavelength Division Multiplexed Networks

R. Durantis[1], *G. Anglaret*[1], *C.J. Hugues*[2], *and G.W. Fehrenbach*[3]

[1]OPTOFRA, Division de Framatome, Tour Fiat, F-92084 Paris la Défense Cedex 16, France

[2]Thorn Emi, Central Research Laboratories, Dawley Road, Hayes, Middlesex UB3 1HH, UK

[3]Degussa AG, Zweigniederlassung Wolfgang, Rodenbacher Chaussee 4, Postfach 1345, D-6450 Hanau 1, Fed. Rep. of Germany

Three known sensing principles are investigated so as to match the requirements of wavelength division multiplexed networks : intensity modulation, frequency modulation and luminescence decay.
A spectral bridging technique is proposed for compensation of intensity modulated analogue sensors. A microengineered optical modulator is proposed for interfacing frequency outdevices to the WDM network and the integration of temperature sensors based on luminescence decay time analysis into a WDM sensor-network is discussed.

INTRODUCTION

The networking of optical sensors may find its way through wavelength-division-multiplexing (WDM) techniques.
A three company effort is carried out in an European BRITE programme to bring those techniques out from the pure Research Age.
The general networking principle is to associate one narrow spectral channel to each sensor : the multiplexing and demultiplexing function is performed by polychromators (bulk optics). It is assumed that an optical source with a sufficiently wide spectrum is available so as to provide several channels. That condition is got either from the sensing element itself using luminescence techniques for instance or from a separate optical source whose power is modulated within each sensor as a function of wavelength. In the latter case, sensors act as simple modulators.
First of all, a system topology is proposed with special care of intensity-modulation multimode sensors : fiber microbending sensors are developed in conjunction with the network.
Then, a microengineered modulator is discussed focusing on the sensor-to-network interface which has been identified as the key element in the design.
Finally, temperature sensors based on luminescence decay analysis are studied in the way to use them under WDM conditions.

Springer Proceedings in Physics, Vol. 44 Optical Fiber Sensors
Editors: H.J. Arditty · J.P. Dakin · R.Th. Kersten

INTENSITY MODULATION AND WDM TECHNIQUES : A SPECTRAL BRIDGE BASED TOPOLOGY

General network topology of WDM systems for back-reflection multimode sensors is provided in figure 1. It is worth noticing that this topology is already implemented on marketed products incorporating on/residual/off sensors. Optical power is launched from an LED into the system towards an optical coupler. First-window LEDs are generally used -which means FWHM around 50 nm - with special care of the peak wavelength which must be adapted to the polychromator performance. A remote polychromator allows it to spectrally dispatch the optical power to the sensors wherein modulation is performed.
A number of twenty channels is available from 800 nm to 910 nm. Thanks to the sensor mirrored termination, modulated light comes back through the original optical coupler to local polychromator which demultiplexes the signals to adequate receivers.
Unfortunately, such a topology generally jeopardize sensor accuracy any time it is used with analogue sensors and classical optical components : it is no use reminding of very poor repeatability of optical connections, variable performance of optical passive components and electronic receivers as a function of temperature and ageing... Designers are compelled to get references within the sensors and to neutralize physical-medium-dependant parameters.
A wavelength-type referencing is used : two different wavelength are launched into each sensor through a unique optical link and then separated thanks to an optical filter upstream to the modulating cell according to figure 2 general functional diagram.
Then, line-dependant parameters are mostly neutralized thanks to a simply mirrored gate on the remote polychromator : two extra signals are thus provided. Four measurements for each sensor are finally processed in the way to get a homothetic function of the loss induced by the measurand. This method has already been described (1) and general design is provided in figure 3.
It must be noticed that reference and auxiliary signals must be spectrally situated as close as possible to one another. Intensity modulation raises two fundamental problems which are not easy to deal with : networks act as modal filters and non-propagating modes may be radiated out from the fiber at

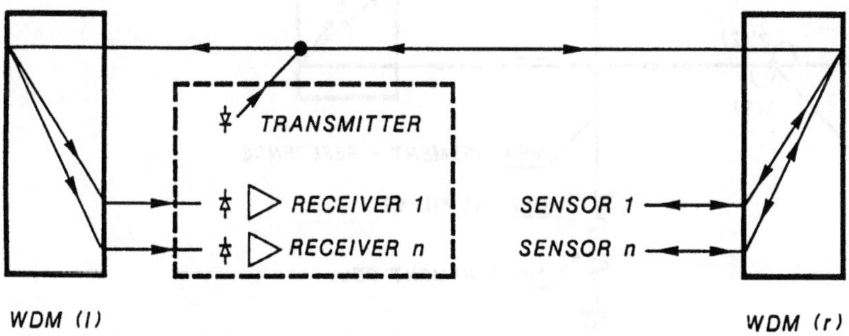

fig.1 GENERAL NETWORK TOPOLOGY

Two spectrum - shifted LEDs launch power into fiber n°1 and fiber n°2.

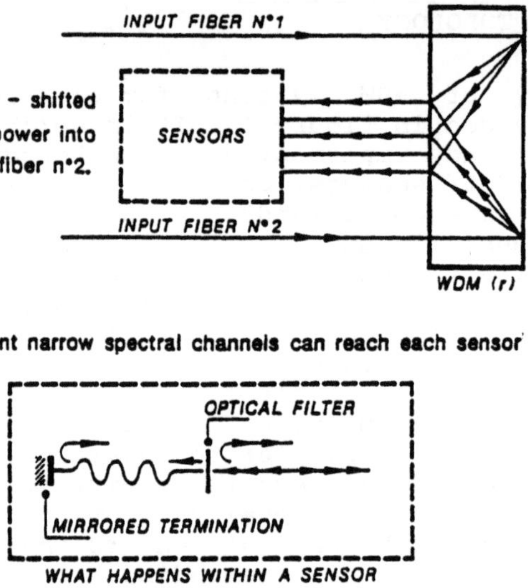

Two different narrow spectral channels can reach each sensor

An adequate optical filter reflects one bandwidth to provide a reference.

fig.2 **GETTING A REFERENCE**

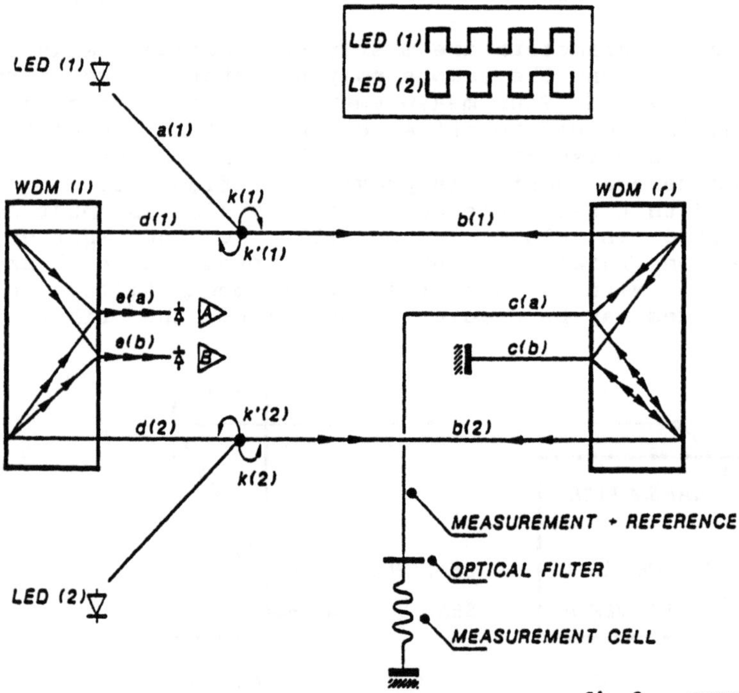

SYSTEM TOPOLOGY

fig.3 **GETTING RID OF LINE - DEPENDENT PARAMETERS**

various distances from the source. First point may be managed thanks to a special selection of optical components. Unfortunately, it is difficult to cope with second one. As a consequence, propagation phenomenons must be as similar as possible for the different spectral channels dedicated to each measurement. It means that associated wavelengths must be situated as close as possible. This is made possible thanks to the narrow spectral gap between two contiguous gates on the polychromator. Working value is six nanometers.

MICROENGINEERED MODULATOR FOR THE WDM SYSTEM

Several workers have proposed the use of hybrid sensors in which electronic sensors are used to capture the measurand information and an optical fibre is used to transmit the information in a digital form (3),(5). The power for the sensors is derived from light also carried by an optical fibre. The WDM topology for the hybrid technology sensors is depicted in figure 4. To minimise the optical energy in the distribution fibre, the electronic sensors used must necessarily have low power requirements. A key element for the hybrid system is the interface between the network and the sensors. An interface which uses an optical modulator design employing microengineered structures is described in the following section. In the design, the modulation technique relies on the electrostatically induced motion of a silicon microengineered structure to modify the reflectivity of the device.

Modulator Design

The enhanced sensitivity of the insertion loss of expanded beam connectors to axial misalignment has been used in the design of sensors (4), compact wavelength division multiplexers (6) and is also used in the modulator design.

The design of the modulator is similar to that of an electromechanical switch proposed for fiber multiplexing (2). It is shown schematically in figure 5 and consists of an array of

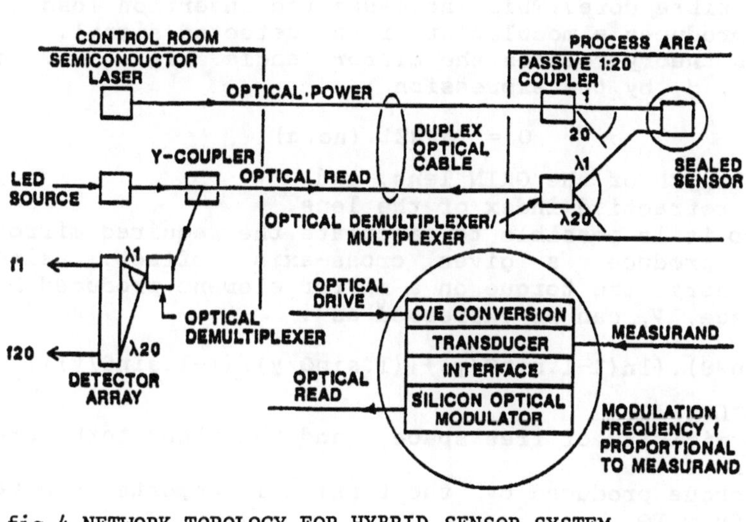

fig.4 NETWORK TOPOLOGY FOR HYBRID SENSOR SYSTEM

fig.5 TORSIONAL MIRROR
MODULATOR

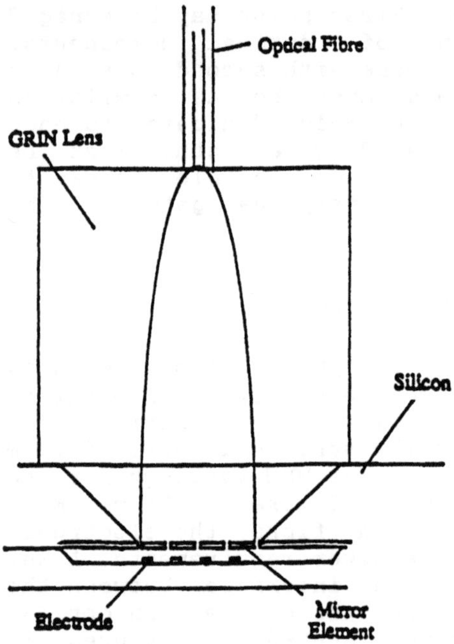

microengineered mirrors torsionally syspended above a set of electrodes. Light from the WDM fibre is collimated by the quarter pitch GRIN lens. The collimated optical beam is incident onto the mirror structure which, in the rest position, retro-reflects the light into the lens. The reflected light is focused by the lens onto the optical fibre where it is recaptured by the fibre core.

The application of a voltage to the electrode structure produces a rotation of the mirror elements. The collimated beam is reflected at an angle to the GRIN lens axis and is focused in the plane of the optical fibre endface but now with a cross-axis offset from the fibre core. This increases the insertion loss of the system and produces a modulation of the detected signal.

Selfoc GRIN lens theory relates the mirror angle shift, Θ, to the axial offset, d, by the expression :

$$\Theta = (d/2Z).(no.\pi)$$

where Z = the length of the GRIN lens, and
no = the refractive index of the lens.

Knowing Z and *no* it is possible to calculate the required mirror deflection to produce a given cross-axis offset. Using electrostatic theory, the torque on a mirror element produced by an applied voltage, *V*, can be expressed as :

$$\Gamma e = A.(\cos\Theta/\sin^2\Theta).(\ln(1-l.\sin\Theta/y)+((l.\sin\Theta/y)/(1-l.\sin\Theta/y)))$$

where $A = (\varepsilon o.LV2)/2$,
εo = permitivity of free space, and the other terms are defined in figure 6.

The restoring torque produced by the torsional supports can be expressed as : $\Gamma r = T\Theta$

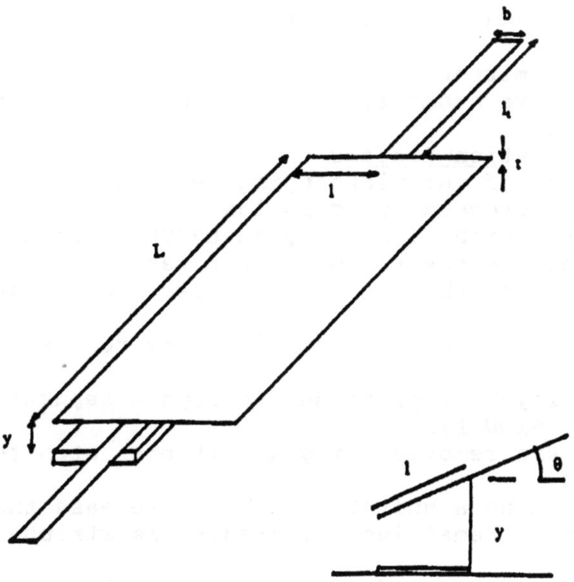

fig.6 DEFINITION OF MODULATOR PARAMETERS

where T = (T.b.t)/(3.(1+v).lt)
In equilibrium these torques are equated and an equation relating the equilibrium voltage to the mirror angle is obtained. When the electrode gap is smaller than the half width of the mirror element, such that the maximum angle that the element can be deflected to is y/l(=θmax), the equilibrium voltage is expressed as :

$$V^2 = 2T\theta \, / (\varepsilon_o L(\ln(1-\theta/\theta_{max}) + (\theta/\theta_{max})/(1-\theta/\theta_{max})))$$

Examination of the relationship reveals that there is a critical voltage, Vc ,beyond which the system becomes unstable and the mirror deflects to its maximum position. This voltage occurs when the mirror angle is approximately 0.44.θmax. In the proposed modulator, the applied voltage will be slightly greater than Vc to obtain the full mirror swing at an effectively reduced drive voltage.
Since the light reflected from the mirror elements is not to be reimaged into an output fibre as in the fibre optic switch, minor distortion of the mirror in the deflected position by the electrostatic force is not important as it will only slightly degrade the modulation depth of the detected signal. Consequently mirror elements with a relatively large value of width to thickness ratio can be used.
As an example, a reflector 1 mm long and 200 µm wide made from 1 µm thick silicon oxynitride, with torsional supports 100 µm long and 50 µm wide, spaced from the counter electrode by 3.5 µm will switch 2° for an applied voltage of approximately 6.5 V.
The resonant frequency of the mirror structure is :

$$f(res) = ½\pi\sqrt{T}/\sqrt{I}$$

where I is the mass moment of inertia of the device which, using the notation of the figure, can be expressed as :

$$I = (d.t/12).(L(21) +2.1t.b)$$

where d = the density of the mirror.
For the structure in the above example, the resonant frequency will be approximately 12 kHz.

Fabrication Process

A process for the fabrication of the microengineered mirrors is illustrated in figure 7. The process is as follows :

(a) Silicon wafers are coated with SiON using an LPCVD process.
(b) One surface of the wafers is covered with photoresist.
(c) The photoresist is patterned through a photomask to expose the SiON film.
(d) The second surface of the wafers is covered with photoresist.
(e) The second photoresist layer is patterned through a separate photomask to expose the SiON film.
(f) The exposed SiON films are removed using a suitable etch to expose the bulk silicon.
(g) The silicon is removed using a suitable etch to release the mirror structures and the remaining photoresist is stripped using a solvent.
(h) The drive electrodes are produced on separate substrates using a single sided etch process and a glod evaporation and patterning stage.
(i) The mirror element and electrode wafers are bonded together and the individual devices are separated and the fibre pigtailed GRIN lens are attached.

Using the above process will allow wafer scale fabrication of the active component of the modulator.

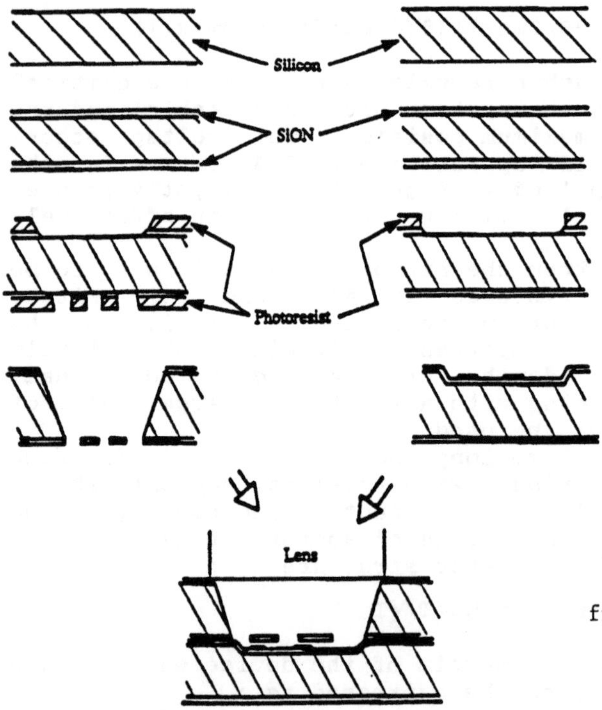

fig.7 TORSIONAL MIRROR MODULATOR FABRICATION

LUMINESCENT TEMPERATURE SENSOR IN A WDM NETWORK

In fiber optic temperature sensor systems based on luminescence decay time analysis, semiconductor light sources can be used for generating the excitation light pulses (figure 8). The absorption bands of the corresponding sensor materials like chromium doped Yttrium-aluminium-garnet (YAG) (7) oder ruby (8) are in the wavelength range around 600 nm.
The optical power and the coupling efficiency to optical fibers of broadband LEDs emitting at 600 nm wavelength are poor compared with special LEDs developed for fiber optic applications.
In order to achieve detected intensity levels in the nanowatt range, thick core (600-1000 µm) high numerical aperture optical fibers have to be used.
Recently laserdiodes with narrow band emission at 670 nm wavelength and 5 mW output power became commercially available. At this wavelength in the low energy wing of the absorption peak at 600 nm wavelength the absorption coefficient of 1 at% Cr:YAG-crytals is below 0.1 cm . To increase the absorption coefficient, Cr:YAG-crystals with Cr-concentrations up to 10 at% have been grown.
First experiments have already shown that sufficient luminescence intensity levels can be detected, using standard 200 µm core fibers with a numerical aperture of 0.2. An estimate of the power level indicates, that it is within the scope of this new excitation scheme to operate 5 temperature sensors (measuring range : -50 to +125°C) in a WDM-System (figure 9).

CONCLUSIONS

It is altogether clear that WDM techniques can accomodate the previously described three sensing principles. They seem to provide good solutions to the problems of referencing intensity

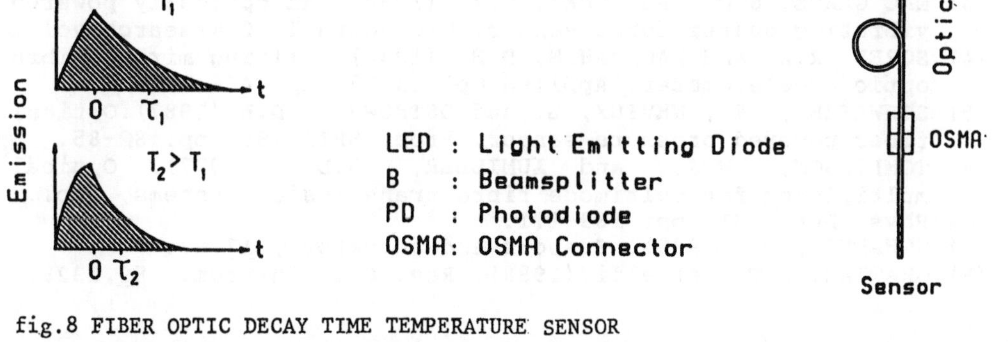

fig.8 FIBER OPTIC DECAY TIME TEMPERATURE SENSOR

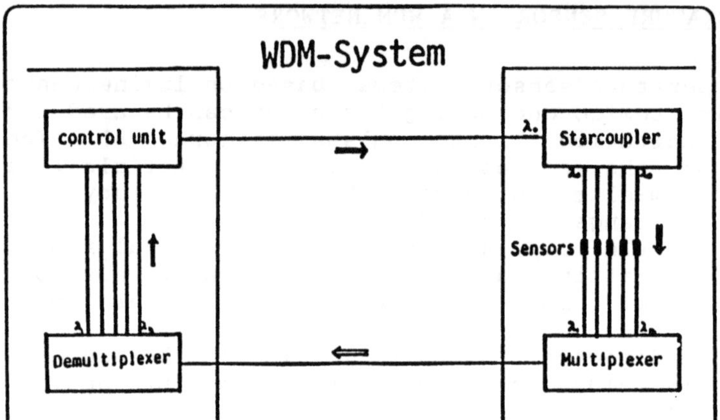

fig.9

modulation sensors and coping with the influence of line parameters upon analogue measurements. Furthermore, good adequacy is predictable between those techniques and microengineered modulators as well as with luminescence-based temperature sensors.

ACKNOWLEDGEMENTS

The above mentioned Research about chromatic multiplexing of multimode fiber sensors is partly carried out thanks to the European Communities Commission through financial support from BRITE programme.

REFERENCES

(1) DURANTIS, R. and ANGLARET, G. (1989). A spectral bridge based method for WDM networks devoted to intensity modulated optical sensors, PROC. EFOC, pp. 408-410.
(2) GUSTAFSSON, K. and HÖK, B. (1987). Fiberoptic switching and multiplexing with a micromechanical scanning mirror, Digest of technical papers- 4th Int. Conf. on Solid State Sensors and Actuators (Tokyo : IEE of Japan), pp. 212-215.
(3) MAC GLADE, S.M. and JONES, G.R. (1984). An optically powered vibrating quartz force sensor, GEC Journal of Research vol.2
(4) SOREF, R.A. and MAC MAHON, D.H. (1984). Tilting mirror fibre optic accelerometer, Applied Optics 23, pp. 486-491.
(5) SCHWEIZER, P., NEVEUX, L. and OSTROWSKY, D.B (1987).Optical fiber powered pressure sensor, Proc. SPIE 798, pp. 82-85.
(6) TOMLINSON, W.J. and AUMILLER, G.D. (1977). Optical multiplexer for multimode fibre transmission systems, Appl. Phys. Lett. 31, pp. 169-171.
(7) SCHÄFFER, W. (1989). Sensors and actuators, 17,27.
(8) GRATTAN, K.T.V et alii (1988), Rev. Sci. Instrum., 59,1328.

A Coherence-Multiplexed Quasi-Distributed Polarimetric Sensor Suitable for Structural Monitoring

V. Gusmeroli, P. Vavassori, and M. Martinelli

CISE Tecnologie Innovative SpA, P.O. Box 12081, I-20134 Milan, Italy

The coherent-multiplexing concept is exploited to develop a quasi-distributed polarimetric-type sensor. The sensing-fiber lengths are marked by microrings which couple the power from the principal propagation mode to the orthogonal one. Hence, the fiber is maintained mechanically and dimensionally intact, suitable for material embedding. A superluminescent diode is used as a source in an experiment involving 5 subsequent sensing-fiber lengths. Preliminary results point out the capability of the quasi-distributed sensor to recover the modal shape of a vibrating structure.

Polarimetric sensors are increasing in importance among coherent fiber-optic sensors. Their relatively low sensitivity is compensated by an intrinsic simplicity of the applicative schemes. Besides, the guide of two orthogonal-polarized modes in the same fiber offers interesting design opportunity. In this context the development of distributed or quasi-distributed polarimetric networks becomes mandatory.
Only the development of such a scheme will permit the polarimetric features to be fully exploited and novel applications to be implemented [1].

The aim of this communication is to present a novel quasi-distributed polarimetric-based scheme and its application to the detection of structure modes. The network is developed following the low-coherent multiplexing concept and using a single-length of fiber. The geometrical and mechanical characteristics of the fiber are maintained intact, so that the sensing-net is suitable to be embedded into composite structures [2].

The developed scheme is reported in Fig.1. A pigtailed superluminescent laser (a coherence length Lc=50μm) is

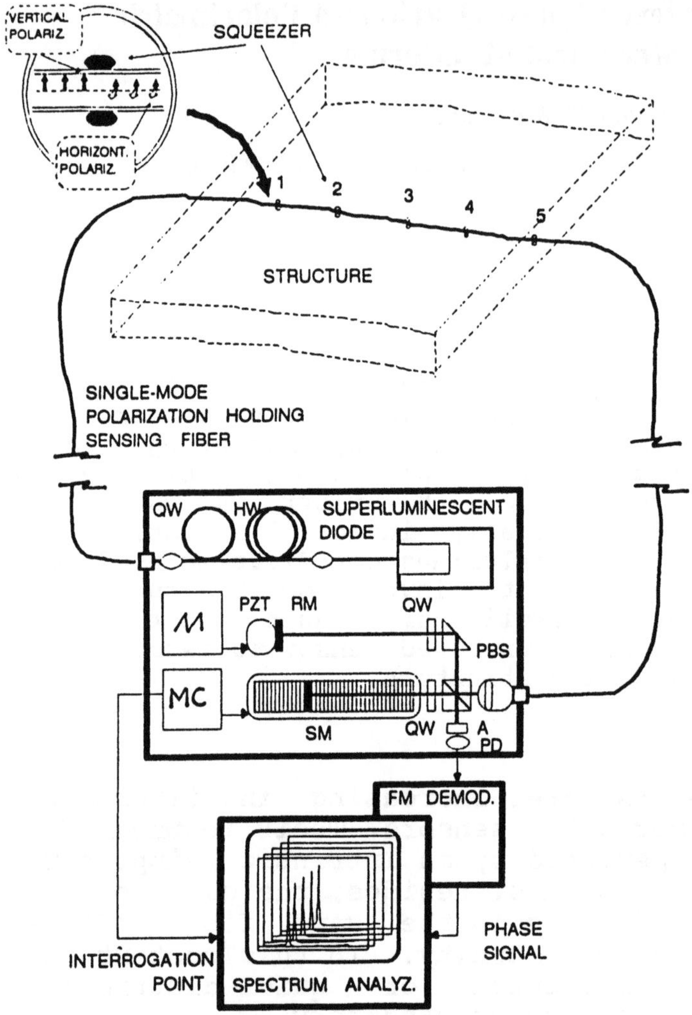

Fig.1 The experimental set-up. The coherence multiplexing technique is applied to a polarimetric sensor. The "squeezer" couples the two propagating modes of the fiber and labels the interrogation points. By moving the mirror SM each sensing length is compensated.

coupled to a polarization-holding fiber (fiber beat length Lb=1mm). By accurately orientating a couple of all fiber retarders (QW,HW), only one of the principal modes of the fiber (e.g. the vertical) is energized. At point 1, a microring squeezes the fiber to couple a calibrated fraction of power from the vertical mode to the horizontal one [3]. The same operation is repeated in each of the four following "squeezed points", i. At the fiber output the principal mode and the coupled signal i show a relative phase delay $\Delta\varphi i=2\pi$ Li/Lb, where Lc is the fiber beat length and Li is the fiber length from the point i to the end. Hence, a series of interferometers, with different values of "unbalancing", are setup along the sensing fiber. When the light returns to the optoelectronics unit, the two guided modes are separated by the Polarizing Beam Splitter (PBS). Thus, the two arms of the external interferometer convey the information from the vertical mode and from the coupled horizontal mode, respectively. If the phase delay accumulated by the two modes is larger than 2π*Lc/λ, where Lc is the coherence length of the source and λ the light wavelength, any interference is not evidenced. Interference is only possible if an unbalance is given to this "external" interferometer which compensates the phase delay $\Delta\varphi$i accumulated by the two modes during the fiber propagation from the point i. By moving mirror SM different values of phase delay are compensated. If the distances beetween succesive squeezed points are setted larger than ΔL=LcLb/λ only the interference from the i-th coupled signal is recovered.

This scheme implements the concept of "quasi-distributed" sensor without use of any special "marker" on the fiber, such as couplers, retroreflective splices or fiber rotators. The address of the multiplexing is obtained by a "soft" and "non-invasive" action, without any need for fiber cutting or for coating removal. As a consequence, the mechanical and size properties of the fiber remain practically intact. This feature can be conveniently appreciated in the context of the structure monitoring, where low-invasivity and mechanical integrity of the sensors are wanted requirements.

An experiment of structure monitoring is simulated by using the proposed scheme. Without any generality loss, let us assume to want to recover the first resonant mode (frequency, amplitude and modal-shape) of a clamped-free structure. A normal polarimetric sensor embedded into the structure does permit to recover only information about the frequency and integral deformation along the fiber path. The developed quasi-distributed sensor allows the modal shape to be also obtained with high accuracy.
Five interrogation points are distributed along the sensing fiber with a mutual distance ΔL=15cm. The

Tab.1 For each fiber-optic sensing segment the strain excitation value, corresponding to the first resonant mode, is reported (row 1). The expected and measured integral strain values are also given (rows 2 and 3).

Sensing segments	1	2	3	4	5
excitation value	0.4406	0.3023	0.1745	0.0715	0.0111
expected strain integral (normalized)	0.4406	0.7429	0.9174	0.9889	1.0000
measured strain integral (normalized)	0.4429	0.7368	0.9205	0.9894	1.0001

number 1 is close to the clamped-end of the structure, while the number 5 is close to the free-end. The resulting five fiber segments are stretched by means of separeted piezoexciters. The excitation frequency is the same for each exciter, whereas the amplitude is changed according to the values reported in the first row of Table 1. These values are the theoretical expected from a structure vibrating on a "pure" mode and are evaluated following Ref.4.

In correspondence with such excitation, the answer of the polarimetric sensor can be easily evaluated. In our case an average sensitivity of $12*10^4$ rad/m is measured. On the base of the fiber integral behaviour theory developed in Ref.5, the normalized expected values of "integral strain" are reported in the 2nd row of the same Table. These values represent the integral of the strain sensed by the fiber from point i to the end of the structure.
The movement of the mirror SM permits one to interrogate each "squeezer" of the sensing-net and then to measure the integral of the strain from point i to the end of the structure. The detected heterodyne

signal (mirror RM, moved by a sawthoot signal, gives the necessary carrier to the scheme) is demodulated and sent to a spectrum analyzer.

The results of the measurements are summarized in Tab.1, Figs.2 and 3. In Fig.2 the rms integral strain is reported in a function of the frequency for the different interrogation points. The same values, normalized, are reported in the third row of Tab.1 and compared with the theoretical curve in Fig.3. An excellent agreement between the experimental data and the expected values turns out. This result points out that not significant cross-talk is present in the developed quasi-distributed network. Besides, it confirms the excellent features of the low-coherence multiplexing method in terms of spatial resolution (ΔL=15cm) and precision. The experimental data permit to recover the first derivative of the mode shape (with spatial resolution limited by the number of interrogation points) and, from this, the wanted mode shape.

In conclusion, a quasi-distributed polarimetric network has been experimented. The novel scheme permits to join the advantage of a low-coherence

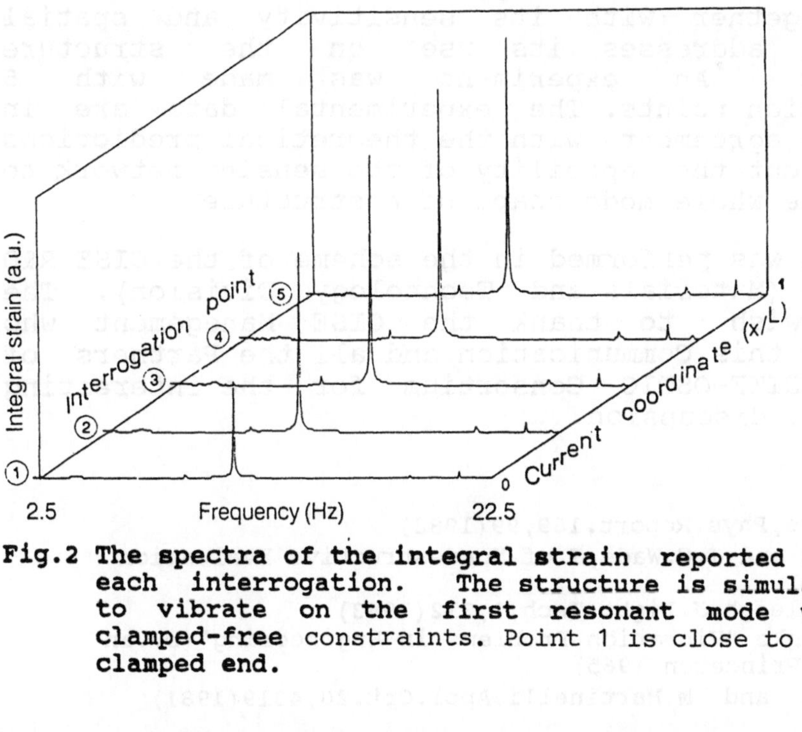

Fig.2 The spectra of the integral strain reported for each interrogation. The structure is simulated to vibrate on the first resonant mode with clamped-free constraints. Point 1 is close to the **clamped end.**

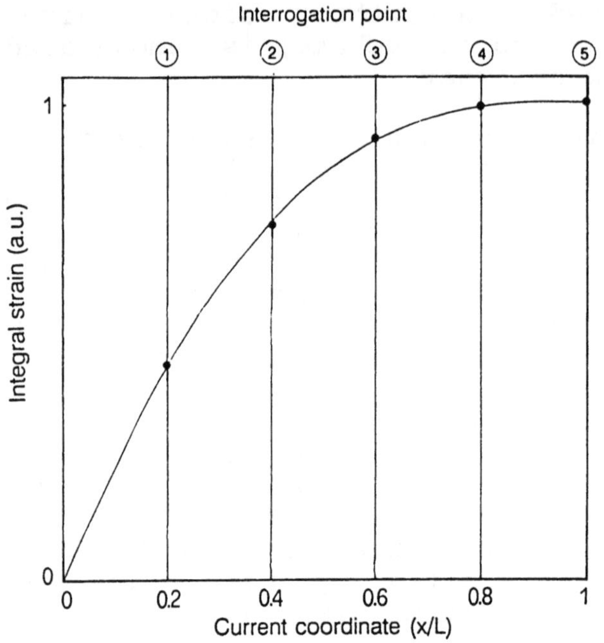

Fig.3 A comparison between the expected integral strain (curve) and the experimental results (dots).

multiplexing to the maintained integrity of the sensing fiber. The very-low invasiveness of the sensor together with its sensitivity and spatial resolution addresses its use on the structure monitoring. An experiment was made with 5 interrogation points. The experimental data are in excellent agreement with the theoretical predictions and point out the capability of the sensing network to recover the whole mode shape of a structure.

This work was performed in the scheme of the CISE R&D activities (Materials and Technology Division). The Authors wish to thank the CISE Manegement who authorized this Communication and all the Partners of the EEC-BRITE-OSTIC Consortium for the interesting and helpful discussion.

REFERENCES

[1]A.J.Rogers,Phys.Report.169,99(1988)

[2]R.O.Clous and J.C.Wade,J.of Nondestructive Evaluation 4,23(1984)

[3]S.C.Rasheleigh,J.Light.Tech.1,312(1983)

[4]S.Timoshenko "Vibration Problems in Engineering"(D.Van Nostrand,Princeton,1965)

[5]C.Liguori and M.Martinelli,Appl.Opt.20,4319(1981)

Quasi-Distributed Fiber Optic Sensor System with Subcarrier Filtering

M.J. Marrone [1], *A.D. Kersey* [1], *A. Dandridge* [1], *and C.A. Wade* [2,*]

[1]Naval Research Laboratory, Optical Techniques Branch, Code 6570, Washington, DC 203745, USA
[2]Sachs-Freeman Associates, Landover, MD 20785, USA
*Present address: Plessey Research (Caswell) Ltd., Northants, UK

Abstract: A quasi-distributed fiber optic sensor system based on an array of discrete differential-delay elements is described. A prototype system is demonstrated with a network of three temperature sensors using subcarrier filtering to interrogate the individual sensors. An analysis of the frequency response and crosstalk behavior of the system is presented.

Introduction: All-optical-fiber sensor systems combine the advantages of fiber sensing and fiber telemetry. The advantages of fiber telemetry can be enhanced by using multiplexing techniques to allow many sensors to be serviced by a single source and detector [1]. Such sensor networks are attractive for monitoring a particular measurand at a number of fixed positions in a remote location. Recently, a multiplexing scheme was demonstrated in which a fiber temperature sensor and a fiber pressure sensor were cascaded with subcarrier filtering used to address the separate sensors [2]. In this paper, we describe a network of three temperature sensors using the subcarrier filtering technique and present an analysis of the frequency response and crosstalk behavior of the system.

Principle: Each sensor in the network is a transversal filter which consists of two fibers of unequal length connected in parallel. In response to an RF intensity-modulated source the recombined light at the output of such a filter exhibits a series of minima when the differential delay in the two fiber paths corresponds to a half-integral number of cycles of the modulation frequency. The normalized frequency response of a single sensor is given by

$$g_i(f) = |\cos(\pi \Delta\tau_i f)| \tag{1}$$

where f is the frequency of the modulation and $\Delta\tau_i$ is the differential delay [2]. For a linear array of sensors the frequency response of the combination is given by

$$G(f) = \prod g_i(f) \tag{2}$$

which is simply the product of the frequency response functions of the individual elements in the array. The purpose of this work is to demonstrate a multisensor system based on the subcarrier filtering scheme and to investigate the interaction or crosstalk between the sensors resulting from the interrogation technique.

Experiment: Three fiber-optic differential-delay filters configured as temperature sensors were used as the elements in the system. Each sensor consisted of a length of single-mode fiber wound on a brass cylinder and connected in parallel with a short reference fiber by means of a pair of 3 dB couplers [3]. The temperature sensitivity of the device is obtained from the dependence of the optical path difference on the thermal expansion of the brass mandrel. In turn, the optical path difference determines the location of the minima in the frequency response of the filter. The experimental arrangement for the three-sensor system is shown in Figure 1. The SLD source was modulated by a voltage-controlled oscillator (VCO). In Figure 2, curves (a), (b) and (c) show the frequency response from 0 to 20 MHz for each sensor independently operated with the source and detection system. At 25°C the net path differences of the three sensors were 58.88, 45.97 and 22.55m with their lowest frequency minima appearing at 1.714, 2.196 and 4.476 MHz, respectively. The measured frequency response of the three-sensor network is shown in curve (d) of Figure 2 from 0 to 20 MHz and in curve (e) from 10 to 16 MHz. Curves (f) and (g) of Figure 2 are computer-generated plots of equation (2) for the three-sensor combination based on delay times of 291.7, 227.7 and 111.7ns, respectively calculated from an accurate determination of the lowest frequency minimum of each sensor operated alone. The calculated frequency response of the combination is in excellent agreement with the corresponding experimental results of curves (d) and (e).

The temperature calibration of the individual sensors and the measurement of the crosstalk were made as follows. The VCO was tuned to a convenient minimum in the frequency response of the sensor. In this case, the 4th minimum of S1 (12.001 MHz), the

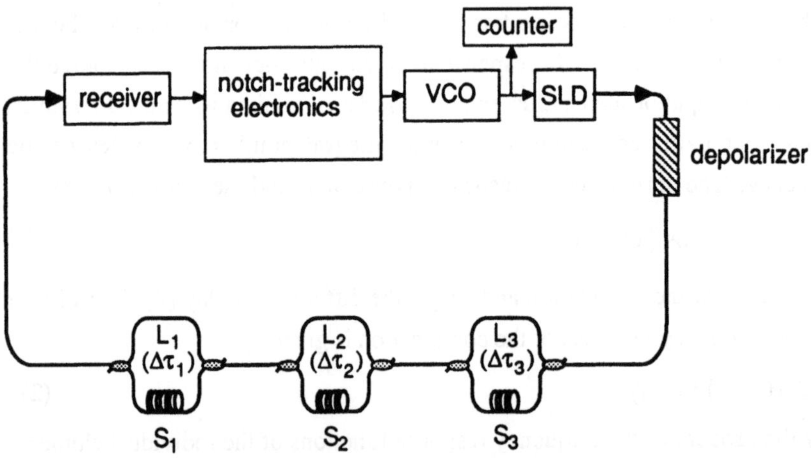

FIGURE 1. Experimental multiplexing arrangement

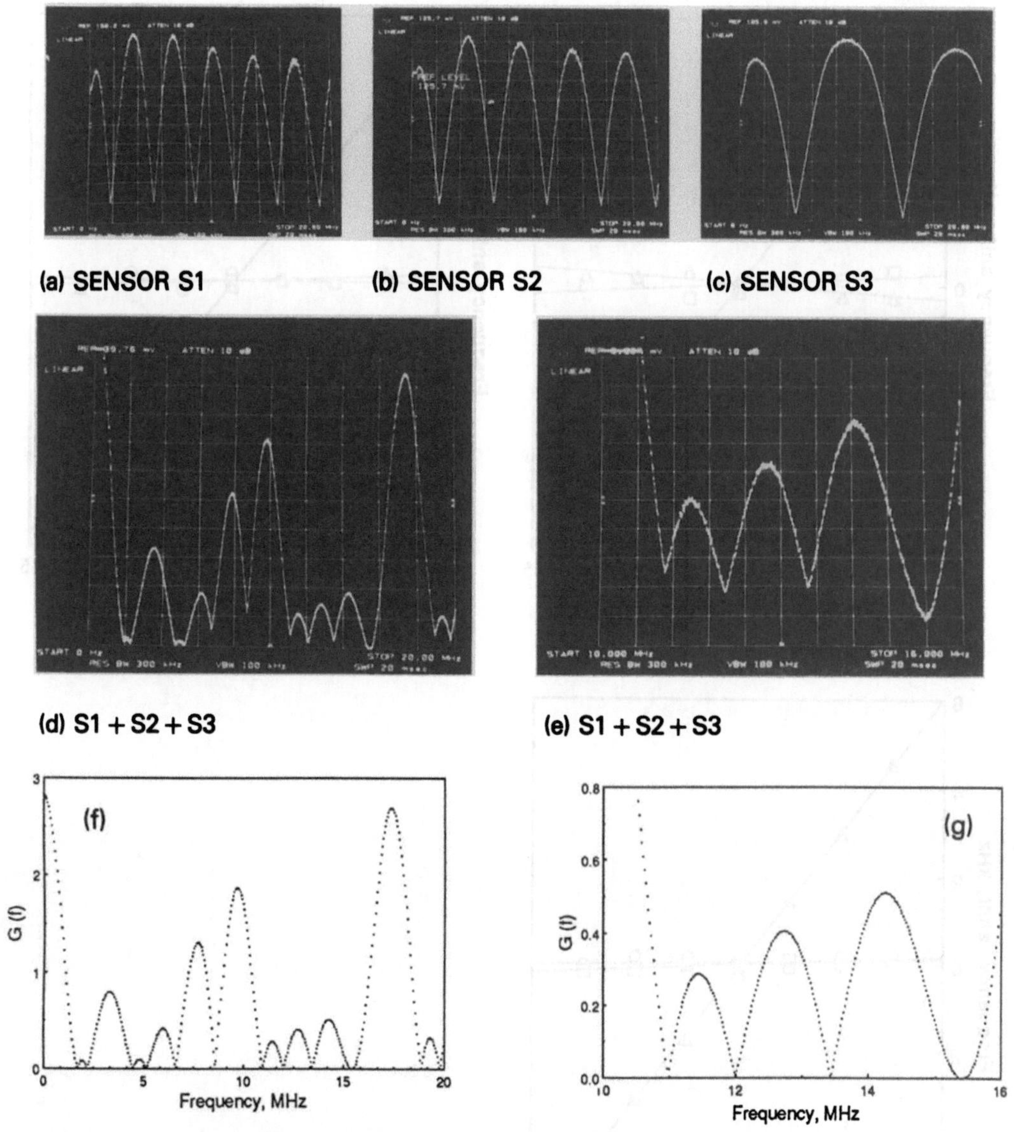

FIGURE 2. Frequency response. (a) - (e), measured response. (f) and (g), calculated response.

3rd minimum of S2 (10.981 MHz) and the 2nd minimum of S3 (13.429 MHz) were chosen because of their relative frequency differences. A square-wave generator provided a 300 Hz signal to dither the VCO by ± 100 kHz about its center frequency. Notch-tracking electronics were used in a feedback loop to lock the VCO at a frequency which minimized the amplitude modulation at the 300 Hz dither frequency. One sensor was immersed in a temperature-controlled water bath while the other two sensors were maintained at constant temperature. Each sensor was interrogated by tuning the VCO to the nominal minimum of its

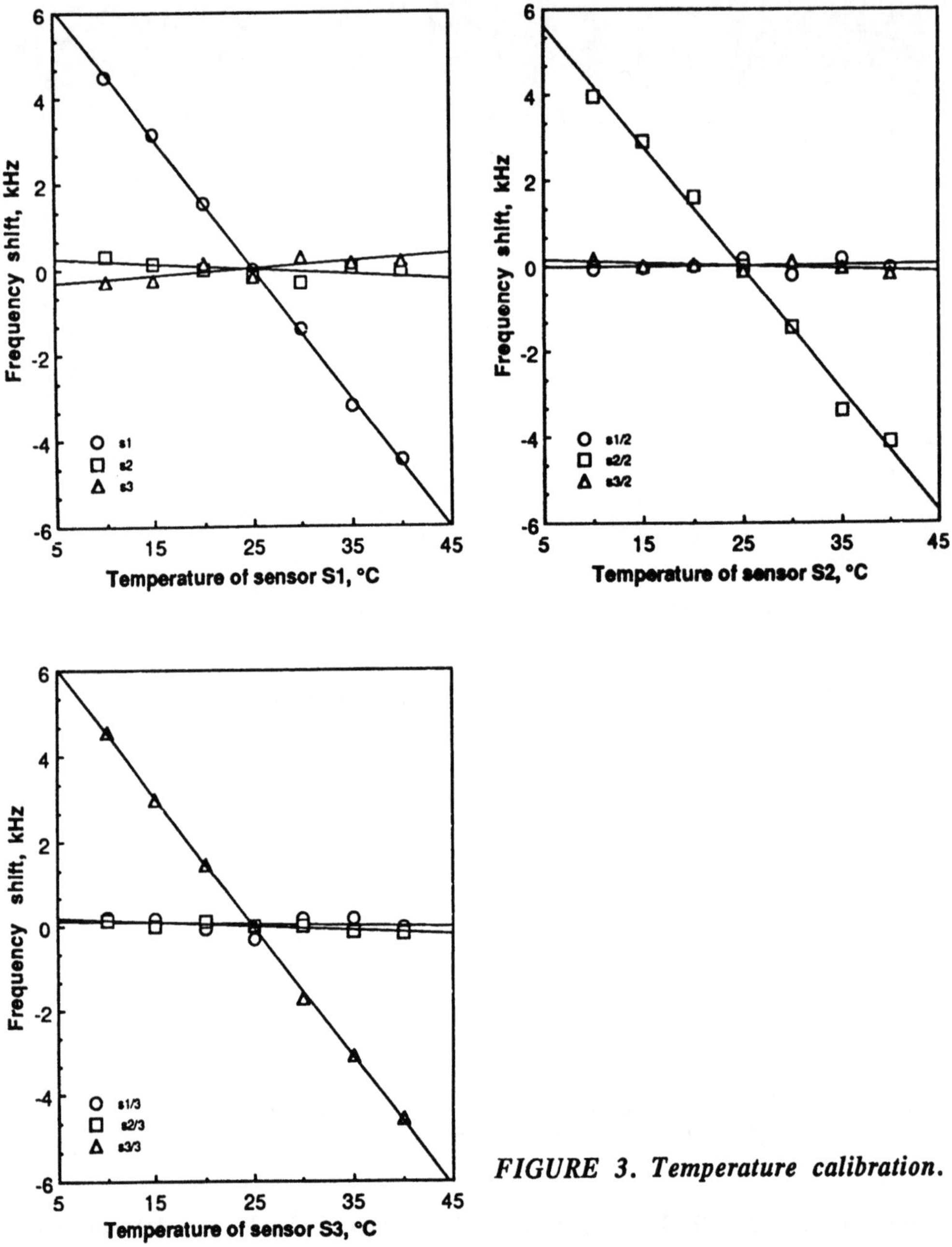

FIGURE 3. Temperature calibration.

notch frequency, locking the notch-tracking electronics and recording the frequency from the VCO counter. Figure 3 shows these data for the 3 sensors in terms of the frequency shift relative to the 25°C value versus the temperature of the water bath. The curves represent a linear fit to the data. The slopes yield thermal calibration constants of -0.303 kHz/°C for S1, -0.284 kHz/°C for S2 and -0.305 kHz/°C for S3. An estimate of the sensor crosstalk can be made from the data of the two sensors held at constant temperature. For example, in Figure 3(a), the slopes for S2 and S3 are -0.012 kHz/°C and +0.016 kHz/°C, respectively. By taking ratios with the calibration curves of S2 and S3 in Figures 3(b) and 3(c), the crosstalk in Figure 3(a) corresponds to +0.04°C in S2 per °C change in S1 and -0.05°C in S3 per °C change in S1. Similarly, the crosstalk behavior of S3 in Figure 3(b) is 0.02°C per °C change in S2. Qualitatively, the fact that the dependence of S3 on S2 is less than the dependence of S3 on S1 is consistent with the fact that the separation of the S3 notch (13.43 MHz) with the S2 notch (10.98 MHz) is more than its separation with the S1 notch (12.00).

In analysing the crosstalk behavior, it is noted that for a single sensor, the shape of each notch as described by equation (1) is symmetric about its minimum value. However, in the case of cascaded sensors, the product of their individual responses may yield a combined frequency response where the notches are asymmetric to varying degrees depending on the location of the notches from the other sensors. In such a case, the dithering technique would result in errors in determining the minimum frequencies of the notches. This represents a system-induced crosstalk mechanism between the sensors. Accordingly, an attempt was made to increase the crosstalk effect by increasing the dither frequency deviation from ± 100 kHz to ± 300 kHz. Figure 4 shows the data for an S1 temperature calibration with the 300 kHz dither frequency deviation used for all 3 sensors. In this case, the slope for S3 is 0.033 kHz/°C which represents a crosstalk effect of approximately twice the magnitude of that observed for the 100 kHz dither frequency deviation. The data for S2 in Figure 4 shows a

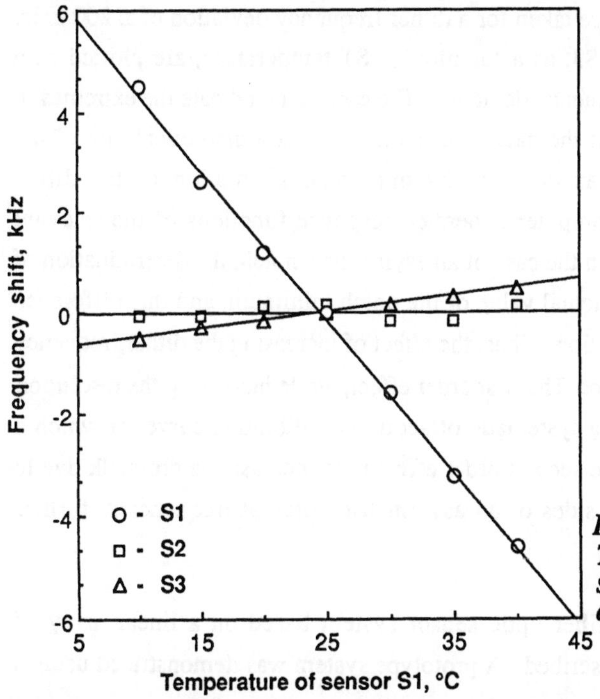

FIGURE 4.
Temperature calibration for sensor # 1; ± 300kHz dither deviation

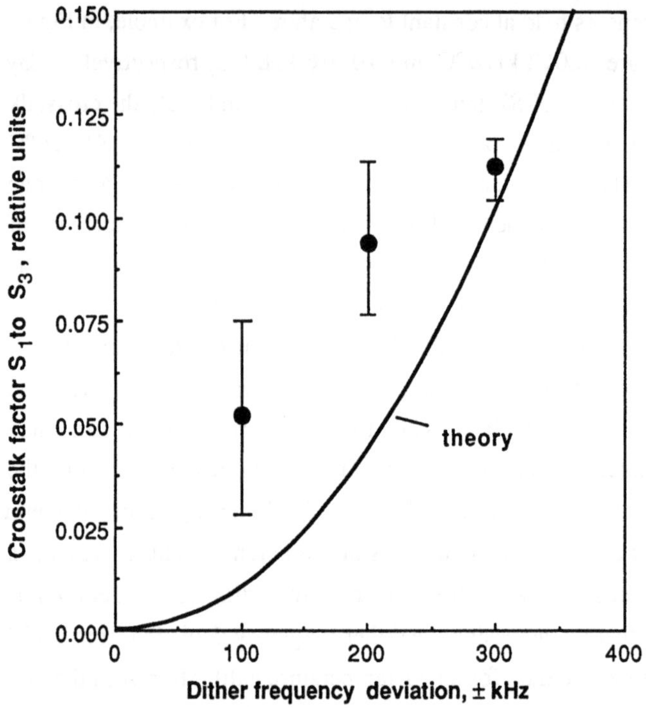

FIGURE 5. Crosstalk factor for sensor # 3 verus dither frequency deviation

slope less than 0.001 kHz/°C which does not indicate any significant crosstalk behavior between S1 and S2 Similar data were taken for a dither frequency deviation of ± 200 kHz. In Figure 5, the relative slopes of S3, as a function of S1 temperature, are plotted as a crosstalk factor versus the dither frequency deviation. The error bars indicate the extremes in the determination of the slopes from the data. The solid curve is a theoretical plot of the relative crosstalk factor. This was obtained by numerical simulation of the dither interrogation approach based on computer-generated response functions of the relevant sensors. This modelling shows that in the case of an asymmetric notch, the determination of the null frequency differs from the actual value of the notch minimum, and this difference increases with dither frequency deviation. Thus, the effect of increasing the dither frequency deviation on a given sensor is twofold: The first order effect, while increasing the resolution of the measurement, is to introduce a systematic offset in the calibration curve for which a simple correction can be added. The second order effect is to increase the crosstalk due to other sensors by sampling the two sides of an asymmetric notch at frequencies further removed from the minimum.

Summary: A quasi-distributed fiber optic sensor system based on a linear array of differential-delay filters has been described. A prototype system was demonstrated using a network of three fiber optic temperature sensors. We have verified that the frequency response of the array is the product of the response functions of the individual components in the array. The system-induced crosstalk mechanism between the sensors has been identified and the calculated crosstalk is shown to be in agreement with the measured value.

References:

1. A. D. Kersey and A. Dandridge, "Distributed and Multiplexed Fiber Optic Sensors", Proc. of 5th Int. Conf. on Optical Fiber Sensors (OFS'88), p.60, New Orleans, Jan. 1988.

2. C. A. Wade, M. J. Marrone, A. D. Kersey and A. Dandridge, "Multiplexing of Sensors Based on Fiber-Optic Differential Delay RF Filters", Electron. Lett., 24, p.1557, 1988.

3. C. A. Wade, A. D. Kersey and A. Dandridge, "Temperature Sensor Based on a Fiber-Optic Differential Delay RF Filter", Electron. Lett., 24, p.1305, 1988.

Bragg-Grating Temperature and Strain Sensors

W.W. Morey, G. Meltz, and W.H. Glenn

United Technologies Research Center, Silver Lane,
East Hartford, CT06108, USA

Bragg–grating sensors have been produced in germanosilicate optical fibers by exposing the core, through the side of the cladding, to a coherent 244 nm two–beam interference pattern. The temperature and strain sensitivity are reported and new techniques for intrinsic quasi–distributed measurements are described.

INTRODUCTION

Recently, we have shown that Bragg gratings can be formed in germanosilicate fiber by exposing the core, through the side of the cladding, to a coherent UV interference pattern with a wavelength selected to lie in the oxygen vacancy defect band of germania.[1] This discovery offers a new means for making accurate, quasi–distributed measurements of temperature and strain either by monitoring the shift in the Bragg wavelength of a tandem array of gratings, each tuned to a distinct wavelength, or by detecting the change in the resonant frequencies of pairs of gratings that form independent Fabry–Perot cavities. Alternatively, the signals from a set of identical grating sensors could be separated through use of conventional optical time or frequency domain methods with a considerable improvement in signal–to–noise ratio over Rayleigh scattered returns.

In this paper, we describe our holographic technique for "writing" Bragg–grating sensors in the core of a fiber and present measurements of the wavelength spectrum of the Bragg filter and its temperature and strain response. Our experiments show that the grating sensors appear to be permanent and stable at high temperatures.

EXPERIMENTAL PROCEDURE

A grating is formed by exposing a short length of bare, photosensitive germanosilicate fiber to a pair of overlapping coherent UV beams. The exper-

Springer Proceedings in Physics, Vol. 44 **Optical Fiber Sensors**
Editors: H.J. Arditty · J.P. Dakin · R.Th. Kersten

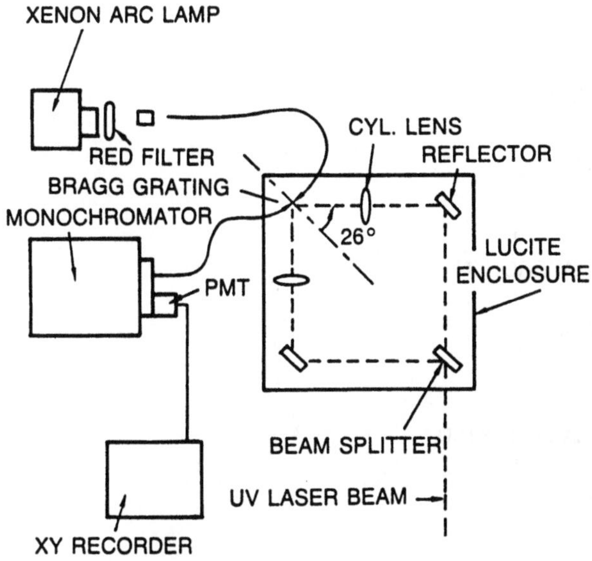

Figure 1. Schematic diagram of the experimental arrangement for writing Bragg gratings. In an alternate set-up (not shown) the arc lamp is replaced by a tunable laser diode and a photodetector is used in place of the monochromator to measure the grating filter spectrum

imental arrangement is shown in Fig. 1. A tunable excimer pumped dye laser, operated at a wavelength in the range 486–500 nm, is used with a frequency doubling crystal to provide a UV source that lies in the 244 nm band and has an adequate coherence length. The UV radiation is split into two equal intensity beams and recombined to produce an interference pattern within the core, normal to the fiber axis. The intensity of the pattern is increased by focusing the beams on the fiber with a pair of cylindrical lenses.

A filtered Xenon arc source or a tunable laser diode is used to measure the reflection and transmission spectra of the grating. The reflected signal is monitored by inserting a beam splitter at the fiber input and the reflectivity is measured by comparing the reflected signal level to the power reflected, at a wavelength near to but out of the filter band, from a mirror placed at the output end of the fiber.

Gratings were written with 244 nm pulsed radiation, having an average power in the range of 4–24 mW. Several different fibers were used, with core diameters from 2.2–2.6 μm and with numerical apertures (NA) from 0.17–0.24, corresponding to GeO_2 doping from 5–12.5 mole percent. Bragg gratings were formed with center wavelengths of 577–591 nm in (a) 6.6 mole percent germanosilicate core, silica clad fiber; (b) in elliptical core, polarization maintaining fiber and (c) in very high germania content fiber (NA = 0.24) containing a small amount (1/2 – 1 mole percent) of phosphorous. In addi-

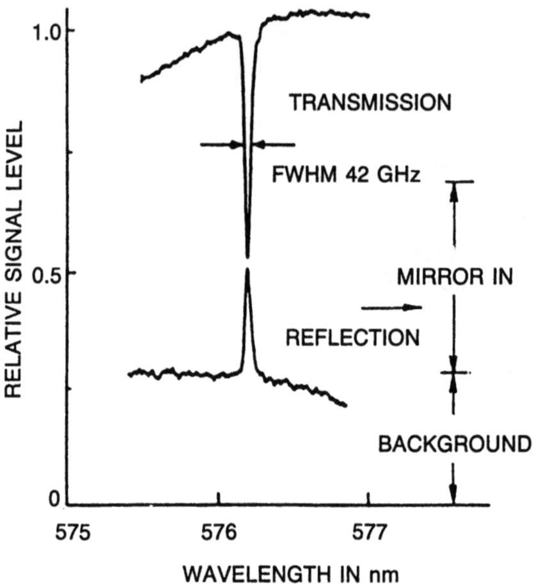

Figure 2. Transmission and reflection spectra for a 4.4 mm long Bragg grating filter. A one meter narrow-band monochromator with a resolution of 0.02 nm was used with a filtered arc lamp source to measure the in-fiber filter characteristics. The measured full width at half maximum (FWHM) is corrected for the monochromator spectral response broadenting.

tion, gratings have been written for the 820 nm laser diode band in both the elliptical core fiber and the strongly germania–doped fiber.

When a periodic index modulation is formed in the fiber core, a narrow notch (or peak) appears in the transmission (reflection) spectrum. The center of the peak or notch occurs at the Bragg wavelength $\lambda = 2n\Lambda$, where Λ is the grating period and n is the mode index. Figure 2 shows the transmission and reflection spectra of a 576.2 nm grating formed after a 5 minute exposure to a 244 nm interference pattern with an average power of 18.5 mW. The length of the exposed region is estimated to be between 4.2 and 4.6 mm, as deduced from inspection of a witness burn spot in a paper target. The observed full–width half maximum line width of the grating is 42 GHz.

TEMPERATURE AND STRAIN SENSITIVITY

An estimate of the temperature and strain sensitivity of an in–fiber grating sensor can be derived from the Bragg condition which relates the wavelength of the reflected radiation at line center λ_o to the grating period Λ:

Bragg condition:

$$\lambda_o = 2n\Lambda$$

$$\therefore \delta\lambda_o/\lambda_o = \delta n/n + \delta\Lambda/\Lambda = (1-p_e)\epsilon + (\alpha+\zeta)\Delta T$$

where:

$$p_e = (n^2/2)[p_{12} - \nu\ (p_{11}+p_{12})] \cong 0.22$$

and α = expansion coefficient, ζ = thermooptic coefficient. Placing the estimated constants in the above equation gives:

$$\delta\lambda_o/\lambda_o = 0.78 \times 10^{-6}\epsilon(\mu \text{ strain}) + 8.86 \times 10^{-6}\ \Delta T(°C)$$

Linewidth, laser stability and signal–noise–ratio considerations determine the minimum detectable change in temperature and strain. For example, an 8 mm grating has a linewidth of about 23 GHz and with an expected S/N of 60 dB.

$$(\delta\lambda_o/\lambda_o)_{min} = (\Delta\lambda/\lambda_o)/\sqrt{S/N} \cong 4.6 \times 10^{-8}$$

$$\text{and}\ \Delta T_{min} \cong 4.6 \times 10^{-3}°C, \qquad \epsilon_{min} \cong 0.046 \text{ microstrain}$$

The Bragg gratings formed by our holographic technique with 244 nm radiation appear to be permanent and stable at high temperatures. A grating in the 6.6 mole percent germanosilicate fiber was heated to 500 °C and maintained at that temperature for 18 hours without change in its reflectivity or line shape. The only variation was an expected shift in line center due to the combination of thermal expansion and a change in refractive index with temperature and stress relief. The features of the Bragg grating wavelength spectrum remain unaltered as the fiber is heated. A theoretical correction for strain relief, which changes the index through the photoelastic effect, improves the agreement between the measured and computed temperature shift values.

The sensitivity to longitudinal strain was also measured by applying a tensile load along the fiber axis with a calibrated spring arrangement. A nearly linear response is observed. The predicted response is somewhat less, but within 17 percent of the measured sensitivity. Uncertainties in the values of the photoelastic coefficients of the fiber core could easily account for the discrepancy.

QUASI–DISTRIBUTED SENSING TECHNIQUES

The Bragg gratings that we have made are 4–10 mm in length. The longer gratings give a higher reflectivity for the same exposure and a narrower spectral width than the shorter grating. To form the grating, the buffer coating is removed from a section of fiber that is slightly greater than the length of the grating to be exposed, and the fiber is held in place in a special fixture. After the exposure, typically a minute, the fiber is recoated with a buffer coating for protection of the surface. This relatively easy operation can be repeated many times over at different locations along a single length of fiber to form a quasi–distributed Bragg grating sensor. The gratings can be made with different Bragg wavelengths by changing the wavelength of the exposing UV radiation or the angle of the interfering beams. The grating sensor signals can be separated by straight forward wavelength division multiplexing methods.

To cover the full range of a single grating strain sensor, approximately 20,000 microstrains would require a measurement of a grating wavelength change of 10 nm. A grating temperature sensor with a range of 500 deg C would require a measurement of 3 nm shift. Laser diode sources can be tuned over ranges of 40 nm to 50 nm.[2] As a consequence, 4 or 5 strain sensors or up to 10 temperature sensors can be addressed with a single laser diode source. A larger number of sensors would require coupling several laser diodes each with a different wavelength tuning range to the sensing fiber.

Another technique that holds promise for detecting a large number of grating sensors is to use time or frequency division techniques in conjunction with a single laser diode source which is tuned over the sensing wavelength range. Nearly identical gratings could be exposed at intervals along the fiber with relatively low reflectivity to reduce the effect of multiple reflections. Time or frequency division multiplexing techniques could be used to separate the individual grating sensor signals, and a wavelength tunable laser diode source could make the temperature or strain measurement for each sensor. The effect of multiple reflections between pairs of sensors is reduced by approximately the reflectivity of the sensor squared. As a consequence, the reflectivity of the individual sensor elements should be reduced to typically 0.1% or less to suppress interference between sensors. For time division multiplexing, the laser is rapidly pulsed, and the signals returned from the gratings are separated by their time of arrival. The individual sensor measurements are made by separately wavelength tuning the laser. With a frequency division multiplexing technique, the laser diode source is rapidly scanned in frequency and heterodyne detection is used with electronic filters to separate the reflected signals from the individual grating sensors. Again, the wavelength of the laser would be tuned along with the frequency chirp to make the sensor measurements.

CONCLUSIONS

The ultraviolet photorefractive sensitivity of germania–doped silica provides a new holographic technique for forming Bragg index gratings in the core of an optical fiber. Each grating can be used as a localized temperature or strain sensor which can be simply multiplexed in a number of ways. The advantages of a Bragg grating sensor are [1] only a short sensor length (typically a few mm long) is required, [2] the spectral width of the return is very narrow and [3] the reflectivity can be made quite large. Our measurements show a nearly linear response to both temperature and strain that is in good agreement with predicted sensitivities. Laser diode sources can be used with a single fiber containing an array of grating sensor to determine the spatial profile of temperature or strain variations in a composite material or along a complex structure.

1. G. Meltz, W. W. Morey and W. H. Glenn: *Formation of Bragg Gratings in Optical Fibers By a Transverse Holographic Method*. Opt. Lett. 14, August 1, 1984 (to be published).

2. J. Mellis et al: *Miniature Packaged External–Cavity Semiconductor Laser with 50 GHz Continuous Electrical Tuning Range*. Elect. Lett. 24, 988 (1988).

Part XIV

Distributed Sensors

Rare-Earth-Doped Fibres for Sensors

D.N. Payne

Optical Fibre Group, The University of Southampton, SO9 5NH, UK

Rare-earth-doped fibre lasers and amplifiers have been demonstrated at several wavelengths in the visible and near infrared regions. As potential sources for optical fibre sensors they have the advantage of wide tunability, high pulse peak-power and complete compatibility with optical fibres.

The incorporation of rare-earth dopants into the core of single-mode fibres [1,2,3] has led to a new class of both active and passive fibre devices including lasers, in-line optical amplifiers, absorption filters and distributed sensors. Several of these devices have reached a state of development where they are finding uses in telecommunication and fibre sensor applications. There is also considerable interest in the traditional laser area where the fibre laser is seen as a route to compact, tunable sources operating at wavelengths from the visible to the infra-red.

The small core size of the single-mode fibre allows high pump intensities for modest (~mW) pump powers. Moreover, the intensity can be maintained over long lengths and this leads to ultra-low lasing thresholds [4] and even permits CW diode-laser-pumped operation of three-level lasers [5]. In conjunction with the long fluorescent lifetime of rare-earths in glass, the high pump intensity allows high-gain

Springer Proceedings in Physics, Vol. 44 — Optical Fiber Sensors
Editors: H.J. Arditty · J.P. Dakin · R.Th. Kersten

(>30dB) operation of fibre amplifiers with excellent saturation properties[6]. In addition, compatibility with existing fibre components is excellent, allowing all-optical fibre circuitry to be assembled with both active and passive components. This is particularly beneficial for the fibre amplifier, where splicing of the active fibre into the link virtually eliminates troublesome Fresnel-reflection feedback which normally limits the gain in optical amplifiers.

A particular attribute of fibre lasers is that the optical-damage threshold of silica is significantly higher than that of the materials used in semiconductor laser fabrication and it is therefore possible to obtain higher peak-power pulses from fibre lasers using the techniques of Q-switching and mode-locking. Recently [7], pulses at a wavelength of 1.060μm with a peak power in excess of 100W and a duration of 15ns have been obtained from a Q-switched 15cm length of Nd^{3+}-doped fibre for a pump power of only 12mW. Similar performance can be obtained from an Er^{3+}-doped fibre, suggesting exciting prospects as a practical source for distributed sensors based on OTDR's, non-linear optics and photonic switching. The broad tunability should also prove useful in wavelength-multiplexed sensor networks.

Perhaps the first applications of rare-earth-doped fibres in sensors will be as a superfluorescent source for use in the fibre gyroscope[8]. The potential advantage here is that the broad emission spectrum is stable with fibre temperature and this provides a constant gyro scale factor.

Developments in rare-earth-doped fibre temperature sensors, lasers and amplifiers will be reviewed with particular emphasis on their applications in optical sensing.

REFERENCES

1. C.J. Koester & E. Snitzer:
"Amplification in a fibre laser",
Appl. Opt., Vol. 3, pp. 1182-1186, 1964.

2. J. Stone & C.A. Burrus:
"Neodymium-doped silica lasers in end-pumped fibre geometry",
Appl. Phys. Lett., Vol. 23, pp. 388-389, 1973.

3. R.J. Mears, L. Reekie, S.B. Poole & D.N. Payne:
"Neodymium-doped silica single-mode fibre lasers",
Electron. Lett., Vol. 21, pp. 738-740, 1985.

4. L. Reekie, R.J. Mears, S.B. Poole & D.N. Payne:
"Tunable single-mode fibre lasers",
IEEE J. Lightwave Technol., Vol. LT-4, pp. 956-960, 1986.

5. L. Reekie, I.M. Jauncey, S.B. Poole & D.N. Payne:
"Diode laser pumped operation of an Er^{3+}-doped single-mode fibre laser",
Electron. Lett., Vol. 23, pp. 1076-1078, 1987.

6. R.J. Mears, L. Reekie, I.M. Jauncey & D.N. Payne:
"Low-noise Erbium-doped fibre amplifier at 1.54μm",
Electron. Lett., Vol. 23, pp. 1026-1028, 1987.

7. W.L. Barnes, J.T. Lin, L. Reekie, D.J. Taylor, I.M. Jauncey, S.B. Poole & D.N. Payne:
"Q-switched and single polarisation operation of a Nd^{3+}-doped fibre laser",
IEE Colloquium (Devices), London, June 1988.

8. K. Liu, M. Digonnet, H.J. Shaw, B.J. Ainslie & S.P. Craig:
"10mW superfluorescent single-mode fibre source at 1060nm",
Electron. Lett., Vol. 23, No. 24, pp. 1320-1321, 1987.

An Improvement of the Accuracy in the Distributed Fiber Temperature Measurement Using Raman Backscattering

Y. Tanabe[1], *A. Takada*[1], *K. Ikawa*[2], *and N. Bando*[2]

[1]Research Center ASAHI Glass Co., Ltd., 1150 Kanagawa-ku, Yokohama, Japan
[2]ASAHI Glass Electronic Products, R&D Center Co., Ltd., 1160 Kanagawa-ku, Yokohama, Japan

A new method of deriving the temperature distribution from the Raman backscattering ratio is proposed. This method can improve the accuracy of measurements. The temperature dependence of the attenuation ratio between anti-Stokes and Stokes lines is fully considered. By this method we reached the accuracy of 1 °C over 1 km length.

1. Introduction

The use of the Raman backscattering ratio to measure the temperature distribution along a conventional silica-based optical fiber has been recently reported [1–4]. It is, however, necessary to improve the temperature measurement accuracy before this technique can be put to practical use. Inaccuracy comes from the temperature dependence of the attenuation ratio between the anti-Stokes and Stokes components. We propose a new method to derive the temperature distribution from the ratio of Raman backscattered components, where temperature-dependent properties of the attenuation ratio are fully considered.

2. Basic Principles

The Raman ratio thermometry relies on monitoring the anti-Stokes and Stokes backscattered signal as a function of distance along the fiber. The amplitude of the signals, unlike the Rayleigh line, show significant temperature dependence, due to the Gibbs or Boltzmann distribution among the various vibrational states of the vitreous macromolecule. In particular, if the ratio between the anti-Stokes and Stokes lines is measured, an absolute temperature measurement can be made. The ratio $R(T)$ is given by

$$R(T) = \exp\left(-hcv/kT\right) \qquad (1)$$

where v is the wavenumber separation of the measured anti-Stokes and Stokes lines from the pump wavelength, h is Planck's constant and T is the absolute temperature of the core of the section of the fiber from which scattered light is received. But in actual optical fibers there are differences of attenuation constants at the anti-Stokes wavelength and Stokes wavelength. Furthermore, these differences are temperature dependent. They make it difficult to make the precise temperature distribution measurement by Raman ratio thermometry in the optical fiber.

3. Design of the Distributed Temperature Sensor

A schematic diagram of our system is shown in Fig. 1 and a photograph in Fig. 2. Our system uses the 850 nm GaAs multiple quantum well diode laser. It is operated at a peak power of

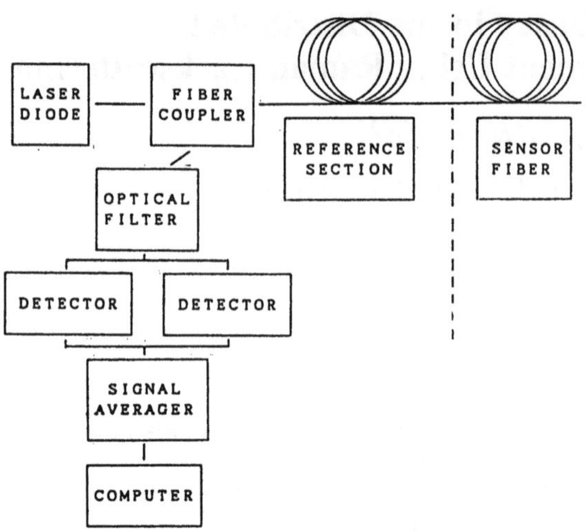

Fig. 1. Schematic diagram of the distributed fiber temperature sensor system

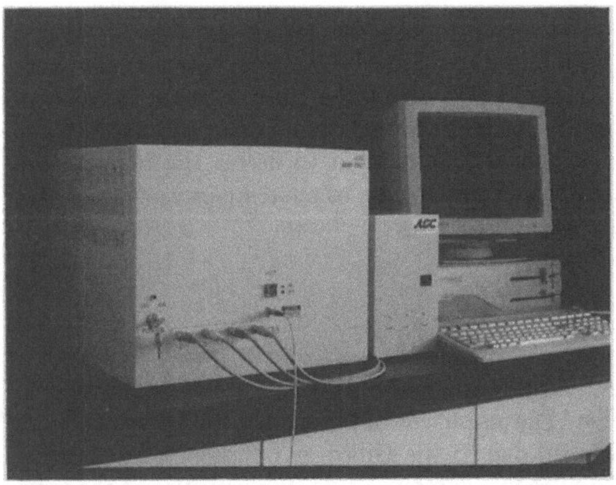

Fig. 2. Distributed fiber temperature sensor system

20 W with 100-ns pulsewidth and 1.5 Hz repetition rate. The power launched into the 100/125-mm multimode sensing fiber is ~ 1 W. Raman backscattered light is separated from Rayleigh backscattering by an optical coupler and optical filters. The spectral bandwidth of the filter is 15 nm FWHM. The Raman backscattering is detected by the photomultipliers. The signal is acquired by the dual high speed A-D converters and averaged 65 000 times in a multichannel signal averager.

4. Temperature Measurement

4.1. Calibration of Measurement Using the Reference Section

It is difficult to measure the absolute ratio between the anti-Stokes and Stokes components, because only an ideal detection system will have equal sensitivities. By installing the reference section of the sensor fiber, which is at a known temperature θ, the unknown temperature T can be found from the relationship [2]

$$\frac{1}{T} = \frac{1}{\theta} - \frac{k}{hcv} \ln \left(\frac{R(T)}{R(\theta)} \right) , \tag{2}$$

where $R(\theta)$ is the Raman backscattering ratio from the reference section. The installed reference section includes a 250 m reference fiber. We measure the temperature of this section by a semiconductor temperature sensor. By installing the reference section we can compensate for the drift of the relative sensitivity during detection of anti-Stokes and Stokes signals.

4.2. Temperature Measurement by the Uncorrected Equation

First we measured the temperature distribution by using (2), which does not consider the temperature-dependent attenuation ratio. The result is shown in Fig. 3. The measured temperature distribution shows constant slope. Laser light launched into the sensor fiber will attenuate when propagating along the long optical fiber. We can compensate for the effect of this attenuation by taking the ratio between the anti-Stokes and Stokes components. But it is impossible to compensate for the effect of the attenuation difference between these components by taking the ratio. This is why the measured temperature distribution shows a slope. We must adopt a new equation to compensate for this effect.

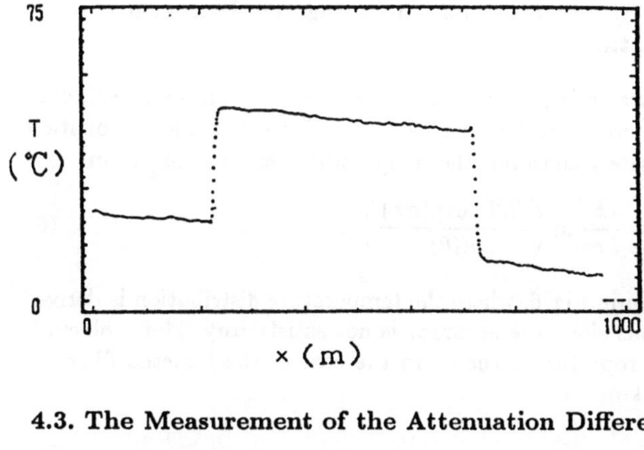

Fig. 3. Temperature distribution derived from (2) including no corrections

4.3. The Measurement of the Attenuation Difference

In an actual system the attenuation difference α is temperature dependent. We have measured α in the cases of the jacketed fiber and fiber cable. If all parts of the sensor fibers have a constant temperature, the amplitude of anti-Stokes and Stokes backscattered light can be written

$$I(x) = A(T) \exp [a(T)x] , \tag{3}$$

where x is distance and $a(T)$ is the temperature-dependent attenuation constant of the anti-Stokes or Stokes component. We take the logarithm of both sides:

$$\ln I(x) = \ln A(T) + a(T)x . \tag{4}$$

From the measured gradients of these lines at several known temperatures we can determine $a(T)$. Then the temperature-dependent attenuation difference $\alpha(T)$ is

$$\alpha(T) = a_S(T) - a_{AS}(T) . \tag{5}$$

The results are shown in Figs. 4 and 5.

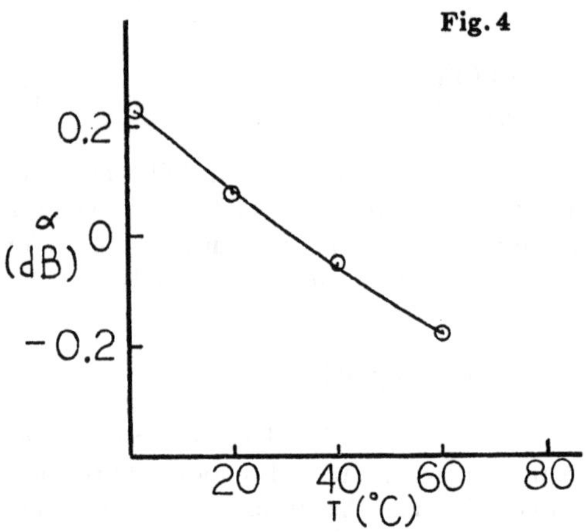

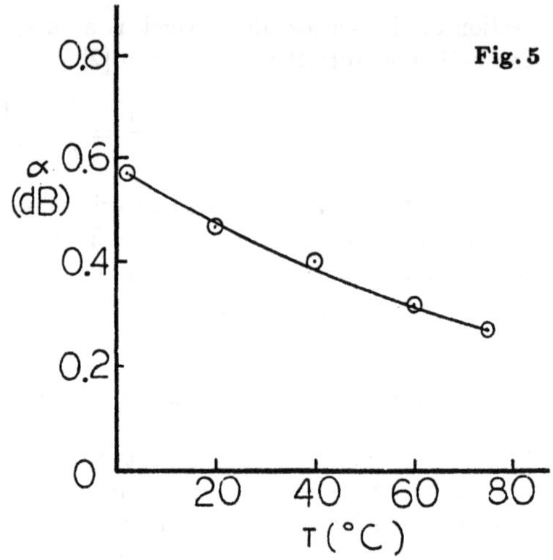

Fig. 4. Temperature dependence of the attenuation ratio α in the fiber cable

Fig. 5. Temperature dependence of the attenuation ratio α in the jacketed fiber

4.4. Temperature Measurement by the Equation Including the Temperature-Independent Attenuation Ratio

Because the attenuation of the backscattering is frequency dependent, the anti-Stokes/Stokes ratio changes with distance. A new term therefore must be added to (2). If the attenuation difference is assumed to be approximately constant, the temperature can be found from

$$\frac{1}{T} = \frac{1}{\theta} - \frac{k}{hcv} \ln \left(\frac{R(T) \exp(\alpha x)}{R(\theta)} \right) . \tag{6}$$

The results of measurement are shown in Fig. 6, where the temperature distribution is derived from (6). Although there is no obvious slope the accuracy is not satisfactory. The problem is due to the temperature-dependent properties of the α. In the case of the jacketed fiber the accuracy is 4.3 °C at a distance of 1 km.

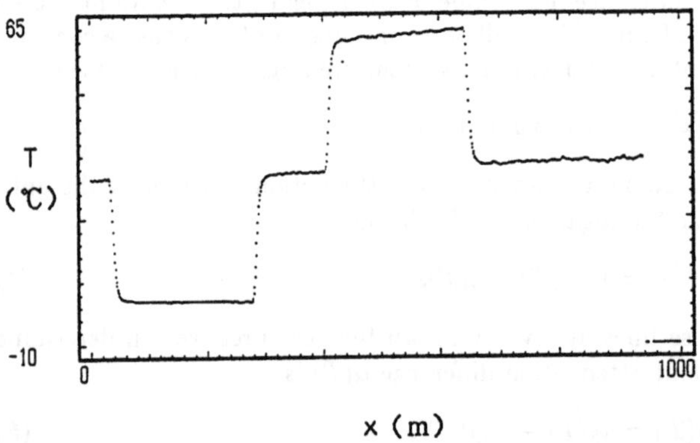

Fig. 6. Temperature distribution derived from (6) including temperature-independent α

5. Proposed Method

5.1. Proposed Equation Including the Temperature-Dependent Attenuation Difference

We propose a new equation to derive the temperature distribution from the Raman backscattering ratio. This equation can resolve the problem with equations (2) and (6). The new equation is

$$\frac{1}{T(x)} = \frac{1}{\theta} - \frac{k}{hcv} \ln \left[\frac{R(T)}{R(\theta)} \exp \left(\int_0^x \alpha(T(\tau)) d\tau \right) \right] . \tag{7}$$

This equation includes the temperature-dependent attenuation difference term on the right side. Because this equation includes the unknown temperature distribution $T(x)$ on both sides, it appears at first sight that it is impossible to solve it. But we can get $T(x)$ by the successive method and convergence method.

5.2. Successive Method

We can get $[T_n]$ by the successive method. The integration actually becomes a summation. Equation (7) can be rewritten as

$$\frac{1}{T_n} = \frac{1}{\theta} - \frac{k}{hcv} \ln \left[\frac{R_n(T)}{R(\theta)} \exp \left(\sum_{p=1}^{n-1} \alpha(T(\tau_p)) \delta x \right) \right] . \tag{8}$$

First we can calculate T_1 by using (2), which does include the attenuation difference term, because the location of T_1 is very close to the starting point. It is scarcely influenced by the temperature-dependent attenuation difference α. Secondly, we calculate T_2 from this calculated T_1. Then, T_3 from T_2, T_1, T_4 from T_3, T_2, T_1, etc. The result of this successive method is shown in Fig. 7. In this example the 1 km long fiber is divided into 1000 points. It is obvious that an accurate temperature distribution measurement is available from this successive method.

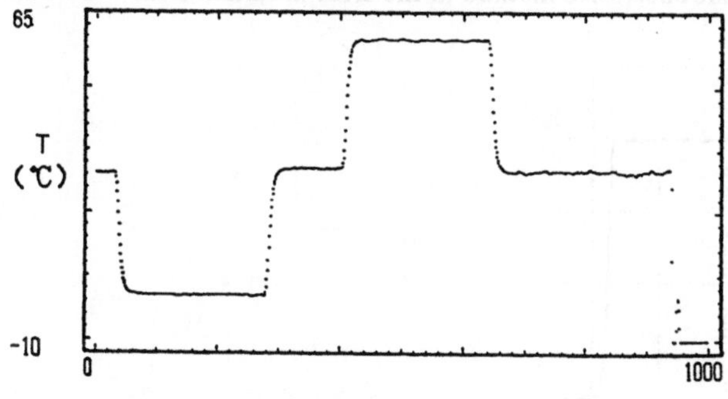

Fig. 7. Temperature distribution derived by the successive method

5.3. Convergence Method

The second solution method is based on the convergence properties of the $[T_n]$ under (9)

$$\frac{1}{T_n} = \frac{1}{\theta} - \frac{k}{hcv} \ln \left[\frac{R_n(T)}{R(\theta)} \exp \left(\sum_{p=1}^{n} \alpha(T(\tau_p))\delta\tau \right) \right] . \tag{9}$$

In this method we can start from the uncorrected temperature distribution $[T_n^0]$, which is calculated from (2). We can consider (9) as a kind of transformation, which is written

$$[T_n^r] = \mathbf{F}\,[T_n^{r-1}] . \tag{10}$$

First, by the substitution of $[T_n^0]$ into the right side of (9), we get $[T_n^1]$ from the left side of (9). Second, this $[T_n^1]$ is resubstituted into the right side of (9). These operations are iterated. The criterion of convergence is defined by

$$\epsilon^r = \sum_{n=1}^{1000} (T_n^r - T_n^{r-1}) < \delta . \tag{11}$$

These transformations are repeated until the ϵ become lower than the predetermined constant δ. Typical values of the calculated ϵ are as follows:

$$\epsilon^1 = 26.7$$

$$\epsilon^2 = 2.4$$

$$\epsilon^3 = 0.13$$

$$\epsilon^4 = 0.534 \times 10^{-2}$$

$$\epsilon^5 = 0.534 \times 10^{-3}$$

$$\epsilon^6 \sim 0$$

A typical result of temperature distribution measurement by this convergent method is shown in Fig. 8. The convergence method is a means of looking for the self-consistent solution of (7). This method has an advantage over the successive method in the error accumulations, we think.

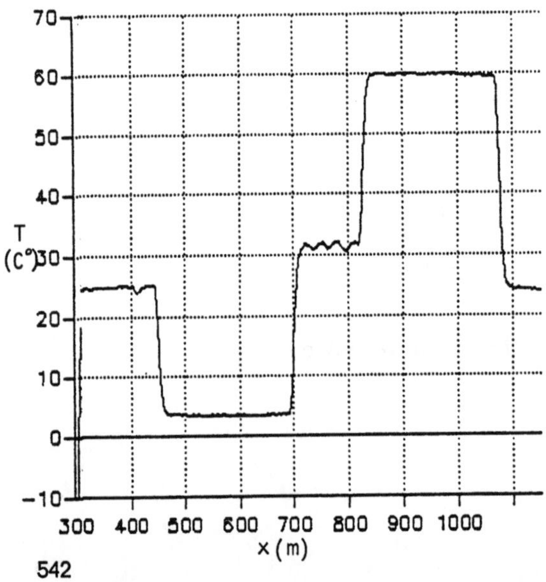

Fig. 8. Temperature distribution derived by the convergence method

6. Conclusions

We propose a new method of deriving the temperature distribution from the Raman backscattering ratio. The temperature-dependent attenuation ratio between the anti-Stokes and Stokes lines is fully considered in the new equation. We can solve this equation by the successive and convergence methods. By our methods we reach an accuracy of about 1 ° C over 1 km of fiber.

References

1. A. H. Hartrog: *IEEE J. LT*-1, 498 (1983)
2. J. P. Dakin, D. J. Pratt, G. W. Bibby, J. N. Ross: *Fiber Optic and Laser Sensor III*, SPIE-566, 249 (1985)
3. J. P. Dakin, D. J. Pratt, G. W. Bibby, J. Ross: *Electron. Lett.* 21, 569 (1985)
4. A. H. Hartrog, A. P. Leach, M. P. Gold: *Electron. Lett.* 21, 1061 (1985)

A Fiber-Optic Distributed Temperature Sensor with High Distance Resolution

K. Ogawa[1], *Y. Ozawa*[1], *H. Kawakami*[2], *T. Tsutsui*[2], *and S. Yamamoto*[2]

[1]Engineering Development Center, The Tokyo Electric Power Co., Inc., 1-1-3, Uchisaiwai-cho, Chiyoda-ku, Tokyo 100, Japan

[2]Cable Research Laboratory, Hitachi Cable Ltd., 5-1-1, Hitaka-cho, Hitachi-shi, Ibaraki-ken, 319-14, Japan

We have developed a fiber optic distributed temperature sensor for temperature supervision of underground power transmission systems. The sensor can detect the temperature distribution at 1m intervals with a measurement accuracy of ±1 °C over a distance of 2 km. The performances have been confirmed through field tests.

1 Introduction

Many authors have recently reported their development work on distributed temperature sensors as an optical fiber itself.[1)-3)]

Fiber optic sensors are immune from electromagnetic noise and are chemically and physically stable. With these advantages, it is expected that distributed temperature sensors will be applied to long-distance power transmission lines and also to many other fields where temperature supervision is required.

"Distance resolution" is one factor that has a determining influence on the perfomance of the distributed-temperature sensor, it mainly consists of sampling distance and response distance. "Sampling distance" is the length of the measurement interval along the fiber ; at each interval, the temperature can be measured. "Response distance" is as follows : When the temperature distribution has a sharp-shaped change along the length of the sensor, the temperature detected is a little relaxed at the point. In this paper, we define the distance where the relaxation phenomenon is observed as the response distance.

While the sampling and response distances are desired to be short, the data reported have been so far in excess of 1 m for the sampling distance and 5 m for the response distance. It is required, however, that the response distance be less than 1 m to detect hotspots in power transmission facilities.

We succeeded in reducing the response distance and the sampling distance to 1 m each, and to achieve a measurement accuracy of ±1°C over a measurement distance of 2 km.

2 Temperature Sensing with Raman Scattering

When light is coupled into an optical fiber, Raman scattering is produced. It is known that the intensity ratio of the two components of Raman scattering "Stokes light" and "anti-Stokes light" is a function of temperature, and thus reveals the temperature. [4)]

This characteristic can be demonstrated by injecting laser light (514nm) into a length of an optical fiber. The fiber (200/250GI, 100m) was set to a fixed temperature in a chamber, and then the intensity of scattering light was measured. The results are shown in Fig. 1.

Fig. 1 shows the intensity ratio of Stokes and anti-Stokes light as a function of temperature. The ratio increases with temperature, and agrees closely with the theoretical calculation result (see equation (A12) in the appendix).

The implications of this result are that if the two components of Raman scattering light generated at each point along the fiber can be detected, then the temperatures at the same points can be measured. Therefore, by measuring Raman scattering distribution in an optical fiber using the optical time domain reflectometry (OTDR) method, it is possible to measure the temperature distribution along the optical fiber.

The configuration of this distributed temperature sensing system is shown in Fig. 2. A pulsed light is injected from a laser diode into a fiber sensor through an OWDD (optical wavelength division demultiplexer). A portion of the scattering light in the fiber continuously comes back to the injection end as

Springer Proceedings in Physics, Vol. 44 Optical Fiber Sensors
Editors: H.J. Arditty · J.P. Dakin · R.Th. Kersten

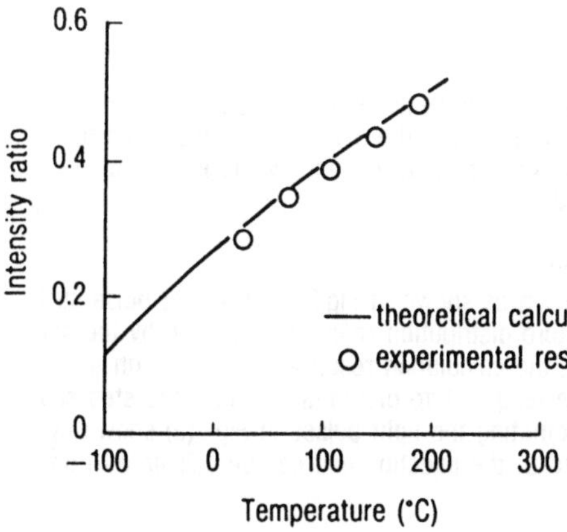

Fig.1 Temperature characteristic of anti-Stokes/Stokes intensity ratio

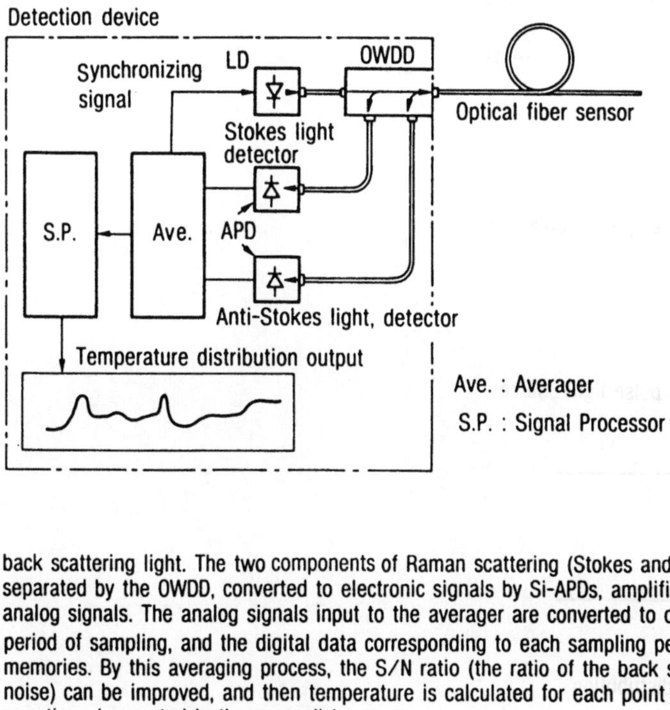

Fig.2 System configuration

back scattering light. The two components of Raman scattering (Stokes and anti-Stokes light) are separated by the OWDD, converted to electronic signals by Si-APDs, amplified, and input to an averager as analog signals. The analog signals input to the averager are converted to digital data according to the period of sampling, and the digital data corresponding to each sampling period is stored in the averager memories. By this averaging process, the S/N ratio (the ratio of the back scattering light intensity to the noise) can be improved, and then temperature is calculated for each point along the fiber using theoretical equations (presented in the appendix).

3 Issues for High Distance Resolution

It is necessary to shorten both the sampling and the response distances. Shortening these distances, however, causes the S/N ratio to deteriorate, which results in larger measurement error. That is, the distance resolution is to be improved over the required measurement distance and also the S/N ratio is to be improved to maintain the required temperature accuracy.

3.1 Sampling Distance

Back scattering light is measured for every sampling period. The relation between distance x (from the injection end to the backscattering light generating point) and time t (the time interval between the light pulse injection and the back scattered light reception) is as follows:

$$x = (v/2)\cdot t \quad \cdots\cdots\cdots (1)$$

where v is the velocity of the light in the optical fiber.

Similarly, the length ΔLs corresponding to sampling time interval Δts becomes $(v/2)\cdot \Delta ts$, and the temperature of the optical fiber is measured at this period. This interval is the sampling distance, and to make it shorter, it is necessary to shorten the sampling period. As our target distance value is 1 m, a sampling time of 10 ns is required (assuming that $v=0.2$m/ns).

3.2 Relation between Pulse Width and Response Distance

In an optical fiber with a step-shaped temperature change such as shown in Fig.3(b), if a light pulse such as shown by the solid line in Fig.3(a) is injected, the temperature distribution calculated is given by the solid line in Fig.3(c). For the section where temperature is constant, the calculation result agrees well with the actual temperature distribution, but at the boundary where the temperature distribution makes the step-shaped change, a relaxed temperature distribution is observed. Approximating the light pulses in Fig.3(a) shown by the dotted line, the intensity of Raman scattering light received at the injection end can be calculated from theoretical equations in the appendix(A1-A11) as follows:

$$Ia(t) = P_0 \cdot tw \cdot Ka \cdot \gamma a(X) \cdot fa(T^*) \quad \cdots\cdots\cdots (2)$$

$$Is(t) = P_0 \cdot tw \cdot Ks \cdot \gamma s(X) \cdot fs(T^*) \quad \cdots\cdots\cdots (3)$$

$$Ia(t)/Is(t) = Ka/Ks \cdot 10^{-(\alpha_a-\alpha_s)\cdot x} \cdot \beta(T^*) \quad \cdots\cdots\cdots (4)$$

$$T^* = \int_{x-\Delta/2}^{x+\Delta/2} T(u)du / \Delta L \quad \cdots\cdots\cdots (5)$$

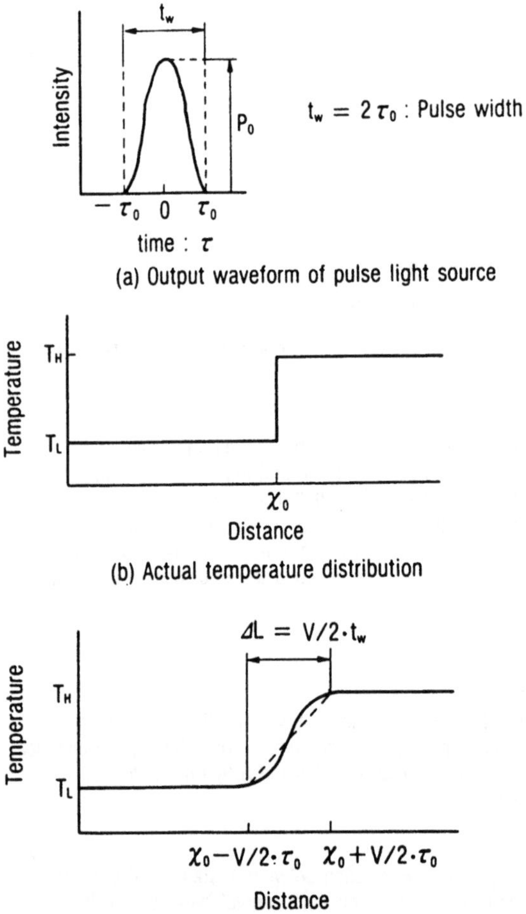

(a) Output waveform of pulse light source

(b) Actual temperature distribution

(c) Calculated results of temperature distribution

Fig.3 Relation between response distance and pulse width

If scattering intensities Ia(t) and Is(t) are measured, $\beta(T^*)$ is obtained from Eq. (4). T^* is obtained from Eq. (A10) in the appendix, but as can be seen from Eq. (5), T^* is the temperature averaged in the distance ΔL which corresponds to pulse width tw ($\Delta L = v/2 \cdot tw$). Thus, in the case of a step-shaped temperature distribution, the relaxed temperature distribution as the dotted line in Fig.3(c) is obtained in the length of ΔL on the boundary. This ΔL is the response distance, and the shorter the distance is, the better the distance resolution is.

To make the response distance less than 1 m as our development target, it is necessary to reduce the pulse width to less than 10 ns. And, also, O/E converters and electronics circuits in amplifiers etc. in the sensing system are required to have a sufficient bandwidth to provide the frequency response needed to keep up with the pulse width of 10 ns.

3.3 Temperature Measurement Accuracy

We have discussed some of the factors that are necessary to develop a sensing system with short response distance. The following factors, however, cause the S/N ratio to deteriorate:

(1) The intensity of received back scattering light decreases proportionally as the pulse width becomes narrower (see Eqs. (2) and (3))

(2) Noise becomes larger with increase of the frequency bandwidth of electronic circuits.

(3) Since the noise due to the running frequency of the digital averager is in inverse proportion to the sampling period, the noise (from averager to O/E converter) increases to the extent the sampling period is shortened.

Deterioration of the S/N ratio causes measurement error to increase in magnitude, and the distance over which the temperature can be measured within a given temperature accuracy is decreased.

Therefore, the following factors also have to be considered to achieve high distance resolution capability:

(1) Increase of peak power of laser diode.

(2) Decrease of power loss in optical system, especially in OWDD

(3) Reduction of circuit noise, especially in O/E converter and amplifier.

(4) Increases of averaging times and speed of averager.

4 Development of a Distributed Temperature Sensor with High Distance Resolution

4.1 Short-Pulse Light Source

A laser diode of 900nm wavelength and of 10W peak power is used as the light source. The target value of 10 ns was realized as shown in Fig. 4.

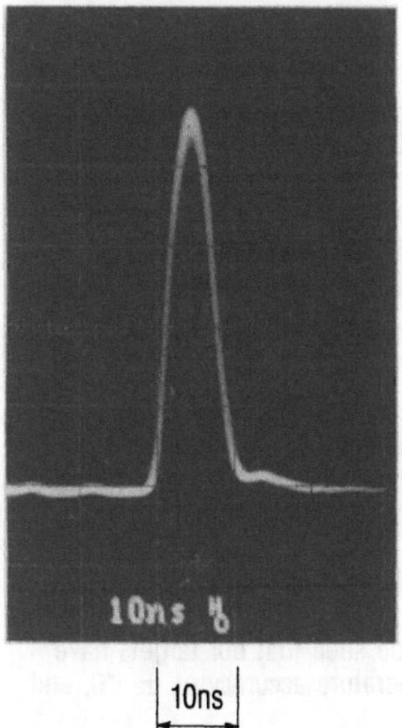

Fig.4 Waveform of light source

4.2 Optical Wavelength Division Demultiplexer

Using some kind of interference filters which selectively transmit or reflect Rayleigh scattering light and Raman scattering light (Stokes and anti-Stokes light), a new OWDD was developed. The OWDD performs very well. From light source to detector over the fiber sensor, the total coupling loss for Stokes light is 6.0 dB and for anti-Stokes light is 5.5 dB.

5 Laboratory and Field Experimental results

Then, we describe experimental results of the distributed temperature sensor based on the developed short-pulse light source and OWDD.

5.1 Response distance

One section of optical fiber sensor was set in a constant temperature chamber and heated to a stable temperature. The relationship between length of the heated sections and amplitude of detected temperature changes is shown in Fig. 5.

When a heated section is longer than 1 m, the measured temperature change agrees with the actual temperature change within the error of ± 1 °C, verifying that the response distance is 1 m.

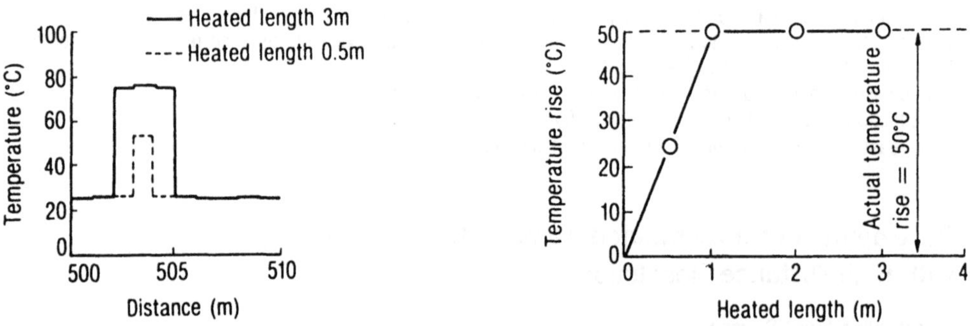

(a) Temperature distribution measured by developed sensor

(b) Relation between heated length and temperature rise

Fig.5 Evaluation of response resolution

5.2 Long-Distance Measurement

A 2000 m fiber sensor was arranged so three sections along the fiber's length were placed in constant temperature chambers. The temperatures were then maintained at −20°C, 90°C, and 150°C respectively. The temperature distribution measurement results are presented in Fig. 6. From the figure it can be seen that for the measured temperature distribution along the 2000 m length, response distance and temperature accuracy are within 1 m and ±1°C, respectively. Next, using the same fiber sensor, we evaluated temperature accuracy through the fiber. The results show that the measurement accuracy of ±1°C is obtained over a length of 2 km.

5.3 Field Test Results

Field tests of this system were conducted in a real underground transmission line. The temperature distribution is presented in Fig. 7. The open circles in the figure give experimental values obtained by a conventional spot-type temperature sensor, and it can be seen that these values closely correspond to the results of the distributed temperature sensor. Through the field tests, it has been confirmed that extremely accurate temperature distribution can be measured by the developed system.

5.4 Total Performance

The performance of the developed sensor is summarized in Table 1. It can be seen that our targets have all been satisfied : sampling distance of 1 m, response distance of 1 m, temperature accuracy of ± 1°C, and measurement distance of 2 km have been obtained.

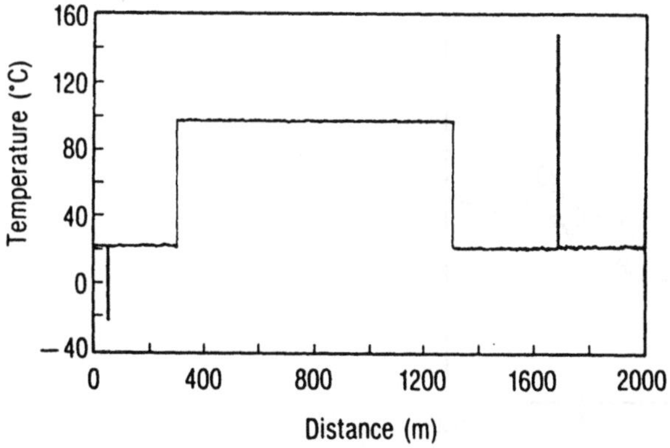

(a) Temperature distribution of whole length

Fig.6 Temperature distribution measurement examples

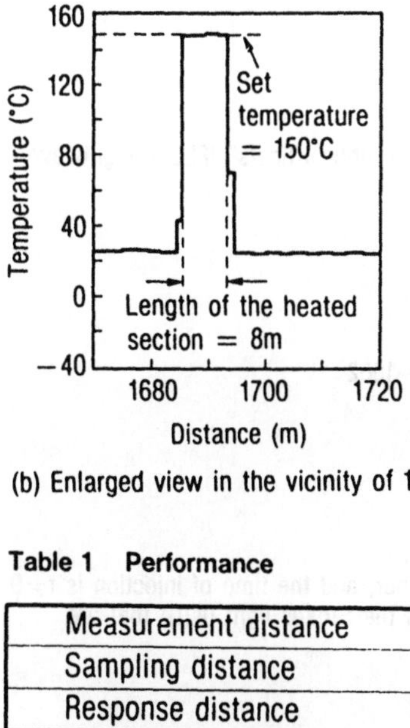

(b) Enlarged view in the vicinity of 1,700 m

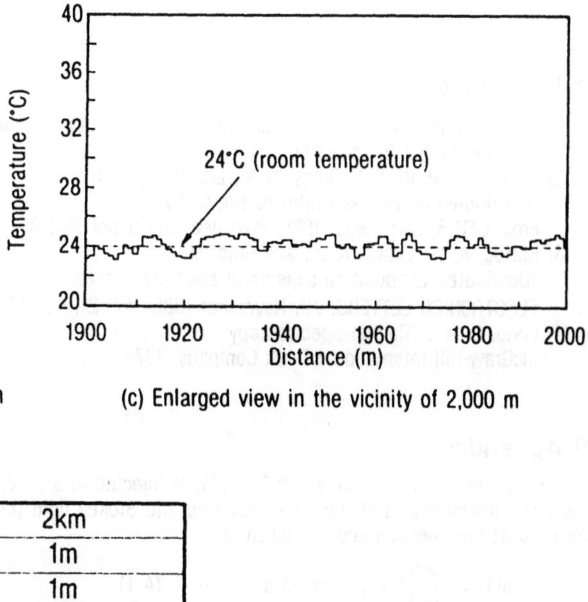

(c) Enlarged view in the vicinity of 2,000 m

Table 1 Performance

Measurement distance	2km
Sampling distance	1m
Response distance	1m
Temperature accuracy	±1°C
Temperature resolution	0.1°C
Temperature range	−50 — +150°C

6 Conclusion

A distributed temperature sensor with high distance resolution has been developed. Laboratory and field tests show that the developed system has good perfomances for practical use.

(1) We have developed the sensor with high distance resolution of 1 m by using a light source of pulse width less than 10 ns.

(2) We have developed a new OWDD (optical wavelength division demultiplexer) to keep the temperature measurement accuracy of less then ± 1 °C in spite of the narrow pulse width of 10 ns.

(3) The developed sensor can detect the temperature distribution with a distance resolution of 1 m, and a measurement accuracy of ± 1°C over a measurement distance of 2 km.

Further research will be concentrated on extending the measurement distance to about 10 km.

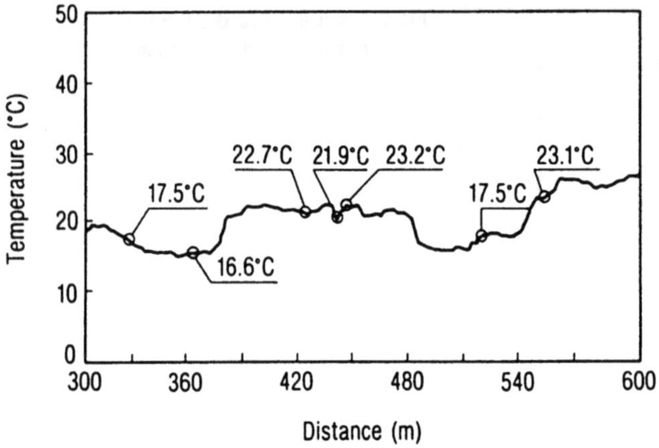

Fig.7 Field-test results (the open circles represent temperatures recorded using a spot-type temperature sensor)

7 References

(1) Hartog, A. H. : "A distributed temperature sensor based on liquid-core optical fibers", IEEE J. Lightwave Technol.,1983LT-3, pp498-509

(2) Dakin, J.P., Pratt, D.J., Bibby, G.W., and Ross, J. N. :
"Distributed anti-Stokes ratio thermometry.",
Proc. OFS 3, San Diego, 1985, Post deadline paper PDS-3

(3) Hartog, A. H., Leach, A. P., and Gold, M. P. :
"Distributed temperature sensing in solid-core fibres.",
ELECTRONICS LETTERS, 7th November 1985, Vol. 21, No. 23, p1061~1062

(4) Long, D. A. : "Raman Spectroscopy"
McGraw-Hill International Book Company, 1977

8 Appendix

If a light pulse $p(\tau)$ as shown in Fig. 3(a) is injected into an optical fiber, and the time of injection is t=0, the time characteristics of the back scattered anti-Stokes light (Ia(t)) and the Stokes light (Is(t)) that are received at the injection end are given by

$$Ia(t) = \int_{-\tau_0}^{\tau_0} P(\tau) \cdot ga(x)\, d\tau \quad \cdots\cdots\cdots(A\ 1)$$

$$Is(t) = \int_{-\tau_0}^{\tau_0} P(\tau) \cdot gs(x)\, d\tau \quad \cdots\cdots\cdots(A\ 2)$$

$$x = (v/2) \cdot (t + \tau) \quad \cdots\cdots\cdots(A\ 3)$$

$$ga(x) = Ka \cdot \gamma a(x) \cdot fa(T) \quad \cdots\cdots\cdots(A\ 4)$$

$$gs(x) = Ks \cdot \gamma s(x) \cdot fs(T) \quad \cdots\cdots\cdots(A\ 5)$$

$$\gamma a(x) = 10^{-(\alpha_r + \alpha_a)x} \quad \cdots\cdots\cdots(A\ 6)$$

$$\gamma s(x) = 10^{-(\alpha_r + \alpha_s)x} \quad \cdots\cdots\cdots(A\ 7)$$

$$fa(T) = \beta(T) / \{1 - \beta(T)\} \quad \cdots\cdots\cdots(A\ 8)$$

$$fs(T) = 1 / \{1 - \beta(T)\} \quad \cdots\cdots\cdots(A\ 9)$$

$$\beta(T) = \exp(-hv\nu / kT) \quad \cdots\cdots\cdots(A10)$$

$$T = T(x) \quad \cdots\cdots\cdots(A11)$$

Here,

K ; a constant determined by optical fiber and wavelength of light source

γ ; transmission loss of optical fiber

α ; optical fiber transmission loss coefficient

x ; distance from injection end
f ; generation probability of Raman scattering
h ; Planck's constant
v ; velocity of light in optical fiber
ν ; Raman shift wave number
k ; Boltzmann's constant
T ; absolute temperature

Sufix a, s, and r indicate anti-Stokes light, Stokes light and Rayleigh light respectively

When the length of response distance is very short and α_a is nearly equal to α_s, the following is given:

$$Ia(t) / Is(t) = (Ka / Ks) \cdot \exp(-hv\nu / kT) \cdots \text{(A12)}$$

The intensity ratio of anti-Stokes light Ia(t) and Stokes light Is(t) measured as back scattering light is in proportion to β (T) and a function of absolute temperature T.

Exploitation of Stimulated Brillouin Scattering as a Sensing Mechanism for Distributed Temperature Sensors and as a Means of Realizing a Tunable Microwave Generator

D. Culverhouse, F. Farahi, C.N. Pannell, and D.A. Jackson

Physics Department, University of Kent at Canterbury, Canterbury, Kent, CT27NR, UK

It is demonstrated that Stimulated Brillouin Scattering(SBS) generated in a monomode optical fibre may be exploited for distributed temperature sensing.

The accuracy of the SBS frequency shift measurement is shown to be significantly improved by heterodyning the SBS signal from the sensing fibre with that generated in a reference fibre.

INTRODUCTION

Distributed fibre optic temperature sensors have been developed which exploit various mechanisms such as back-scattered Rayleigh Scattering in liquid filled fibres[1], doped fibres[2], back-scattered Stimulated Raman Scattering(SRS)[3] (in this reference the sensor measured distributed stress rather than temperature) and back-scattered Spontaneous Raman Scattering[4] of which the latter appears to be the most promising technique, despite the fact that the scattering cross-section is relatively small. The effect of this small scattering cross-section is to limit the thermal and spatial resolution obtainable whilst at the same time requiring relatively long measurement times. In this paper we propose the exploitation of another scattering mechanism, namely Stimulated Brillouin Scattering, for distributed Temperature Sensing (DTS) as the Stimulated Brillouin Scattering cross-section is several orders of magnitude greater. In addition to demonstrating that SBS is a viable mechanism for DTS we also show that by heterodyning the ≅30GHz SBS signal from the sensor with another SBS signal of similar frequency generated in a fibre held at constant temperature, the frequency shift may be recovered electronically rather than optically, thus greatly simplifying the measurement system whilst at the same time yielding greater accuracy.

Stimulated Brillouin Scattering (SBS) occurs for relatively low input powers in monomode optical fibres as the confinement of the radiation to the vicinity of the fibre core (typical area 20pm^2) results in a high optical power density[5]. For applications such as optical communications SBS is seen as a very deleterious effect as it can limit the maximum optical power transmitted by the fibre and hence decrease the distance

Springer Proceedings in Physics, Vol. 44 Optical Fiber Sensors
Editors: H.J. Arditty · J.P. Dakin · R.Th. Kersten

between repeaters. SBS can be used to advantage in optical fibres, for example to produce amplification[6].

Stimulated Brillouin Scattering can be described as a 3 wave interaction involving the incident light (the pump), a generated acoustic wave and the scattered light wave (the Stokes line). Classically, the pump creates a pressure wave in the medium due to electrostriction and the resultant variation in density changes the optical susceptibility. Thus the incident wave pumps the acoustic wave which scatters it, and this scattering generates the Stokes wave. This wave is down-shifted in frequency by an amount equal to the acoustic frequency ν_A which is given in the plane wave approximation by:

$$\nu_A = \frac{2V_A n}{\lambda} \quad \mathrm{Sin}\ \theta/2 \tag{1}$$

where θ is the angle through which the light is scattered (θ = 180° in the case of full back-scattering such that $\sin\theta/2$ = 1 in this case), V_A the velocity of the acoustic wave, n the refractive index of the fibre core and λ the vacuum wavelength of the incident beam.

In normal spontaneous Brillouin Scattering (Sp.BS) there is both a Stokes and anti-Stokes line where the modulus of the frequency shift is equal to ν_A. In both cases the frequency of the scattered light, ν_A, is a function of temperature. The dominant effect producing this dependence is the change of refractive index with temperature. The change in the velocity of the acoustic wave with temperature amounts to only 7% for the range 20°C to 1620°C[7]. Hence from equation (1) with $\sin\theta/2 = 1$, we have

$$\frac{d\nu_A}{dT} = \frac{2}{\lambda}\left[V_A\frac{dn}{dT} + n\frac{dV_A}{dT}\right] \tag{2}$$

where T is the temperature. Thus the change in shift of the Brillouin line with temperature provides a possible mechanism which can be exploited for a distributed temperature sensor.

If SBS is to prove viable for DTS measurements, it is necessary to establish that a change in frequency shift associated with a change in temperature is measurable and that signals corresponding to different temperature regions along the 'effective length' of the fibre can be simultaneously determined. The effective length[8] of the sensing (or reference) fibre is shorter than its physical length as the pump power is continuously depleted along the fibre due to the SBS process such that at a specific distance from the injection point in the fibre there is insufficient pump power above threshold to produce the SBS signal.

In a typical monomode optical fibre, $\nu_A \cong$ 30GHz and $d\nu_B/dT$ = 5.5 MHz/°C, hence optical resolutions of $\geq$ 5MHz are required for a 1°C temperature resolution in a DTS. Although resolutions of this order of magnitude are achievable with high contrast confocal scanning Fabry-Perot interferometers, absolute optical frequency measurements with this degree of accuracy are difficult to achieve over protracted periods.

An alternative technique for signal recovery which offers significantly higher resolutions [9] is optical heterodyning. As SBS is a coherent process the generated Stokes line is extremely narrow offering very high heterodyning efficiencies. The heterodyne signal could in principle be produced by mixing the SBS signal with the original pump line, however the resulting signal at ≅30GHz would be difficult to process electronically. A more effective solution which is demonstrated here is to produce a 'reference' SBS signal in a second optical fibre held at constant conditions and to beat this signal with the SBS emanating from the sensing fibre such as to generate a signal at a low frequency readily accessible to conventional electronics.

EXPERIMENTAL

The experimental configuration shown schematically in figure 1 was used (1) to ascertain the dependence of the Brillouin Shift with temperature, and (2) to test the feasibility of heterodyning two SBS signals arising from two separate fibres.

Light from a single frequency argon ion laser operating at 514.5nm was launched, via a beam splitter, into two 500m lengths of monomode fibre of slightly different composition. An optical isolator comprising a Glan Thompson Prism and a quarter wave plate was used to prevent any instabilities in the laser output frequency being caused by back-scattered light entering the laser cavity. The back-scattered SBS signals from both fibres were combined on the first beam splitter and intensity divided at the second beam splitter to enable the optical and heterodyne spectra to be measured. The optical measurements were performed with a high resolution confocal Fabry-Perot (CFP) of free spectral range 2GHz, and the heterodyne spectra analysed with an electronic spectrum analyser.

OPTICAL MEASUREMENTS

In the case of the optical studies, the transmitted pump line from one of the fibres was also detected via the CFP; this signal served as a frequency reference. The frequency shifts of the back-scattered Brillouin lines were recorded whilst the temperature difference applied to the fibres was varied by cooling one spool to 0°C and heating the second to 35°C.

The change in frequency of each Brillouin line from the stationary pump line is 2.67MHz per °C (for fibre 1) when heated to ≅ 35°C and ≅ 6.4MHz per °C (for fibre 2) when cooled to 0°C. The difference in the frequency shifts is accounted for by the difference in physical properties of the two fibres.

A second experiment was performed using the arrangement shown in figure 2, where only a single fibre (fibre 2) was

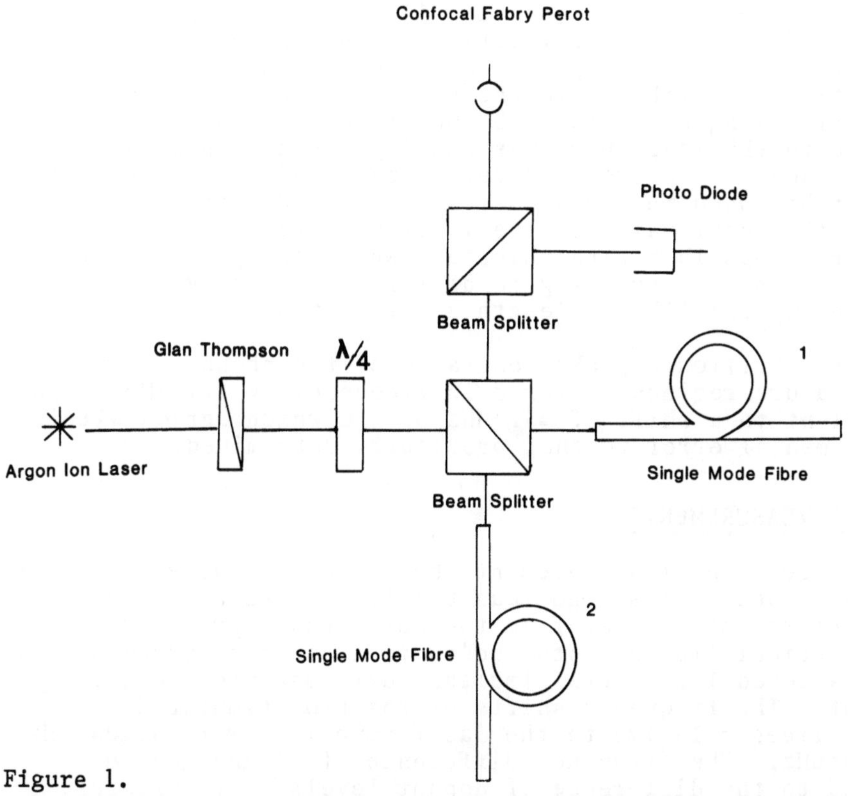

Figure 1.

Experimental arrangement, Stimulated Brillouin Shifted Signals from the two 500m spools of single mode fibre were analysed with a scanning CFP with a FSR of 2GHz. The beat frequency was detected by a photo-detector and investigated with an electronic RF Spectrum Analyzer.

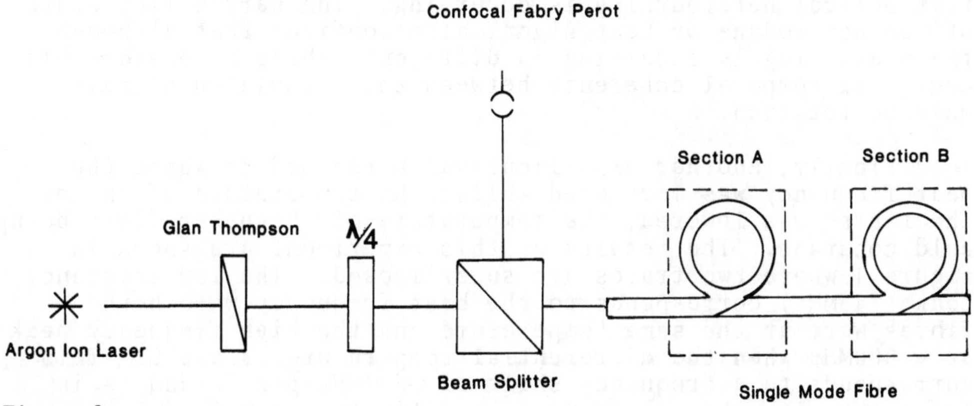

Figure 2.
Experimental configuration to study the effect of temperature variation along a single fibre.

used. Here, 150m of fibre (section A) was cooled to 0°C and the remaining 350m of the fibre (section B) heated to ≅30°C. During this experiment the laser power was adjusted such that both fibre sections contributed to the scattering process; that is, the effective length and the subsequent generation of phonons attributed to the laser pump was equally divided between sections A and B. The results verify that SBS can occur along a single fibre when the frequency shifts are different and depend on the local value of the fibre temperature. These results are compatible with Shibata[10] where the presence of Brillouin peaks corresponding to different refractive index profiles along the fibre's length are reported.

The two Brillouin peaks generated by the phonons in the two temperature regions differed in frequency by 171.4MHz. This is equivalent to a shift of ≅5.7MHz per °C which agrees within the experimental error to that previously determined.

HETERODYNE MEASUREMENTS

The experimental arrangement shown schematically in figure 1 was again used to test the feasibility of generating an optical heterodyne carrier from two SBS signals produced in separate optical fibres. Both the optical and heterodyne spectra of the scattered light were simultaneously measured in this experiment. The frequency shifts of the two stimulated Brillouin lines relative to the laser pump line were 32.042GHz and 32.796GHz. The frequency difference of 754MHz may be attributed to the difference of dopant levels in the fibres.

The two back-scattered SBS signals were also detected on a high speed PIN photodiode, and the spectrum of its output current was analysed with an HP 8590A RF Spectrum Analyser. Figure 3 shows the resulting power spectrum (the fibres being maintained under the same room temperature conditions), where a strong (<1MHz wide) peak at 760MHz is easily identified. The frequency of this peak clearly corresponds to the difference frequency between the two SBS signals thus one may conclude that optical heterodyning is occurring. The narrow line width of the heterodyne or beat signal also confirms that although the scattering is occurring in different fibres a considerable degree of temporal coherence between the stimulated signals must be retained.

Finally, another experiment was performed in which the beat frequency was monitored whilst the temperature of one of the fibres was lowered, the temperature of the other fibre being held constant. The results of this experiment are shown in figure 4 where two traces are superimposed. The low frequency peak (748MHz) corresponds to the beat frequency when both fibres were at the same temperature and the high frequency peak at ≅ 800MHz when the differential temperature is ≅10°C. This corresponds to a frequency shift of ≅ 5.2MHz per °C and is in agreement with the temperature sensitivity determined in the first experiment. It would thus appear that this phenomenon can be exploited to realise firstly, a high resolution distribu-

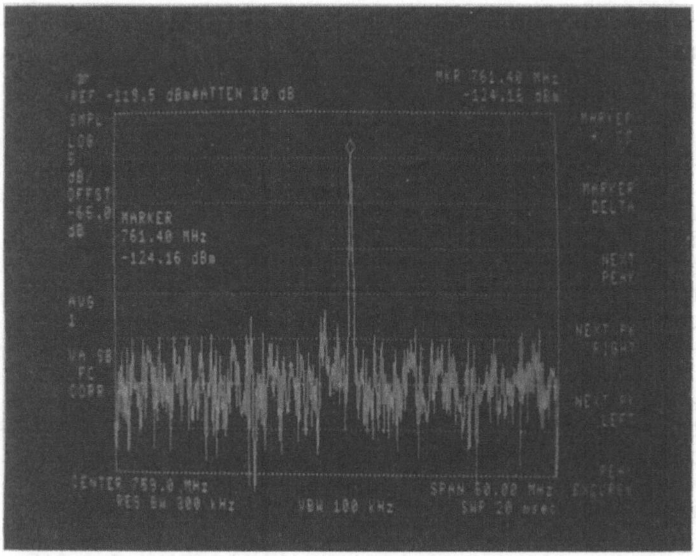

Figure 3

The heterodyne frequency at 761.4MHz generated at the photo-diode with both fibres held at room temperature.

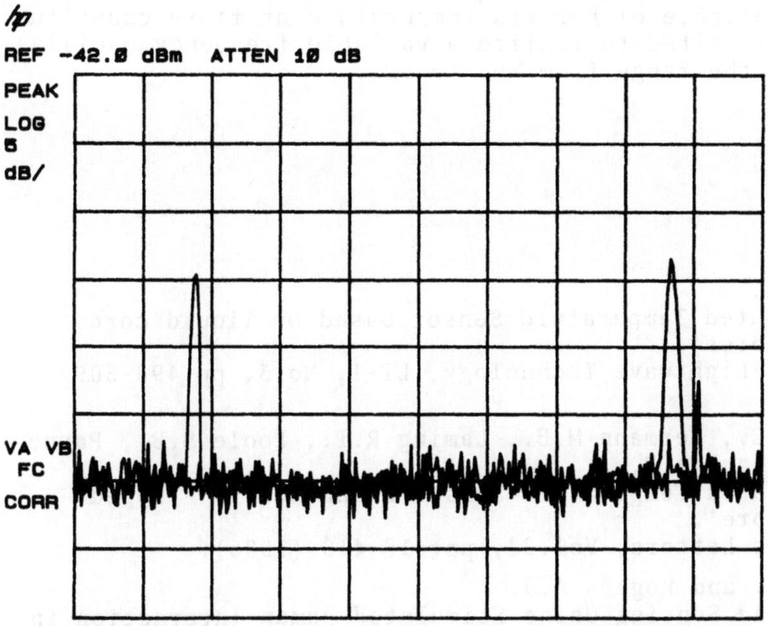

Figure 4

The lower frequency (748MHz) corresponds to the beat frequency when both fibres were at room temperature. The higher frequency peak (800MHz) corresponds to the beat note when fibre 1 was heated by 10°C.

ted temperature sensor and secondly, a tunable microwave generator with a large frequency range.

SUMMARY

We have demonstrated that SBS can be generated along a length of single mode fibre where the Brillouin frequency shifts correspond to the temperature of specific sections of the fibre. Hence it should be possible to exploit SBS for distributed temperature sensing. It has also been verified experimentally that a strong heterodyne beat signal with a narrow line width can be produced by mixing two SBS signals generated in separate optical fibres. This beat frequency is shown to vary rapidly with temperature and to depend on the physical properties of the fibre. Hence a DTS system may be optimized by selecting appropriate fibres to generate a frequency accessible to conventional electronic signal processing.

In practice distributed temperature sensing based on SBS would be based on solid state lasers operated in the pulsed mode in order to determine the spatial distribution in the fibre sensor. Research is currently under way in these laboratories to realise such devices.

Finally, the very large frequency range of the signal generated by heterodyning SBS signals from different fibres, which is controllable either via temperature or fibre constituents, may be exploited to realize a variable frequency oscillator tunable in the range from MHz to GHz.

REFERENCES

1. Hartog A.F.
   'A Distributed Temperature Sensor based on liquid core optical fibres'.
   Journal of Lightwave Technology, LT-1, No.3, pp.498-509 (1983).

2. Farries M.C., Fermann M.E., Laming R.I., Poole S.B., Payne D.N. and Leach A.P.
   'A Distributed Temperature Sensor using No.1^{3+} - doped optical fibre'.
   Electronics Letters, Vol.22, pp.418-419 (1986).

3. Farries M.C. and Rogers A.J.
   'Distributed Sensing Using Stimulated Raman Interaction in a monomode optical fibre'.
   Proceedings of the 2nd International Conference on Optical Fibre Sensors, Stuttgart, Paper 4.5., pp.121-132 (September 1984).

4. Dakin J.P.
   'Temperature measurements using Intrinsic Fibre Sensors'.
   Journal of Optical Sensors, Vol.1, No.2 pp 101-114 (1986).

5. R.H.Stolen,
'Non-linearity in Fibre Transmission'.
Proceedings of IEEE, Vol.68, No.10 (October, 1980)

6. Atkins C.G., Cotter D., Smith D.W. and Wyatt R.
'Application of Brillouin Amplification in Coherent optical Transmission'.
Electronic Letters, Vol.22, No.10, pp.556-558 (8 May, 1986)

7. Bucaro J.A. and Dardy H.D.
'High temperature Brillouin Scattering in fused quartz'.
Journal of Applied Physics. Vol.45., No.12 (December, 1984).

8. Smith R.G.
'Optical power handling capacity of low loss optical fibres as determined by stimulated Raman and Brillouin scattering'.
Applied Optics Letters, Vol.11, No.11 (November, 1972).

9. Cummins H.Z. and Swinney H.L.
'The Theory of Light Beating Spectroscopy'.
Progress in Optics 8, p.133 (1970).

10. A.Shibata et al.
'Brillouin Gain Spectra for single mode fibres having pure silica, GeO_2 doped and P_2O_5 doped cores'.
Optics Letters, Vol.12, No.4 (April, 1987).

A New Technique for a Fiber Distribution Sensor

K. Nakamura, N. Kagi, and S. Yoshida

The Furukawa Electric Co. Ltd., Yokohama R&D Laboratories,
2-4-3 Okano Nishiku Yokohama, Japan

We devised a new method for distribution sensing using the ESA (Excited State Absorption)of rare-earth doped fibers. This method involves a detected signal which is a forward transmitted continuous wave, and so it is possible to obtain a higher S/N ratio than with an OTDR method. The dynamic range is controllable by pumping intensity.

Introduction

Optical fiber sensors are the subject of considerable interest for the measurement of a great number of physical quantities, as extensively reviewed in ref.(1). Especially, distributed sensing is a method utilizing the strong advantage of low loss optical fibers. Some distributed fiber sensors are based on optical time domain reflectometry (OTDR), which is a well-known technique for observing the length dependence of physical parameters along the sensing fiber (2).

Because OTDR detects a weak back-scattered signal which is generated by forward travelling short time pulses, the detectable power is low and the S/N ratio and dynamic range are also small. This leads to long time averaging and a high power laser source must be used. The problem of OTDR is that only a small fraction of the input power is available.

We devised a new method for distributed sensing utilizing a ESA (Excited State Absorption)of a rare- earth doped fiber. Because this method is able to detect a cw forward travelling wave, a highly efficient sensing system with a high S/N ratio was realized.

Springer Proceedings in Physics, Vol. 44 Optical Fiber Sensors
Editors: H.J. Arditty · J.P. Dakin · R.Th. Kersten

Principle

Fig.1 shows an arrangement of this distributed sensing method. A pumping laser source (A), driven by a pulse generator, is coupled into the rare-earth doped sensor fiber through a beam splitter. Another source (B), used for the sensing signal, operated as continuous wave, is coupled into the other fiber end. That is, the excitation and signal sources are counter propagating. The signal output from the sensor fiber is reflected by the beam splitter and detected by a photo detector.

This method uses the ESA (Excited State Absorption) of a rare-earth doped fiber (3). ESA means the absorption by a stable excited state.

In Fig.2, we show an energy diagram for the Er^{3+}ion. E_g, E_a and E_b in this figure show a ground state, lower excitation state and upper excitation states respectively. GSA (Ground State Absorption) appears as an absorption from E_g to E_b, and ESA is generated by the absorption from E_a to E_b, when the state E_a, which has a long life time, is excited.

Fig.3 shows an attenuation loss spectra of the Er^{3+} doped fiber. The lower spectrum, (a), indicates the ESA by the 1st excited state, and the upper spectrum, (b), shows the GSA.

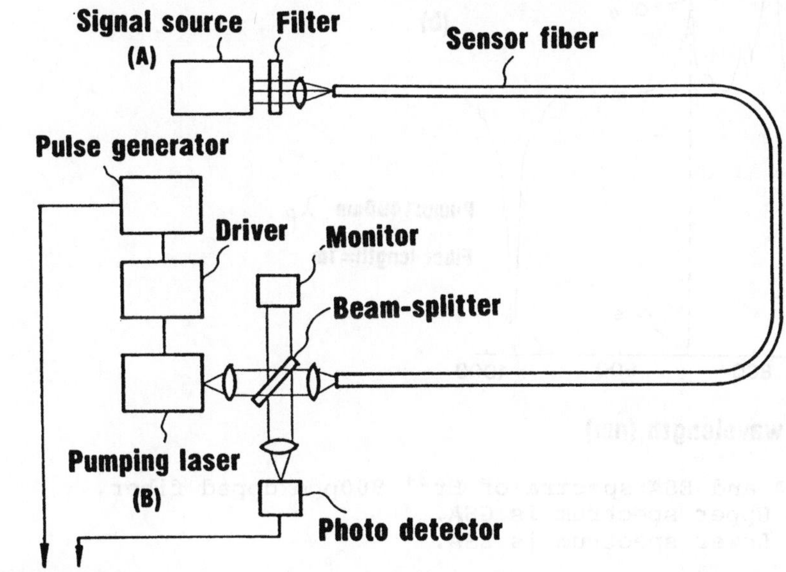

Fig.1 The arrangement of this sensing method.

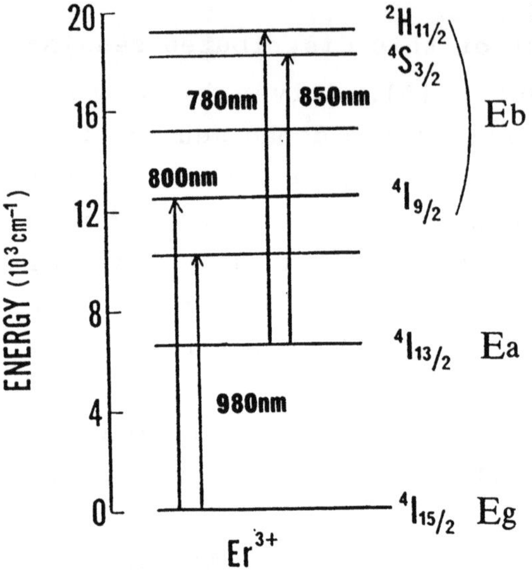

Fig.2 The energy diagram of Er^{3+} ion.

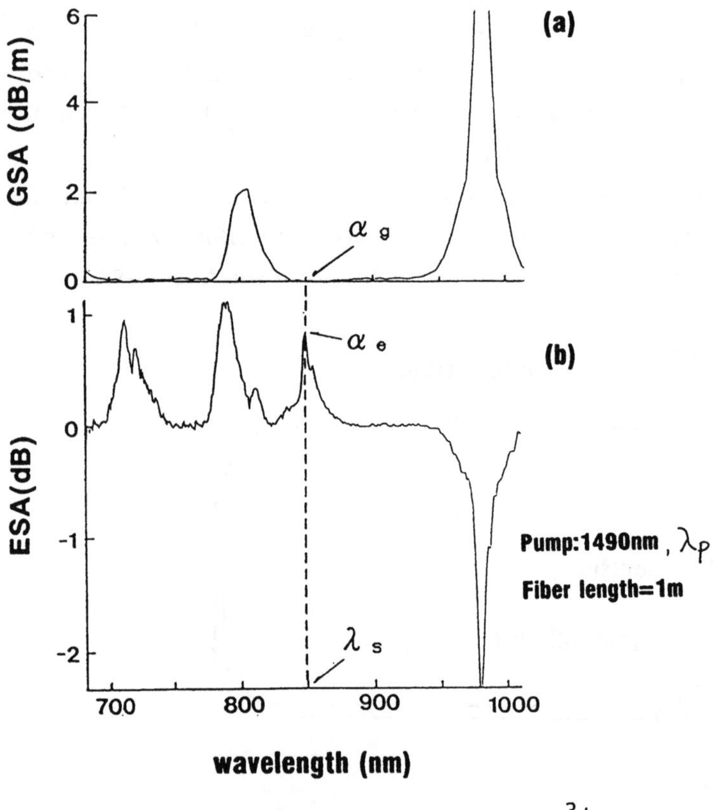

Fig.3 The GSA and ESA spectra of Er^{3+} 900ppm doped fiber.
(a) Upper spectrum is GSA.
(b) Lower spectrum is ESA.

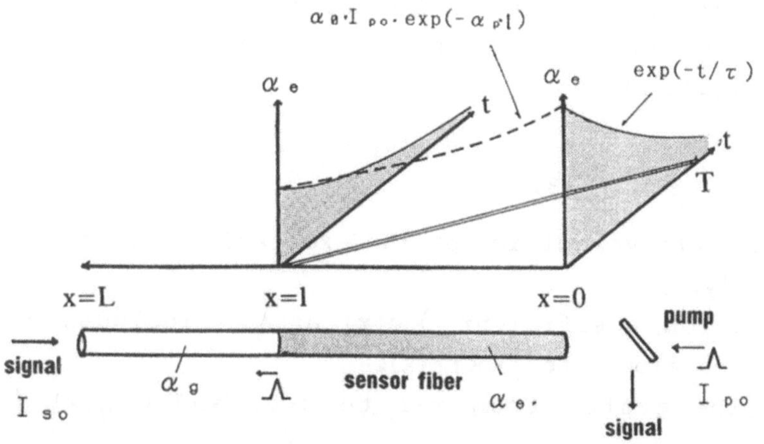

Fig.4 The concept of this method. I_{so} and I_{po} are input signal and input pumping intensities, respectively. When the pumping pulse is located at x=l, the absorption loss due to ESA is generated in a region from x=0 to x=l.

In the GSA, as shown in this figure, the attenuation (α_g) at λ_s is small, depending on only scattering loss, and the output signal I_s, passing through the fiber of L in length, is expressed by

$$I_s = I_{so} \cdot \exp(-\alpha_p \cdot L) \qquad (1)$$

where I_{so} is an input signal. When pumping is carried out by an excitation source at λ_p, because of ESA, the attenuation loss at λ_s is changed to $\alpha_o \cdot I_p$, when I_p is the excitation intensity.

We indicate a conception of this method in Fig.4. As shown in this figure, once the excitation pulse is coupled into the sensor fiber ,the detected signal intensity I_{so} from the same end varies as function of time (t) because the absorption of the sensing wave changes with the propagating excitation pulse, due to ESA.

The signal intensity characteristic at x=x, when the excitation pulse is located at x=l (0<x<l), is given by

$$-dI_s = (\alpha_e + \alpha_g) \cdot I_s \cdot (-dx) \qquad (2a)$$

$$\alpha_e = \alpha_a \cdot I_{po} \cdot e^{(-\alpha_p x)} \cdot e^{(-t_1/\tau)} \qquad (0<x<l) \qquad (2b)$$

$$= 0 \qquad (l<x<L) \qquad (2c)$$

where I_{po} is the initial excitation intensity, τ is a life time of level E_a, αp is the attenuation coefficient of the excitation, and $t_1=(1-x)\,V$.
V is defined as

$$V = 1/v_p + 1/v_s \qquad (3)$$

where v_p, v_s are light velocities of the excitation and signal waves, respectively.

The attenuation coefficient $\alpha o(x)$ at λs includes a sensing information along the fiber (x).

Eq.(2) is integrated from x=l to x=0, and $\alpha_0(x)$ is expressed by

$$\alpha_0(l)=(1/I_p)\cdot\exp(-\alpha_p l)\{(-V/\tau)(\log(I_s/I_{s0})+\alpha_0 l) - \frac{d}{dl}(\log I_s/I_{s0})\} \qquad (4a)$$

Using $T = l\,V$ and $d/dl = d/dt \; dt/dl = V \; d/dt$,

$$\alpha_0(l)=(1/I_p)\cdot\exp(-\alpha_p T/V)\{(-V/\tau)(\log(I_s/I_{s0})+\alpha_0 T/V) -V\frac{d}{dV}(\log I_s/I_{s0})\} \qquad (4b)$$

Using this formula, from the change of $\alpha_o(l)$ we can monitor an interesting physical quantity, (e.g. temperature) along the fiber and identify the position by time T.

Fig.5 (a) shows a case when $\alpha_o(l)$ = constant, i.e., the sensing fiber lies in a homogeneous environment. If ESA increases as the temperature increases and if the temperature at x=l is higher than that at other places, we will observe the time domain signal as shown in Fig.5 (b). We calculate the change of $\alpha_0(l)$ from the change of the signal output(I), and obtained the temperature at x=l.

We examined the temperature dependence of ESA in Er^{3+} doped fibers. The ESA spectra showed no spectral shift for a temperature change from +30°C to +90°C, but the attenuation level has temperature dependence, as shown in Fig.6. As a result of Fig.6, we estimated the sensitivity of this fiber to be 2.45mdB/C in the 900ppm doped Er^{3+} fiber in the temperature region from 30°C to 90°C. This variation is not large, however, the method advantage of high S/N ratio enables the sensing.

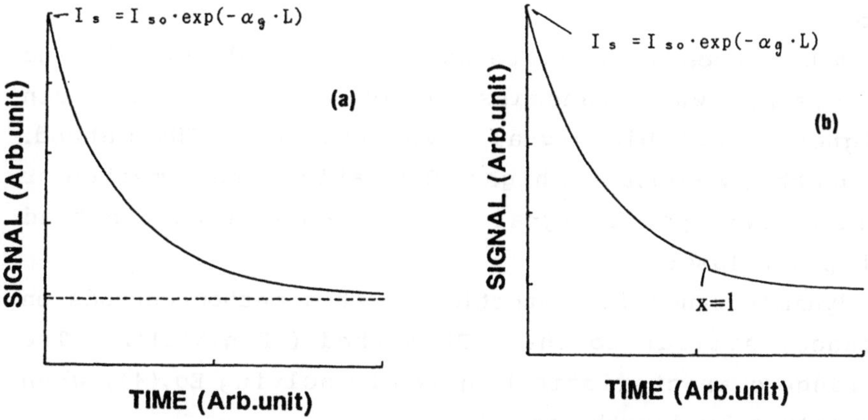

Fig.5 The signal output characteristics.
(a) The signal output in a case of a homogeneous detectable parameter (e.g. temperature).
(b) When the parameter varies at a certain place,the signal output changes at the time corresponding to the place.

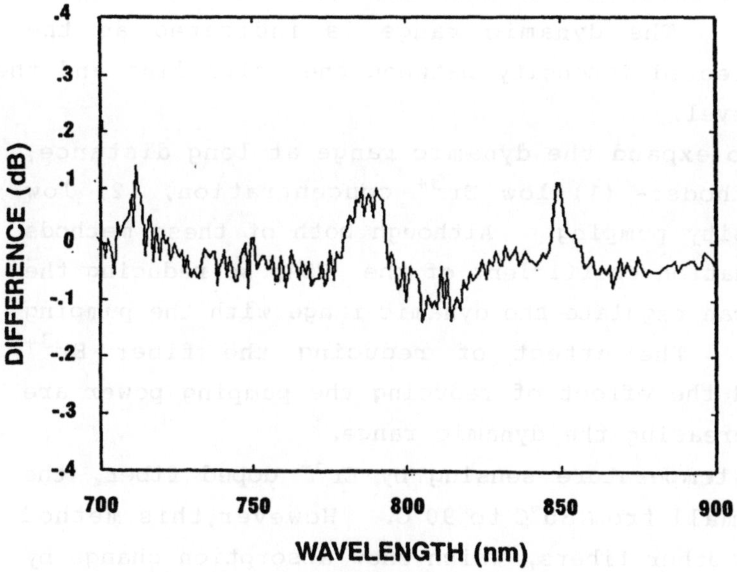

Fig.6 Temperature dependence of ESA in a Er^{3+} doped fiber. The ordinate shows the difference of ESA from 30°C to 90°C.

Discussion

This method uses a CW source as the detectable signal, and the source is a forward transmission wave. As a result, we can get a higher detectable signal power than the OTDR method. This means that we obtain a higher S/N ratio by this method or we can use a lower power signal source such as a LED, instead of a high power laser.

The dynamic range for detection in this method depends on the distance, similar to the OTDR method (Fig.5(a)). The dynamic range characteristic is given by solving Eq.(4), when $_0(l)$ is constant by length, as

$$I_s/I_{s0} = e^{(-\alpha_g \cdot l)} \cdot \exp\{\alpha_0 \cdot I_p \cdot \tau\,(e^{(-V \cdot l/\tau)} - e^{(-\alpha_p \cdot l)})/(V - \alpha_p \cdot \tau)\} \quad (5)$$

and the solution is shown in Fig.5. The solid line in Fig.5 (a), is approximated by $\tau \gg T = l \cdot V$. If $l=0$, $I_s = I_{so} e^{-\alpha g l}$, i.e., the initial output signal is that obtained for no excitation. The dashed line in this figure shows the case of $l \rightarrow \infty$ or $T \rightarrow \infty$. The dynamic range is indicated as the difference in detected intensity between the solid line and the detector noise level.

In order to expand the dynamic range at long distance, we have two methods:- (1) low Er^{3+} concentration, (2) low excitation intensity pumping. Although both of these methods reduce the attenuation coefficient of the ESA, so reducing the sensitivity, we can regulate the dynamic range with the pumping intensity. The effect of reducing the fiber Er^{3+} concentration and the effect of reducing the pumping power are equivalent in increasing the dynamic range.

Considering temperature sensing by Er^{3+} doped fiber, the sensitivity is small from 30°C to 90°C. However, this method can be applied to other fibers, which show absorption change by an excitation, for example Ho^{3+} or Nd^{3+} doped fiber, whose absorption is expected to be temperature dependent (4).

Conclusion

We developed a the new technique for distributed sensing by using an optical fiber. In this method, two optical sources, one used as the excitation to generate the ESA, and the other used as the monitoring signal, travel in the sensing fiber in opposite directions. The monitoring signal output varies, with the variation beginning at the moment when the excitation pulse is coupled into the fiber. We observed the time domain characteristic of the signal. The time is used to identify the position in the fiber. We introduced a characteristic formula. The sensing information is included in the ESA feature. That is, when the ESA has a temperature dependence, we obtained the distributed temperature along the fiber. This method uses a cw and forward travelling source as the detectable signal, implying that we can get a higher S/N ratio than an OTDR system. We examined the temperature dependence of the ESA of Er^{3+} doped fiber and estimated the temperature sensing ability. Other than a Er^{3+} doped fiber, this method can be applied to any fiber whose attenuation changes when excitation pumping and environment physical quantities change.

Referneces

(1) T.G.Giallorenzi, et al: IEEE J.Quantum Electron. QE-18 pp.629-665 (1982)

(2) M.K.Barnoski and S.M.Jensen : Appl.Opt.15 pp.2112-2115 (1976)

S.D.Personick : Bell Syst.Tech.J.56 pp.355-367 (1977)

A.H.Hartog and D.N.Payne : Proc.8th European Conf. on Optical Commun.(France),pp.215-220 (1982)

(3) R.I.Laming et al: Opt.Lett.13 pp.1084-1086 (1988)

(4) M.C.Farries et al: Electron.Lett.22 pp.418-419 (1986)

White Light Interferometry for Distributed Sensing on Dual Mode Fibers

G. Kotrotsios and O. Parriaux

Centre Suisse d'Electronique et de Microtechnique (C.S.E.M.) S.A.,
Recherche et Développement, Maladière 71, CH-2000 Neuchâtel 7, Switzerland

Abstract. *A new interferometric method for distributed sensing in the transmission mode over dual mode fibers is presented. Intermodal delay time, proportional to the distance of mode coupling transducers from the fiber end, is measured by a Michelson interferometer exterior to the sensing fiber. Centimetric spatial resolution is demontsrated over a fiber length of 75 m.*

1. Introduction.

Distributed sensing in the transmission mode (1) is an interesting alternative to reflectometry(2), particularly because of its considerably higher dynamic range. The penalty to be payed for this is a lower spatial resolution of the sensor system in comparison with reflectometric techniques as well as an inherent difficulty in identifying and demultiplexing several transducers activated simultaneously on the fiber line. A new transmission technique is proposed here, permitting a considerable improvement of the spatial resolution, high dynamic range and real time detection. "White light" interferometry between the LP-modes of a dual mode fiber is used for locating and identifying the various transducers installed along a single fiber line.

2. Demultiplexing of mode coupling transducers.

The fundamental LP_{01} mode of a dual mode fiber is excited by wave trains coming from a low coherence source. It propagates up to the mode coupling transducer with very low cross-talk with the second order mode, LP_{11}(1). Mode coupling is provided by a microbending transducer(4) of suitable spatial period. Then, part of the wave trains are carried by the LP_{01} mode and the coupled part by the LP_{11} mode down to the fiber end. As these modes do propagate with different group velocities, the wave trains reach the fiber end with a time delay $\Delta\tau$ (figure 1). The measurement of $\Delta\tau$ provides the location of the mode coupling event if the intermodal group delay time per unit length in the fiber is known.

The amplitude of the field emerging from the fiber end is given by:

$$\mathcal{E} = a\, E_{01}(x,y,\omega)\, e^{-j\beta_{01}L_i} + b\, E_{11}(x,y,\omega)\, e^{-j\beta_{11}L_i} \qquad (1)$$

Springer Proceedings in Physics, Vol. 44 Optical Fiber Sensors
Editors: H.J. Arditty · J.P. Dakin · R.Th. Kersten

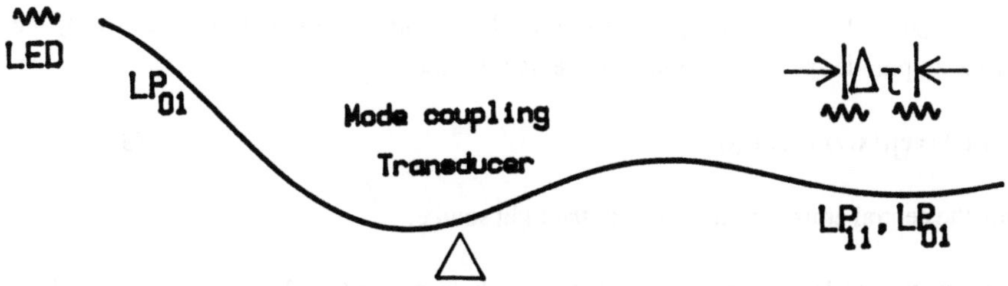

Figure 1. Wave trains propagation in a dual mode fiber after a mode coupling transducer.

with L: distance of the transducer from the fiber end, β_{j1},E_{j1}: propagation constant and modal field of the LP_{j1} (j=0,1), and b^2: normalized power coupled to the second order mode at the mode coupling transducer.

A Michelson interferometer at the fiber end is used for the time delay measurement. This interferometer acts as an autocorrelator, providing a variable delay between its arms, as depicted in figure 2.

After splitting and recombination in the interferometer the resulting field becomes:

$$\mathcal{E} = (a\, E_{01}(x,y,\omega)\, e^{-j\beta_{01}L} + b\, E_{11}(x,y,\omega)\, e^{-j\beta_{11}L})(\, 1 + e^{-j\tau\omega}) \qquad (2)$$

where τ: variable delay between the two arms of the interferometer.

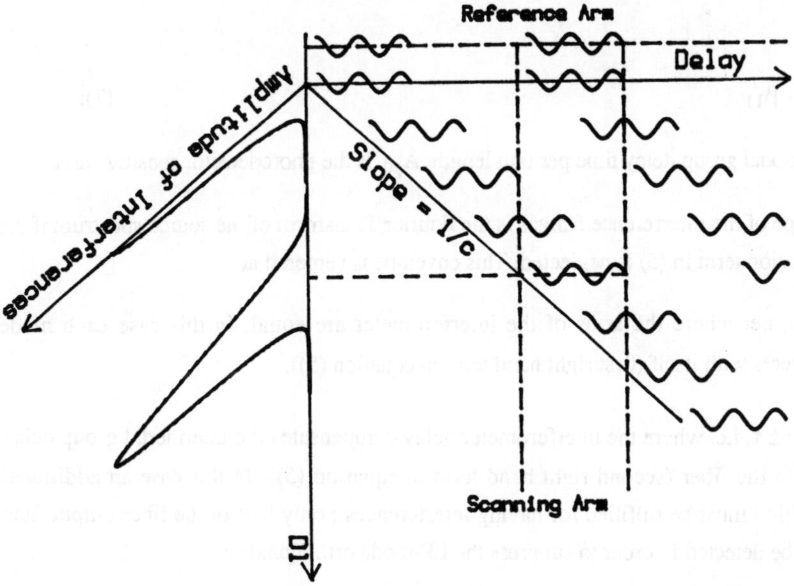

Figure 2. Conditions of interference of the fields propagated by a dual mode fiber in a variable delay interferometer.

The wavelength and the spatial dependance of the field amplitudes may be dissociated by the introduction of the amplitude spectrum $F(\omega-\omega_o)$ of the light source:

$$E_{j1}(x,y,\omega) = e_{j1}(x,y)\, F(\omega-\omega_o) \tag{3}$$

ω_o being the central emission frequency of the light source.

Furthermore, the mode propagation constants may be expanded in Taylor's series around ω_o as:

$$\beta_{j1} = \beta_{j1}(\omega_o) + \beta_{j1}'\,(\omega-\omega_o) + \beta_{j1}''\frac{(\omega-\omega_o)^2}{2} \tag{4}$$

Introducing (3),(4) into (2) and considering that the detected power is obtained by integrating $\mathcal{E}\mathcal{E}^*$ over the detection sensitive area and over the source's spectrum, one obtains for the oscillating component of the detected power:

$$P_{ac}(\tau) \propto g(\tau) \int_{Apd} \left[a^2 e_{01}(x,y)\, e_{01}(x,y) + b^2 e_{11}(x,y)\, e_{11}(x,y)\right] dA +$$

$$\left\{ g(\Delta\beta' L-\tau) + g(\Delta\beta' L-\tau) \right\} ab \int_{Apd} [\, e_{11}(x,y)\, e_{01}(x,y)]\, dA \tag{5}$$

where

$$g(\tau)\text{: Fourier transform of the function } F(\omega-\omega_o) \exp(-j\Delta\beta'' L\frac{(\omega-\omega_o)^2}{2}), \tag{6}$$

and

$$\Delta\beta' = \beta_{01}' - \beta_{11}' \qquad (7):$$

is the intermodal group delay time per unit length. A_{pd} is the photodetector sensitive area.

The envelope of the interference fringes is the Fourier Transform of the source spectrum if the second order dispersion term in (5) is neglected. This envelope is centered at :

a. $\tau = 0$, i.e. where the arms of the interferometer are equal. In this case each mode interferes with itself (first right hand term in equation (5)).

b. $\Delta\beta' L = \pm\,\tau$, i.e. where the interferometer delay compensates the intermodal group delay time in the fiber (second right hand term in equation (5)). In this case an additional condition must be fulfilled for having interferences : only half of the fiber output field must be detected in order to suppress the LP-mode orthogonality.

The effect of the neglected term in (6) is to broaden the envelope width and to reduce the fringe contrast[5].

Measurement of the air path difference Δl between positions of the mobile mirror where interferences take place provides τ ($=\frac{2\Delta l}{c}$) and therefore the position of the transducer, L ($=\frac{2\Delta l}{c\,\Delta\beta'}$).

If more than one transducer are activated simultaneously, additional interefence envelopes appear as well as cross term interferences corresponding to distances between transducers. These terms can be neglected if the power coupled to the second order mode at each transducer is a small fraction of the fundamental mode power.

3. Experimental set-up and measurements.

A Philips light-pen laser in the low injection current regime is used as a low coherence source. The FWMH spectral width is 30 nm and the central emission wavelength is 830 nm. The dual mode fiber is a Cabloptic telecommunication fiber single mode in the second window. For this fiber at the dual mode regime the intermodal beat length is 365 µm and the intermodal group delay time is 1.7 ns/km at the source central wavalength (fiber V-number =3.55).

The demonstrated readout system consists of two types of fiber transducers: First a lateral offsetted butt-joint between an exciting single mode fiber and the dual mode sensing fiber and, second, a microbending transducer composed of a pair of 30 theeth jaws with a spatial period equal to the intermodal beat length, i.e., 365 µm. The Michelson interferometer is made of bulk optic elements. The position of the scanning mirror is controlled by a 0.1µm resolution stepping motor.

Figure 3 shows the interferogram obtained by scanning the mobile arm of the interferometer with only the butt joint transducer activated.

The calibration of the displacement of the mobile mirror relative to the transducer position for a given type of fiber consists in measuring the intermodal group dealy time per unit length, $\Delta\beta'$, over a known fiber length. In the present fiber the group delay is measured to be 1.76±0.01 ns/km (corresponding to an air path difference of 1.12 mm), which is in accordance with the same figure measured previously: 1.7 ns/km(1). The form of the interference enveloppe is approximately gaussian.

Activating the microbending transducer at a distance of 2.18 m from the fiber end results in an additional interferogram when scanning the variable arm of the interferometer as shown in figure 4.

Figure 5 illustrates the effect of shifting the microbending transducer further down at 1.17 m from the fiber end. The interferogram shifts accordingly closer to the origin,i.e., where the interferometer is balanced.

From figures 4 and 5 the position of the transducer is found to be 2.23 m and 1.22m from the fiber end respectively.

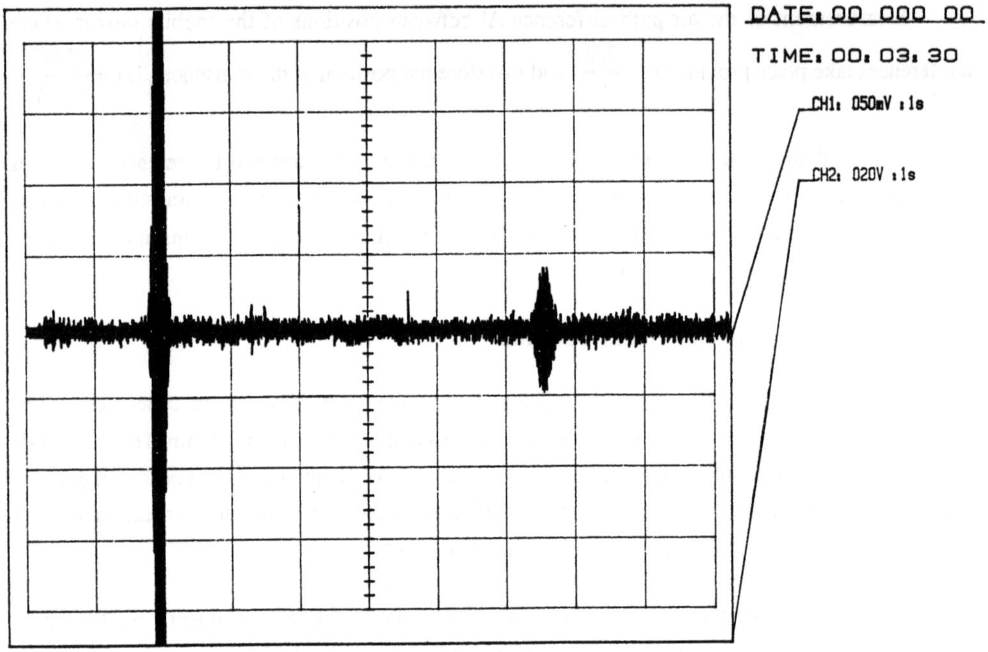

Figure 3. Interferogram obtained when varying the mirror position. Only one mode converter is activated at a distance of 4.25 m from the fiber end.

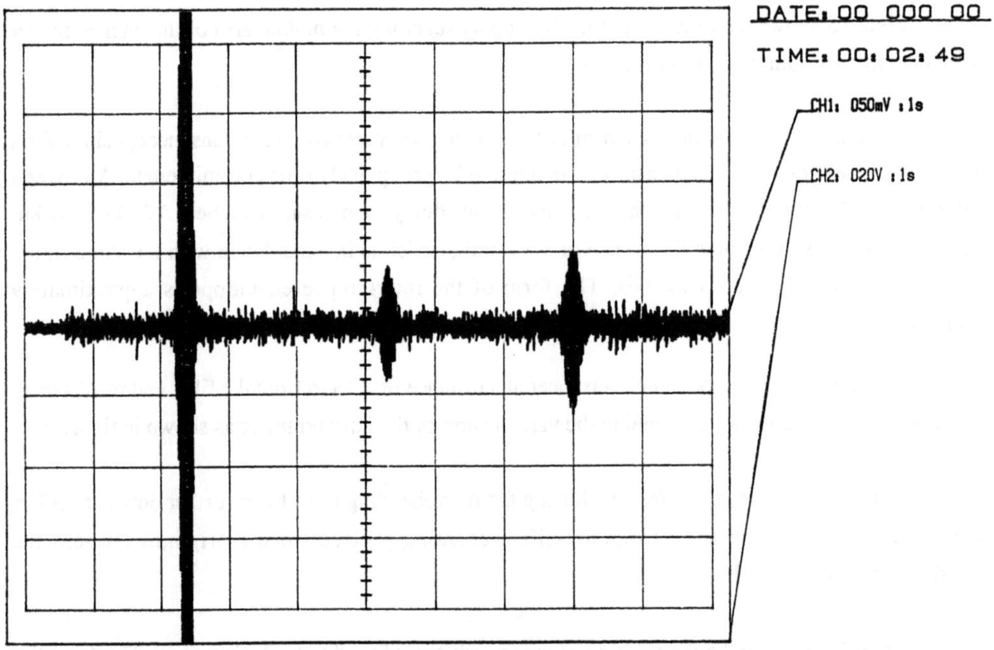

Figure 4. Interferogram as in figure 3 with a microbending transducer activated at 2.18 m from the fiber end.

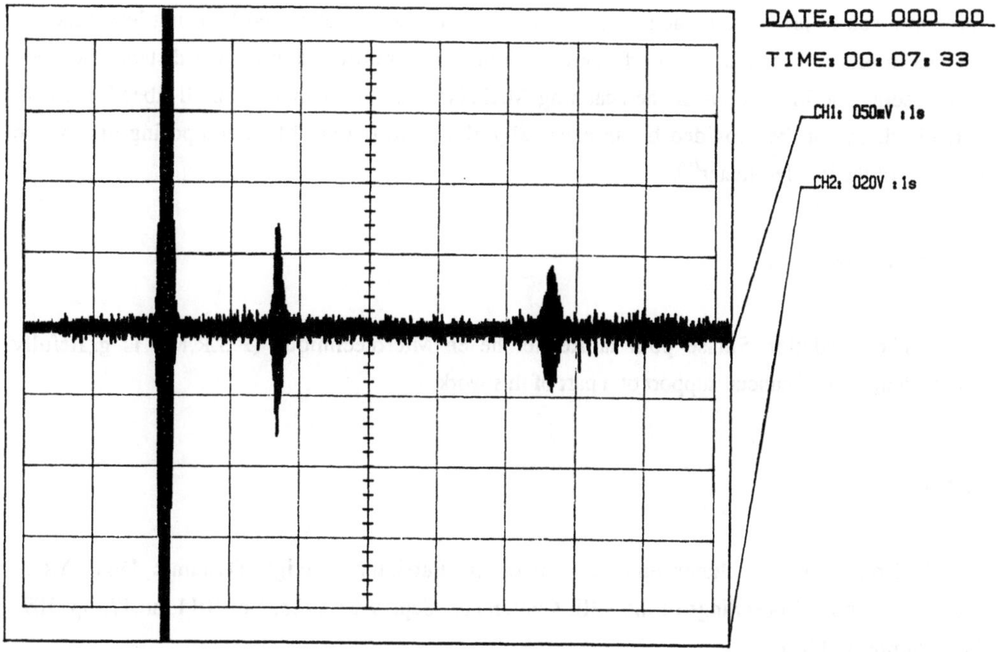

Figure 5. Interferogram as in figure 3 with the microbending transducer activated at a distance 1.17m from the fiber end.

4. Discussion.

The full temporal width of the interference envelope in the present system is measured from figures 3-5 to be of the order of 0.4 ps, which corresponds to a differential mode propagation length in the dual mode fiber of 0.23 m (as $\Delta\beta'$=1.76 ns/km). This is the minimum distance between adjacent transducers with the source used in this experiment. The total fiber sensor length depends on the mobile mirror excursion length. In the demontrated system the latter was 2 cm, allowing the readout of a 75 m long sensing fiber. The number of transducers which could be multiplexed is limited by the spectral width of the source and, more critically, by the LP_{01} mode power available for each transducer. The spatial resolution depends on the accuracy with which the center of the interferogram can be defined. The error in the definition of this point can easily be reduced to less than 10% of the interferogram's width, allowing a spatial resolution of 23 mm.

Random changes of the orientation of the LP_{11} mode lobes may result in fading effects. However using a broadband incoherent source evades this problem.

5. Conclusion.

A novel distributed fiber optic sensing technique in the transmission mode using optical interferometry between delayed LP-modes of a dual mode fiber was discussed. A spatial resolution of some mm can in principle be achieved. Further improvement is expected by using less coherent sources in order to reduce the interferogram width and therefore the minimum distance between transducers. The integration of the scanning Michelson interferometer in an all-fiber form with variable delay can be provided by an elastically elongated optical fiber, composing one of the branches of the interferometer(5).

Acknowledgements

The Fondation Suisse pour la Recherche en Microtechnique (F.S.R.M.) is gratefully acknowledged for financial support of a part of this work.

References

[1] G. Kotrotsios, P. Dénervaud, L. Falco, O. Parriaux, " High Dynamic Dual Mode Transmitometry", Proceedings of the SPIE Conference, Sep.19-23 1988, Vol.1011, no.17, pp. 109-115, Hamburg, FRG.

[2] A.R. Mickelson, O. Klevhus, M. Eriksud, "Backscatter Readout from Serial Microbending Sensors", IEEE J. of Lightwave Tech., Vol. LT-2, no.5, pp.700-709, (1984).

[3] K. Kurosawa, S. Hattori, T. Yoshino, "Distributed Fiber Optic Sensor using Forward Traveling Light in Polarisation Maintaining Fibers", Proceedings of the SPIE Conference, vol. 798, no. 04, March 30- Avril 3 1987, The Hague, The Nederlands.

[4] J.N. Blake, B.Y. Kim, H.J. Shaw, "Fiber-optic coupler using periodic microbending", Optics Lett., Vol.11, no.3, pp.177-179, (1986).

[5] L. Thevenaz, J.-P. Pellaux,J.-P. Von Der Weid, " All-Fiber Interferometer for Chromatic Dispersion Measurements", IEEE J. of Lightwave Tech., Vol. LT-6, no.1, pp.1-7,(1988).

Index of Contributors

Springer Proceedings in Physics

Managing Editor: H. K. V. Lotsch

Springer Proceedings in Physics

Managing Editor: H. K. V. Lotsch

Volume 1 Fluctuations and Sensitivity in Nonequilibrium Systems [illegible]

[illegible]

Volume 13 [illegible]

Volume 14 Magnetic Properties of Low-Dimensional Systems [illegible]
Editors: L. M. Falicov [illegible]

Volume 15 Gas Flow and Chemical Lasers [illegible]

Volume 16 Photons and Continuum States of Atoms and Molecules
Editors: N. K. Rahman [illegible]

Volume 17 Quantum Aspects of Molecular Motions in Solids
Editors: A. Heidemann [illegible] M. Prager, D. Richter, and T. Springer

Volume 18 Electro-optic and Photorefractive Materials [illegible]

Volume 19 Lasers and Synergetics [illegible] R. Graham and A. Wunderlin

Volume 20 Primary Processes in Photobiology [illegible] T. Kobayashi

Volume 21 Physics of [illegible]
and [illegible]

Volume 22 Semiconductor Interfaces: Formation and Properties
Editors: G. Le Lay [illegible]

Volume 23 Magnetic Excitations and Fluctuations II [illegible] U. Balucani, S. W. Lovesey,
M. G. Rasetti, and V. Tognetti

Volume 24 Recent Topics in Theoretical Physics [illegible] Editor: H. Takayama

Volume 25 Excitons in Confined Systems [illegible] Editors: R. Del Sole, A. D'Andrea, and
A. Lapiccirella

Volume 26 The Elementary Structure of Matter [illegible] Editors: J.-M. Richard, E. Aslanides, and
N. Boccara

Volume 27 Competing Interactions and Microstructures: Statics and Dynamics
Editors: R. LeSar, A. Bishop, and R. Heffner

Volume 28 [illegible] Editors: [illegible]

Volume 29 Polymer Motion in Dense Systems Editors: D. Richter and T. Springer

MIX
Papier aus verantwortungsvollen Quellen
Paper from responsible sources
FSC® C105338

If you have any concerns about our products,
you can contact us on
ProductSafety@springernature.com

In case Publisher is established outside the EU,
the EU authorized representative is:
Springer Nature Customer Service Center GmbH
Europaplatz 3, 69115 Heidelberg, Germany

Printed by Libri Plureos GmbH
in Hamburg, Germany